Refrigeration Principles, Practices and Performance

Refrigeration Principles, Practices and Performance

Chris Langley

THOMSON
DELMAR LEARNING

Australia Canada Mexico Singapore Spain United Kingdom United States

Refrigeration Principles, Practices and Performance
Chris Langley

Vice President, Technology and Trades ABU:
David Garza

Vice President, Technology Professional Business Unit:
Gregory L. Clayton

Publisher, Technology Professional Business Unit:
David Koontz

Director of Learning Solutions:
Sandy Clark

Managing Editor:
Larry Main

Acquisitions Editor:
James DeVoe

Senior Product Manager:
John Fisher

Marketing Coordinator:
Mark Pierro

Director of Production:
Patty Stephan

Production Manager:
Andrew Crouth

Content Project Manager:
Andrea Majot

Technology Project Manager:
Linda Verde

Editorial Assistant:
Tom Best

COPYRIGHT © 2008 Thomson Delmar Learning, a division of Thomson Learning Inc. All rights reserved. The Thomson Learning Inc. logo is a registered trademark used herein under license.

Printed in Canada
1 2 3 4 5 XX 09 08 07

For more information contact
Thomson Delmar Learning
Executive Woods
5 Maxwell Drive, PO Box 8007,
Clifton Park, NY 12065-8007
Or find us on the World Wide Web at
www.delmarlearning.com

ALL RIGHTS RESERVED. No part of this work covered by the copyright hereon may be reproduced in any form or by any means—graphic, electronic, or mechanical, including photocopying, recording, taping, Web distribution, or information storage and retrieval systems—without the written permission of the publisher.

For permission to use material from the text or product, contact us by
Tel.   (800) 730-2214
Fax   (800) 730-2215
www.thomsonrights.com

Library of Congress Cataloging-in-Publication Data:

Refrigeration principles, practices, and performance/Chris Langley.
   p. cm.
Includes index.
ISBN 1-4180-6097-6
   1. Refrigeration and refrigerating machinery. I. Title.
TP492.L37 2006
621.5'6—dc22   2007000699

NOTICE TO THE READER

Publisher does not warrant or guarantee any of the products described herein or perform any independent analysis in connection with any of the product information contained herein. Publisher does not assume, and expressly disclaims, any obligation to obtain and include information other than that provided to it by the manufacturer.

The reader is expressly warned to consider and adopt all safety precautions that might be indicated by the activities herein and to avoid all potential hazards. By following the instructions contained herein, the reader willingly assumes all risks in connection with such instructions.

The publisher makes no representation or warranties of any kind, including but not limited to, the warranties of fitness for particular purpose or merchantability, nor are any such representations implied with respect to the material set forth herein, and the publisher takes no responsibility with respect to such material. The publisher shall not be liable for any special, consequential, or exemplary damages resulting, in whole or part, from the readers' use of, or reliance upon, this material.

Dedication

A very special thank you to my wife Noreen. Without her love and technical support, this work would not have been possible.

Table of Contents

Preface xv

Chapter 1   Fundamentals 1
    Heat and Temperature 1
    Specific Heat 3
    Latent Heat 3
    Change of State 5
    Vapor 5
    Gas and Vapor 5
    Energy 5
    Pressure 6
    Partial Pressure 7
    Heat Transfer 7
    Scope of Refrigeration 8

Chapter 2   The Compression System of Refrigeration 10
    Evaporation 10
    Pressure Cycle 14
    Temperature Cycle 16
    Heat Cycle 17
    Liquid–Vapor Cycle 19
    Standard Ton Conditions 21

Chapter 3   Refrigerants: Physical and
                  Refrigerating Properties 22
    Requirements 22
    Physical Properties 25
    Critical Pressure 28
    Historical Development of Refrigerants 29
    Water 32
    Ammonia 32
    Halocarbons 34
    Conditions Other Than Standard Ton 36
    Changing Refrigerants 37
    Handling Refrigerants 38
    Refrigerant Cylinder Colors 39

Chapter 4 Liquid Feed Devices or Expansion Valves 40

Function of the Expansion Valve 40
Hand Expansion Valve 41
Low-Side Float Valve 42
High-Side Float Valve 44
Capillary Tube 45
Automatic Expansion Valve 46
Thermostatic or Thermal Expansion Valve 48
Solid-State or Thermal-Electric Expansion Valve 54
Liquid Solenoid Valve with Electric Float 55
Liquid Solenoid Valve with Thermal Differential Sensor 56
Liquid Solenoid Valve with Variable-Capacitance Probe 57
Pneumatic High-Side Float 58

Chapter 5 Evaporators 60

Requirements for Evaporators 60
Types of Evaporators 61
Flooded Evaporator 62
Dry Evaporator 64
Improved Designs 66
Convection and Forced-Draft Evaporators 67
Evaporator Feeds 69
Surge Drum 69
Recirculated or Overfeed System 70
Oil in Evaporators 71

Chapter 6 Compressors 74

How a Compressor Works 74
Historical Development: Compressor Types 80
Construction Details 85
Other Types of Compressors 93
Compressor Sizes 103

Chapter 7 Condensers 105

Requirements 105
Operation 105
Air-Cooled Condenser 106
Air- vs. Water-Cooled Condensers 108
Double-Pipe Water-Cooled Condenser 108

Shell-and-Tube Condenser *109*
Shell-and-Coil Condenser *110*
Water Supply; Cooling Towers *111*
Evaporative Condenser *114*

Chapter 8 Flow Equipment **116**
Liquid Receiver *116*
Pipes and Piping *116*
Hand Valves *119*
Filters and Strainers *121*
Driers and Dehydrators *122*
Oil Traps *123*
Automatic Purgers *124*
Sight Glasses *125*
Heat Exchangers *126*

Chapter 9 Electrical Controls and Control Valves **128**
Backpressure Control *128*
Temperature Control *130*
High-Pressure and High-Temperature Cut-outs *131*
Low-Oil Cut-out *131*
Low-Water Cut-out *132*
Float Switch *132*
Control Applications *132*
Solenoid Valve *133*
Evaporator Pressure-Regulating Valve *135*
Crankcase Pressure-Regulating Valve *136*
Crankcase Heater *137*
Check Valve *137*
Water-Regulating Valve *137*
Pressure-Relief Devices *138*
Computers and Refrigeration *139*
Energy Management *140*

Chapter 10 Lubrication **142**
Oil Types *142*
Requirements *143*
Oil Foaming *143*
Oil Selection *144*

Table of Contents

Chapter 11 Defrosting 147
- *Defrost Cycle* 147
- *Hot-Gas Defrost* 147
- *Thermobank System* 149
- *Water-Spray Defrost* 149
- *Electric Heater Defrost* 149
- *Defrost Controls* 151
- *Brine Spray* 151

Chapter 12 Compressor Drives 154
- *Requirements* 154
- *Electric Drive* 154
- *Potential Relay* 156
- *Contactor and Motor Starter* 157
- *Variable-Frequency Drive* 160
- *Steam Drive* 161
- *Diesel or Gasoline Engine* 161

Chapter 13 Food Preservation 163
- *Natural Ripening and Enzyme Action* 163
- *Bacteria and Molds* 164
- *Other Changes* 164
- *Cold Storage* 165
- *Freezing* 165

Chapter 14 Operating 167
- *Maintaining Temperature* 167
- *Economy* 168
- *Safety* 172
- *Automatic Refrigeration* 178
- *General Maintenance* 178

Chapter 15 Servicing 181
- *Variations from Refrigeration Requirements* 181
- *Food Storage Problems* 189

Chapter 16 Refrigerated Enclosures 191
- *Insulation* 191
- *Moisture Proofing* 196

Openings and Doors 197
Cabinet Construction 197
Tank and Pipe Insulation 198
Excessive Heat 198

Chapter 17 Instruments and Meters 200

Stem Thermometer 200
Dial Thermometer 204
Recording Thermometer 205
Pressure Gage 205
Transducer 208
Thermocouple 208
Thermistor 208
Electric Instruments 208

Chapter 18 Heat Calculations 210

Design Temperature 210
The k Factor 211
R Value 212
Usage Factor 214
Condensing Unit Sizing 215
The K Factor 215
Evaporator Sizing 216
Boxes Larger than 1500 Cubic Feet 219
Insulation Thickness 223
Condenser Calculation 223

Chapter 19 Humidity in Refrigeration 229

Relative and Absolute Humidities 229
Humidity Measurement 230
Effect of Evaporator Temperature 234
Electronics and Humidity Measurement 236

Chapter 20 Compressor Calculations I 237

Refrigerant Tables 237
Compressor Size 238
Compressor Horsepower 241
Complete Analysis of Expansion Valve and Evaporator 242
Pressure–Enthalpy Diagram 243
Sizing Compressors 245

Chapter 21 — Compressor Calculations II 252

Capacity Variation 252
Effect of Changing Suction Temperature 253
Effect of Changing Condensing Temperature 254
Detailed Standard Ton Conditions 254
Effect of Subcooling 256
Effect of Superheated Suction Vapor 257
Wet Compression 257
Changing Refrigerant 258
Heat Exchangers 260
Two-Stage Systems and Booster Systems 263
Cascade Systems 268

Chapter 22 — Refrigerant Lines and Pressure Drops 271

Effect of Pressure Drop 271
Calculating Velocity and Pressure Drop 273
Effect of Changing Flow Conditions 277

Chapter 23 — Brine in Refrigeration 282

Brine Chemistry 284
Corrosion Control 285
Brine Concentration 286
Congealing Tanks and Eutectic Plates 290

Chapter 24 — Liquid Cooling 292

Beverage Cooling Requirements 292
Water Coolers 292
Beer Coolers 295
Charged Water, Soda Water, and Seltzer 296
Milk Coolers 296
Brewery Refrigeration 297
Refrigeration in Wine Making 299

Chapter 25 — Complete Systems 301

Commercial Multiplexing 301
Cold Storage 305
Locker Plants 307
Air Conditioning 308
Heat Pumps 311
Ice Making 313

Chapter 26 Carbon Dioxide—Dry Ice **318**
Carbon Dioxide Refrigeration *318*
Manufacture of Dry Ice *318*
Advantages and Disadvantages of Dry Ice *320*
Uses of Dry Ice *321*

Chapter 27 Altitude and Its Effects **323**

Chapter 28 Absorption Systems **327**
Continuous Absorption System *328*
Diffusion Absorption System *329*

Chapter 29 Refrigeration Codes **332**

Appendix **334**

Index **401**

Preface

During many years of teaching refrigeration to practical engineers and technicians, I have never found a text that I consider to be wholly suitable. For practical people, such a text must be written in everyday English, yet it must not be oversimplified or leave out basic theory. This book is an attempt to fulfill these requirements.

A brief review is first given of the basic physical principles necessary to understand refrigeration cycles. The student is then given a detailed physical description of compression refrigeration. Each piece of equipment used is described, stressing its part in the complete refrigeration cycle. Illustrations are drawn from all branches of equipment: domestic, commercial, marine, and industrial. Where there are differences among these branches, they are pointed out. No attempt has been made to separate them by subject. Pictures and diagrams are used wherever possible to help provide a picture of the processes explained.

After the student thoroughly understands the physical process of refrigeration, calculations show how to choose the proper size equipment. These calculations are kept as simple as possible so that anyone who can solve a simple formula should be able to master them. Still, simplification has been done, not by omission, but by using everyday terms and illustrations.

Many tables and charts have been prepared that compare conditions and provide shortcuts for calculations. Useful tables and similar data are gathered in the Appendix so that they can be easily accessed for reference. Questions at the end of each chapter stress the key points in the chapter and give the student practice in solving problems. Answers to all numerical problems are supplied in the Instructor's Guide.

This work is a compilation of much knowledge I have collected both from practical experience and from the association and help of other people over more than 30 years. It is impossible to acknowledge all the sources of information collected here or to give credit to all the people, books, and magazines that have supplied information that has gone into this book.

I hope this book will be of value to beginners and to students of refrigeration as well as to practical people in the field, whether they are selling, installing, servicing, or operating large or small equipment.

Supplements

An Instructor's Guide providing answers and solutions to the end-of-chapter questions is available: ISBN 1-4180-6098-4.

Acknowledgments

I would like to thank Eugene Silberstein for performing a technical edit on the manuscript and providing me with detailed feedback, suggestions, and recommendations.

I would like to express appreciation to the following people for their input as reviewers of this edition:

Robert Chatenever, Oxnard College, Oxnard, CA
Terry Rogers, Midlands Technical College, West Columbia, SC
Eugene Silberstein, Suffolk Country Community College, Brentwood, NY

About the Author

Chris Langley graduated from Kean College of New Jersey in 1996 with a bachelor's degree in industrial engineering technology. He completed his master's studies in 2003 at Stevens Institute. He has over 30 years of experience in all facets of industrial utilities, including 18 years of hands-on experience with ammonia systems. Mr. Langley has also taught HVAC, stationary engineering, and refrigeration engineering in two of New Jersey's vocational-technical schools for 18 years as of this writing. He holds a New Jersey double gold engineer's license.

Basic engineering concepts have remained reasonably unchanged over the centuries. What has changed in recent years is the technology used to control refrigeration systems. These changes have allowed us better energy efficiency, easier troubleshooting, and more automatic operation of equipment, to name a few advances. I have also included information about one of the more popular halofluorocarbon refrigerants, R-134a. There are so many refrigerants in use, both old and new, to discuss them all would require a whole separate text. I encourage the student to keep current by attending seminars, talking to manufacturers and distributors, and using the Internet. DuPont, as well as the many organizations mentioned in Chapter 29, have always been extremely helpful to technicians/engineers, and I encourage the student to actively seek them out and participate in their offered activities.

Disclaimer

This book can only be considered as a general guide. The author and publisher have no liability nor can they be responsible to any person or entity for misunderstanding, misuse, or misapplication that may cause loss or damage of any kind, including material or personal injury, or alleged to be caused directly or indirectly by the information contained in this book.

Refrigeration Principles, Practices and Performance

Chapter 1 Fundamentals

To understand the principles of refrigeration it is necessary to understand:

1. The mechanical equipment and components used to produce refrigeration
2. The physical laws being exercised by the fluids within this mechanical equipment
3. How the mechanical equipment controls the behavior of the fluids, while still adhering to the physical laws that the fluids must obey

Therefore, so that we can obtain a full understanding of our subject, we shall review the important physical laws involved.

Heat and Temperature

The first and most important subject to take up is heat and its effects. You may ask: Why study heat when we are interested in refrigeration, or the production of cold? The answer is that cold is in fact defined by the absence of heat. If we remove heat from a substance, we cool it. So, in order to understand the refrigeration process, we must know something about heat transfer.

It is important to understand the difference between heat and temperature. *Temperature* is the intensity of heat, or how strong the heat is—how hot an object or substance is. In the refrigeration industry, the most commonly used temperature scale is the Fahrenheit scale. Fahrenheit temperatures are abbreviated using the letter "F." Temperature scales other than Fahrenheit and their relationships are shown in Figure 1-1.

Figure 1-1 shows us how to convert from Fahrenheit to celsius and back to Fahrenheit and compares four temperature scales: Rankine, Fahrenheit, kelvin, and Celsius. The Rankine scale is the absolute Fahrenheit scale, and the Kelvin scale is the absolute celsius scale. Absolute scales are useful in calculating ratios, which are mathematical relationships between like quantities. Example 1 illustrates a simple ratio calculation.

EXAMPLE 1 How much more dense (increase in mass per unit volume) is air at 50°F than at 100°F?

First, convert both temperatures so they are absolute temperatures. In this case, we use the Rankine scale because we are given Fahrenheit units. Since 100°F = 560°R and 50°F = 510°R, [1 − (510/560)] × 100 equals approximately 8.93%.

This kind of information is useful in combustion formulas. Combustion equipment performs better in the winter than in summer because the air is more dense (contains more oxygen per unit volume) in the summer.

Consider two containers on a stove, one containing 1 pint of water and the other containing 8 pints, or 1 gallon, of water. The containers are the same size, so they absorb the same amount of heat in the same length of time. Suppose conditions are such that the single pint of water comes to a boil in 5 minutes. We will find that the 8-pint kettle will

2 CHAPTER 1 *Fundamentals*

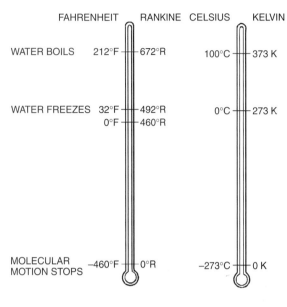

TEMPERATURE FORMULAS

$°F = \frac{9}{5}°C + 32; °C = \frac{5}{9}(°F - 32);$

$°R = \frac{4}{9}(°F - 32); °F = \frac{9}{4}°R + 32;$

$°C = \frac{5}{4}°R; °R = \frac{4}{5}°C$

Figure 1–1 Various temperature scales.

take 8 times as long, or 40 minutes, to come a boil.[1] If both pots are boiling, the water in both kettles will be at the same temperature. The boiling temperature of water is 212°F.[2] It can be shown that it takes 8 times more heat to bring 1 gallon of water to the boiling temperature than it does to bring 1 pint of water to the boiling temperature.

If the gallon of water is heated for only 20 minutes, rather than the 40 minutes required to bring it to the boiling temperature, the temperature of the water will be at some level below the boiling temperature. A thermometer will show the pint to be much hotter than the gallon. The gallon, however, will contain more heat than the boiling single pint of water because it has absorbed heat from the stove for a longer period of time. If the gallon heated 20 minutes is removed from the stove, it will stay warm longer than the pint heated to the boiling point. The gallon will give up more heat to warm something (or someone), even though it was not as hot to start with. That is why measuring temperature is not enough: We must be able to measure the *quantity* of heat in addition to the *level* of heat. The quantity of heat depends on the initial and final temperatures of the substance, the quantity of substance, and the type of substance.

Heat is a form of energy. For engineering work, including refrigeration, it is measured in *British thermal units,* better known by the abbreviation Btu. *The Btu is the amount of heat necessary to raise the temperature of 1 pound of water 1 degree Fahrenheit.*

Returning to the previous problem, let us consider 1 pound of water instead of 1 pint.[3] Let us assume the water is at 60°F. To raise its temperature to 212°F, it must be heated:

$$212 - 60 = 152°F$$

Since it takes 1 Btu to heat 1 pound of water 1 degree, it will take 152 Btu to heat 1 pound of water 152 degrees, or to the boiling point. If a second container has 8 pounds of water in it,[4] each pound will require 152 Btu to heat it to the boiling point. Therefore 8 pounds of water will require

$$8 \times 152 = 1216 \text{ Btu}$$

Eight pounds of water heated only 20 minutes will absorb half as much heat, or

$$1/2 \times 1216 = 608 \text{ Btu}$$

This will raise its temperature half of the 152 degrees necessary to raise it to the boiling point. Its final temperature is therefore

$$(1/2 \times 152) + 60 = 136°F$$

---

[1] This assumes no variations in losses. Such losses would change these figures slightly.
[2] At atmospheric pressure at sea level (14.696 psia).
[3] One pound of water is almost a pint.
[4] Eight pounds of water is 0.96 gallons.

Therefore, the kettle with 1 pound of water heated to 212°F has 152 Btu of heat added to it, and the larger kettle, which is heated to 136°F, has 608 Btu added.

Specific Heat

If we put 1 pound of iron on the stove and check its temperature with a thermometer, we will find that it reaches 212°F in less than 1 minute. However, if is removed from the stove, it will cool off more quickly than the water did. It takes much less heat to raise the temperature of iron than to raise the temperature of water a given amount. Also, to reduce its temperature, we need to remove less heat. Experiment shows that it takes only 0.118 Btu of heat to raise the temperature of 1 pound of iron by 1°F. This is called the specific heat of iron. The *specific heat* of a substance is the amount of heat necessary to raise the temperature of 1 pound of that substance by 1°F. Since it takes 1 Btu to raise the temperature of 1 pound of water by 1 degree, the specific heat of water is 1.

Figure 1-2, as well as Table A-1 and Table A-2 in the Appendix, gives some useful specific heats. To find the total heat necessary to raise the temperature of any substance, multiply its weight times its specific heat times the temperature change:

$$H = WS(t_1 - t_2)$$

Food	Specific Heat
	Btu per pound
FRUITS AND VEGETABLES	
Most fruits, berries, vegetables	0.92
MEAT	
Beef	0.77
Fish, poultry	0.82
Veal	0.70
Lamb, pork	0.66
DAIRY PRODUCTS	
Butter, cheese	0.64
Eggs	0.76
Milk, cream	0.92

Figure 1-2 Specific heats of some common foods. (For complete table, see Appendix Table A-2)

where

H = heat in Btu
W = weight in pounds
S = specific heat, from Figure 1-2, Table A-1, or Table A-2
t_1 = original temperature in °F
t_2 = final temperature in °F

Note: There is no such thing as a negative Btu. Always use absolute values (distance from zero). To heat something, we add heat; to cool something, we remove heat.

EXAMPLE 2 How much heat is required to heat 25 pounds of lead from 70°F to 200°F?

$$H = WS(t_1 - t_2)$$
$$= 25 \times 0.030(200 - 70)$$
$$= 97.5 \text{ Btu of heat}$$

Also, if 97.5 Btu of heat are removed from 25 pounds of 200°F lead, it will reduce the temperature to 70°F. That is, the formula will work for heating or for cooling.

EXAMPLE 3 How much refrigeration is required to cool 400 pounds of beef from 85°F to 36°F?

$$H = WS(t_1 - t_2)$$
$$= 400 \times 0.77(36 - 85)$$
$$= -15{,}100 \text{ Btu}$$

The (−) sign can be taken as requiring cooling instead of heating—that is, removing heat instead of adding it.

Latent Heat

Let us go back to our pint of water. If it is left on the stove after it comes to a boil, its temperature will remain the same. However, it will gradually boil away; that is, it will slowly change from water to steam. And if it is taken from the stove, it will stop boiling. Simply stated, heat is required to boil water. If the water came to a boil in 5 minutes, it will take about 38 minutes to boil dry. So it takes much more heat to evaporate a pound of water already at the boiling temperature than to heat it to that temperature. Experiment shows it takes 970 Btu of heat to evaporate 1 pound of water that is already at 212°F. This is known as the latent heat of evaporation or the latent heat of vaporization. "Latent" means hidden. The heat is hidden as far as the temperature is concerned because the steam is still at 212°F. Thus, the *latent heat of vaporization* is the amount of heat necessary to change a liquid to a vapor with no change in temperature.

If this vapor is collected in a coil or steam radiator instead of boiled off into the room, it will give 970 Btu of heat back to the room as it condenses. Thus, this latent heat is available to be used for heating if the steam is condensed back to water.

If we put 1 pound of ice on the stove and melt it, we will find that it takes almost as long to melt the ice completely as to heat the pound of water from 60°F to 212°F. Furthermore, the water around the ice will not heat up as long as it is in contact with unmelted ice. A thermometer will show both the ice and the water to be at 32°F during the entire melting period. Thus, we have another hidden or latent heat. Experiment shows that it takes 144 Btu[5] of heat to melt 1 pound of ice. This is the latent heat of fusion. Thus, the *latent heat of fusion* is the heat necessary to change 1 pound of a solid to a liquid without a change in temperature.

Since ice must absorb heat to melt, this heat must be supplied by some other source. As we said earlier, to remove heat from something is to cool it. Therefore, the melting ice will cool whatever supplies this heat.

EXAMPLE 4 In Example 3, we figured it took 15,100 Btu to cool 400 pounds of beef from 85°F to 36°F. How much ice would be required to do this cooling?

Each pound of ice will absorb 144 Btu. Thus it will take 15,100/144 = 104.9 pounds of ice to do the job, neglecting losses.

This is a typical refrigeration problem. The refrigeration can be accomplished by ice or by mechanical means. In fact, mechanical refrigerators are usually compared to ice to describe the amount of cooling they are capable of providing. The cooling capacity of small machines is now rated directly in Btu per hour cooling capacity. At one time, their cooling capacities were rated in "ice melting equivalents," better known as IME.

EXAMPLE 5 How many Btu per hour of refrigeration can a machine rated at 10 IME produce?

Since 1 pound of ice absorbs 144 Btu, 10 IME are equivalent to

$$10 \times 144 = 1440 \text{ Btu per hour}$$

Larger-sized machines are rated in tons of refrigeration. One *ton of refrigeration* is the same amount of refrigeration as is produced by 1 ton of ice melted each 24 hours. Since 1 pound of melting ice at 32°F has a latent heat of 144 Btu, 1 ton of ice can absorb 2000 × 144 = 288,000 Btu. That is, it will take 288,000 Btu of heat to melt 1 ton of ice. Since a 1-ton machine will absorb 288,000 Btu in 24 hours, it will absorb 288,000 Btu per day or 12,000 Btu per hour or 200 Btu per minute.

EXAMPLE 6 One ton of refrigeration is equivalent to how many IME per hour?

One ton of refrigeration is equivalent to 12,000 Btu per hour. Therefore, 12,000/144 = 83.3 IME.

Besides latent heat of evaporation and latent heat of fusion, latent heat of sublimation is another term that you may encounter. Certain substances, such as dry ice, which is ice made of carbon dioxide gas, evaporate directly from the solid or ice form to a vapor. Such evaporation from the solid form is known as sublimation. Evaporation by any other name, such as sublimation, still requires heat. Thus, *latent heat of sublimation* is the heat necessary to change 1 pound of a solid to a vapor. It may be considered as the sum of the latent heat of fusion and the latent heat of evaporation. For dry ice this latent heat is 246 Btu per pound. Its sublimation point is −109°F.

EXAMPLE 7 How much dry ice is required to cool the beef in Example 3?

---

[5] 143.33 Btu has been established as a more accurate value, but the whole number 144 Btu has been adopted by the trade. Practical figures as commonly used are given throughout this book.

Since 15,100 Btu of cooling is required, it will take

$$15{,}100/246 = 61.3 \text{ pounds of dry ice}$$

to cool the beef.

Note that in Example 4, it took 105 pounds of water ice to do this same job. However, other factors, such as cost and required temperatures, may be more important than weight alone. These factors will be be discussed in more detail in Chapter 26.

Change of State

So far, we have discussed water in the form of a solid (ice), a liquid, and a vapor (steam). Many other substances can be changed from solid to liquid or from liquid to vapor by applying heat. Latent heat must be added to change a solid to a liquid or to change a liquid to a vapor. Heat must be absorbed, or taken away, to change a vapor to a liquid or to change a liquid to a solid. The amounts of heat will vary greatly with different substances. They will also vary as the pressures exerted on the substances change. As we also noted with the example of dry ice, changes directly from a solid to a vapor are possible. Any such change of a substance from solid, liquid, or vapor form to any other of these forms is known as a *change of state*.

Vapor

When steam or vapor is in contact with the liquid from which it has evaporated, or when it is at its evaporating temperature, it is known as *saturated vapor*. If it is boiling quietly so that only vapor, no liquid, is present, it is known as dry saturated vapor. However, if it is boiling violently so that droplets of liquid spatter into the vapor, or if an attempt is made to cool the vapor so that some of it will recondense, it is known as *wet vapor*. Wet vapor may be droplets of liquid carried through a pipe with a flow of vapor, or it may form as a fog in the vapor.

If more heat is applied to a boiling liquid, the liquid boils faster. The temperature of a vapor from this liquid cannot be raised above the boiling point as long as liquid is present. However, if the vapor is carried away from the liquid in a pipe or otherwise and then heated, nothing prevents a rise in temperature. A vapor so heated above its boiling temperature is known as *superheated vapor*.

Thus, vapor may be wet, dry saturated, or superheated. A *wet vapor* contains some unevaporated liquid. A *dry saturated vapor* has no liquid mixed with it, but is exactly at evaporating temperature. A *superheated vapor* has been heated above its boiling temperature.

Gas and Vapor

We have spoken of evaporated liquid as a vapor. This is the correct term for it. Many times, we hear such a substance incorrectly called a gas. A *gas* is actually a highly superheated vapor. The principal difference between a vapor and a gas is that a gas will follow the so-called *gas law*, which states that the volume of a gas varies inversely with the absolute pressure and directly with the absolute temperature. Any such change can be calculated using formulas given in any elementary book of physics or chemistry. However, as a gas approaches its condensing temperature, using these formulas leads to errors. The closer the temperature of the gas is to the condensing temperature, the greater these errors become. *Vapor is the gaseous substance on which the "gas formulas" will not work.* Such information, when desired, must be taken from tables which are records of tests. No exact dividing line can be given between a vapor and a gas. *For refrigeration purposes such a substance is a vapor, and tables must be used to find its properties.*

Energy

Heat is a form of energy. *Energy* is ability to do work. It may be in the form of mechanical energy, such as that delivered by a steam engine or electric motor. It may be chemical energy, such as the energy in gasoline or gun powder. It may be electrical energy, as delivered by a generator or battery; or it may be heat energy, which can be used to drive internal combustion engines or steam engines.

Notice that heat energy is changed to mechanical energy in the last-mentioned case. Chemical energy can be changed to heat energy by combustion. Electric energy can be converted to mechanical energy in an electric motor, or to heat energy in an electric heater. These are only a few of the possible changes of the form of energy. However, energy can neither be created nor destroyed.[6] We can only convert it from one form to another.

Since energy is neither created nor destroyed, there are always the same relationships between the units of the different forms of energy. That is, 746 watts of electrical energy equals 1 horsepower of mechanical energy. One horsepower of mechanical energy equals 2545 Btu per hour, or 42.42 Btu per minute of heat energy. A complete table of energy conversion equivalents is given in Table A–4 in the Appendix.

Pressure

All refrigeration systems contain pressure which must be controlled. Most of this pressure is vapor pressure. Any gas or vapor enclosed in a container will press or push out equally in all directions. This pressure may be greater than or less than atmospheric pressure. If it is greater than atmospheric pressure, it is measured by the amount of pressure or push on each square inch of surface. The unit used is pounds per square inch, or psi. The pressure usually indicated is *gage pressure,* abbreviated psig. This is the pressure shown on an ordinary pressure gage. It is important to remember that such a gage measures the difference between the pressure in the system and atmospheric pressure. Thus, a change in atmospheric pressure without a change in the system will give a different gage reading. Normally, the variation in atmospheric pressure at any one place is so small that it cannot be detected on the gage. However, a pressure specified for sea level will give a different condition at an altitude, because altitude affects atmospheric pressure.

The average atmospheric pressure at sea level is 14.7 psi.[7] Therefore, we can get the total pressure in any system at sea level by adding 14.7 to the gage pressure. Such a pressure is called an *absolute pressure,* abbreviated psia. It is the pressure in the system measured above a perfect vacuum, that is, measured above no pressure at all. The same results will always be obtained at the same absolute pressure. This is why absolute pressures are used so much. If a gage pressure is specified, it is assumed to be at sea level unless otherwise stated. A more complete description, with the variations to be expected at different altitudes, will be given in Chapter 27.

If the pressure in the system is less than atmospheric, it is usually measured in inches of vacuum (the pressure of a column of mercury so many inches high). Here again, gage pressure is compared to atmospheric pressure. A 30-inch[8] vacuum will be a perfect vacuum at sea level where the barometric pressure is 30 inches. A 10-inch vacuum will be 10 inches below atmospheric pressure, or one-third of a perfect vacuum. This is sometimes called a gage pressure of 10 inches of vacuum, despite the contradiction of the terms "pressure" and "vacuum." However, it should be remembered that any partial vacuum has some absolute pressure.

If it is necessary to change a vacuum to an absolute pressure, the vacuum value is subtracted from 30 inches. Thus, our 10-inch vacuum has an absolute pressure of $30 - 10 = 20$ inches of mercury absolute pressure.

On the other hand, absolute pressures are nearly always given in psi, so it is necessary to convert our inches to psi. Remember that atmospheric pressure is 14.7 psi and also 30 inches, so 14.7 psi = 30 inches. From this we can set up a proportion:

$$\frac{\text{Pounds} = 14.7}{\text{Inches} = 30}$$

---

[6] Even in nuclear reactions, mass is converted to energy. The energy is contained in the mass.
[7] This varies with the barometric pressure; 14.696 psi is the accurate mean pressure.
[8] 29.92 inches is the accurate mean.

Thus, to convert our 20 inches of mercury absolute pressure to psi:

$$\frac{\text{Pounds}}{20} = \frac{14.7}{30}$$

$$\text{Pounds} = \frac{20 \times 14.7}{30} = 9.8 \text{ psia pressure}$$

Alternatively, the direct ratio of psi to inches as given in Table A–5 in the Appendix may be used:

$$0.49 - 20 = 9.8 \text{ psia}$$

A rough approximation of a conversion can be obtained by remembering that 2 inches equals 1 psi. This will usually give an answer as near as can be read on a gage. For instance, our 20 inches of mercury absolute pressure then becomes 10 psi absolute. This is very close to the 9.8 psia obtained above.

Partial Pressure

Consider two containers of equal volumes connected by a line and a pump, as shown in Figure 1–3. Let container A be filled with a light gas X to a pressure of 50 psia. Let container B be filled with a heavy gas Y to a pressure of 25 psia. Now, if the pump is operated to force all of gas Y into container A with gas X, the result will be a pressure of 75 psia. That is, X still exerts its 50 psia and Y adds its 25 psia to this. Furthermore, despite the fact that Y is heavier than X, it will be found that there is no tendency for Y to fall to the bottom, or for X to rise to the top.[9] Each gas fills the entire container and exerts its individual pressure on all inside surfaces of the container. This is known as *Dalton's law of partial pressure*.

Heat Transfer

Heat may be transferred or moved from one point to another by three methods: conduction, convection, and radiation. *Conduction* is heat transfer by contact.

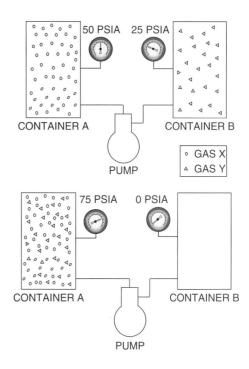

Figure 1–3 Dalton's law of partial pressure.

If one end of a metal bar is heated, the other end soon warms up. Each atom or particle that is warmer than its neighbor passes some of its heat to this neighbor by contact or conduction. Heat is transferred from the warm to the cold side of a wall by conduction. Heat gets through the walls of a metal pipe by conduction.

Convection is heat transfer by the movement of a fluid, usually water or air. Air blowing over a product and back to cooling coils picks up heat from the product by conduction. This heat is carried to the coils by convection. Air convection may be natural (by gravity circulation) or forced (by a blower). In other cases, water or brine, a salt solution, may be used to transfer heat from a product to cooling coils.

Radiation is heat movement through space by shining, just as light is radiated. The sun radiates

---

[9] These gases are blended. In practice, heavy gases will sink, as when hydrochlorofluorocarbons (HCFCs), chlorofluorocarbons (CFCs), or hydrofluorocarbons (HFCs) fill a basement or crawl space and displace oxygen. This poses a life-threatening situation.

heat to the earth. A person near a fire feels warmth from the fire. One near a cold surface feels chilled as heat from his warm body is radiated to the colder surface. A product near a coil in a room above 32°F may be frozen as it radiates its heat directly to the colder coil. Radiant heat can be blocked just as light can be blocked by anything that forms a shadow. Radiant heat is the only form of heat that can travel through a vacuum (such as space).

Scope of Refrigeration

Refrigeration equipment can be roughly divided into four general classifications: domestic, commercial, industrial and marine.

Domestic refrigeration is used for home kitchens. It is nearly always a cooling cabinet with its mechanical equipment built in to give a completely self-contained unit. The cabinets may vary in size from 2 or 3 cubic feet up to 20 or more cubic feet.

Commercial refrigeration is used for holding food for sale in a retail store. This includes the general run of grocery refrigerators, meat boxes, refrigerated showcases, ice cream cabinets, beverage coolers, and small-sized air conditioning equipment. The refrigerated compartment may be a cabinet similar to a domestic box, but usually larger, called a reach-in cabinet; or it may be a walk-in cooler. It may be self-contained, or the mechanical equipment may be in a basement or other convenient place.

Industrial refrigeration is used in the manufacturing or processing of foods or other products requiring refrigeration. This includes ice plants, cold storage plants, dairies, food freezing, meat packing plants, breweries, large air conditioning equipment, and so on. The amount of refrigeration required is large. The cooling may be done in refrigerated rooms, or in insulated warehouses which are entirely refrigerated. The required mechanical equipment is usually grouped in a central engine room.

Marine refrigeration is most closely associated with industrial refrigeration. We find this type of equipment on fishing boats, processor/tenders, naval ships, liners, cargo ships, and so on. The main difference between industrial and marine refrigeration is that industrial systems are built on solid ground. Marine equipment must be designed to work on a vessel tossing and rolling on the open ocean. Special design considerations must be taken to ensure that evaporators can function without flooding and that condensers maintain their liquid seal.

There are no distinct dividing lines between these classifications. A large domestic box might also be sold to a grocery store or restaurant and be called commercial refrigeration. A dairy might supply refrigeration to retail cabinets from the same engine room that supplies refrigeration for milk cooling or ice cream freezing. However, this classification gives us a general idea whether we are working with small, medium, or large equipment.

Questions

1. Why is a knowledge of heat and its effects necessary in a study of refrigeration?
2. How many Btu of refrigeration must be used to cool 1 gallon of milk from 80°F to 45°F?
3. A 1-quart glass milk bottle weighs 1.44 pounds. How much heat must be removed per gallon of milk to cool the milk bottles from 80°F to 45°F?
4. A 1-quart cardboard milk carton weighs 3 ounces. How much heat must be removed per gallon of milk to cool the cartons from 80°F to 45°F?
5. If the cooling in Questions 2, 3, and 4 is done with ice, how much ice will be required per gallon of milk, neglecting other losses?
6. Repeat Question 5 using dry ice.
7. Explain where conduction is an aid in a refrigeration system.
8. Explain where conduction should be reduced as much as possible in a refrigeration system.

9. Explain where convection is an aid in a refrigeration system.
10. Explain where radiation is an aid in a refrigeration system.
11. Explain where radiation should be reduced as much as possible in a refrigeration system.
12. What is the difference between a wet vapor and a superheated vapor?
13. Are formulas or tables best for finding the properties of refrigerant vapors? Explain your answer.
14. Convert 25 psig to psia.
15. What is the difference between domestic, commercial, industrial, and marine refrigeration?

Chapter 2
The Compression System of Refrigeration

Evaporation

Let us return to our example in Chapter 1 about water boiling on a stove. As long as the water is boiling, it is absorbing heat. In order for a substance to absorb heat, however, another substance must be rejecting heat. To supply or give up heat is to cool. Therefore, as long as heat is absorbed by the boiling water, whatever gives up that heat is cooled off. In our boiling water example, the flame from the stove is cooled. The double boiler, Figure 2-1, uses this fact. As long as water is in the lower container, anything in the upper container will not be heated above 212°F—in spite of a flame that may be 2000°F. More fire or a hotter flame causes more rapid boiling, or the evaporation of a greater quantity of liquid, but no rise in temperature. Put the other way around, the more evaporation there is, the more the cooling effect on the flame.

So far we have always spoken of the boiling temperature of water as 212°F. However, if we were to boil our water on the top of a mountain 10,000 feet high, we would find that our water would boil at 193°F. Again, nothing we could do in applying more heat would cause the water to get hotter than 193°F. Perhaps you have heard of a camping trip in the mountains where eggs could not be hard boiled, or beans could not be cooked. The reason is that boiling water at such elevations is not hot enough to do the required cooking.

This phenomenon occurs because atmospheric pressure is lower at higher altitudes. At 10,000 feet the atmospheric pressure is only 10 psi, not 14.7 psi. A barometer, which measures atmospheric pressure, will read 20.4 inches instead of 30 inches. A 20.4-inch vacuum would be a perfect vacuum at this elevation. The lower the pressure, the lower the boiling temperature will be. If we check the effect of a greater pressure, we find the boiling temperature to be higher. In a steam boiler with the pressure at 100 psig, the temperature of the boiling water is 338°F. *So, a change of pressure will change the boiling temperature.* This is due to the vapor pressure pushing down on the surface of a liquid.

Figure 2-2 shows the boiling temperature of water at various pressures. From this figure, we can see that at a pressure of 0.178 psia, or a gage pressure of 29.56 inches of mercury, water can be boiled at 50°F. As long as this vacuum can be maintained, the water temperature cannot be raised above 50°F. The boiling water absorbs heat, which removes heat from, or cools, the heat source. As long as there is air—or anything else warmer than 50°F—around the water, boiling or evaporation will continue. The only difficulty is maintaining the required vacuum. As the water boils, the steam or vapor formed must be removed by a pump or other device to prevent the pressure from rising. Such a pump must have an enormous capacity, because at these very low pressures the vapor occupies an enormous volume. At atmospheric pressure, the steam occupies approximately 1600 times the volume of the water from which it was generated. However, there are refrigeration systems that use water as the refrigerant.

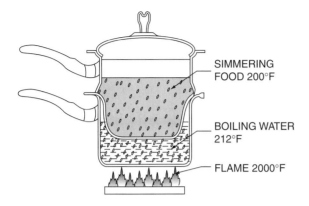

Figure 2–1 The double boiler.

To avoid some of the difficulties of using water for refrigeration, we need a liquid that evaporates more easily. If we spill a little ether on our hands, our hands become quite cool. There is enough evaporation without boiling to absorb considerable heat. In this case, the heat absorbed from our hands cools the hands. If we boil the ether at atmospheric pressure, we will find that it boils at 94.3°F. Figure 2–3 shows the boiling temperature of ether at various pressures. From this figure we can see that we can boil ether at 15°F if we maintain a 25.-inch vacuum over it. The vapor pressure above the fluid at 25.5 inches is still more than 2 psia.

Again, anything warmer than 15°F will supply heat to the ether at a 25.5-inch vacuum and cause it to boil. Whatever supplies this heat is cooled, and the temperature of the heat source will drop closer to 15°F. This makes it possible to produce satisfactory refrigerating temperatures without such extreme vacuums. Ether was once used as the heat transfer medium in refrigeration applications, but other fluids that produce better results have since been developed.

One of the oldest and best-known substances that has been used as an effective refrigerant is ammonia. Figure 2–4 shows the pressure–temperature characteristics of ammonia. At atmospheric pressure, ammonia boils at 28°F below 0°F. A cylinder of the liquid at room temperature, say 85°F, will be under a pressure of 152 psig. The liquid in the cylinder always maintains the pressure corresponding to the temperature found on the chart. When a substance is saturated, its temperature will always correspond to a specific temperature.

If the valve on a cylinder of ammonia is opened to allow vapor to escape, the liquid that remains in

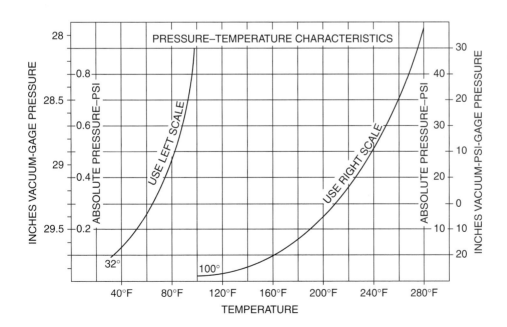

Figure 2–2 Water: pressure–temperature characteristics.

12 CHAPTER 2 *The Compression System of Refrigeration*

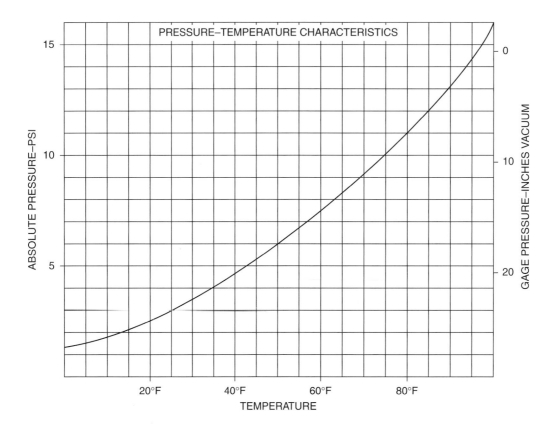

Figure 2–3 Ether: pressure–temperature characteristics.

the cylinder will evaporate or boil off more vapor to replace it. Vapor pressure above the liquid is reduced. Vapor from the liquid will be generated to replace this lost vapor pressure. However, evaporation requires heat, which is drawn from the liquid itself, so the liquid remaining in the cylinder will become somewhat cooler. Therefore the pressure in the cylinder will become slightly lower. This action could be continued by leaving the valve open until the pressure drops to atmospheric. The temperature of the remaining liquid will then be reduced to −28°F as shown in Figure 2-5. At any time before the pressure reaches atmospheric, the pressure–temperature relationship will be that shown in Figure 2-4.

If the cylinder is turned upside down and the valve opened, liquid will come out of the valve. However, the boiling temperature of liquid ammonia at atmospheric pressure is −28°F. Therefore, as it emerges from the valve, some of it will evaporate instantly, or *flash* into vapor. Heat to evaporate this liquid is taken from itself, as well as the air surrounding the released liquid, so the unevaporated liquid is immediately chilled to −28°F. If the liquid in the cylinder was at 85°F, about 21 percent of it would have to evaporate to cool the rest to −28°F. The rest of the liquid would then be available for refrigeration. The air surrounding this vaporizing liquid is also being cooled.

If the liquid ammonia is fed into a long coil, Figure 2-6, the −28°F liquid will cool the metal walls of the coil to nearly that temperature. The coil could then be used to cool a room or food

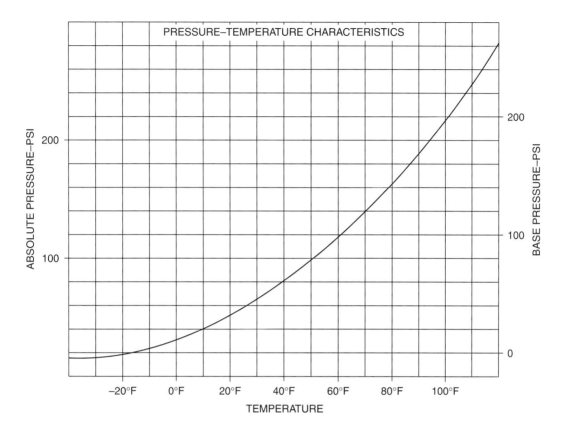

Figure 2–4 Ammonia: pressure–temperature characteristics.

products. As the coil absorbs heat from the room or food, this heat will evaporate the rest of the liquid in the coil. Such a coil is called an *evaporator coil* or simply an *evaporator*.

Refer back to the three methods of heat transfer described in Chapter 1. The air movement throughout the room most closely defines convection. As the product "touches" the air and the air in turn "touches" the coil, heat is transferred by conduction. Radiation heat transfer occurs throughout refrigerated spaces. It is here that the refrigerant, as the ammonia is called, is evaporated by absorbing heat. The valve should be adjusted so that just enough liquid is allowed to flow into the coil that it will all be evaporated by the time it reaches the end of the coil. The valve used to make this adjustment is called an *expansion valve*. As the liquid boils, it absorbs latent heat (no temperature change).

Latent heat quantities are much greater than sensible heat quantities.

Such a system will produce refrigeration as long as ammonia is present in the cylinder. Once the ammonia supply is depleted, the cooling effect will stop. In addition, if the ammonia were allowed to escape to the air, it would create a public nuisance, and replacing the ammonia supply would be costly. For these reasons, it is desirable to have a cycle that repeats itself. In such a system, the refrigerant supply is not depleted and the refrigeration effect is provided as long as the system is in operation. The additional system components that are required to create this repeating cycle are a compressor, a condenser, and any associated motor drive. These components act as a reclaiming plane that takes the evaporated vapor and converts it back to a liquid so it may be reused.

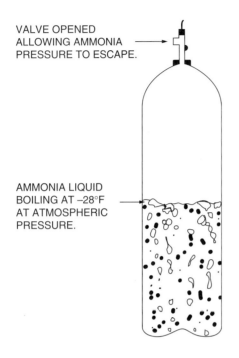

Figure 2-5 Boiling liquid ammonia.

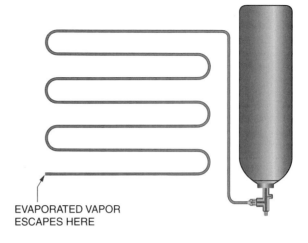

Figure 2-6 Liquid ammonia fed to a coil.

This should be stressed: The compressor and its motor in no way produce cold directly. The compressor is used to remove the vapor from the coil as fast as it is evaporated. It then compresses this vapor back to 152 psig. The mechanical energy used in this compression ends up as heat energy in the vapor. This heats the vapor to 290°F as it leaves the compressor. This hot, high-pressure, superheated vapor then goes to a condenser. The condenser is an assembly of pipes having water or air, or both, flowing over them. The air or water absorbs heat from the pipes, which first cools the vapor to 85°F. This is the *saturation temperature* of the ammonia at the given pressure. The vapor cannot be cooled below this temperature and remain a vapor at that pressure. So further cooling condenses it back to a liquid. At that point it drains back to a cylinder, where it is ready to be used again. This storage cylinder is called the *liquid receiver*. The complete cycle is shown schematically in Figure 2-7. The air or water cooling the condensing refrigerant is an example of *sensible heat*.

Pressure Cycle

From the preceding discussion and Figure 2-7, we see that refrigerant circulates continuously through the system. From the expansion valve it travels through the evaporator, then back through the suction line to the compressor. This portion of the system is at the low pressure required by the evaporator. This low pressure will vary depending on the temperature required. It can be found opposite the required evaporator temperature in Figure 2-4. It is important to recognize that the pressure will be the same from the expansion valve up to the compressor. This region is called the *low-pressure side*, or more simply, the *low side*, of the system. Because the evaporator itself is the most important part of the low side, the term *low side* is often used to mean an evaporator coil. The pressure in the low side is sometimes called the *back pressure*. That is, it is the pressure on the back side of the compressor. It is also sometimes called *suction pressure*.

The compressor takes the low-pressure vapor and builds it up to a pressure high enough to condense. The compressor discharge line, the condenser, the liquid receiver, and the liquid line are all at this high pressure. The discharge line connects the compressor to the condenser(s) and the liquid line supplies liquid refrigerant to the evaporator(s) from

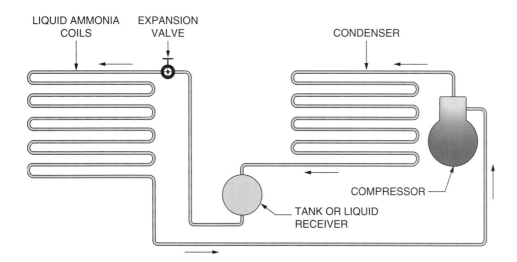

Figure 2–7 Schematic of refrigeration system.

the storage vessel. These parts of the system are collectively called the *high side*. Actually, the compressor crankcase usually contains low-pressure vapor. The entire compressor is sometimes considered part of the high side, but not always.

Thus the complete system is divided into a high side and a low side, Figure 2–8. The expansion valve and the compressor are the division points. A pressure gage anywhere in the high side will give the high pressure. A gage anywhere in the low side will give the low pressure. Gages are usually put at or near the compressor, but again, it is important to remember that these gages show the pressure found throughout the system. These indicated pressures are all saturation pressures.

In the preceding discussion we have spoken of ammonia evaporating at atmospheric pressure, which is −28°F. Temperatures this cold are sometimes used, but higher temperatures are more commonly encountered. A 15°F temperature coil is more common. From Figure 2–4, or from tables, we can find that we must maintain a 28.4-psig pressure on the low side to allow the ammonia to boil at 15°F. As in the previous case, we still have the same pressure throughout the low side, but now at 28.4 psig. This need not change the condensing

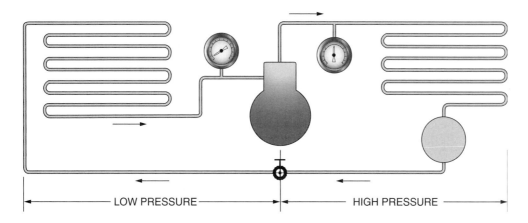

Figure 2–8 High and low sides.

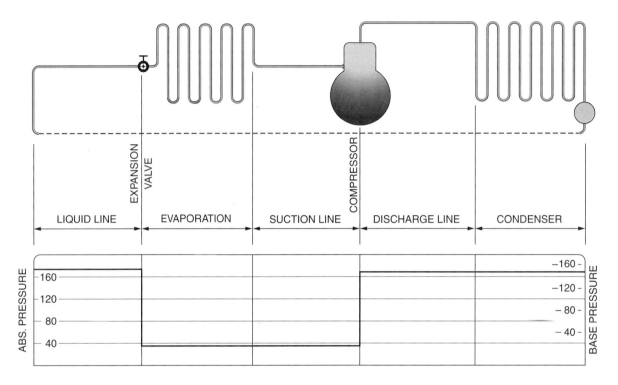

Figure 2-9 Pressure cycle of refrigerating system.

pressure, which remains at 152 psig. These latter pressures are shown graphically in Figure 2-9.

Temperature Cycle

The temperature cycle is shown graphically in Figure 2-10. If the pressure in the evaporator is such that the refrigerant boils at 15°F, any part of the evaporator in which liquid is present will be at this temperature. It is important to remember that as the end of the coil is approached, its temperature does not rise gradually as the coil absorbs heat. The heat boils the liquid within the coil. This liquid boils at the same temperature in all parts of the coil, because it is all at the same pressure. Once the last bit of liquid in the coil is evaporated, only vapor is present.

With no liquid present, the vapor can and does become heated above its boiling or saturation temperature. Therefore, superheated vapor—that is, vapor heated above its boiling temperature—usually leaves the coil. It may pick up 4°F of superheat, heating it from 15°F to 19°F. As the vapor flows back through the suction line to the compressor, additional superheat will be absorbed. If the line is well insulated, this may not amount to more than 7°F, heating the suction gas to a temperature 26°F. Thus, entering the compressor we have a 26°F vapor at a pressure of 28.4 psig. Notice that the temperature of 26°F and the pressure of 28.4 psig do not correspond to each other on the pressure-temperature chart (Figure 2-4). This is because the refrigerant is superheated and the pressure-temperature relationship holds only for saturated refrigerants. The compressor increases the pressure of the superheated vapor from 28.4 psig to the compressor discharge pressure of 152 psig. At these conditions, the heat of compression will heat this vapor up to 200°F. Thus, the vapor leaving the compressor is at 152 psig pressure and 200°F temperature.

At this high temperature, some cooling will be encountered in the discharge line, the amount

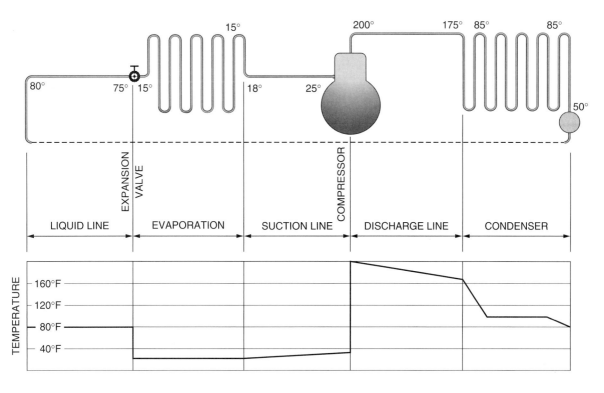

Figure 2-10 Temperature cycle of refrigerating system.

depending on the length of the line as well as the temperature of the air surrounding the discharge line. If we assume that the refrigerant cools to 175°F going to the condenser, we have 152 psig vapor at 175°F entering the condenser. The first few coils of the condenser cool the vapor to 85°F, the condensing temperature. Then any further heat removal condenses the vapor to a liquid. As the mixture of vapor and liquid progress through the condenser coil, there is progressively less vapor and more liquid present. However, as long as a vapor is present, the liquid cannot be cooled below 85° F. If sufficient liquid collects in the lower coils so it does not contact vapor, it may be subcooled below 85°F. *Subcooling* is defined as cooling the liquid below its saturation temperature.

Let us assume that the liquid is subcooled 5 degrees. This will reduce its temperature to 80°F. Thus an 80°F liquid, still at 152 psig pressure, will collect in the liquid receiver. From here it will flow through the liquid line to the expansion valve. There may or may not be a slight change of temperature of the liquid as it flows through pipes before reaching the expansion valve. As the refrigerant exits the expansion valve, the temperature changes instantly as the pressure is reduced from 152 psig to 28.4 psig (saturated pressure–temperature relationship). Naturally, temperatures will be somewhat different in different systems, but this represents a typical example. Remember: The pressure-temperature relationship applies only to saturated fluids—not those that have been superheated or subcooled.

Heat Cycle

The pressure and temperature cycles can be checked directly in a system with gages and thermometers. However, there is another important cycle that cannot be measured directly. That is the heat cycle as shown in Figure 2-11. The heat cycle can be

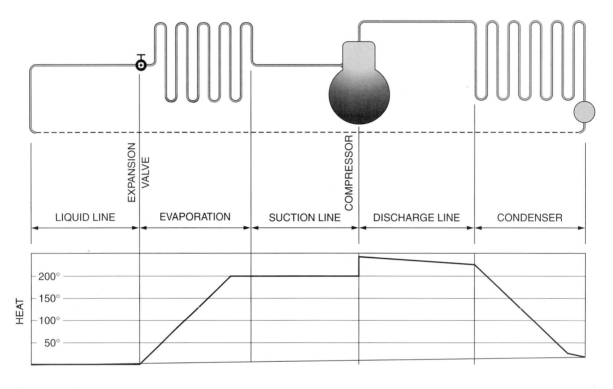

Figure 2–11 Heat cycle.

found from tables once the pressure–temperature information is obtained. We will show how this is done in Chapter 20. Information is given here on the heat cycle for the above system with a 15°F evaporator and an 85°F condenser to complete the picture.

The heat absorbed may be calculated per pound of ammonia circulated, or per ton of refrigeration required. One ton of refrigeration is the cooling effect that takes place when 200 Btu per minute are transferred. Therefore, if we figure our system on this basis, 200 Btu per minute must be absorbed by the evaporator. This heat is absorbed gradually over the whole evaporator surface. In the suction line, 2.5 Btu more are picked up.

The vapor is then drawn to the compressor, where it is compressed from 28.4 psig to 152 psig. The mechanical energy used in this compression is converted to heat energy in the vapor itself. This adds 33.2 (from the thermodynamic tables) Btu of heat to the vapor. Thus we now have a total of 235.7 Btu in the vapor leaving the compressor. This vapor goes to the condenser. Here the 235.7 Btu must be removed to liquefy it and make it ready for use again in the evaporator.

Notice that the evaporator and compressor add heat to the fluid. The condenser takes heat from the fluid. The total heat taken from the fluid by the condenser must equal all the heat absorbed from the evaporator and from the compressor. Thus the entire system is nothing more than a heat pump. Heat is pumped from the evaporator at a low temperature level (15°F) to the condenser at a high temperature (85°F). At the higher temperature, it may be disposed of by air or water cooling. The air or water used to cool the condenser allows the condenser to reject the heat that was absorbed into the system through the evaporator and suction line and the heat that was generated in the compressor.

Also notice that, to absorb 200 Btu of heat in the evaporator, only 33.2 (from the thermodynamic tables) Btu of mechanical power are necessary. More refrigeration is produced than mechanical power is involved. This is not a form of perpetual motion, or getting something for nothing. The refrigeration is produced by the evaporating liquid, not by direct action of mechanical power. The mechanical power is used to "pump" the heat to a higher level so it may be rejected from the system. In terms of efficiency, we must treat a total of 235.7 Btu in the condenser to obtain 200 Btu of refrigeration in the evaporator. Thus the efficiency is

$$\frac{200}{235.7} \approx 0.85 \text{ or } 85\%$$

Thus 85 percent of the total heat handled is utilized directly in cooling. However, because mechanical power is more costly, and therefore more important than condenser cooling, this actual efficiency is not used. The ratio of refrigeration produced to mechanical energy required is used instead. For this case that ratio is

$$\frac{200}{33.2} = 6.02$$

Because efficiency can range from 0 to 100 percent and this result is greater than 100 percent, we cannot classify this figure as efficiency. The term used is *coefficient of performance,* or COP. Thus, a coefficient of performance of 6.02 means that 6.02 times as much refrigeration is produced as the heat equivalent of the mechanical power consumed.

Liquid–Vapor Cycle

The liquid–vapor cycle, Figure 2–12, was described at the beginning of the chapter, but will be reviewed here. The liquid in the receiver, usually subcooled, is fed to the expansion valve. As the refrigerant leaves the evaporator, its pressure is reduced. This evaporates some of it, 13 percent in this case, to chill the other 87 percent of the remaining liquid to 15°F. This 13 percent of evaporated vapor is called *flash gas.* The air passing over or through the evaporator gradually evaporates the rest of the saturated liquid

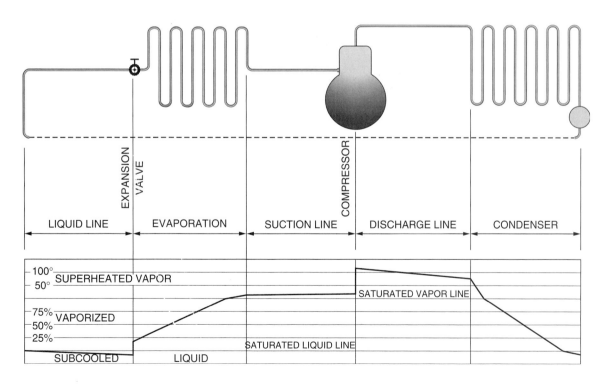

Figure 2–12 Liquid–vapor cycle.

CHAPTER 2 The Compression System of Refrigeration

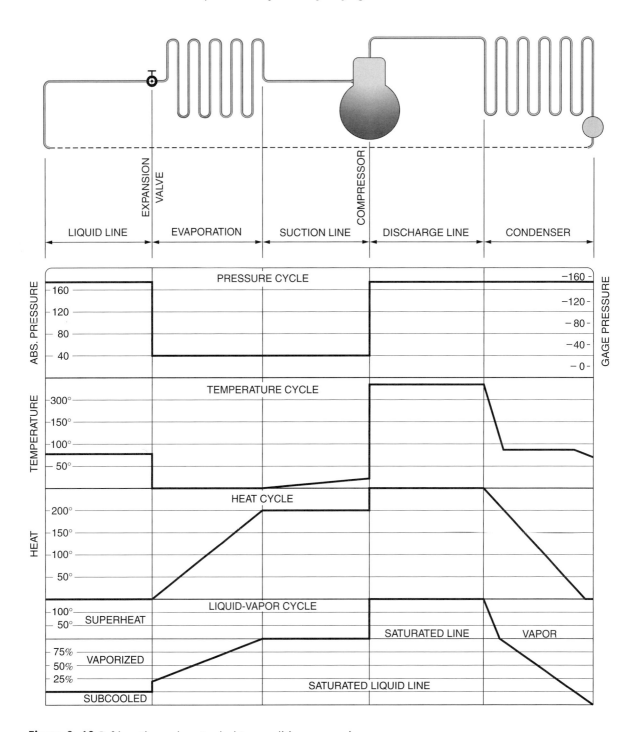

Figure 2-13 Refrigerating cycles: standard ton conditions, ammonia.

until the refrigerant becomes a saturated vapor. At or near the end of the evaporator, the last of the liquid is evaporated. Any additional heat absorbed superheats the vapor. Only vapor should return to the compressor. The compressor raises the pressure and superheat of the vapor. From there, it goes to the condenser, where first the superheat is removed to form a saturated vapor, then this saturated vapor is condensed to a saturated liquid.

Standard Ton Conditions

The charts we have been using apply only to an ammonia system operating at the given temperatures. Figure 2–13 combines a similar set of data for a slightly different condition. Here we have a 5°F evaporator, no superheat in the evaporator, and 9°F of superheat in the suction line. The condenser is at 86°F, and the liquid leaving the condenser is subcooled 9°F. This condition is considered a standard or base with which to compare other conditions or systems. It is called *standard ton conditions,* and is frequently referred to. For simplicity, the 9°F of superheat in the suction line and the 9°F subcooled liquid is sometimes omitted.

In Chapter 1, the information was given that 200 Btu per minute is the equivalent of 1 ton of ice per 24 hours, and is called one ton of refrigeration. *One ton of refrigeration at standard ton conditions* is the amount of refrigeration produced with a 5°F evaporator and an 86°F condenser.

Questions

1. a. What is the boiling temperature of water at 29 inches vacuum?
   b. What is the boiling temperature of ether at 16 inches vacuum?
   c. What is the boiling temperature of ammonia at 20 psig pressure?
2. a. In what part of the refrigeration system is the actual cooling produced?
   b. What is the purpose of the compressor and the condenser?
3. Draw a schematic picture of the complete refrigeration system.
4. a. Which parts of the system are included in the high-pressure side?
   b. Which parts of the system are included in the low side?
5. a. Which parts of the system are cold in relation to the other parts of the system?
   b. Which parts are cool to the touch but not frosted?
   c. Which parts are warm to the touch?
   d. Which parts are hot?
6. a. Which parts of the system add heat to the refrigerating fluid being circulated?
   b. Which parts take heat from this fluid?
7. What is the relation of the total quantity of heat picked up by the refrigerating fluid to that lost from this fluid?
8. What is the meaning of the term *coefficient of performance?*
9. a. In what parts of the system are there only liquid?
   b. In what parts are there only vapor?
   c. In what parts are there both liquid and vapor?
10. What is the complete definition of a ton of refrigeration at standard ton conditions?

Chapter 3

Refrigerants: Physical and Refrigerating Properties

We have mentioned that the evaporation of water, ether, and ammonia have been used to produce cooling, or refrigeration. Each of these and many other fluids may be used for the same purpose, but some give better results than others. What is required of a good refrigerant? Why are some refrigerants used for certain applications, and other refrigerants are used elsewhere? No one refrigerant has all the desirable qualities for all applications, so the refrigerant selected must provide the best balance between desirable and undesirable qualities for the required application.

Requirements

A good refrigerant should have the following properties:

1. It should produce maximum refrigeration per cubic foot of vapor pumped.
2. It should have a condensing pressure suitable for the type of compressor to be used.
3. It should have an evaporating pressure suitable for the type of compressor to be used.
4. It should be stable.
5. It should have no effect on metals.
6. It should have no effect on oils.
7. Its critical temperature should be well above its condensing temperature.
8. It should be nonpoisonous and nonirritating.
9. It should be nonflammable.
10. It should be available at a reasonable price.
11. It should be detectable.
12. A minimum of power should be required to compress it.
13. Its freezing point should be well below the evaporator temperature.

These requirements are discussed in the following paragraphs:

1. *It should produce maximum refrigeration per cubic foot of vapor pumped.* A refrigerant should give the maximum amount of refrigeration for a given size or investment in equipment. In using different refrigerants for the same application, the physical size of the evaporator and the condenser size remain more or less the same, but a compressor of given size can pump only a given volume of refrigerant vapor. When the proper amount of refrigerant is evaporated to do the required amount of cooling, the vapor has a certain volume, which will vary considerably for different refrigerants. The resulting volume of the superheated vapor that is produced as a result of heat absorption in the evaporator and the suction line is comprised of the latent heat of the refrigerant (the amount of heat necessary to evaporate 1 pound of it), its volume per pound, and the number of pounds of refrigerant circulated through the system per unit time.

2. *It should have a condensing pressure suitable for the type of compressor to be used.* The condensing pressure depends on the saturation pressure of the refrigerant at ambient atmospheric temperatures. This pressure should not be too high. Although a

reciprocating compressor can be designed for any required pressure, pressures above 300 psig require heavy (which means expensive) equipment and piping.

Centrifugal compressors cannot build up a large pressure without an excessive number of stages. Therefore, they require a very low condensing pressure. Most centrifugal compressors use a refrigerant with a condensing pressure between zero and 10 psig. At low condensing temperatures, the condenser may operate under a slight vacuum (see Chapter 6).

Figure 3-1 gives the temperature–pressure relationships for some common refrigerants. These are given to a logarithmic scale so that both high and low values can be shown. Thermodynamic tables are much easier to read than graphs. We have included selected thermodynamic refrigerant tables in Appendix A-15.

3. *It should have an evaporating pressure suitable for the type of compressor to be used.* For reciprocating compressors that are not hermetically sealed, it is desirable to have an evaporator pressure very near, but slightly above, atmospheric. The rotating shaft of the compressor must come out through the crankcase wall with a packing or seal to prevent leakage of refrigerant out or air in, see Figure 3-2, but this seal cannot always be made leakproof. If the seal has about the same pressure on each side of it, the leakage will be minimum. Because air or moisture entering the system can cause significant problems, it is more desirable to lose a small amount of refrigerant than to allow air to leak into the system. (Problems associated with moisture are discussed in later chapters.) Therefore, the crankcase pressure, which is usually the suction pressure, should be slightly above atmospheric. Because different jobs require different evaporator temperatures, this leads to the use of different refrigerants to approach the ideal of a suction pressure that is slightly above atmospheric pressure. Suction pressures for any given temperature can be found in Figure 3-1 or the refrigerant tables in Appendix A-15.

Hermetically sealed compressors do not have a shaft seal, so they can be operated at a vacuum or at high suction pressures with no trouble.

Centrifugal compressors work best at high vacuums or very low absolute pressures. Therefore, low pressure refrigerants are commonly used in conjunction with these compressors.

4. *It should be stable.* Stability means that the refrigerant must remain in its original chemical form. For instance, ammonia is a chemical combination of nitrogen and hydrogen. If it should separate into these gases, neither would make a good refrigerant. Many vapors or gases that are not made of a single chemical element have a breakdown temperature above which they are not stable. That is, they are apt to separate into the elements of which they are made, so the breakdown temperature for the refrigerant selected should be well above the operating temperatures of the system. Some of the newer blends will fractionate (the individual refrigerants will separate from each other) under the right conditions. However, the individual fluids maintain their original chemical identity.

5. *It should have no effect on metals.* The refrigerant chosen must not corrode or otherwise react with the metals normally used in refrigeration systems. For instance, an absorption system using sulfuric acid as an absorber was once tried. Some of the problems introduced by using such a fluid can easily be imagined. Absorption systems are discussed in later chapters.

6. *It should have no effect on oils.* The refrigerant chosen must have no harmful effects on properly selected lubricating oils. Some refrigerants may react with some oils to cause gumming, sludging, or varnish formation. Some refrigerants may also mix with and thin some oils, to a point that the oil becomes a poor lubricant. Ethyl chloride, for example, which was once used as a refrigerant, acted as a solvent, thinning the lubricating oil.

7. *Its critical temperature should be well above its condensing temperature.* All vapors or gases have a temperature above which it is impossible to liquefy them, regardless of the pressure applied. This is the critical temperature. It is important that this temperature be higher than any possible condensing temperature to which the refrigerant will be exposed during the course of system operation. Notice the difference between the critical temperature and the breakdown temperature, mentioned in connection with stability. The vapor may be heated above the critical temperature by the

24 CHAPTER 3 *Refrigerants: Physical and Refrigerating Properties*

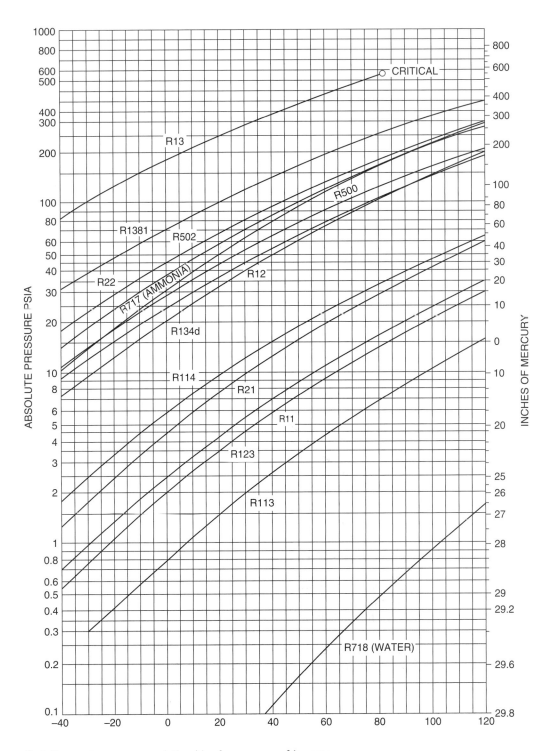

Figure 3–1 Temperature–pressure relationships for common refrigerants.

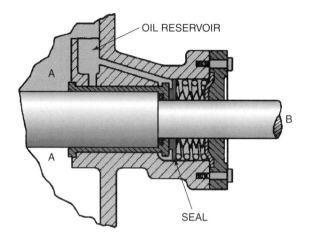

Figure 3–2 Why low pressure, but not a vacuum, is wanted.

heat of compression. If it can be cooled below this temperature before condensing, no harm will be done: It will condense at any temperature below its critical temperature. However, if it is heated above its breakdown temperature at any point in the system, irreparable damage may occur.

8. *It should be nonpoisonous and nonirritating.* Refrigerants must be handled by operators and service technicians. If the refrigerant is poisonous or highly irritating, a hazard is created whenever the refrigerant comes in contact with service personnel or is released into the atmosphere. Imagine the panic if an ammonia leak was to occur in an auditorium or other crowded location.

9. *It should be nonflammable.* A flammable refrigerant would also create a hazard, both when handled by operators and in case of leakage.

10. *It should be available at a reasonable price.* Sufficient quantities of the refrigerant should be available at a reasonable cost. At this cost, it must be of sufficient purity to cause no refrigeration difficulties. Recent developments have changed both the costs and availability of many refrigerants.

11. *It should be detectable.* Some simple means of determining whether leaks are present in the system, and of finding exactly where those leaks are located, should be possible.

12. *A minimum of power should be required to compress it.* The different refrigerants may require different amounts of power to compress them. Obviously, the less the power is required, the lower will be the operating costs of the refrigeration plant.

13. *Its freezing point should be well below the evaporator temperature.* If flow is to be maintained in the system, a refrigerant that freezes at temperatures encountered in the evaporator cannot be used. Except where water is used as a refrigerant, this did not have to be considered a few years ago, and still does not on ordinary jobs. However, on those special jobs that require temperatures 100°F or more below 0°F, this must be considered.

Physical Properties

Figure 3–3, Figure 3–4, and Figure 3–6 list some important physical properties of common refrigerants. All fluids used as refrigerants have been given a number with the prefix "R." Most refrigerants are commonly known by this number. Refrigerants are sometimes referred to by their chemical name or formula.

Items *1, 2,* and *3* give the common name (number), the chemical name, and the chemical symbol.

Item *4* gives the boiling point at atmospheric pressure. This factor alone gives some indication of the pressure characteristics. A low boiling temperature means that boiling can be obtained in the evaporator without a vacuum. However, a positive pressure is necessary to condense the refrigerant. A very low boiling temperature, such as 50 to 100°F below zero, indicates that considerable pressure is necessary in the evaporator at ordinary refrigerating temperatures. An excessively high condensing pressure would probably be required. On the other hand, a boiling point of 50 to 100°F above zero indicates that a vacuum is needed in the evaporator, and the condensing pressure will be near atmospheric. Boiling temperatures around that of water (212°F) indicate the need for an exceedingly high vacuum in the evaporator, and some vacuum in the condenser. Compare the evaporating and condensing pressures of water to those of R-22 or R-123.

Items *5* and *6* give the condensing pressure and evaporator pressure for standard ton conditions, that is, an 86°F condenser and a 5°F evaporator.

Figure 3-3 Physical properties of water (Refrigerant 718).

1.	Common name	Refrigerant 718
2.	Chemical name	Water
3.	Chemical symbol	H_2O
4.	Boiling temp. at atmospheric press	212°F
5.	Cond. press. at 86°F	28.67 in
6.	Evap. press. at 5°F	29.67 in[1]
7.	Latent heat at 5°F, Btu/lb	1071[1]
8.	Volume at 5°F, cu ft/lb	2444[1]
9.	Displacement at 5°F, 86°F, cfm/ton	477[1]
10.	Critical temp.	706.1°F
11.	Freezing temp.	32°F
12.	Least detectable odor	None
13.	Irritability	None
14.	Toxicity	None
15.	Flammability	None
16.	Toxic in flame	None
17.	Sp. gr., liquid, 5°F. water = 1	1.00[1]
18.	Sp. gr., vapor, atm. air = 1	NA
19.	Power, 5°F, 86°F	NA[2]
20.	Compress. temp. at 5°F, 86°F	NA[2]

[1] For 40°F instead of 5°F.
[2] Used in steam jet or absorption. Mechanical power and compression temperature have no meaning.

Item 7 gives the latent heat per pound of refrigerant at 5°F. This is the amount of heat necessary to evaporate 1 pound of the refrigerant.

Item 8 gives the volume, or actual cubic feet of space, that would be filled by the vapor from 1 pound of refrigerant evaporated at the pressure shown in item 6.

Item 9 combines items 7 and 8 (with a correction for flash gas), to show the actual amount of vapor that must be pumped per minute to produce 1 ton of refrigeration, or 200 Btu per minute. These figures give a direct comparison of the sizes of compressors necessary to produce the same amount of refrigeration. Figure 3-7 shows graphically the volumetric requirements for various compressors operating with different refrigerants.

Item 10 gives the critical temperature, and *Item 11* gives the freezing point. The condensing temperature must always be below the critical temperature, and the evaporator temperature must always be above the freezing point.

Item 12 gives the minimum amount of the refrigerant mixed with air that is detectable by odor. This value may be given in percent or in parts per million (ppm). Note that it takes 10,000 ppm to make 1 percent.

Item 13 gives the percent usually found to be irritating. *Item 14* gives the percent and time of exposure necessary for a fatal dose. The actual hazard of a refrigerant is usually a combination of these. A refrigerant that is easily detected by smell, and particularly one that is irritating, may not be as hazardous as a less toxic one that gives no particular warning of its presence.

If a vapor can reach 10 percent concentration before it creates a health hazard, it can be considered to be relatively nontoxic. At 10 percent or greater concentrations, however, enough air is displaced that normal breathing does not supply sufficient oxygen. Therefore any gas, toxic or not, will cause discomfort. Water is not a poison, but if you were shut up in a room filled with it, you would not live long. Similarly, this could be true of any substance other than air. One other thing to consider with respect to hazards is the weight of the refrigerant vapor. This is shown in *item 18*. If the vapor is heavier than air, any leakage could very easily collect in large concentrations in low places such as basements or holds of ships. Once the vapor is allowed to concentrate, positive or mechanical ventilation is necessary to eliminate it.

Item 15 gives some information regarding the flammability and explosion hazards of the refrigerant. Interstate commerce regulations only distinguish flammable from nonflammable refrigerants, but there are also variations with respect to the hazards involved. Flammable vapors will burn with more or less limited admixtures of air. These mixtures must not be too rich or too lean. Naturally, the narrower the limits of burning of a mixture that will burn, the less chance there is of exactly that mixture being

Figure 3–4 Physical properties of Refrigerants 13, 134a, 123, and 717.

		Refrigerant 13	Refrigerant 134a	Refrigerant 123	Refrigerant 717
1.	Common Name	Refrigerant 13	Refrigerant 134a	Refrigerant 123	Refrigerant 717
2.	Chemical name	Monochlorotri-fluoromethane	Ethane, 1,1,1,2-tetrafluoro	Dichloro-trifluoro-ethane	Ammonia
3.	Chemical symbol	$CCIF_3$	CH_2FCF_3	CHC_2CF_3	NH_3
4.	Boiling temp. at atmospheric press.	−114.6°F	−15.7°F	81.7°F	−28.0°F
5.	Cond. press. at 86°F	—[1]	96.626 psig	1.3 psig	154 psig
6.	Evap. press. at 5°F	177 psig	9.0751 psig	25.1 in	19.6 psig
7.	Latent heat at 5°F, Btu/lb	44.5	88.298	78.9	565
8.	Volume at 5°F, cu ft/lb	0.190	1.8969	13.986	8.15
9.	Displacement at 5°F, 86°F, cfm/ton	2.07[2]	6.07	44.32	3.44
10.	Critical temp.	83.9°F	213.9°F	362.63°F	271°F
11.	Freezing temp.	−294°F	−103.3°F	−161°F	−108°F
12.	Least detectable odor	20%	Odorless	20%	53 ppm
13.	Irritability	None	None	Eyes	700 ppm
14.	Toxicity	20% 2 hr	None[3]	>50 ppm 8 hr	0.5% 30 min
15.	Flammability	None	None	None	13.1%-26.8% @ 54 psig
16.	Toxic in flame	Yes	No	Yes	No additional
17.	Sp. gr., liquid, 5°F. water = 1	1.18	1.208 at 77°F	1.46	0.66
18.	Sp. gr., vapor, atm. air = 1	5.7	3.6	5.3	0.74
19.	Power, 5°F, 86°F	1.98[2]	1.035	0.946	0.99
20.	Compression temp. at 5°F, 86°F	110°F	100°F	90°F	210°F

[1] Above critical temperature.
[2] At 5°F and 70°F.
[3] Concentrations over 75,000 ppm may cause cardiac irregularities.

present. After they are ignited, some vapors burn much more rapidly than others. Also, the burning increases the temperature, which increases the pressure. High-speed burning plus the pressure increase is what causes explosive violence. Item 15 gives the limiting proportions of flammable mixtures, and the maximum pressures built up by ignition.

Item 16 gives information regarding another hazard. Some refrigerants that are themselves nonpoisonous and nonflammable break down to poisonous compounds when they are exposed to high temperatures. Thus, such a refrigerant leaking into a room with any high temperature source could be hazardous. Unvented gas stoves or electric heating elements could cause such a breakdown. However, the toxic products are very irritating so will tend to drive occupants out of a room before dangerous amounts can be breathed. This breakdown

Figure 3–5 Refrigerating Properties of Common Refrigerants.

	Refrigerant 718 Water	Refrigerant 22	Refrigerant 717 Ammonia
1. Maximum refrigeration per cu ft	(Note 1)	Good	Good
2. Reasonable condensing pressure	Poor	Good	Good
3. Reasonable evaporator pressure	Poor	Good	Good
4. Stability	Good	Good	Good
5. No effect on metals	Good	Good	Good
6. No effect on oil	Good	Good	Good
7. High crit. temp.	Good	Good	Good
8. Nonpoisonous; nonirritating	Good	Good	Poor
9. Nonflammable	Good	Good	Good
10. Availability; cost	Good	Fair	Good
11. Ease of finding leaks	Fair	Fair	Good
12. Power required	Good	Good	Good
13. Freezing point	Poor	Good	Good

[1]Suitable only for steam jet or lithium bromide absorption.

is also a definite hazard to fire department personnel in case of fire in the building.

Item 17 gives the specific gravity of the liquid at 5°F compared to water, that is, the ratio of its weight to the weight of the same volume of water. There is some variation in the weight of the liquid with temperature, but for most refrigerants this will not run over 10 percent for a change from 0°F to 100°F.

Item 18 gives the specific gravity of the vapor at atmospheric pressure compared to the weight of air. This is the condition of the vapor when it escapes from a system. The weight of the vapor varies almost directly with absolute pressure.

Item 19 gives the horsepower required to produce 1 ton of refrigeration at standard ton conditions. This is the power required to actually compress the vapor, without including any of the losses involved. Actual power including losses will run up to 50 percent higher than these figures. However, for the same size jobs, the power losses will be approximately the same for the different refrigerants.

Item 20 gives the compression temperature reached by the vapor at standard ton conditions. That is the temperature of the compressor discharge. High temperatures require water-jacketed cylinders, and give rise to lubrication troubles due to oil breakdown.

Figure 3–5 and 3–6 combines the factors discussed before in an overall comparison of the various refrigerants.

Critical Pressure

Critical pressure may also be an issue when specifying refrigerants. Above its critical pressure, a particular refrigerant will not evaporate no matter how much heat is applied. This is not a problem in most applications. Nevertheless, in some special or extreme applications, critical pressure must at least be considered. The critical pressure for ammonia is 1657 psia. The critical pressure for water is 3206.2 psia, the critical pressure for R-134a is 588.9 psia, and the critical pressure for carbon dioxide is approximately 1057 psia. It is interesting to note that the critical pressure of carbon dioxide corresponds to 88°F (only 2°F above the standard ton

Figure 3-6 Physical properties of less common refrigerants.

1. Common Name	Refrigerant 13B1	Refrigerant 21	Refrigerant 40	Refrigerant 113	Refrigerant 114
2. Chemical name	Bromotri fluoromethane	Dichloromono fluoromethane	Methyl chloride	Trichlorotri fluoroethane	Dichlorotetr. fluoroethane
3. Chemical symbol	$CBrF_3$	$CHCL_2F$	CH_3CL	$C_2CL_3F_3$	$C_2CL_2F_4$
4. Boiling temp. at Atmospheric press.	−72.0°F	48.1°F	−10.8°F	117.6°F	38.4°F
5. Cond. press. at 86°F	247.1 psig	16.5 psig	80.0 psig	13.9 in	22.0 psig
6. Evap. press. at 5°F	63.2 psig	19.2 in	6.5 psig	27.9 in	16.1 in
7. Latent heat at 5°F, Btu/lb	44.1	109.3	180.7	70.6	62.0
8. Volume at 5°F, cu ft/lb	0.38	9.13	4.47	27.0	4.23
9. Displacement at 5°F, 86°F, cfm/ton	2.62	20.4	5.95	102	20.14
10. Critical temp.	152.6°F	353°F	289°F	417°F	294°F
11. Freezing temp.	−270°F	−211°F	−144°F	−31°F	−137°F
12. Least detectable odor	20%	20%	None	20%	20%
13. Irritability	None	None	None	None	None
14. Toxicity	20% 2 hr	10% 30 min	2% 2 hr	5% 1 hr	20% 2hr
15. Flammability	None	None		None	None
16. Toxic in flame	Yes	Yes	Yes	Yes	Yes
17. Sp. gr., liquid, 5°F. water = 1	1.5	1.46	0.99	1.66	1.46
18. Sp. gr., vapor, atm. air = 1	7.2	3.75	2.04	6.5	3.75
19. Power, 5°F, 86°F	1.03	0.94	0.96	0.97	1.05
20. Compression temp. at 5°F, 86°F	124°F	142°F	172°F	86°F	86°F

condensing temperature). Critical pressures may be found on DuPont's website as necessary.

Historical Development of Refrigerants

A very rough timeline for the development of refrigerants is as follows:

- The ancients used natural ice to cool drinks and food. They also knew about the cooling effect of evaporation and that a mixture of ice and salt would lower the freezing temperature of water.
- Underground rooms have long been built to store food using natural ice. Eight to ten feet into the ground, the temperature is reasonably constant all year.
- In the early 1800s, Michael Faraday successfully liquefied ammonia. Prior to this, ammonia had been considered to be a "fixed gas"—that is, the scientific community of the time did not think it could be liquefied.
- Carbon dioxide and sulfur dioxide were used as refrigerants.

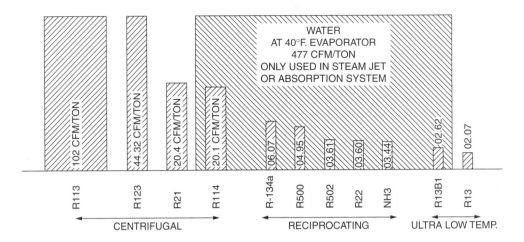

Figure 3-7 Graphic displacement of common refrigerants.

- In 1931, Thomas Midgely and C. F. Kettering developed R-12 (a chlorofluorocarbon compound, or CFC).
- During the mid-1900s, a number of other CFCs were developed.
- In the late 1900s, hydrofluorocarbons (HFCs) were developed (they include no chlorine).

The first refrigerants used were known chemicals that could be easily liquefied and evaporated. Ammonia was the first chemical to be used extensively for refrigeration. It proved so successful that it has been the most important industrial refrigerant ever since. It was also used in the first commercial butcher shops, retail dairies, and so on.

The hazards of ammonia led to the search for something safer for certain applications. Carbon dioxide became the accepted refrigerant for hospitals, prisons, and passenger ships up until the introduction of halocarbon refrigerants. However, the excessive pressure, low critical temperature, and high power requirements prevented the general adoption of carbon dioxide in other applications. Carbon dioxide is now used in the United States only for the production of dry ice (liquid carbon dioxide) (see Chapter 26).

Ammonia, though widely used in commercial refrigeration, requires too high a pressure, and too cumbersome equipment, to be applied to anything approaching domestic applications. So other chemicals were investigated by those experimenting with a "mechanical ice box." Many substances were tried, but sulfur dioxide became the favorite, closely followed by methyl chloride. These two substances were successful enough to be applied to early commercial equipment.

Both these refrigerants, however, presented hazards that, some designers felt, held back the full development of domestic and commercial refrigeration and air conditioning. Sulfur dioxide, when dissolved into water, forms sulfuric or sulfurous acid. It can contaminate food and ruin flowers. Methyl chloride is famous (or infamous) for its "methyl drunk" effect. So efforts were undertaken to find a refrigerant that had all the qualities desired of a good refrigerant. A whole new series of chemicals was discovered and developed, the halogenated hydrocarbons or halocarbons, which had most of the properties desired. Different members of the series had boiling points from 100°F or more below zero to over 200°F above zero. Therefore, one could be found with pressure characteristics that would fit almost any refrigeration application.

Such refrigerants were first called by their initial trade name, Freon, followed by a designating number. As other manufacturers began to supply these chemicals, however, they gave them their own trade names. Much confusion resulted.

Headed by the American Society of Refrigeration Engineers (now called the American Society of Heating, Refrigeration and Air Conditioning Engineers, or ASHRAE), the industry decided to call each of these compounds a "Refrigerant," followed by an identifying number. The "Freon" or DuPont numbering system was adopted for the halocarbons.

This is why we may see two or more different names for the same chemical. It is perfectly proper to call a halocarbon by its trade name, or simply as "Refrigerant" followed by the proper number. The older refrigerants are, at present, still most commonly identified by their chemical names. "Halocarbon" is a shortened term for "halogenated hydrocarbon." Fully halogenated refrigerants (CFCs) have been phased out because of their harmful effects on the environment.

Halocarbon refrigerants have entirely replaced sulfur dioxide and methyl chloride in new domestic and commercial equipment and have been widely used in virtually all refrigeration applications for many decades.

Freon is the trade name for refrigerants manufactured by DuPont. It is as much a household word as other trade names such as Klixon, Thermos, and Amprobe. We all know what these "devices" are and rarely stop to think of them as marketed by their respective companies.

The original refrigerants sold by DuPont and others fell into two "families" of chemicals. Chlorofluorocarbons, the CFCs, contain the chemical elements chlorine, fluorine, and carbon. R-11 and R-12 fall into this group, as do as their azeotropes R-500 and R-502. An azeotrope is a particular kind of mixture, and an azeotropic refrigerant is a mixture of more than one refrigerant. *Do not attempt to mix refrigerants yourself*. These are the more common CFC refrigerants, although there are many more in this family.

Hydrochlorofluorocarbons, or HCFCs, contain the chemical elements hydrogen, chlorine, fluorine, and carbon. R-21, R-22, R-123, as well as other refrigerants, fall into this family. R-22 is also part of the azeotropic mixture R-502.

Hydrofluorocarbons, HFCs, comprise the relatively new family. These refrigerants contain the chemical elements hydrogen, fluorine, and carbon.

Note that there is no chlorine in these refrigerants. This is by design. The most common and well-known refrigerant in this family is R-134a. R-507 is also in this group. It is azeotropic at temperatures well below zero but will fractionate at higher temperatures. R-125 is a less popular HFC.

The CFC and HCFC refrigerants were in wide use globally for many decades. Then, in June 1974, two chemists at the University of California proposed a theory. They claimed that certain chlorine-based chemicals were responsible for ozone depletion in the atmosphere and that further study was warranted. Although the scientific community was (and still is) divided on this theory, Rowland and Molina's "discovery" immediately gained the attention of the press.

Subsequent news releases stated that chlorine-based chemicals were undeniably responsible for ozone depletion. Many solutions were proposed and debated, which led to the ban of chlorine-based propellants in aerosol products. Take note here that aerosol products are intentionally released to the atmosphere.

The debates went on and led to the well-publicized Montreal Protocol of 1987. In the Montreal Protocol, 17 developed nations agreed to a significant reduction in the production of CFCs. This spawned further debate. One extreme idea was to ban all CFCs and HCFCs by the year 2000. Others claimed that HCFCs were the solution to ozone depletion and not part of the problem. Some individuals retracted their suggestions for reasons unknown.

We ultimately came to an accelerated ban on the CFC family and an extended schedule to eliminate the HCFC family. Legislation changes so quickly that sometimes it is hard to keep current. At this writing, equipment using HCFC refrigerants will not be built after the year 2010, and HCFCs will be banned by the year 2020.

One relatively new rating for refrigerants is the ozone depletion potential, or ODP. R-11 has a rating of 1 (the highest rating for refrigerants). Some halogens (used in fire blanketing systems) have even higher ratings. The other refrigerants have ODPs of less than 1. The HFC refrigerants have ODPs of zero. Space constraints prohibit printing an all-inclusive list of refrigerant properties.

In any event, as we mentioned before, HFCs conspicuously do not contain chlorine. R-134a is the overwhelming choice for applications that used to be dominated by R-12. R-12 equipment may be retrofitted with R-134a. However, caution must be exercised. The old equipment must be cleaned and evacuated thoroughly, and the proper oil must be used. R-12 and R-134a and their oils are completely incompatible. Responsible engineering calls for custom designing equipment to use a specific refrigerant and sticking with that refrigerant.

Water

Comparing the list of requirements of a good refrigerant with the properties of water suggests that water is a very poor choice for a refrigerant. The enormous volume of its vapor is such that using an ordinary compressor is impossible. High vacuums are required in both the low- and high-pressure sides of the system. The required vacuum on the low side is so low or deep that it is difficult to maintain. Leaks are difficult to find. They only become evident when air leaking into the system prevents proper operation. To check for a suspected leak, the system must be charged with compressed air and all joints brushed with soap suds. The 32°F freezing point of water makes it impossible to use for ordinary applications.

However, for certain applications above 32°F, such as chilling large quantities of water and for air conditioning, it satisfactorily fulfills some very rigid requirements. First, water is absolutely safe. It has no odor or poisonous properties, and it is not flammable. If water is accidentally released in a crowded building, it will not cause panic. It is cheap and can be obtained anywhere. It has the highest latent heat of vaporization compared to other refrigerants.

The problem of handling the excessive volume of the low-pressure vapor has been solved in two ways. Steam ejector systems and lithium bromide absorption systems can create the high vacuums needed and handle the large volumes of vapor produced. Water is therefore quite commonly used as a refrigerant in certain applications.

Ammonia

Ammonia (also called Refrigerant 717 or R-717) was the first refrigerant to be used successfully for mechanical refrigeration, and still is one of the leading refrigerants in terms of quantity produced and sold, and in terms of the tonnage of refrigeration produced by it. It used to be produced from nitrates, but since World War I, the bulk of it has been produced by chemically combining nitrogen from the air with hydrogen taken from water.

Ammonia is colorless in both liquid and vapor form. It has a well known, strong irritating odor. It boils at −28°F at atmospheric pressure. It is very highly soluble in water. Household ammonia is 2 or 3 percent ammonia dissolved in water. Commercial aqua ammonia is 28 percent ammonia dissolved in water. The specific gravity of pure liquid ammonia is 0.66, and that of the vapor is 0.74. Thus, the liquid is lighter than water, and the vapor is lighter than air.

The very high latent heat of ammonia more than offsets its light weight. It gives more refrigeration than R-134a for the same displacement. Its low boiling point requires a pressure on the low side for ordinary evaporator temperatures, 19.6 psig at 5°F, but makes typical low-temperature applications possible without having to operate in the vacuums range. Its condensing pressure gets high enough, 154.5 psig at 86°F, that proper care must be taken in designing equipment to hold it. Extra heavy pipe is usually used on the high side to give a sufficient factor of safety.

Pure ammonia has been proven stable at the pressures and temperatures existing in a refrigeration system. However, there seems to be some catalytic action in the presence of oil and other impurities. This causes some breakdown at compressor discharge temperatures and pressures. This breakdown forms noncondensable gases, which must be purged off.

Below 200°F there seems to be very little of this action. Between 250 and 300°F the increase in noncondensable gas becomes very great. Exact temperatures at which this happens cannot be specified because it varies for different systems, probably due to varying amounts of oil carryover, carbon present, moisture, and other factors. Some of the noncondensable gas may also come from oil breakdown or cracking (separating) occurring at discharge conditions. Regardless of where noncondensables come from, they indicate that discharge temperatures should not go above 250°F if such a limit is possible.

Ammonia when dry has no effect on metals, but it will attack and destroy copper or copper alloys in the presence of moisture. Because ammonia combines so quickly and completely with any moisture present, even moisture in the air, it has proven very difficult to keep dry in systems the size of the usual ammonia refrigeration equipment. Parts of the system must periodically be opened to the air for inspection or maintenance. This difficulty has led to the design of systems containing no copper alloys. Special bronzes can be made that are corrosion-proof against ammonia and water. These bronzes have been introduced into ammonia equipment in a limited way.

Pure ammonia will not have any effect on a properly refined lubricating oil, but with an acid oil or any moisture present, it may sludge badly. Although the ammonia has no effect on a good oil, compression temperatures must be watched. Ammonia has a higher compression temperature than other refrigerants, 210°F at standard ton conditions. Higher temperatures, at higher compression ratios, will cause the oil to break down, which means carbonization, sludge, and varnish formation. Ammonia compressors must be water-jacketed to limit high discharge temperature, but this will not wholly prevent oil breakdown. A water jacket is an area around the compressor cylinder(s) through which water flows to help cool the cylinder wall(s).

The critical temperature and freezing point of ammonia impose no limitations on ordinary systems. However, in special applications where temperatures of $-100°F$ or lower are required, some other refrigerant must be used because of ammonia's freezing point is $-107.9°F$.

The irritating and toxic properties of ammonia are additional negative features. When ammonia combines with moisture, which it does readily, it forms a powerful caustic, ammonium hydroxide. This, in contact with human flesh, can cause first-, second-, or even third-degree burns, depending on the concentration and length of exposure. This is especially true of the moist surfaces of the eyes, nose, mouth, and lungs. Any moist part of the body is instantly irritated, and with increased exposure, burned by it. This action takes place to a lesser degree on the outside skin because the skin contains moisture. Any sweaty surface of the body can be badly burned. However, the percent of ammonia that is highly irritating to breathe is much smaller than the percent that can burn or be fatal, so it is its own warning agent. Ammonia is deadly in certain concentrations.

The irritating odor or ammonia, on the other hand, gives instant warning of a leak. For a positive leak test, the white smoke from a mixture of ammonia and sulfur dioxide is used. The sulfur dioxide is supplied by burning sulfur sticks. These are made by dipping long slivers of wood in melted sulfur. A previously prepared supply is usually kept on hand, or sulfur sticks may be obtained from suppliers of ammonia. In the presence of ammonia, burning sulfur gives off white smoke.

Ammonia is flammable, but within rather narrow limits of concentration in air. As refrigeration plants contain oil, this will aggravate a fire. However, ammonia is one of the cheapest refrigerants, and it is so well established that it can be obtained all over the civilized world. The power required to produce refrigeration with ammonia is about average. It is not more than 2 percent higher than that required for the lowest-powered refrigerant at standard ton conditions.

Although ammonia is not as safe as some other refrigerants, in large plants it is common to have it under the supervision of engineers. Small leaks can be immediately repaired. In case of a bad leak, the engineer is present. He knows immediately how to isolate the leaking part, and make the necessary repairs.

In automated plants that are not under the constant supervision of an engineer, a leak warns anyone in the neighborhood that attention is required. Service personnel can then be called before enough refrigerant is lost to cause faulty refrigeration.

From the standpoint of refrigeration produced, ammonia is one of the best refrigerants available. Its hazardous attributes are the only factors against it. It predominates in large-capacity industrial plants. The average ice plant, cold storage plant, ice cream factory, wholesale creamery, or any other application with similar refrigeration requirements would not, in the past, consider any other refrigerant.

Halocarbons

Many synthetic refrigerants have been developed from methane and ethane. This class of refrigerants is called halogenated hydrocarbons or halocarbons because of the presence of the halogen elements, bromine, chlorine, or fluorine. Various combinations of these elements with carbon and sometimes hydrogen give different characteristics, many of which exactly fit certain refrigeration requirements.

Those refrigerants that contain some fluorine are more stable or chemically less active. This means they are nonflammable and relatively nontoxic. All modern halocarbon refrigerants contain fluorine. However, even nonflammable halocarbons will break down to irritating and toxic gases when in contact with flame or hot surfaces.

General properties of all halocarbon refrigerants are quite similar except for their boiling temperatures. They are similar to ether (also a halocarbon) in having a sweet, ethereal odor. Most of them have such a slight odor that it is noticeable only in a large concentration. Both the liquid and vapor are clear and colorless. The liquid is quite heavy, usually about 1.5 times as heavy as water. Vapors are from three to six times the weight of air. With such a heavy, dense liquid and vapor, pipelines, passages in compressors, and valve openings must be of liberal size, and have a minimum of bends or other restrictions. Otherwise, excessive pressure drops might occur.

All halocarbons will mix with oil. At temperatures found within refrigeration systems, they will mix in any proportion. A few separate out, as oil separates from water, at low temperatures encountered in evaporators. All halocarbons thin the oil as they mix with it. Most compression equipment is now designed with this in mind, to keep mixing at a minimum.

Halocarbons will dissolve any natural rubber material, so this must be kept in mind when selecting gaskets or packing. Synthetic, oil-resisting rubber of the neoprene or chloroprene types will hold it.

Leaks may be checked with a halide torch or an electronic leak detector. Different electronic leak detectors are available for different refrigerants. For systems that operate at a vacuum, a charge of air must be added to bring the pressure above atmospheric. A mixture of air and refrigerant escaping from a leak will still be indicated on the halide torch or electronic detector. Keep in mind that phosgene gas is produced when halocarbons are burned. Halide torches should only be used in well-ventilated spaces, as phosgene is quite deadly.

Refrigerant 30 (Carrene 1) was one of the first synthetic halides developed exclusively for refrigeration purposes. It was developed for centrifugal compressors, and used exclusively in this country for the first centrifugal refrigeration systems. These were applied entirely to air conditioning work.

These early centrifugal compressors are now considered obsolete, but many of them are still in operation and require this refrigerant. The following discussion of halocarbons includes those most commonly used for refrigerants, and a few of the less common ones. Many more such chemicals are available. A complete list of all fluids that are being used as refrigerants can be found in the ASHRAE *Handbook of Fundamentals,* published by the American Society of Heating, Refrigerating and Air Conditioning Engineers, as well as on DuPont's website.

Refrigerant 11 is a low-pressure refrigerant suitable for either low- or high-temperature applications in a centrifugal compressor. It used to be the most common refrigerant in centrifugal compressors. Virtually all current stocks of R-11 are the result of recovery/reclaim. It also an excellent solvent and cleaning fluid and was once used as such. R-11 is a CFC.

Refrigerant 12 is a moderate-pressure refrigerant suitable for reciprocating compressors. This was the first of the halocarbons and began to replace other refrigerants such as sulfur dioxide and methyl chloride. It used to be the most common refrigerant found in domestic and commercial refrigeration and small air conditioning systems. Most of the current stocks of R-12 are the result of recovery/reclaim. R-12 is also a CFC.

Refrigerant 13 is a high-pressure refrigerant with a very low boiling point, $-114.6°F$ at atmospheric pressure. Its high pressure and low critical temperature make it unsuitable for ordinary applications. However, it is an excellent refrigerant in the low-temperature end of a cascade system, producing refrigeration at $-100°F$ or lower. R-13 is a CFC.

Refrigerant 13B1 is another refrigerant that is suitable for low temperatures, but not for common applications. Its boiling point at atmospheric pressure is −72.0°F. Therefore, it is not as easy to get ultralow temperatures as with R-13, but condensing pressures are not so high.

Refrigerant 21 is another refrigerant whose pressure and volume characteristics make it suitable only for centrifugal compressors. It is a little higher pressure than R-11, so it does not require quite as high a suction vacuum and will produce more refrigeration for the same volume pumped as R-11. R-21 is an HCFC.

Refrigerant 22 is quite similar to R-12 but has a higher pressure. Its pressure characteristics are somewhat similar to those of ammonia. R-22 is used in commercial and small industrial applications, and in some air conditioning systems. A smaller compressor can be used with this refrigerant than with R-12, but this advantage must be balanced against higher condensing pressures—as high as 300 psig and sometimes more with air-cooled condensers. R-22 is also an HCFC. Its discharge temperature is the highest of the halocarbon refrigerants.

Refrigerant 40, or *methyl chloride,* is an older refrigerant that at one time had considerable use in commercial applications. In its pure state, methyl chloride has no effect on most metals of compressor construction, *but it must never be used in contact with zinc or aluminum.* In contact with these metals, it breaks down to form spontaneously combustible gas. If it is contaminated with moisture, it will cause electrolytic action, which will dissolve copper from tubing and deposit it on wearing steel parts. Methyl chloride will dissolve natural rubber, so rubber must not be used in gaskets or packing.

Methyl chloride is toxic and slightly asphyxiating. It has a rather sickly sweet ether odor, but small amounts cannot be detected by smell. Small quantities can be breathed with no ill effects. Greater amounts will cause drowsiness, mental confusion, and nausea. It has about the same effect as too much alcoholic stimulant without the stimulation, and has been called a "methyl drunk." Greater concentrations can cause asphyxiation and even death.

Because of the toxic and fire hazards of methyl chloride, it is no longer used in new equipment. But much of it is still found in older systems.

Refrigerant 113 is a low-pressure (high-vacuum) refrigerant suitable only for centrifugal compressors at air conditioning temperatures. R-113 is a CFC.

Refrigerant 114 is a low-pressure refrigerant that has been used in centrifugal compressors in large-size applications, and in rotary compressors for domestic sizes. Its pressure is a little higher than most centrifugal refrigerants. Its volume is too great to be suitable for a reciprocating compressor. R-114 is also a CFC.

Refrigerant 123 is an HCFC and is suitable for centrifugal applications. These machines are typically large tonnage and used for air conditioning or process loads. The chilled water supply temperature is usually above water's freezing temperature. Otherwise, some type of antifreeze solution must be employed. R-123 is not a new refrigerant, but it has recently became popular as a replacement for R-11. This is, of course, a result of the ban on CFC refrigerants, but as we mentioned, HCFCs are on their way out too.

Some older machines that used R-11 have been retrofitted to use R-123. These two refrigerants have similar boiling temperatures and are reasonably compatible. However, as with any retrofit, the old system must be thoroughly cleaned and evacuated before charging with the replacement refrigerant. Make sure the right oil is used. Also keep in mind that refrigeration systems perform best with their original refrigerant. Performance and efficiency are almost invariably compromised as the result of a retrofit.

Refrigerant 134a is an HFC that has recently been employed in a wide variety of applications. Its boiling temperature is slightly higher than that of R-12. We now find R-134a in automobiles, homes, places of business, and industrial plants. It works well in all types of compressors. Some older machines that used R-12 have been retrofitted with R-134a. However, keep the already-mentioned caveats in mind. It is very important to remember that HFCs and CFC/HCFCs and their lubricants are completely incompatible and must not be mixed. The right gaskets must also be used. Serious damage to equipment can occur if retrofit instructions are not followed carefully.

Refrigerant 500 is an azeotropic mixture of R-12 (a CFC) and R-152a (a HFC). An azeotropic mixture is a particular chemical mixture of fluids that cannot be separated by boiling or evaporation.

Therefore, the fluids evaporate and condense as if they were a single fluid, without any separation. Azeotropic mixtures were developed after the halocarbon family was discovered. The aim was to find mixes of refrigerants that would have better properties than the base refrigerants. Azeotropes are the result of careful and time-consuming experimentation. Individuals should not attempt to "invent" their own mixtures.

Refrigerant 500 has pressure and volume characteristics between those of R-12 and R-22. Its principal use has been in hermetically sealed, reciprocating, commercial compressors.

Refrigerant 502 is an azeotropic mixture of R-12 and R-115. Its pressure characteristics are very similar to those of R-22. However, its discharge temperature is much lower than that of R-22. Because of this lower discharge temperature, its use has reduced oil breakdown and has increased the life of hermetic motor windings. Comparing R-502 with R-12, they have almost the same discharge temperatures (101°F and 99°F, respectively), but R-502 has a boiling temperature approximately 28°F lower than that of R-12.

Refrigerant 503 is an azeotropic mixture that contains CFC and HFC refrigerants. It consists of 40.1 percent R-23 and 59.9 percent R-13. It boils at −126°F at atmospheric pressure. Because of this very low boiling temperature, it works well in the low stage of cascade systems. Cascade systems are covered in a later chapter. The critical temperature of R-503 is 67°F, so it cannot be used in a single-stage system because the refrigerant vapor would not condense in typical summer temperatures. Additionally, a single-stage system employing R-503 would have a prohibitive compression ratio. However, R-503 is noncorrosive, nonflammable, and considered to be relatively nontoxic.

Refrigerant 401A is a near-azeotropic blend that was developed to retrofit systems containing R-12 or R-502. It contains R-22, R-152a, and R-124. It works well in systems that operate between 0 and 32°F. It does not have fixed evaporating or condensing temperatures but experiences what is known as *glide*. So, for a given fixed pressure, we may see a range of up to 8°F in which the refrigerant will evaporate or condense. We must find the average temperature of this range to properly determine evaporating and condensing temperatures. Alkylbenzene oil is recommended when retrofitting to this refrigerant.

Refrigerant 404A is used primarily in medium- and low-temperature equipment. It is also classified as near-azeotropic and is a blend of R-134a, R-143a, and R-125. This refrigerant may be specified for new equipment but may also be specified as a replacement for R-502. Systems using R-404A require POE (see Chapter 10) oil. It has a relatively low glide and a small potential to fractionate.

Refrigerant 407C is a near-azeotropic blend that was developed to retrofit systems containing R-22. It may also be specified in new systems. It contains R-32, R-134a, and R-125. It may be found in refrigeration and air conditioning applications. As with any retrofit, be sure that the manufacturer's instructions are followed precisely. This refrigerant requires POE oil. Its temperature glide may be over 10°F.

Conditions Other Than Standard Ton

Standard ton conditions have been used for all comparisons that would be affected by a change of temperature. This is to give a fair comparison of each refrigerant with the others. On the whole, operating conditions for different refrigerants will change by about the same amount with changing temperatures. There are a few cases, however, in which some properties which are nearly the same will change relative positions for higher or lower evaporator temperatures. Also note that conditions such as condenser and evaporator pressures, displacements, horsepower, and compression temperatures will change greatly with changing temperatures. So what is marked good or fair for standard ton conditions might be poor for other, widely varying conditions. And some items marked poor could be good at other conditions. Any more complete comparison would have to be worked out for the exact conditions prevailing, but the data given will serve as a good comparison of systems operating under average conditions.

Changing Refrigerants

The question is sometimes asked: Can the refrigerant in a system be changed without affecting the operation of that system? Can obsolete refrigerants be replaced with something more modern?

Before making any such changes, it is best to consult the manufacturer of the refrigeration system. The manufacturer is best qualified to evaluate any contemplated changes. If manufacturer's recommendations are unavailable, the following points should be kept in mind:

1. The system will probably give maximum efficiency with the refrigerant for which it was designed. Sizes of lines, ports, and passages in the compressor, valve clearances, and the pressure applied to the valve springs are all designed to give the best possible results based on the volume and density of the fluid to be handled. A different refrigerant will have a different volume and density, so all the conditions so carefully designed for will be changed. It would be like having a suit tailored to an individual, then selling it to someone else. Even though it might fit approximately, it could never fit another person as well as the one for whom it was originally tailored.

The result in the case of refrigerants is that the losses in the system, particularly in the compressor, are apt to be greater. In some cases, they will be considerably greater. Sometimes manufacturers do design a compressor that can be adapted to use two or more refrigerants, with or without minor changes such as valve clearance and valve springs. Such compressors could be changed over with no ill effect if other factors are considered.

2. All refrigerants have different refrigerating effects for a given displacement. Therefore, if no other changes are made, there will be a variation in refrigerating capacity. This variation, neglecting losses mentioned above, will be proportional to the data given under item 9 of Figure 3–3 and Figure 3–4 and in the graphic illustration of Figure 3–7. If a change such as from R-22 to R-134a is made, the capacity will be reduced. This will reduce the load on the condenser and driving motor. On the other hand, the opposite change (from R-134a to R-22) will increase the capacity, which will overload the motor and the condenser. The overloaded condenser would raise the head pressure, which would further overload the motor.

In open-type, belt-driven compressors this variation in capacity is usually compensated for by changing the operating speed of the compressor. This is done by changing the size of the motor pulley. When the unit is to be slowed down, a smaller motor pulley is required. Reducing the size of the drive pulley will reduce the speed of the compressor crankshaft.

3. Any change in refrigerant of course requires that the entire system be thoroughly cleaned. This includes draining and flushing out the compressor oil. The new oil must be suitable for the new refrigerant to be used.

4. Expansion valves or float valves should be exchanged or recalibrated for the new refrigerant. A thermostatic expansion valve is always made for a specific refrigerant, and will give best operating characteristics with this refrigerant. All pressure-regulating valves or pressure switches must be readjusted to the new pressure conditions imposed by the new refrigerant.

5. Ammonia cannot be used in a system designed for halocarbon refrigerants. Copper tubing, brass and bronze fittings, and bronze bearings are used extensively in designs intended for halocarbons. These metals are all attacked by ammonia. Ammonia operates at a higher pressure than many halocarbon refrigerants, and may prove harder to hold. Because of the high discharge temperatures, ammonia compressors are either water-jacketed or are designed to be cooled by cold suction vapor. Most halocarbon compressors use only air cooling on the cylinder and head.

6. Refrigerants 12, 22, 134a, 500, 502, and 507a are the only common refrigerants with characteristics near enough to those of ammonia to be considered as possible substitutes for the latter. However, it is not good practice to change an ammonia system without a complete change of all equipment. The halocarbons are such excellent solvents that they will loosen gums, sludges, and scales left by ammonia that ordinary cleaning processes will not touch. Therefore, such a change would lead to a

constant repetition of plugged valve orifices, filters, and screens. Also, the halocarbons are so much heavier than ammonia that excessive pressure drops could be expected in the piping, ports, and passages designed for the much lighter ammonia.

Handling Refrigerants

The high-vacuum refrigerants, that is, those with boiling points at or near atmospheric temperatures, or with condensing pressures near atmospheric, are stored, handled, and shipped in drums similar to oil drums. It should be remembered that the pressure of any refrigerant rises rapidly with a rise in temperature. Such refrigerants should not be stored near a heat source, such as a heater or boiler.

Also, there is always some expansion of the liquid with a temperature rise. Therefore the drum should not be filled completely full, because any rise in temperature will expand the refrigerant and cause it to bulge or even burst the drum. Drums should not be filled to more than 80 percent capacity.

All higher-pressure refrigerants are handled in special containers that are made to conform with Interstate Commerce Commission regulations. This makes it possible to ship the refrigerant in the same container in which it is stored or sold. The U.S. Department of Transportation oversees the transport and storage of refrigerants.

Small quantities of the most common halocarbon refrigerants are sold in 1- and 2- pound disposable containers. A special valve is available that can be attached to the container. Some refrigerants are available in disposable containers of up to 25 pounds capacity.

Larger quantities of halocarbon refrigerants or ammonia are sold in steel drums as shown in Figure 3–8. These drums can be returned for credit when emptied. Drums are available that hold up to 150 pounds of some refrigerants. For different refrigerants, the weight capacity varies according to the density of the refrigerant. Refrigerants are also sold in tankcar lots to manufacturers or to cold storage plants that can use such large quantities.

Refrigerant drums contain either a fusible plug or a safety valve as a protection against excessive pressure. A fusible plug protects against overpressure

Figure 3–8 Cylinders and drums of various refrigerants (color coded). *Courtesy National Refrigerants.*

from fire or other excessive heat. A safety valve protects against overpressure from any source.

During service operations, refrigerant is sometimes pumped from a system into empty refrigerant drums for storage. It is of utmost importance that these drums not be overfilled. The design capacity of the drum must be known and not exceeded. Liquid in a drum filled nearly full will expand to fill the cylinder with a small rise in temperature. Then even a slight additional rise will exert a hydrostatic pressure in the drum that no steel would hold. The result would be an explosion. No drum should ever be filled without first determining that it is empty. Then it should be weighed both before and after filling, to ensure that its capacity is not exceeded.

The technicians who handle different sized cylinders soon get to know the amount of the different refrigerants for which each one is safely rated. However, for a new refrigerant, a new size cylinder, or a new person, the amount should be checked from the supplier or other source. The law says never fill a cylinder to more than 80 percent of its liquid capacity. You can determine the expansion of any liquid refrigerant by dividing the specific volume at the final temperature by the specific volume at the starting temperature. For example, the specific volume of liquid ammonia at 100°F is 0.02747 cubic feet/pound (from Table A–15), and the specific volume of liquid ammonia at 0°F is 0.02419 cubic feet/pound.

$$\frac{0.0247}{0.02419} = 1.10321$$

So, liquid ammonia will expand approximately 10 percent as it heats up from 0°F to 100°F.

Refrigerant Cylinder Colors

Here we list the refrigerants by family (along with their azeotropes) and the color of the cylinder in which they are stored. The colors have been standardized over the years to reduce confusion and prevent accidents.

- CFCs:
  R-11 orange
  R-12 white
  R-13 light blue
  R-13 B1 coral
  R-113 purple
  R-114 dark blue
  R-500 yellow (azeotrope)
  R-502 light purple (azeotrope)
  R-503 aquamarine (azeotrope)

Note: Although R-503 is in the CFC family, it is an azeotropic mixture of R-13 (CFC) and R-23 (HFC). Once again, we caution against "experimenting" with refrigerant mixtures.

- HCFCs:
  R-22 light green
  R-123 light gray
  R-124 deep green
  R-401A coral red (zeotrope)
  R-401B mustard yellow (zoetrope)
  R-401C blue-green (zoetrope)
  R-402A pale brown (zoetrope)
  R-402B green-brown (zoetrope)
  R-404A orange (zoetrope)
  R-406A light gray-green (zoetrope)

- HFCs:
  R-23 light gray
  R-125 tan
  R-134a light blue
  R-407A bright green (zoetrope)
  R-407B peach (zoetrope)
  R-407C chocolate brown (zoetrope)
  R-410A rose (zoetrope)
  R-507 teal (azeotrope)
- Ammonia silver

The intent is to simplify identification of refrigerant cylinders. Always read what is on the service cylinders; not everybody has the same color perception. Note that orange appears twice, and there are fine shades of blue and green.

Questions

1. Which properties that are desirable in a refrigerant are most important to produce dependable refrigeration?
2. Which properties are important to produce economical refrigeration?
3. Which properties are most important from a safety point of view?
4. Why is R-134a so popular in domestic and commercial applications?
5. Do HFC refrigerants contain chlorine? Why or why not?
6. What other refrigerants are also commonly used in commercial applications?
7. In what type of application might we find R-503?
8. Why is ammonia not used in common commercial applications?
9. Why is ammonia not used in directly air conditioned spaces?
10. Why has R-12 or R-22 not displaced ammonia in most industrial plants?
11. In what types of installations is R-11 to be found?
12. List some of the problems involved when the refrigerant in the system is replaced with another.
13. Why is it important to weigh the refrigerant being transferred to any refrigerant drum?

Chapter 4

Liquid Feed Devices or Expansion Valves

The refrigeration industry commonly calls the valve or device that feeds refrigerant to the evaporator an *expansion valve*. Many technical personnel, however, object to the use of this term. True, some of the liquid changes to vapor at this point, and expands in volume as it does so. However, as pointed out in Chapter 2, only a small part of the liquid changes to vapor as it goes through this valve. The remaining liquid contracts as it chills. Also, when float valves and capillary tubes are used for this purpose, the term "expansion valve" becomes an even less accurate description yet. For these reasons, the device is sometimes called an *evaporator feed device, refrigerant feed device, metering device, liquid-feed device,* or *refrigerant control device*.

Function of the Expansion Valve

To understand clearly the function of an expansion valve, let us first consider an analogy. Consider a tank that supplies water to a coil placed inside the firebox of a furnace (Figure 4-1A). The flow of water from the tank to the coil is regulated by a control valve. This valve is set so that water from the tank flows into the coil at exactly the same rate as it is evaporated, or boiled away, by the furnace. It should be adjusted so that some water will run all the way through the coil, but the last drop of water should be evaporated to steam before it rolls out of the end of the coil. If the valve is not open wide enough, all the water is evaporated before it reaches the end of the coil. The last few lengths of pipe are then useless in that they evaporate no water. If the valve is opened too wide, unevaporated water leaves the coil. This wastes water, and if dry steam is required, the excess water might cause damage.

Figure 4-1B shows how this analogy fits a refrigeration system. The liquid receiver corresponds to the water tank, and the liquid refrigerant, to the water. The regulating valve is the expansion valve. The evaporator coil corresponds to the pipe coil in the furnace. The liquid receiver does not have to be as high as the evaporator coil because the pressure of the high side will push the liquid through the expansion valve to this low-pressure coil. The box, at 40°F, will be hot compared to the boiling temperature of the refrigerant. Therefore, it will evaporate the liquid refrigerant in the coil just as the water is boiled in the analogy. The expansion valve must be regulated so that some liquid refrigerant will reach the last pipe before evaporating. Too little refrigerant will not utilize the entire coil. Too much will return liquid to the compressor, where it may cause damage.

If warm air or warm food enters the box, the greater amount of heat will evaporate all the refrigerant before it reaches the end of the coil. Therefore, the expansion valve will have to be opened wider if the entire coil is to be used. This will more rapidly remove the extra heat than if only a part of the coil is effective. As the heat is extracted from the warmer goods, less heat will be

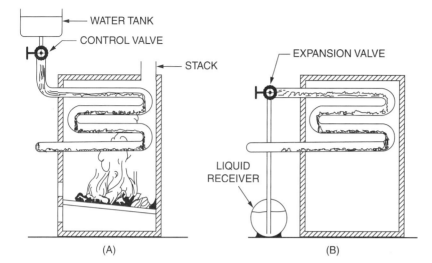

Figure 4-1 Evaporator analogy.

left to evaporate the refrigerant. Therefore, the valve will have to be gradually closed to compensate for this decrease in heat load. That is, the expansion valve must regulate the flow of refrigerant according to the heat load. The expansion valve does not control temperature.

Referring back to the refrigeration cycle charts, Figure 2-13, we note the following conditions at the expansion valve:

1. The pressure drops instantly from high pressure to low pressure at the expansion valve needle.
2. The temperature drops instantly from the condensing temperature to the evaporating temperature at the valve needle.
3. There is no change in the heat content of the refrigerant as it passes through the expansion valve. This is because the heat removed to cool the liquid becomes part of the refrigerant vapor.
4. The refrigerant approaches the expansion valve as a liquid. Part of it, 14 percent at the conditions illustrated, changes instantly to a vapor at the needle. This 14 percent is known as *flash gas* and is discussed in Chapter 20.

Hand Expansion Valve

The hand expansion valve, Figure 4-2, is nothing more than a needle valve, or in large sizes a plug valve, which makes fine adjustments or fine control of the flow possible. Obviously, an attendant or engineer must always be present to make whatever adjustments are necessary with this type of valve.

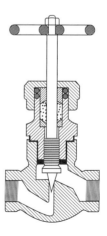

Figure 4-2 Hand expansion valve.

CHAPTER 4 — Liquid Feed Devices or Expansion Valves

Figure 4-3 Hand expansion valve bypassing automatic valve.

At one time the hand expansion valve was all that was available for refrigeration plants. It was the only expansion valve available with the early ammonia systems. Now, some form of automatic valve is standard. However, a hand expansion valve is often used as a bypass around the automatic valve to make operation possible in case of failure or repairs to the automatic valve, as shown in Figure 4-3.

Low-Side Float Valve

One of the easiest of the automatic feed devices to understand is the low-side float valve. This type of valve was at one time widely used in both domestic and small commercial systems. Now, however, *its* only common use is in large commercial and industrial systems. Figure 4-4 shows the float mechanism itself. Figure 4-5 shows how it can be connected inside the liquid header of an ammonia evaporator.

Figure 4-6 shows how it can be mounted outside the accumulator in a separate chamber. The latter method of installation is more common, as it can be more easily serviced.

In all these cases, the liquid refrigerant level is maintained at the float level. As liquid is evaporated, the level drops and the connecting linkage opens the valve to allow more liquid to flow into the evaporator. As the entering liquid raises the level to the required point, the float is lifted and this closes the valve. Shut-off is complete and positive when there is no evaporation of refrigerant, as during the off cycle. The valve is called a low-side float valve because the float ball and mechanism is in the low-pressure side of the system. During high loads, the float will drop and allow more refrigerant to flow to the evaporator. Under light or no load, the float will rise and restrict or stop the flow of refrigerant.

The principal advantages of the low-side float valve are that it gives excellent automatic control and it maintains the proper refrigerant level regardless of load conditions, load changes, off cycles of the compressor, or any other such operating variables. Another advantage is that any number of evaporators can be operated in the same system. Each float valve will only pass the refrigerant required for its evaporator, no more and no less, regardless of conditions. Therefore, there is no limit to the number of low-side, float-operated evaporators that may be properly operated in the same system. Each will draw only its required refrigerant.

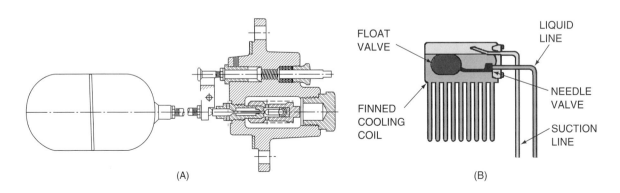

Figure 4-4 (A) Low-side float valve for ammonia and (B) Low-side float. *Courtesy of H. A. Phillips & Co.*

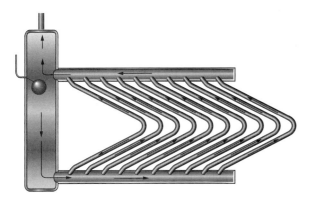

Figure 4–5 Low-side float valve inside liquid header. *Courtesy of York Corp.*

The principal disadvantage of this valve is that it will not work satisfactorily in a dry expansion coil (Chapter 5). Also, if it is improperly applied to a flooded evaporator (Chapter 5), it will give erratic operation or will trap oil. The chamber containing the float must be located so that the boiling action will not make the float bobble up and down, interfering with smooth operation.

Following are some of the difficulties that may be encountered with a low-side float valve. The ball may spring a leak from corrosion, from cracks, or from solder joints. The obvious result of such

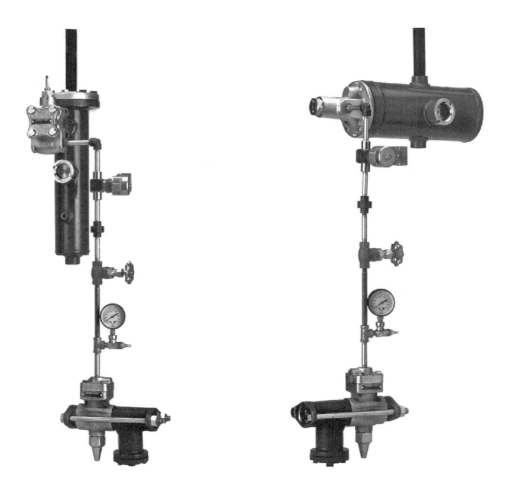

Figure 4–6 Low-side float valve in separate chamber. *Courtesy of H. A. Phillips & Co.*

defects is that the float will sink and open the valve. This will flood the coil and allow refrigerant to flow back to the suction line. This may lead to serious compressor damage.

The needle or seat, or both, may wear. They may wear smoothly so that they still hold refrigerant when the valve is shut off tight. However, to compensate for the wear, the float must rise higher to push the needle farther into the seat. The higher float means a higher liquid level. This may cause floodback (liquid refrigerant carried to the compressor) under the more violent boiling caused by heavy loads. Or the needle and seat may wear unevenly or even become pitted by wire drawing[1] of the refrigerant. Then the valve will leak continuously.

Too much refrigerant may have no effect on an evaporator with a low-side float valve. This is because the refrigerant is admitted only as required. However, if there is insufficient refrigerant, the liquid level in the evaporator cannot be maintained high enough to close the valve completely. During the running cycle, bubbles of warm, high-pressure vapor will be admitted to the evaporator. This must be removed by the compressor. This uses up part of the useful compressor capacity, yet still consumes power. The compressor operating on this excess warm vapor may run hot. If the refrigerant shortage is acute enough that a great deal of warm vapor enters the evaporator with what liquid can be condensed, the evaporator itself will begin to warm up. Such a condition can increase to a point at which the compressor will only recirculate vapor, producing a negligible cooling effect.

With automatic operation, slight refrigerant shortages allow high-pressure liquid or vapor to enter the evaporator during the off cycle. This will rapidly increase the pressure and warm up the evaporator. This makes for short off cycles, thus increasing the running time, and the cost of operating the system. Under a slight shortage with backpressure controls, such short cycling has been known to make an evaporator too cold, by operating the compressor when refrigeration is not needed.

Because the float of a low-side float valve system must have the proper buoyancy to operate the needle at the proper level, the floats must be calibrated for the refrigerant used. Floats for halocarbon refrigerants and for ammonia have such different calibrations, as well as being made of different materials, they cannot be interchanged.

High-Side Float Valve

Another form of float valve control is the high-side float valve. This differs from the low-side float valve in that the float chamber is on the high-pressure side of the valve needle, and a rising float opens the valve (Figure 4–7). Thus, this valve dumps the liquid to the evaporator as rapidly as it is condensed, but does not allow any uncondensed vapor to pass through the valve. Such a system makes the evaporator, instead of the liquid receiver, the refrigerant storage part of the system. This kind of valve is sometimes used without a liquid receiver.

One problem peculiar to early designs of high-side float valves was that they sometimes became air-bound. If air or other noncondensable gas was in the system, the liquid flowing to the float valve chamber carried this gas with it. Because the valve would pass only liquid, the gas collected. As the valve chamber filled with gas, the gas pressure prevented more liquid from flowing into the chamber. If no liquid entered, the float valve would not open, and the system would fail to function. The remedy was to "purge" the valve—that is, open a plug provided in the top of the chamber and let the air or gas blow out. Once all the air was eliminated from the system, the trouble would not recur. With ammonia valves, a bleeder tube is inserted in the valve to allow any noncondensable gas to bleed off to the evaporator (Figure 4–7).

Because the evaporator is the storage chamber for the refrigerant, this valve is very sensitive to high or low refrigerant charges. Too much refrigerant in the system will flood the evaporator (fill it to overflowing). With too little refrigerant, the valve will work satisfactorily, but the refrigeration and the frost level on the evaporator will be low.

---

[1] Wire drawing is a flow of high velocity through a nearly closed valve. The velocity becomes so great through the small orifice under pressure that it may actually erode or cut the metal.

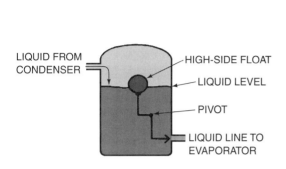
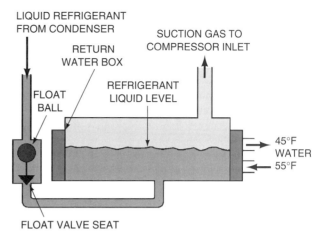

Figure 4-7 High-side floats.

Some evaporators with high-side float valves have become very erratic with low refrigerant charges.

The main use of the high-side float valve was in early domestic units. It has been used in some self-contained commercial applications, both with low-pressure refrigerants and with ammonia, but its use is very restricted even in these applications.

Capillary Tube

The capillary tube is not a valve at all, but functions as an expansion valve in domestic systems and in some small commercial systems. This fact shows more than ever that the purpose of the expansion valve is to divide the high- and low-pressure sides of the system while permitting flow to the low side. The capillary tube is nothing more than a coil of several feet of very fine tubing, usually having an orifice of about 0.03 to 0.06 in. The pressure of the high side of the system is used up in forcing the liquid through the long restricted passage. This device is sometimes called a *restrictor tube*. The length of tube used gives just the right restriction to allow the proper amount of liquid to trickle through to the evaporator. The only adjustment possible is in the length of tube used, and once this is installed, it is fixed. The three factors that affect flow through the capillary tube are the length of the tube, its bore, and the pressure drop across it.

However, a properly sized capillary tube with the proper refrigerant charge is somewhat self-regulating, or self-adjusting. It is installed without a liquid receiver. If insufficient refrigerant passes through, the liquid refrigerant backs up into the condenser. This restricts the inside volume of the condenser available for the gas discharged from the compressor. The head pressure rises and forces a greater flow through the capillary tube. If all the liquid is drained from the condenser (as happens during the off cycle), there is still sufficient restriction to give a separation between the high and low pressures.

Naturally, such a system is not as efficient in operation as an expansion valve that adjusts itself to the refrigerant flow required. However, for domestic and small, self-contained commercial units, the operating cost is relatively low. Therefore, the savings possible with a better device are insignificant. The capillary tube is cheaper than any other evaporator feed device, and has nothing to wear out. It is now used in all domestic systems. It is being applied to more and more hermetically sealed commercial equipment as well.

In many ways, the capillary tube works like the high-side float valve. It passes condensed refrigerant to the evaporator. Thus, the liquid storage is in the evaporator. One essential difference between the operation of a capillary tube and that of a high-side float valve is that with a capillary tube the pressure balances during the off cycle. That is, the high pressure in the high side bleeds through the tube until the pressures on the high side and the low side are the same. After all the condensed liquid has passed, the warm vapor goes over to the low side.

This warm vapor is then chilled and condensed by the cold liquid in the evaporator. To prevent too large a loss from this process, the volume of the evaporator is usually large compared to the volume of the high side of the system. This is another reason for omitting the receiver from this type of system. Then there is not enough of this warm vapor compared to the cold liquid to make any appreciable difference in the evaporator temperature during the normal off cycle.

The principal trouble to be expected with such a small tube is that it may become plugged with dirt. This difficulty was experienced in early applications. Also, moisture in the refrigerant may plug the tube with ice. However, an ultraclean environment has been found to be essential to the manufacture of any successful refrigeration unit, and this has eliminated the difficulties once encountered with the capillary tube.

Like the high-side float valve, the capillary tube is critical to the amount of refrigerant charged. Too much will fill the evaporator and flood back to the compressor. Too little will improperly fill the evaporator, leading to oil clogging and erratic operation.

A change of refrigerant most probably will not work properly. The length and diameter of the tube chosen must balance the pressure forcing the refrigerant through, the quantity of refrigerant required, and the viscosity of the refrigerant. Any change will upset this balance.

Automatic Expansion Valve

The automatic expansion valve is nothing more than a pressure-reducing valve, as shown in Figure 4–8. The sensitive element, the flexible diaphragm, has the evaporator pressure on the lower side of it,

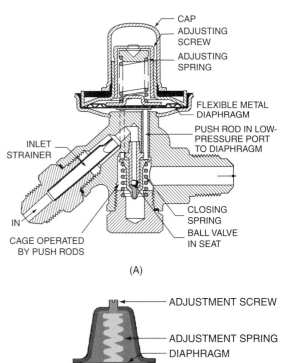

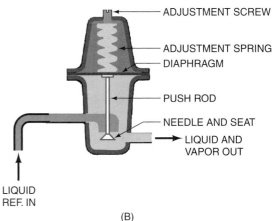

Figure 4–8 Automatic expansion valves. *Courtesy of Controls Company of America.*

and atmospheric pressure above it. These pressures are balanced and adjusted with springs. For a given pressure setting, the valve opens just enough that the flow through it balances the vapor removed by the compressor. If the evaporator pressure starts to decrease, the pressure below the diaphragm is reduced, and the adjusting spring pushes down. This motion is transferred to the cage and needle by push rods to push the needle out of the seat, causing more refrigerant to flow into the evaporator. This raises the evaporator to the desired level.

Any increase in pressure similarly raises the diaphragm, and the closing spring closes the valve. When the compressor stops for an off cycle, there is an immediate rise in pressure because the compressor is no longer removing the evaporating vapor. This rise in pressure closes the valve during the off cycle. A more commonly used automatic expansion valve uses only one spring as the opening pressure, and the evaporator pressure itself as the closing pressure.

The constant-pressure type of valve has advantages and disadvantages. Its biggest advantage is that it will maintain a constant evaporator temperature. Its disadvantage is that it does not respond well to varying loads. When the temperature surrounding the evaporator coil rises due to a greater heat load, the liquid refrigerant in the coil is all evaporated before any of it reaches the end of the coil. Thus, the last coils of the evaporator are useless for extracting heat from the room.

However, when the temperature has risen, this is just the time when all of the coil is most needed. Under such conditions, the room comes down to the proper temperature much more slowly than if the entire coil were actively refrigerated. On the other hand, if there is insufficient heat in the room to evaporate all the refrigerant entering the coil, there is nothing to prevent the unevaporated liquid from flooding back through the suction line to the compressor.

The action of the automatic expansion valve can best be illustrated by following what happens when refrigeration is started on a warm coil fed with this valve (Figure 4-9). During the off cycle, the valve remains closed. As the compressor is started, the reduction in pressure opens the valve enough to allow a limited amount of refrigerant to flow. This small quantity is all evaporated by the first section of warm evaporator tubing, *A–B*. The evaporation of this liquid refrigerates or cools section *A–B*, so the liquid following can flow through without all evaporating. The following liquid then strikes the warm section *B–C* and is evaporated, chilling *B–C* as it does so. Thus, the liquid will gradually work its way toward the end of the coil. However, some heat from the room will pass through the refrigerated section of the

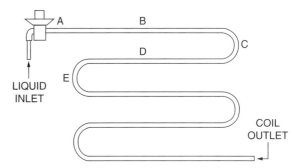

Figure 4–9 Action of automatic expansion valve on evaporator.

coil at all times. This also evaporates some of the liquid refrigerant. If the coil is long enough, or the heat passage through it is great enough, a point will be reached where the heat flow through the coil is sufficient to evaporate all of the liquid refrigerant. If this happens, liquid will never reach the end of the coil.

On the other hand, if the heat in the room is insufficient to evaporate all the liquid refrigerant, liquid will gradually work its way back to the suction line. The suction line then becomes part of the evaporator and produces refrigeration where it is not needed, but which must be paid for in power bills. If the suction line is short, the liquid may reach the compressor and cause damage. Of course, such conditions could be corrected by readjusting the valve. However, the expansion valve is expected to work automatically, so in practice an average setting must be found that uses as much of the coil as possible under heavy loads, but does not flood back under light loads. A thermostat is used and so adjusted that the machine turns off before it floods back.

Besides not responding well to a varying refrigeration load, the automatic expansion valve has one other disadvantage: It does not work well with more than one evaporator in the system. Figure 4-10 shows why: If evaporator A is operated at a higher temperature, which means a higher pressure, than evaporator C, the higher-pressure vapor from evaporator B will back up through evaporator D to evaporator C and close the other valve. Thus, coil C will get no refrigerant

48 CHAPTER 4 *Liquid Feed Devices or Expansion Valves*

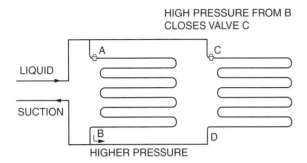

Figure 4–10 Why an automatic expansion valve will not multiplex well.

and will not refrigerate. Under limited conditions both coils could be operated at exactly the same pressure, but if one valve had its pressure set just a little too high, or if due to a slight change in adjustment due to wear, stickiness, change of temperature of valve parts, etc., the other coils will be starved. Such applications have been used, but they must be accurately adjusted and closely watched, and are not recommended.

Because of its constant temperature characteristics, the automatic expansion valve is much used in certain applications, particularly in liquid cooling systems in which it is important to maintain a liquid bath at a constant temperature. The valve is also simpler, which means cheaper, and possibly a little more trouble-free than valves that are more responsive to the load.

Because of its relative simplicity, there is little that can go wrong with this valve. Needles and seats may erode so that they leak during the off cycle, but present-day materials have largely eliminated this problem. Also, gradual wear will necessitate the valve moving farther into the seat to shut off. This will raise the evaporator pressure and tend to cause floodback if uncorrected, but it can be corrected by readjustment. Diaphragms have been known to crack or corrode through, but again, modern valves have largely eliminated this problem.

An excessive refrigerant charge will have no effect on this valve. A low charge will allow some high-pressure vapor to enter the evaporator. Peculiarly, this will sometimes cause a slug of liquid back to the compressor because a large bubble of high-pressure vapor can sweep everything before it as it travels through the evaporator to the compressor.

Each valve usually has considerable range of pressure through which it can be adjusted. Its operation is usually equally satisfactory at any required evaporator temperature. For very cold temperatures, below –20°F, the manufacturer should be consulted for recommendations.

Thermostatic or Thermal Expansion Valve

The thermostatic or thermal expansion valve is sometimes called a DX or a TX valve. It is an automatic expansion valve with an additional device to correct the adjustment for the load on the evaporator, as shown in Figure 4–11. It is actually more automatic in its operation than the so-called automatic expansion valve. To understand how the thermostatic expansion valve works, refer to the temperature cycle of Figure 2–13. The expansion valve and evaporator section is reproduced to a larger scale in Figure 4–12 under three different conditions.

The condition of Figure 4–12A is found with a starved evaporator; that of Figure 4–12B with an evaporator operating nearly flooded; and that of Figure 4–12C with it full flooded. There are two important differences in the different conditions illustrated. The first and most important is that in Figure 4–12A, only about 60 percent of the evaporator surface is wet on the inside. Because the part that is dry does practically no work, only about 60 percent of the evaporator is doing useful work—that is, cooling the room. About 95 percent of the evaporator is used in Figure 4–12B, which means only a 5 percent loss in the full usage of the evaporator. The condition of Figure 4–12C gives fullest use of the evaporator surface. However, if there is liquid at the end of the coil, as in this case, any sudden surge in evaporation might push some of the liquid on through to the

Thermostatic or Thermal Expansion Valve

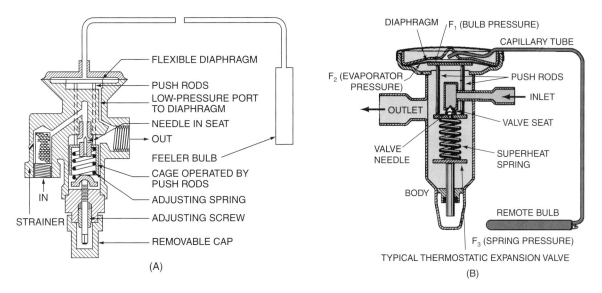

Figure 4–11 Thermostatic expansion valves.

compressor. Therefore, it is too critical a point at which to attempt to operate this type of evaporator.[2] The condition of Figure 4-12B causes relatively little loss in capacity, so for safety's sake it is the one used.

The second difference to be noted in Figure 4-12 is the difference in the temperature of the vapor leaving the evaporator. In Figure 4-12A, 20°F of superheat is picked up in the dry section of the coil. In Figure 4-12B, there is 10°F of superheat,

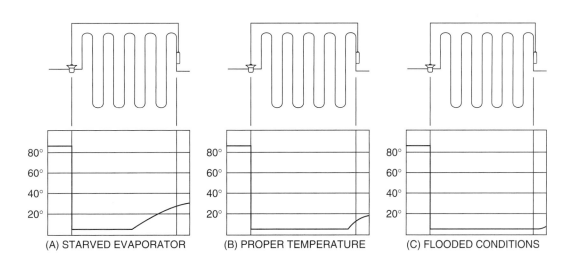

Figure 4–12 Comparison of suction temperature under various evaporator conditions.

---
[2] See Chapter 5 for a more complete discussion of evaporator types.

and in Figure 4–12C, there is none. Notice that this is a difference in temperature between the active part of the coil and the end of the coil. Differences such as these will exist regardless of the temperature at which the coils operate. This fact is made use of to adjust the expansion valve to operate the coil at the condition of Figure 4–12B.

The thermostatic expansion valve in Figure 4–11 is an automatic expansion valve with a temperature-sensitive element attached. The feeler bulb is attached to the last coil of the evaporator to "feel" the temperature change of Figure 4–12. The bulb is charged with a fluid having characteristics similar to those of the refrigerant in the system. The pressure created by the refrigerant in the bulb is transmitted to the top of the flexible diaphragm. An adjusting spring and screw are provided to balance the valve to the evaporator. Once this is set, changes in the load on the coil are automatically adjusted for by the changing pressure on top of the diaphragm. The preliminary adjustment should be such that approximately 10°F of superheat balance the coil.

Figure 4–13 shows how this 10°F superheat gives a pressure difference that actuates the valve. The conditions given are for Refrigerant 134a. The 5°F evaporator has a pressure of 9 psig. This pressure is applied to the under side of the diaphragm, and remains constant as long as the evaporator temperature does not change. The 10°F superheat in the suction line heats the feeler bulb to 15°F. This causes a pressure of 15 psig, which is applied to the top of the diaphragm.

The difference of 6 psi is balanced by the adjusting spring. When there is 6 psi more pressure on top of the diaphragm than on the bottom, the flow of the refrigerant through the valve is just sufficient to maintain the condition of Figure 4–12B. The valve is properly adjusted to allow the right amount of refrigerant into the coil to balance the load.

The condition of Figure 4–12A is an extreme condition, but if it did occur, the 20°F of superheat would heat the feeler bulb to 25°F. This would cause a pressure of 22.1 psig on top of the diaphragm. The pressure below would remain at 9 psig. Thus there is 22.1 − 9 = 13.1 psi greater pressure above the diaphragm. This will hold the valve wide open to permit the flow of a maximum amount of refrigerant.

A less extreme case will occur if there is sufficient heat around the coil to evaporate all the liquid, and 15°F of superheat is picked up by the time the refrigerant reaches the bulb. A 20°F bulb will cause a pressure difference of 9.4 psi on the diaphragm and open it wider. If this wider opening allows the refrigerant to reach the point of Figure 4–12B, which gives 10°F of superheat, the valve will balance and stay at this new position. If not, the valve will readjust.

If the heat load drops so that unevaporated refrigerant approaches the end of the coil, the superheat drops. If the superheat drops to 5°F, this will heat the bulb to 10°F, which gives a pressure of 11.9 psig or a difference of 2.9 psi on the diaphragm. Thus, the spring will close the valve a little further, which will reduce the flow of refrigerant.

If the liquid refrigerant reaches the coil end, the temperature will be 5°F, and the pressure will be the same on both sides of the diaphragm. The spring will then close the valve completely and hold it closed until the pressure on top of the diaphragm increases again.

It might seem that the valve should be operated with much less than 10°F of superheat to better use the entire coil. Some superheat is obviously necessary, because if the liquid reaches the control bulb, there will be no way of telling whether it just reached it, or flooded clear back to the compressor. The reason 1°F or 2°F of superheat is not used is

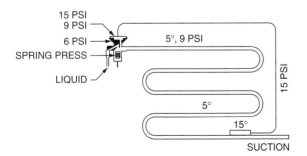

Figure 4–13 Pressures and temperatures in thermostatic expansion valve and evaporator.

because this would mean a difference of less than 1 psi between normal operation and a completely closed valve. It is practically impossible to make a valve that is sensitive enough to react to such a small change, but that will not fluctuate too widely under changing load conditions. That is, it would be either completely closed or wide open most of the time, instead of following the load smoothly. Manual adjustment makes it possible to change the superheat setting at which the valve will operate. About 10°F of superheat has been found to work best in practice for natural-convection coils. As little as 5°F may be used on some blower coils and some freezer coils.

A variation in the construction of the thermostatic expansion valve is the external equalizer. A valve with an external equalizer is recommended for large evaporators in which there is considerable pressure drop through the evaporator. Any such pressure drop will be maximum for heavy loads, and will be little or nothing for light loads. Figure 4–14 illustrates extreme conditions under heavy loads. The maximum pressure drop in this case is 4 psi. Assume that the valve is set for an average superheat of 10°F under a light load as in Figure 4–13. The pressure below the diaphragm is 9 psig and above it is 15 psig, giving a difference of 6 psig. Under a heavy load, Figure 4–14, the pressure at the end of the coil is 9 − 4 = 5.8 psi. This gives an evaporating temperature of −3°F.[3] However, 15°F at the end of the coil is necessary to get the 6 psi difference. Thus, to balance the valve, the total superheat setting is actually 15 − (−3) = 18 F. This will starve the coil, but a heavy load condition is just the time when maximum flooding is required.

The external equalizer takes the pressure that is applied under the diaphragm from the same point as the temperature of the superheat bulb is taken, as shown in Figure 4–15. Therefore, any change in the pressure in the coil is compensated for. The 10°F superheat is added to the temperature of the evaporating liquid at the end of the coil. It is always constant, regardless of any change of pressure through the coil.

Because active refrigeration is not produced during the off cycle, the high conductivity of the metal evaporator tube tends to equalize the temperature throughout the coil. This keeps the difference in temperature between the active section of the coil containing liquid and the bulb less than 10°F, which closes the valve. This shuts it off tight during the off cycle so it will not flood the evaporator.

The big advantage of the thermostatic expansion valve is that, when used properly, it gives completely automatic evaporator flow control. It responds readily to variations in load, and balances the entire coil without allowing any floodbacks.

Its principal disadvantage is its complexity, which means that it is not only more costly, but that it has more parts that can go wrong.

However, if this type of valve is properly selected for the job, and properly installed, it gives very little trouble. The right valve must be chosen for the job. It must be of the proper size. Too small a valve will not allow enough refrigerant to pass to fill the entire coil during heavy loads. Too large a valve will cause "hunting," that is, a valve that opens too wide when an increasing load calls for

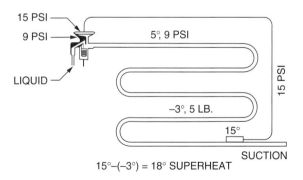

Figure 4–14 Pressure drop in large coil.

---

[3] It has been stated that the evaporator temperature is constant through the entire evaporator. Normally it is, but there are special cases when there may be sufficient variation to alter operating conditions somewhat. In this case, note that the temperature gets colder as the end of the coil is approached.

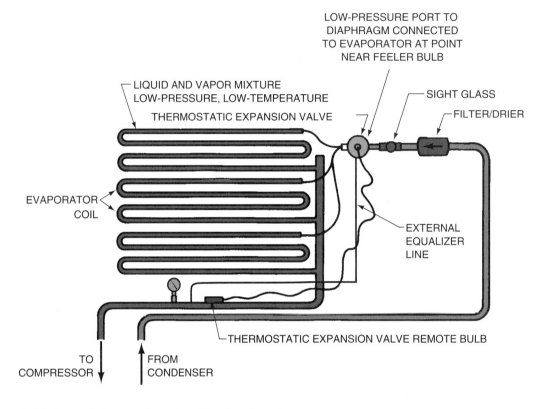

Figure 4-15 Thermostatic expansion valve with external equalizer.

more refrigerant. This wide opening causes flooding, which chills the control bulb and closes the valve. Thus, the valve alternately opens too wide, then closes too much, and does not settle down readily to a proper balance.

Some valve manufacturers supply different sized orifices for different loads on valves to give different ratings between the different sizes manufactured. Figure 4-16 gives generic information demonstrating how the difference in pressure

Valve No.	Orifice Assembly No.	Design Pressure Difference (Lb.)						
		30	40	50	60	70	80	90
ABC	111B	0.35	0.45	0.55	0.60	0.65	0.70	0.75
ABC	11B	0.70	0.80	0.90	1.00	1.10	1.10	1.20
ABC	1B	1.40	1.60	1.80	2.00	2.20	2.30	2.40
ABC	2B	2.50	2.90	3.30	3.60	3.90	4.20	4.40
ABC	3B	3.40	3.90	4.40	4.80	5.20	5.60	5.90

Figure 4-16 Capacities (in tons) of one size of thermostatic expansion valve at different pressures, with different orifice assemblies (generic information for illustrative purposes only).

between the high and low sides actually determines how much refrigerant is forced through a given sized orifice. Note that with this valve and a given orifice size, extreme pressure changes may reduce capacities 30 percent or increase them 20 percent from average conditions.

Some manufacturers supply an orifice that goes in the outlet of the valve. This orifice is small enough that at anywhere near full load, most of the pressure drop is across this orifice instead of at the needle and seat. In this way, any erosion due to wire drawing will take place in this secondary orifice instead of on the needle or seat. In some cases, valve capacities are more easily changed by changing this orifice than by changing the needle and seat assembly.

One manufacturer has adapted the thermostatic expansion valve to flooded operation. The feeler bulb has a built-in electric heater. Its design is such that when it is submerged in liquid, the heat does not reach the bulb, so the valve remains closed. When the liquid level drops below the bulb–heater assembly, the heat reaches the bulb and the valve opens.

In some applications of this valve there may not be enough difference in temperature between the room and the coil to supply the required superheat. If, for instance, there is an extra large coil in a room so that a coil only 8°F colder than the room is necessary, then 10°F of superheat cannot be picked up. Such a condition could be remedied in a halocarbon system by using a heat exchanger, Figure 4–17. Here the extra heat is supplied by the warm liquid from the condenser that feeds the expansion valve.

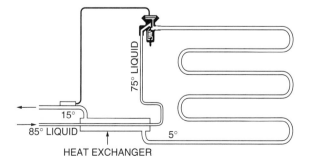

Figure 4–17 Heat exchanger to supply expansion valve superheat.

Heat exchangers are not desirable in ammonia systems. The warmer suction gas causes an undesirable increase in compressor discharge temperature. Some thermostatic expansion valves for ammonia have been designed that will operate on as little as 2°F of superheat.

A similar condition might occur at very low temperatures for a different reason. Normally, the sensing bulb is charged with the same refrigerant as that with which the system is charged. It has been shown how 10°F of superheat at a 5°F evaporator gives 6 psi pressure difference across the diaphragm. But at a −40°F evaporator, the pressure with Refrigerant 134a would be 7.4 psi. This plus 6 pounds would give 13.4 psia pressure on top of the diaphragm necessary to balance the valve. However, the temperature for 13.4 psia is −18 F, or 22°F of superheat. Thus, if no heat exchanger or other form of correction is used, the valve will not work in a room colder than −18°F. Such a wide difference in temperature between the coil and the room is very uneconomical.

Manufacturers have corrected this condition somewhat by using a cross-charged fluid in the sensing bulb. The fluid in the sensing element has thermodynamic characteristics different from those of the refrigerant with which the system is charged. Typically, this fluid has a much lower boiling point than the refrigerant in the system. This special fluid tends to hold the same pressure difference for a given temperature difference whether at high or low temperature. However, there is a limit to the temperature range for which cross-charging can correct. Some manufacturers make special valves for evaporator temperatures below 0°F. It is always safest to depend on the manufacturer's recommendations in such applications.

In liquid-charged sensing bulbs, some liquid is always present regardless of system conditions. These devices normally maintain constant superheat regardless of system conditions. Some thermostatic expansion valves employ gas-charged sensing bulbs. They are similar in operation to their liquid-charged cousins. However, it is possible for all of the liquid in the sensing element to vaporize. When this happens, any further rise in temperature causes no further rise in pressure in the sensing bulb. Therefore, the expansion valve is as wide open as it can get once all of the liquid in the sensing bulb is vaporized. The vapor cross-charged sensing bulb operates similarly to the vapor-charged sensing bulb, with the exception that it uses a fluid with thermodynamic characteristics different from those of the system's refrigerant. These vapor cross-charged devices are used in special refrigeration applications.

Each valve is made specifically for the refrigerant it is to be used with, as well as the size of the project. The bulb is charged with a cross-charged fluid balanced against the refrigerant with which it is to be used. Occasionally, because the wrong valve is available, someone tries to use it and compensate for the difference by readjusting. Such adjustments of a misapplied valve will never give satisfactory results.

In commercial applications, more thermostatic expansion valves are used at the present time than all other types of refrigerant controls put together. Their fully automatic characteristics under all load conditions make them excellently suited for the attention-free automatic requirements of commercial operation. Their adoption in industrial applications was slower than in commercial ones, but they are now being used in ever-increasing numbers. They do have limitations in full flooded evaporators, and in very cold temperature work.

Solid-State or Thermal-Electric Expansion Valve

The previously mentioned devices all respond to either pressure or thermal (temperature) differences. In other words, they are thermal, mechanical, or a combination thereof. Even the hand expansion needle valve is adjusted by the operating engineer as conditions and loads change in the system. Although these types of devices are still in wide use, some liquid-feed devices employ electricity and/or electronics to feed liquid refrigerant to the evaporator.

Solid-state expansion valves, figure 4–18, employ a control voltage transformer, a heat (thermal) motor, and a reverse-acting thermistor. The heat motor is simply a bi-metal device. It responds exactly the

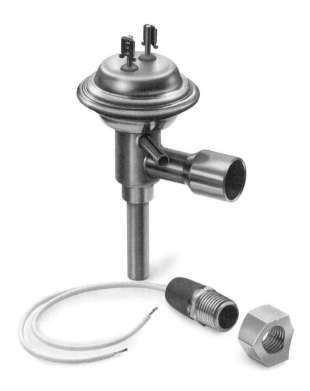

Figure 4–18 Thermal electric valve shown with liquid-sensing thermistor and thermistor suction line adaptor. *Courtesy of Parker Hannifan Corporation.*

same way as the thermal overload device used in hermetically sealed compressors and some thermostats. Typically, two dissimilar metals are riveted or otherwise bonded together. The metals have different coefficients of thermal expansion. Consequently, any thermal change causes the bi-metal strip to bend.

A *thermistor* is an electronic device whose electrical resistance changes with temperature. Some thermistors offer more resistance to the flow of electricity when their temperature increases (positive temperature coefficient). Others offer less resistance to the flow of electricity as their temperature increases (negative temperature coefficient).

In a system that uses the thermal-electric expansion valve, the control voltage transformer typically reduces 120 volts down to 24 volts. These transformers are available to use with virtually any primary voltage. However, the control voltage is almost invariably 24 volts. The 24-volt power is wired in a series with the reverse-acting thermistor and the heat motor of the liquid refrigerant valve. The thermistor is positioned so that is in the direct path of the refrigerant vapor leaving the evaporator.

An increase in suction vapor temperature occurs when the evaporator requires an increase in the rate of refrigerant being fed to it. This increase in suction vapor temperature causes the temperature of the thermistor to increase, which in turn causes a decrease in the thermistor's electrical resistance. Now, more current is allowed to flow to the heat motor of the refrigerant liquid feed valve. The bi-metal strip bends and forces the valve to open. As a result, more liquid refrigerant is fed to the evaporator.

When the evaporator is saturated with liquid refrigerant, the sequence of operations reverses. As the saturated mixture of liquid and vapor refrigerant approaches the thermistor, the vapor temperature and the temperature of the thermistor are lowered. Now, the thermistor's electrical resistance is raised. Less current is allowed to flow to the heat motor and the bi-metal returns or nearly returns to its original shape. The liquid refrigerant valve moves more to the closed position and less liquid refrigerant is fed to the evaporator. These cycles repeat over and over to keep the evaporator filled with the proper amount of liquid refrigerant.

The solid-state expansion valves are relatively new in the family of evaporator liquid-feed devices. The thermistor represents relatively low mass (compared to the thermo-bulb of the thermostatic expansion valve). Also, the thermistor is physically in the path of the refrigerant suction vapor (compared to the thermo-bulb, which is strapped to the outside of the suction line). These two features combine to make the solid state-expansion valve respond much more quickly to changes in evaporator load and conditions. The solid-state expansion valve is capable of working with much less superheat than its cousins. This allows the evaporator to operate with more liquid refrigerant and overall makes for more efficient operation.

Solid-state expansion valves allow refrigerant to flow in both directions. This feature makes them a natural choice for heat pump applications (see Chapter 25). Some solid-state expansion valves have a bleed port. This means that they never completely stop the flow of liquid refrigerant to the evaporator, even when the valve is completely shut. This feature allows the high-side pressure to equalize with the evaporator pressure during the off cycle (capillary tubes do this as well). As a result, a low-starting-torque motor may be used to drive the compressor.

Liquid Solenoid Valve with Electric Float

Solenoid valves have a coil of wire with an iron core running through the center. The iron core is allowed to move back and forth through the coil and does so when the coil is energized or de-energized. This reciprocating motion causes a disk to move in and out of the seat of the valve, thus opening or closing the valve. In some larger valves, the coil and shifting core may operate a pilot valve, which in turn causes the main disc to move in and out of the main seat. A normally closed valve opens when current is applied to the coil. A normally open valve closes when current is applied to the coil.

In an electric float and solenoid combination, voltage (normally 120 volts) is wired in a series with the electric float and the solenoid valve. The down side of the solenoid is connected to ground. The electric float consists of a steel housing (required for ammonia), a float (hollow, sealed steel ball) inside the housing, an iron rod, and a stainless tube into which the iron rod is placed. The iron rod is connected to the sealed steel ball and can ride up and down in the stainless tube. The steel housing is hard-piped to the vessel in which we want to maintain a level of liquid refrigerant. The upper pipe is connected to the vapor portion of the vessel and the lower pipe is connected to the liquid portion of the vessel. Consequently, the liquid level in the vessel will give a corresponding level in the housing. The elevation of the housing corresponds to the desired liquid level.

Outside the stainless tube is a magnet attached to a pivot. If the iron rod is high up in the stainless tube, the magnet is attracted to the iron rod and moves toward the rod until the magnet is "hugging" the stainless tube. When the iron rod moves down inside the stainless tube, the magnet falls away from the stainless tube. The base of the pivot has switch contacts connected to it, and a circuit is made or broken as the magnet either "hugs" the stainless tube or falls away from it. The entire magnet–pivot–switch assembly is normally enclosed in a clear plastic housing for protection.

As liquid refrigerant evaporates inside the evaporator, the liquid refrigerant level in the vessel drops. This vessel might be a recirculator vessel or a surge drum (see Chapter 5). As a result, the liquid refrigerant level drops inside the steel housing and the sealed steel ball also drops. In turn, the iron rod moves to a lower position inside the stainless tube and the magnet falls away from the tube. When this happens, the electrical contacts close and the solenoid valve is energized, causing it to open. The solenoid valve is now feeding liquid refrigerant to the evaporator and the liquid level is rising. When the level, float, and iron rod rise up high enough, the magnet is attracted to the iron rod and "hugs" the stainless tube. The electrical contacts open and the liquid solenoid is de-energized. The liquid refrigerant valve closes and the vessel is satisfied with refrigerant. Refrigerant evaporation continues to take place, and when the level in the vessel drops low enough, the sequence starts all over again.

The contacts in the device just described open on rise. In other words, as the liquid refrigerant level rises, the electrical circuit is broken. These devices may also be wired to close on rise. We may choose to do this if we want to use the float to give an alarm such as a high suction trap level. These devices may also be used to stop a compressor on high suction trap level or stop a pump on low recirculator vessel level.

Occasionally, the steel ball develops a leak. When this happens, it can no longer float on the liquid refrigerant. The housing and float must be replaced. In the meantime, on an emergency basis, the float may be disconnected electrically. Most solenoid valves have an override stem that may be screwed in to open the valve manually and keep the vessel satisfied with liquid refrigerant.

Liquid Solenoid Valve with Thermal Differential Sensor

The thermal differential sensor (TDS) consists of a metal block, an electric resistance heater, a bi-metal strip, and a sealed liquid metal tube. (**Caution:** The heater gets hot enough to burn you.) The metal block is strapped to the external pipe of the vessel in which we want to maintain a level of liquid refrigerant. The pipe is normally 3/4 inch in diameter and is connected to the vapor and liquid portions of the subject vessel. Therefore, the liquid level in the vessel will give a corresponding level in the pipe. The block is at the desired level we wish to maintain. The heater is inside this block, and the bi-metal strip can sense the temperature of the metal block. The bi-metal strip twists and untwists as its temperature changes. As it twists, it tilts the sealed liquid metal tube. The sealed liquid metal tube has contact points inside. If the points are both covered with the liquid metal, the contacts are closed and the circuit is made.

As liquid refrigerant evaporates inside the evaporator, the liquid refrigerant level in the vessel drops. When this happens, the liquid level moves down and away from the metal block. The electric heater raises the temperature of the metal block and the bi-metal strip twists. The liquid metal tube tilts and the contact points are covered with liquid metal. The contact points are wired in a series with the solenoid valve, so the solenoid valve opens. The vessel is now being fed with liquid refrigerant and the level rises. As the liquid approaches the metal block, it overrides the heater and the block's temperature is lowered. The bi-metal "untwists," at least one contact point is uncovered with liquid metal, and the circuit is broken. The solenoid valve is de-energized, the valve closes, and the vessel is now satisfied with liquid refrigerant. Refrigerant evaporation continues to take place, and when the level in the vessel drops low enough, the sequence of events starts over again.

Compare the electric float with the TDS. The electric float is hard-piped to the subject system. If we need to replace the electric float or move it (change elevation), we have to open the refrigeration system. However, the TDS may be removed from the system without having to open it. As with the electric float, the TDS may be wired to open on rise or to close on rise. Therefore, it may be used in any application in which we find the electric float.

Liquid Solenoid Valve with Variable-Capacitance Probe

The variable-capacitance probe is simply an iron rod coated with Teflon and inserted into an iron pipe, Figure 4–19. The iron pipe is normally 3 inches or 4 inches in diameter. The pipe is connected to the vapor and liquid portions of the subject vessel. Therefore, the liquid level in the vessel will give a corresponding level in the pipe. The capacitance of the iron rod varies in direct proportion to the level of the liquid refrigerant on it. A current is applied to the rod and we read how much current we get back from the rod (normally 4 to 20 milliamperes). The control panel normally gives percentage of probe, which can be easily converted to inches, feet, gallons, pounds, or any convenient unit of measure. The control panel also has electric contacts that can be hard-wired to open and close valves, start and stop equipment, and give an alarm, all based on the elevation of the liquid as sensed by the probe. All of these levels are easily adjusted on the control panel, as is the required span.

The *span* is simply how sensitive the control is. If we want to maintain 36 inches of liquid ammonia in a vessel, we can set the control panel to energize the liquid solenoid at 35 inches of liquid and to de-energize the liquid solenoid at 37 inches of liquid. In this case, there is 1 inch of span (1 inch either way of the desired 36 inches). For finer control, one might want to program for 1/2 inch of span. In this case, the liquid solenoid would be energized at 35.5 inches of liquid and de-energized at 36.5 inches of liquid.

These variable-capacitance probes and their control panels are easy to set up and are very

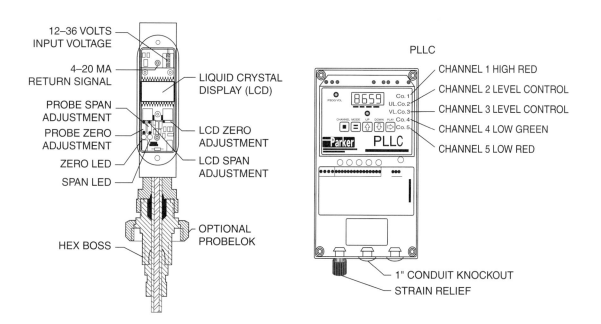

Figure 4–19 Variable-capacitance probe shown with its associated control panel. *Courtesy of Parker Hannifan Corporation.*

user-friendly. One probe can do the job of five electric floats or TDSs. A typical application might involve five "trigger" points, all controlled by one probe. The five points might be:

- Liquid level control (energizes and de-energizes the liquid solenoid as necessary)
- HLA—high level alarm (gives an audible and/or visual alarm on high vessel level)
- HLSD—high level shutdown (shuts down compressor[s] on high vessel level; the HLA is set at a lower level than the HLSD)
- LLA—low level alarm (gives an audible and/or visual alarm on low vessel level)
- LLSD—low level shutdown (shuts down pump[s] on low vessel level; the LLA is set at a higher level than the LLSD)

Pneumatic High-Side Float

One manufacturer's system employs a pneumatically actuated high-side float. The float chamber, Figure 4–20, item 7, allows control air (3–15 psig) to pass through at varying rates according to the level in the condenser section of the chiller. The air is piped to the upper side of the diaphragm of the liquid feed valve, Figure 4–20, item 8. As the level rises in the condenser barrel, the evaporator feed valve opens and feeds liquid refrigerant to the evaporator barrel. When the liquid level falls in the condenser, control air flow is restricted and the evaporator feed valve moves toward the closed position. The entire system is built to keep the first part of the first pass of cooling-water tubes surrounded with liquid refrigerant. Here the liquid refrigerant is subcooled. The remainder of the first pass of cooling-water tubes and the second pass of cooling-water tubes are surrounded by compressor discharge vapor.

This operating description refers to Figure 4–20. The condenser as shown has an integral economizer in the first-pass section. The thermal economizer increases machine efficiency and lowers the required horsepower per ton, thus lowering operating costs. Refrigerant is condensed on the condenser tubes (1), drips down, and is collected on the plate separating the condenser section from the thermal economizer section. The liquid is then directed to the end of the vessel opposite the condenser water inlet. From there, it flows to the thermal economizer through a designated opening (2). It then flows around a series of baffles (3) and out the thermal economizer liquid outlet line (4). The condenser water in the two-pass illustration flows in at (5) through the thermal economizer and condenser first pass, reverses and flows back through the condenser second pass, and exits at (6). The counterflow of entering condenser water and leaving refrigerant subcools the liquid refrigerant to a temperature near that of the entering water temperature, thus increasing the refrigeration effect of the system. The thermal economizer tubes are kept immersed in liquid. The level is maintained by a liquid level control (7), which operates a control valve (8) in the liquid line to the cooler.

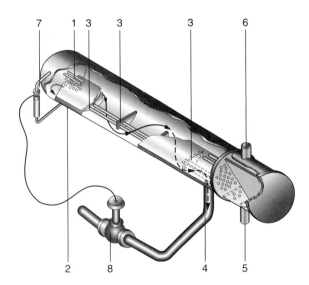

Figure 4–20 Pneumatic high-side float used in conjunction with a thermal economizer. *Courtesy of Carrier Corporation.*

Questions

1. What is the principal purpose of an expansion valve?
2. Can an expansion valve normally be used to control temperature?

3. What are the principal differences between the high-side float valve and the low-side float valve?

4. What are the advantages of the low-side float valve?

5. a. How is the operation of the high-side float valve similar to the operation of the capillary tube?

   b. How does it differ?

6. What are the operating characteristics of the automatic expansion valve?

7. What are the advantages of the thermostatic expansion valve?

8. The thermostatic expansion valve is sometimes called a superheat valve. Why?

9. Can a single model thermostatic expansion valve be applied to all refrigerants?

10. What are some of the problems encountered in operating a thermostatic expansion valve at $-30°F$?

11. Name some liquid-feed devices that use electricity and/or electronics to aid in their operation.

12. Is a liquid solenoid valve normally used in conjunction with some other type of device? Explain.

13. Can one variable-capacitance probe take the place of several other types of devices?

14. What do HLSD, HLA, LLA, and LLSD stand for?

Chapter 5   Evaporators

The evaporator, as the name implies, is the part of the system in which the liquid refrigerant is evaporated. It is variously called the chilling coil, chiller barrel, freezing coil, evaporator, or low side. The evaporation process, that is, the heat absorbed in evaporating the liquid refrigerant, is the purpose of the entire refrigerating system. It is the absorption of this heat that we call cooling. Because the evaporator is relatively simple in design, its importance is often overlooked. However, the whole system depends on the proper operation of the evaporator.

Requirements for Evaporators

There are three principal requirements for good evaporator design:

1. The evaporator must provide a sufficiently large heat transfer surface that can be maintained at a temperature lower than the medium being cooled.
2. The evaporator must have a chamber to hold the liquid refrigerant to be evaporated, and room for the evaporated vapor to separate from the liquid.
3. The evaporator must provide circulation without excessive pressure drop.

1. *The evaporator must provide a sufficiently large heat transfer surface that can be maintained at a temperature lower than the medium being cooled.* The total surface area must be proportional to the required load. The refrigerant could be boiled in a tank, but this would be uneconomical because of the large volume of refrigerant required to fill it and the small exterior surface it offers to absorb heat. The usual way of obtaining the required surface is to construct the evaporator in the form of pipe coils as shown in Figure 5-1. Copper tubing is usually used for halocarbon refrigerants. Its surface is often extended, or enlarged, usually by the addition of sheet metal or aluminum fins (Figure 5-2). Other methods have also been used. Figure 5-3 shows a plate surface evaporator. In this design, plates are set vertically for ceiling coils and horizontally for shelf coils. Refrigerant channels in the plates may be formed by soldering or welding tubing between the two surface plates, or they may be formed by grooves pressed in a second plate that is welded to the first. Iron pipe is used for ammonia.

The use of extended surfaces such as fins is limited to applications where the evaporator coils are used to chill air. Where water or other liquids are cooled by evaporators, fins are not needed. Heat transfer from a liquid is much more rapid than from a gas, such as air. Note how much more rapidly cold water will chill us (remove heat from us) than will cold air.

2. *The evaporator must have a chamber to hold the liquid refrigerant to be evaporated, and room for the evaporated vapor to separate from the liquid.* This separation must be complete so that vapor can be removed by the compressor without drawing slugs of liquid with it. In other words, under heavy

Figure 5–1 Sample pipe coil. *Courtesy of Rempe Co.*

loads (lots of heat), it must not boil over. Even a slight amount of liquid returned to the suction line will reduce the efficiency of the compressor. A large amount may damage the compressor.

3. *The evaporator must provide circulation without excessive pressure drop.* Circulation is essential for maximum heat transfer. If one watches water come to a boil, there is a period when bubbles begin to form on the inside surface of the pot and cling there. These bubbles insulate the liquid from the metal. As noted earlier, liquid transfers heat better than gas (bubbles). If these bubbles could be swept from the surface, actual boiling would commence sooner. The same is true inside an evaporator. The only sure way of knowing that bubbles are not insulating the pipe from the liquid is to have a circulation that sweeps these bubbles from the surface as fast as they are formed. The circulating force is usually provided by the flow of refrigerant through the coils.

Excessive pressure drop through the coil affects proper expansion valve operation, penalizes the compressor capacity, and under heavy loads may cause coil starving (not enough refrigerant). So, although flow or circulation of refrigerant is necessary, it must not be such that friction will be excessive.

Types of Evaporators

Evaporators may be divided into "flooded" and "dry" types. A *flooded evaporator* has some form of tank or header that keeps the inside evaporator

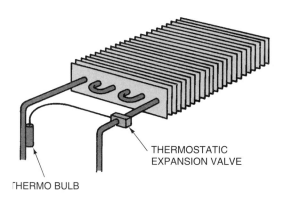

Figure 5–2 Finned coil evaporator.

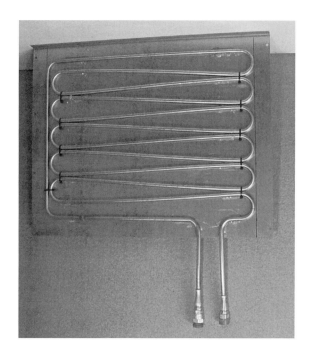

Figure 5–3 Stainless steel evaporator plate. *Courtesy of Ralph Bailey, for Dometic Company.*

surface full, or flooded with liquid. A higher heat transfer rate per square foot of evaporator surface is usually attained with flooded systems, but they require a larger charge of liquid refrigerant to keep them full. A *dry evaporator* is not actually "dry," but it does contain relatively less liquid than a flooded system. The inside of the evaporator is only about one-fourth or one-third full of liquid. However, the boiling action serves to keep most, if not all, of the inner surface splashed or wetted with liquid refrigerant.

Flooded Evaporator

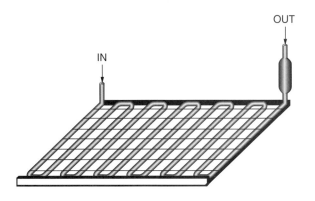

Figure 5–4 Shelf-type evaporator.

Figure 5–4 illustrates the simplest type of flooded evaporator, fed by a capillary tube. A header or accumulator on the outlet allows excess refrigerant to boil off so that it does not return to the compressor via the suction line. Several coils similar to Figure 5–4 may be used as shelves in a domestic freezer cabinet. The inlet is to the bottom shelf. The outlet of the bottom shelf feeds the next higher shelf. This type of feed continues through each shelf to the top. The accumulator is on the outlet of the top shelf.

Figure 5–5(A) illustrates a common form of domestic evaporator. It is made of two sheets of aluminum embossed to form suitable refrigerant passages, then welded together. In this particular evaporator, the passages are arranged to give series flow from the inlet through three sides. The fourth side, shown on the right, allows all flow to enter interconnected parallel passages that act as an accumulator.

Typical industrial flooded evaporators fed with low-side float valves are shown in Figure 4–5 and Figure 4–6. Such evaporators are commonly used for cooling liquids.

Figure 5–6 shows a liquid or brine cooler similar to a horizontal-return tube boiler, but it provides

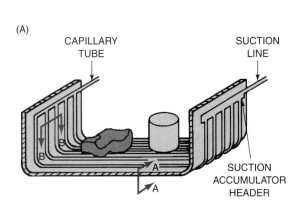

Figure 5–5 Domestic evaporator of formed plates. (A) *Courtesy of Refrigeration & AC Technology (Delmar).* (B) *Courtesy of Noreen.*

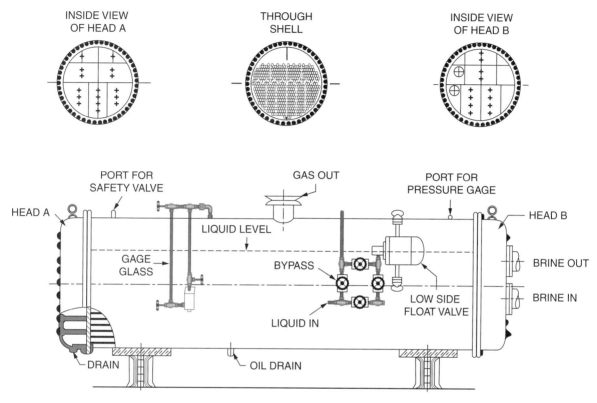

Figure 5-6 Multipass brine cooler. *Courtesy of Henry Vogt Machine Co.*

for several passes of the brine back and forth in these tubes. The refrigerant is in the tank or boiler. The heat flows from the brine or water through the tube walls to the refrigerant. The tubes give the surface necessary to transfer the heat. The top surface of the refrigerant is adequate to allow the bubbles formed by boiling to break through to give proper vapor separation. The bubbles rising from the tubes induce circulation of the liquid refrigerant up through the tube bundles and back down the sides of the evaporator (Figure 5–7).

Figure 5-8 shows a similar evaporator designed for use in the brine tank of an ice plant. Here the brine makes only one pass through the tubes. It is forced through by the action of an agitator in the brine. Figure 5-9 shows an ice maker of similar construction, but set vertically.

Many designs have been evolved to make a trombone or hairpin type of coil like Figure 5-1 operate fully flooded. Figure 5-10 shows how this can be done. An accumulator is used of to keep the coil flooded and to separate the vapor from the liquid. Such an evaporator can be fed by a low-side float

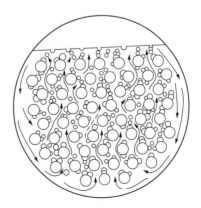

Figure 5-7 Liquid circulation in boiler-type evaporator.

Figure 5-8 Single-pass brine cooler. *Courtesy of Henry Vogt Machine Co.*

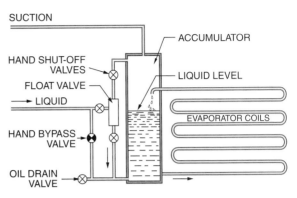

Figure 5-10 Accumulator used on flooded evaporator.

Figure 5-9 Packaged tube ice maker. *Photo courtesy of Noreen.*

valve, a thermostatic expansion valve designed for flooded operation as described in Chapter 4, or a float switch and solenoid valve.

The best industrial evaporators of modern design combine an accumulator with headering (Figure 4-6). Shorter lengths of pipe are used for the evaporator surface. These shorter lengths are connected with headers or feeders, which in turn are supplied with liquid from an accumulator. Return headers carry a mixture of liquid and vapor back to the accumulator. In this design, the mixture of liquid and vapor in the tubes is lighter than the liquid in the accumulator. This induces a rapid flow through the tubes from the bottom to the top to give the circulation required for good heat transfer, yet the headers and tubes are short enough that this good circulation does not set up an excessive pressure drop. Accumulators must be of adequate size or surging may occur, that is, they may become filled to overflowing by the "boiling over" action in the tubes when a sudden heat load is applied.

One design makes use of a liquid ammonia pump to provide forced circulation from an accumulator. Figure 5-11 shows the details of such a system.

Dry Evaporator

In Figure 4-1, an analogy was shown between an evaporator and a water pipe in a furnace. Such a coil is typical of a "dry" expansion evaporator. There

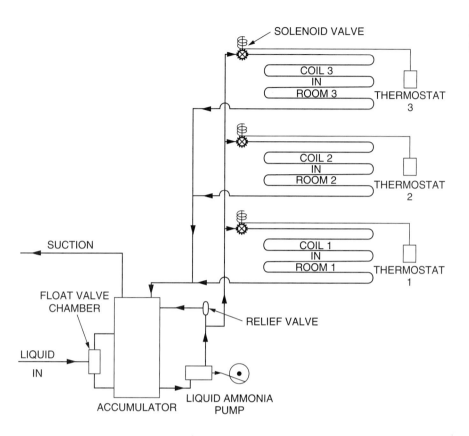

Figure 5-11 Liquid ammonia circulating system.

are actually two different ways such a coil may operate. Some vapor (flash gas) is formed as the liquid goes through the expansion valve, Figure 2-12. Figure 5-12 shows how this vapor forms bubbles in the pipe. As more evaporation occurs, these bubbles increase in size until they fill the tube. From here on, the evaporator is a series of sections filled with liquid alternated with vapor. The sections of liquid decrease in length and the sections of vapor increase in length. These sections travel rapidly from the point of formation to the end of the coil. The coil can still be adjusted so that the last section of liquid evaporates entirely at the approximate end of the coil, but any increase in load will immediately increase the size of the bubbles and push the liquid ahead faster. Some of this liquid may be pushed out so quickly that it is pushed clear through to the compressor before it has time to evaporate. Improved designs, feeding refrigerant to the bottom of the evaporator, and the use of suction accumulators have virtually eliminated this condition. Also, superheating of the bubble takes place between the slugs of liquid. This dries off the inner surface between the liquid slugs, so there are actually

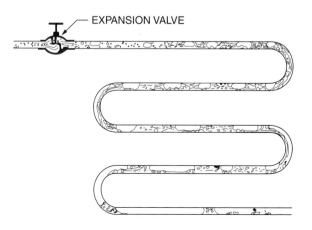

Figure 5-12 Evaporation in small tube evaporator.

"dry" sections of the evaporator. Any dry section of evaporator is an inactive one; that is, it will absorb almost no heat. The circulation in such an evaporator is good, but the separation of liquid and vapor is poor. And if the evaporator is of any great length, the pressure drop can be excessive.

To correct this condition, the coil diameter can be increased as shown in Figure 5-13. Here the liquid flows along the bottom of the tube, while the vapor separates from it and flows through the top of the tube. There should be enough splashing or agitation to keep the entire inner wall of the tube wet, but not enough to throw the liquid together in a slug that can shut off the gas passage completely. Here the liquid flow in the bottom and the vapor flow through the top of the tube provides circulation, but there is enough room for separation of the vapor and liquid. Such an evaporator is best, and can be obtained by proportioning the tube size to the amount of refrigerant circulated and the speed of the flow. However, too slow a speed is also possible. With low-pressure refrigerants, some oil is always carried to the evaporator, which must be swept through to the suction line. A pipe that is too large will slow the circulation to the point that this oil is not carried along, and it will become trapped in any low points in the coil. Also, the right sized evaporator coil for a light load would not be the correct one for a heavy load. Actually, a dry expansion evaporator of about the right size will probably operate as in Figure 5-13 on light loads, and as Figure 5-12 on heavy loads. It may even act as in Figure 5-12 for part of its length and as in Figure 5-13 for the rest.

Both of the foregoing illustrations show top feed. That is, the liquid is fed into the coil at the top, allowed to flow to the bottom by gravity, and the vapor is pumped out at the bottom. Sometimes an up feed is used. With coils of small pipe size, these will operate similar to Figure 5-12. With larger pipe sizes they will operate more as in Figure 5-13 on the horizontal runs, but as in Figure 5-12 in the vertical lifts from coil to coil. Such operation is necessary to move the liquid uphill against gravity. More liquid must be present in the coil to make such operation possible, and to get liquid up the coil high enough to reach the expansion valve bulb. This makes the coil operate more nearly flooded than with a down feed, with the advantages of higher heat transfer. However, it is also more apt to flood back and give erratic operation under varying load conditions. It does not operate as completely flooded as a true flooded evaporator, and if it is not operated sufficiently flooded, it will trap oil badly.

Improved Designs

The first coils designed for dry expansion were entirely of the "trombone" type shown in Figure 5-1. Such coils are still used to a limited extent in all branches of refrigeration, but study of their operation showed that for a coil of any size the difficulties explained in connection with Figure 5-12 were encountered.

One of the first methods used to offset these difficulties was to use an accumulator, as mentioned with regard to flooded evaporators. Accumulators were hooked up in many ways, but always with the same purpose: to keep the coil more nearly flooded, to separate the unevaporated liquid from the vapor, returning the liquid to the coil and the vapor to the suction line.

A second method of giving better operation is *headering* shown in Figure 5-14. Here the circuits are divided so that all the liquid does not have to be fed through one pipe. This smaller amount of liquid

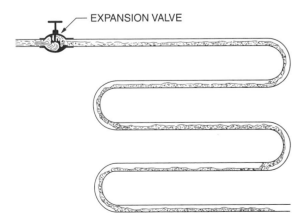

Figure 5-13 Evaporation in large tube evaporator.

Convection and Forced-Draft Evaporators

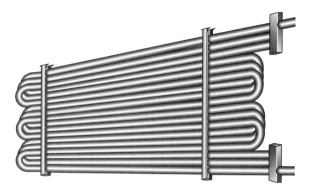

Figure 5-14 Headered pipe coil. *Courtesy of Rempe Co.*

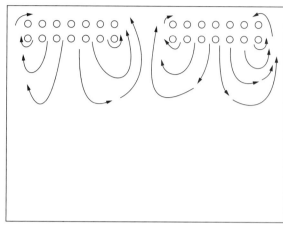

Figure 5-15 Air circulation from coils.

per coil will operate more nearly like Figure 5-13. Notice that the flooded evaporators of Figure 4-5 and Figure 4-6, and the domestic evaporator of Figure 5-5, are all headered. The number of parallel passes from a header is based on the amount of refrigerant circulated, the speed of circulation, and the pressure drop resulting from this amount and speed. Both commercial and industrial evaporators make extensive use of headers.

Convection and Forced-Draft Evaporators

Evaporators used for cooling air may be divided into two general types, convection or natural-draft evaporators, and forced-draft evaporators. To cool an entire room satisfactorily, air must be made to flow from the evaporator to all parts of the room. If natural convection is used, circulation is supplied by the tendency of cold air to drop and warm air to rise. If a bare coil is used, there is considerable interference between the cold air falling and the warm air rising as shown in Figure 5-15. Radiation as well as convection are effective in cooling the room with a bare coil, but radiation is not so effective in a crowded room because the heat from the goods in the room must "shine" directly to the coil. If these goods are "shaded" by other goods, they will be very slow in coming down to temperature if air circulation is poor.

Bare coils such as those shown in Figure 5-15 are common in industrial freezer storage plants.

One method of increasing this circulation is to use *baffles*. These are insulated pans, sometimes with vertical flues, placed so as to separate the chilled, falling air from the warm, rising air. This gives a more positive circulation (Figure 5-16). Note that air from the room is guided up, then through the entire coil, then back down to the room. If coil temperatures get too low, goods directly under the baffle outlet can be damaged by freezing. However, because these goods are shielded (shaded) from the coils by the baffles,

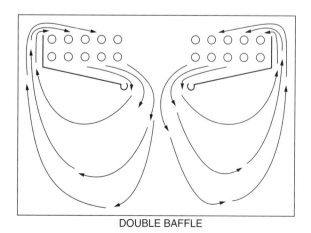

Figure 5-16 Baffles to direct air circulation from coils.

lower coil temperatures can be carried than without baffles before there is danger of freezing. Baffled coils are not used now as much as they once were.

The most effective method of obtaining rapid, positive air circulation is with a forced-draft evaporator. Figure 5–17, Figure 5–18, and Figure 5–19 show common forms of blower coils. Many other designs are also used. Blower coils provide more even temperatures in all parts of the room than other common evaporator systems. Also, it is possible to obtain closer control of humidity with such coils. Ammonia coils like the one shown in Figure 5–19 may either be headered or operated flooded from an accumulator. The latter system is most common.

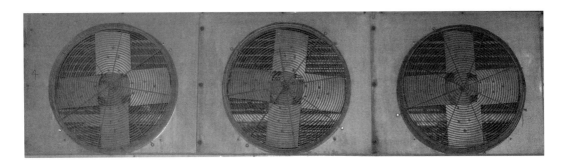

Figure 5–17 Blower-type evaporator to be mounted from ceiling. *Photo courtesy of Noreen.*

Figure 5–18 Wall-type blower coil. *Courtesy of Peerless of America.*

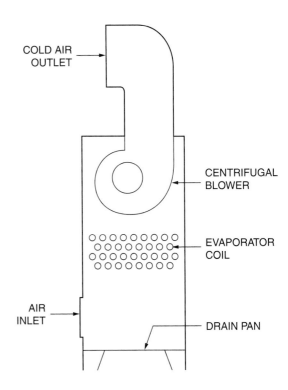

Figure 5–19 Industrial blower coil.

Evaporator Feeds

One very important thing that should be understood about different types of evaporators is the type of expansion valve suitable to feed each. A capillary tube or high-side float valve is suitable only for a flooded evaporator such as that shown in Figure 5–4 or Figure 5–5. The header, or miniature accumulator, takes up any variation in quantity or change in volume of the refrigerant. If such a feed was applied to an ordinary trombone coil, the refrigerant charge of the system would be so critical that a few ounces too much or too little would flood or starve the coil. Also, the proper charge for a heavy load would not be proper for a light load.

The low-side float valve is used in industrial evaporators similar to the ones in Figure 4–5, Figure 4–6, or Figure 5–6 to Figure 5–9. If applied to a trombone coil as in Figure 5–10, an accumulator is necessary as a surge chamber. It has the same purpose as the header in the high-side float evaporator. A low-side float valve cannot be used on a dry expansion coil without some form of header or accumulator.

On the other hand, an automatic expansion valve cannot be used on any of the above-mentioned flooded coils. It will not work because the pressure in no way determines how full the coil or header is of refrigerant. It would starve under heavy loads and flood under light loads. Neither will the thermostatic expansion valve work directly on a flooded coil. The vapor leaves the flooded coil with little or no superheat. This is an advantage for compressor operation, but superheat is necessary to operate this valve. Some flooded coils similar to Figure 5–6 with halocarbon refrigerants have been fed by a thermostatic expansion valve by using a heat exchanger to supply the necessary superheat as shown in Figure 5–22. The necessary heat is supplied by the warm liquid from the liquid line. A more detailed discussion of the use of heat exchangers is given in Chapter 8.

Surge Drum

A surge drum is typically a horizontal cylindrical vessel as shown in Figure 5–20. It is located physically above the flooded evaporator coils that it feeds.

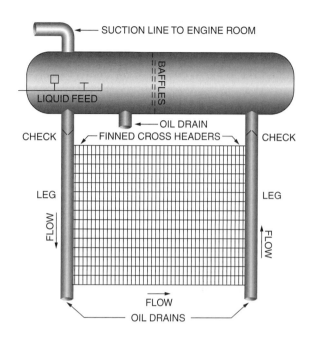

Figure 5–20 Typical surge drum arrangement (drum is usually in a separate room above the refrigerated space).

It is connected to these coils by vertical legs with check valves. These legs are iron pipe and are usually 6 to 8 inches in diameter. The liquid level in this vessel is maintained by one of the methods described in Chapter 4. As long as liquid is maintained in the surge drum, all of the coils located beneath this vessel, including the vertical legs, will be full of liquid refrigerant.

The check valves in the vertical legs allow refrigerant flow in only one direction. This makes the entire assembly more efficient and helps return oil from the lower portions of the coils back to the surge drum. The oil may be easily removed, as it collects in an oil trap located on the bottom of the surge drum. This trap is simply a "blister" welded on to the bottom of the surge drum and is fitted with a drain valve to allow removal of accumulated oil.

Allowing refrigerant to flow in only one direction keeps the unevaporated refrigerant moving down the "down leg" and the mixture of liquid refrigerant and the vapor that is formed after evaporation in the coils moving up the "up leg." This provides recirculation and keeps refrigerant moving through the coils where it is needed. At the top of the "up leg," large slugs of liquid refrigerant are carried up into the surge drum with the refrigerant vapor. To prevent these slugs from being sucked into the suction line (and ultimately into the compressor), baffles are provided inside the surge drum. The baffles give the mixture of liquid and vapor a chance to settle down. The suction line is attached to the surge drum at the opposite end of the vessel from the "up leg." Employing suction traps or accumulators in the suction lines further reduces the chances of liquid refrigerant reaching the compressor(s).

The coils may circulate air naturally, but more often, large-capacity fans are employed to circulate the air. Such an arrangement is called a *blast freezer* and can be used to freeze large amounts of raw product in a very short time. The surge drum is usually located in a separate room above the blast freezer and is heavily insulated to enhance thermal efficiency. The coils and most of the connecting piping are located physically within the refrigerated space and, of course, are bare.

The surge drum is fed liquid refrigerant at or near condensing temperature (or economizer temperature—see Chapter 6). Consequently, flash gas is formed in the surge drum as the refrigerant is fed into the vessel and immediately encounters suction pressure. The surge drum and its connecting suction piping must be sized to accommodate this flash gas. Let's compare this system to a recirculated system.

Recirculated or Overfeed System

In a recirculated system, a *recirculator package* is usually located in the engine room (space permitting). This vessel functions as a surge tank and a suction trap and provides a positive liquid head for the recirculator pumps. A liquid level is maintained in this vessel by one of the methods described in Chapter 4. A suction line (usually relatively short) connects the recirculator vessel to the compressor(s). The flash gas that is generated as the vessel takes in liquid refrigerant is returned immediately to the compressors.

Superheat is added only in the suction line between the recirculator vessel and the compressor. These pieces of equipment are usually located together in the engine room, possibly immediately adjacent to each other. Compare this to the flooded arrangement. After the vapor leaves the surge drum, it may travel several hundred feet and pick up a considerable amount of superheat. In both cases, insulation is essential. The liquid refrigerant in the recirculator vessel is at or near process temperature before it even leaves the engine room area.

Underneath the recirculator vessel are recirculator pumps (usually two or more). These pumps supply refrigerant to evaporators located throughout the plant. These pumps may return liquid refrigerant directly back to the recirculator vessel through back-pressure regulators (see Figure 5–21). This may happen if no evaporators are calling for cooling. The number of evaporators allowed is limited only by plant needs and economics. As mentioned earlier, the refrigerant is already at or near process temperature. No evaporator capacity is compromised as a result of flash gas, and the need for surge drums has been eliminated. A surge drum system usually requires one surge drum per evaporator.

The overfeed ratio is typically 4 to 1. This means that if we feed 4 pounds of liquid refrigerant to the evaporators, we will get 1 pound of refrigerant vapor and 3 pounds of unevaporated liquid back in the suction line. The suction line has to be sloped back toward the engine room as the unevaporated liquid is returned back to the recirculator vessel. If the entire line cannot be sloped, lifting station(s) (typically gas-operated pumps) can be employed. This entire suction line is a mix of saturated liquid and vapor, with virtually no superheat gained. Compare this to a conventional flooded system. The suction lines are long and dry. In some cases, too much superheat may be gained, requiring the use of desuperheaters.

The 4-to-1 overfeed ratio is for design purposes. It costs extra money to pump the liquid around in the system. In these tough economic times, we can reduce the overfeed ratio significantly without compromising system efficiency or performance. We can in fact reduce it as much as we want as long as we are still getting some unevaporated liquid

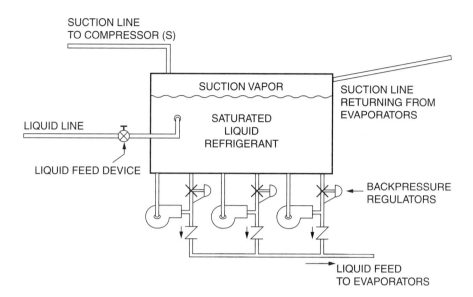

Figure 5–21 Recirculator package.

returning back to the recirculator package through the suction line.

The evaporators in a recirculated system perform better than their flooded cousins. Look at the engineering data for any company that makes evaporators. The same model evaporator will give more tonnage in a recirculated system compared to using it in a flooded system. This is attributed to the fact that the recirculated evaporator receives liquid at or near process temperature. The evaporator used in the flooded system receives liquid at or near condensing temperature (or economizer temperature—see Chapter 6).

Oil in Evaporators

Liquid ammonia and oil separate like oil and water. Ammonia is lighter than oil, so the oil settles as a layer on the bottom. Remember the "blister" on the surge drum. In an ammonia system, most of the oil is taken out of the vapor leaving the compressor with an oil separator (see Chapter 8). More oil separates out in the liquid receiver, where it can be drained off through an oil drain valve placed at a low point as shown in Figure 8–2. A small amount of oil remains in the refrigerant and is carried to the evaporator.

All halocarbon refrigerants dissolve in oil in any proportion at the temperatures existing on the high-pressure side of the system. This fact makes oil harder to separate from a halocarbon refrigerant than from ammonia. Small halocarbon systems seldom use an oil separator because of its poor performance. When a separator is used, less oil is separated from a halocarbon refrigerant than from ammonia. With halocarbon refrigerants, no separation will take place in the liquid receiver. Therefore, a great deal more oil reaches the evaporator than in an ammonia system.

With a dry expansion evaporator, with either ammonia or a halocarbon refrigerant, oil is carried along with the liquid refrigerant until all the refrigerant is evaporated. Then the high-velocity suction gas blows the oil as drops or an oil fog back through the suction line to the compressor. Suction lines must be designed so this oil is not trapped.

A starved dry expansion evaporator will not wash oil to the evaporator outlet where it can be eliminated. Under this condition, oil can collect in the evaporator until there is more oil than refrigerant. That part of the evaporator filled with oil will produce no refrigeration, and is said to be *oil-logged*. Because a halocarbon refrigerant carries more oil to the evaporator than ammonia does, oil logging is more apt to take place in halocarbon evaporators than in ammonia evaporators.

In a flooded ammonia evaporator, oil will settle to the bottom, from which it must be drained periodically. Oil drain valves are put at the low points for this purpose as shown in Figure 5-6.

With R-22 and R-123, the oil will separate from the liquid refrigerant at evaporator temperatures. The oil is lighter than the halocarbon refrigerant, so it floats on top. The oil may exist as a layer or as a foam. R-134a is also heavier than oil.

Oil does not separate from other halocarbon refrigerants. However, as refrigerant liquid is evaporated, an oil-rich solution is formed with the remaining liquid refrigerant. This oil-rich solution is lighter than refrigerant with little or no oil and so will rise to the top. Some foaming usually takes place.

To get oil out of a flooded halocarbon evaporator and return it to the compressor, several systems have been tried. Details of these systems vary, but most revolve around one of two ideas. The first uses a high enough liquid level in the evaporator that some liquid from the surface and any foam that has formed will be swept into the suction line. A large heat exchanger then warms the solution to evaporate the refrigerant from the oil-refrigerant mixture (Figure 5-22). The refrigerant vapor and the oil

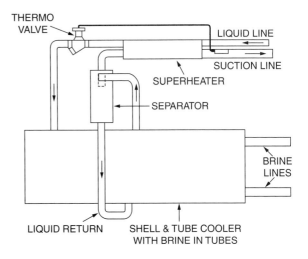

Figure 5-22 Heat exchanger to supply superheat to flooded coil. *Courtesy of Alco Controls Co.*

that will not evaporate is swept through the suction line to the compressor.

The second system carried a lower liquid level so only vapor can enter the suction line. A small bleed-off tube draws the oil-refrigerant solution from the liquid surface as shown in Figure 5-23. This solution

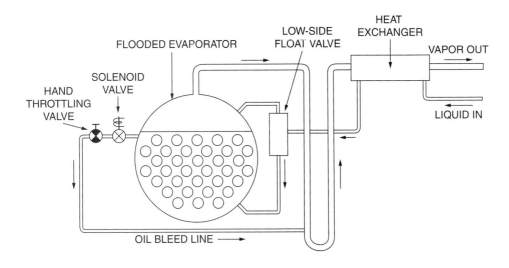

Figure 5-23 Flooded halocarbon evaporator with oil return.

is then mixed with the main suction vapor and sent to a heat exchanger to evaporate off the liquid refrigerant. A needle valve similar to a hand expansion valve is located in the oil bleed line and must be adjusted to bleed off enough liquid that oil does not collect, yet not enough to cause refrigerant flooding in the suction line. A solenoid valve is also located in this line, which closes whenever the compressor stops. Thus, liquid cannot continue to bleed through this line during the time the compressor is not operating.

When properly designed and properly adjusted, either of the above systems will work well on steady loads. The more a refrigerant load fluctuates, the more difficult it is to find a good balance point with either system. Flooding can easily occur with heavy loads, and oil logging can occur with light loads. Refrigerant 22 and Refrigerant 123, because of the blanketing action of the oil layer, tend to require more critical adjustments than other halocarbon refrigerants.

Questions

1. What are the requirements of a good evaporator?
2. What is the difference between a dry and a flooded evaporator?
3. What are the advantages of a flooded evaporator?
4. What is the importance of sufficient evaporator surface?
5. Why is it important to separate the vapor from the liquid in the evaporator?
6. What types of expansion valves are suitable for dry expansion evaporators?
7. What types of expansion valves are suitable for flooded evaporators?
8. What are the advantages of proper headering of evaporators?
9. Why can excessive velocity not be permitted in an evaporator?
10. What are the advantages of forced-draft evaporators?
11. What is the purpose of a surge drum?
12. What is meant by the term "overfeed ratio"?
13. Some designers refuse to use a halocarbon refrigerant in a flooded evaporator. Why?
14. Name two differences in the way oil reacts with ammonia and with a halocarbon refrigerant in the evaporator.

Chapter 6 Compressors

The refrigeration compressor is the heart of the refrigeration system. It removes the vapor from the evaporator and introduces vapor to the high-pressure side of the system. It maintains the low-side pressure at which the refrigerant evaporates, and the high-side pressure at which it condenses. In brief, it supplies the pressure differences necessary to keep the system refrigerant flowing through the system. Many different types of compressors have been used to do this, and many different details tried with individual types of compressors.

Figure 6-1 shows a cross section of a typical reciprocating compressor used in industrial applications, the details of which will be considered later.

How a Compressor Works

Figure 2-9 shows the pressure being instantly raised from the low pressure to the high pressure by the compressor. Figure 6-2 shows this process drawn out over the stroke of the compressor, showing exactly how the pressure is built up. However, this figure only shows what happens on the compression stroke of the compressor. A complete analysis of the suction and discharge strokes follows.

Both the suction and discharge valves are held closed by a combination of spring pressure and vapor pressure. On the down stroke, with both these valves closed, the pressure in the cylinder is lowered as the volume increases. When the pressure in the cylinder becomes less than the pressure below the suction valves, the difference in pressure opens the valves. On the rest of the down or suction stroke, the vapor from the evaporator and suction line flows into this constantly increasing volume.

At the bottom of the stroke, there is no longer an increase in volume to keep the vapor flowing. Therefore, there is no pressure difference, and the springs close the suction valve, trapping the cylinder full of vapor. On the rising stroke, this vapor is compressed from its low pressure to a pressure greater than the discharge pressure on top of the discharge valve. This greater pressure forces the discharge valve open. At this point, compression stops and the rest of the stroke forces this compressed vapor out through this valve. At the end of the stroke, there is no longer a difference of pressure to hold the valve open, so the spring closes it. A small amount of high-pressure vapor is still trapped in the space between the piston and the valve plate. This space is referred to as the *clearance volume*. The vapor in the clearance volume is called *clearance vapor* and must be re-expanded to a pressure slightly lower than the suction pressure before the suction valve opens again.

This action is well shown on a compressor indicator diagram. Such a diagram can be made by an instrument called an indicator, which can be connected to a compressor in operation.

For older, slower machines, indicators were common engine room test devices used to check compressor operation. They do not react fast enough

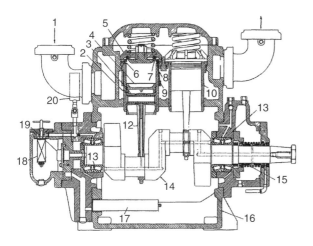

1. SUCTION INLET	11. DISCHARGE OUTLET
2. PISTON RINGS	12. DROP FORGED
3. WRIST PIN	13. ROLLER MAIN BEARING
4. PISTON	14. CRANKSHAFT
5. SPRING LEADED SAFETY	15. DOUBLE SHAFT SEAL
6. SUCTION VALVE	16. CRANKCASE OIL SUMP
7. DISCHARGE VALVE	17. CRANKCASE OIL
8. UNLOADER LIFT PIN	18. OIL FILTER
9. UNLOADER LIFT RING	19. OIL PUMP
10. ALLOY IRON CYLINDER	20. OIL PRESSURE GAGE

Figure 6–1 Cross section of open-type compressor, 20- to 200-horsepower range. *Courtesy of Vilter Mfg. Co.*

to keep up with modern high-speed compressors and so are no longer used. However, they illustrate the principles of compressor operation and the pressure changes that take place in the cylinder.

An indicator is actually nothing more than a recording pressure gage. The fundamentals of an indicator are shown in Figure 6-3. The pressure in the compressor cylinder is applied to the indicator piston. The greater the pressure, the higher the piston is forced up against the spring. This rise can be calibrated so the pressure in the cylinder can be determined by the height to which the indicator piston rises. The piston is connected to a pencil so that, as it rises and falls, it leaves a mark. The paper under this pencil is moved with the movement of the compressor piston.

Figure 6-4 shows an indicator diagram as drawn by an indicator. As the compressor piston goes up on the compression stroke from A' to B', the pressure of the vapor trapped in the compressor cylinder is raised as shown by A–B on the diagram. At B the pressure must be raised above the condensing pressure by the amount F to open the discharge valve against its spring pressure and the vapor pressure against it. Once the valve is opened, the pressure in the cylinder tends to equalize with the discharge pressure. Then the spring starts to close the valve again. This causes the valve to flutter and gives the peculiar rippled effect from B to C in the diagram as the discharge vapor escapes through this fluttering valve. The first hump, at B, is always a little higher than the others because it has to supply the initial push to open the discharge valve. The end of the stroke is reached at C'. This appears on the diagram at C'. Because the stroke is finished, there is no more flow through the valve.

There is still a small amount of high-pressure vapor trapped in the cylinder, represented by E. At the beginning of the suction stroke, this high-pressure vapor must be expanded down to the low pressure before any new low-pressure vapor can enter the cylinder. C'–D' of the stroke accomplishes this. It is shown on the diagram by the line C–D. At D the pressure must be brought below the suction pressure by the amount G to open the suction valve. There is no heavy head pressure against this valve, so a lighter spring can be used to make it hold. With the lighter spring and smaller vapor pressure, there is practically no flutter as there is with the discharge valve. Once the valve opens, the flow of vapor through it is enough to hold it open, and the suction line D–A is steady.

Figure 6-4 is for a hypothetical compressor using ammonia as the refrigerant. However, except for the pressures involved, it is practically the same shape of diagram as would be found for any refrigerant. These diagrams will vary considerably with different operating conditions. They will even be affected by the mechanical condition of the compressor. The actual shape of the curve A–B will tell whether there is leakage within the cylinder, such as past valves, rings, and so on.

Several possible discharge lines are shown in Figure 6-5. This line will always be above the discharge vapor pressure. The higher the line is, the

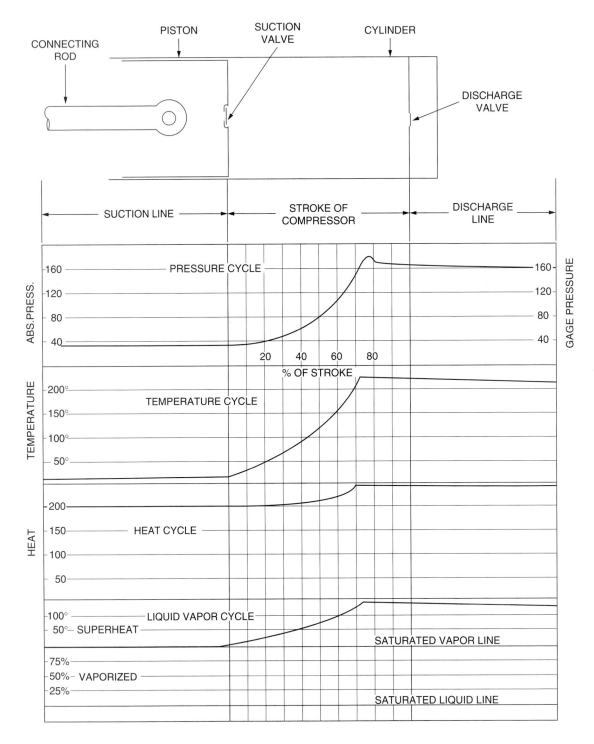

Figure 6-2 Refrigeration cycles detailed at the compressor. Ammonia at standard ton conditions.

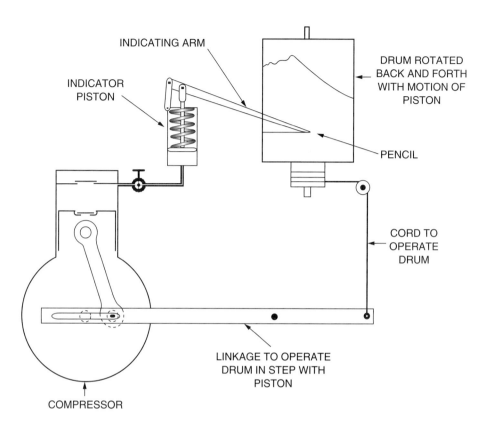

Figure 6-3 Schematic indicator connected to compressor.

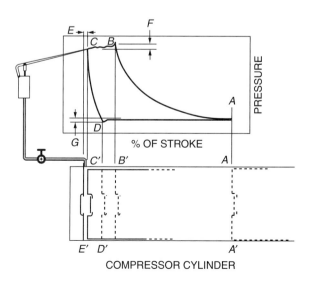

Figure 6-4 Indicator diagram as drawn by indicator.

more power is required to compress against this higher pressure. A heavy discharge valve such as a poppet valve, shown later in Figure 6-21, will give a diagram as in Figure 6-5A. It takes considerable extra pressure to open such a valve. If the spring holding the valve is light, the pressure will drop off as shown. A heavy spring with such a valve will give a curve such as Figure 6-5B. Light plate or disk valves, which are also shown in Figure 6-21, can give a lower discharge line, although this can be offset by too strong a spring as shown in Figure 6-5C and Figure 6-5D.

Figure 6-6 illustrates the complete diagram. The extra dip at D in the suction line is the extra push necessary to open the suction valve. A heavy suction valve such as a poppet valve could cause a larger dip than a plate or feather valve if actuated by spring pressure alone. However, some poppet valves have been so balanced that the inertia of the rising and falling piston helps to snap them open

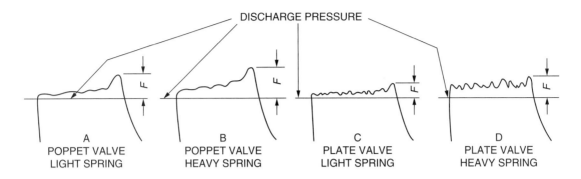

Figure 6-5 Typical indicator discharge lines.

and closed. This should give a curve with almost no dip at the heel. This distance, *G*, is caused by the pressure necessary to hold the valve open against the spring. This pressure will be large with a heavy spring or small with a light one.

The length of the line *H–J* is actually the length of the effective stroke of the compressor figured at the suction pressure. This is the portion of the stroke during which refrigerant vapor is drawn into the compressor's cylinder(s), in spite of the fact that the line *K–L* represents the total actual stroke of the piston. That is, a piston traveling the distance *H–J* with no leakage, loss of pressure, or other losses would do as much work as the actual piston traveling the distance *K–L*. The length of the line *H–J* divided by the length of the line *K–L* is the volumetric efficiency of the compressor. Thus, as shown in Figure 6-7, a heavy suction valve spring can materially reduce the effective stroke and the volumetric efficiency.

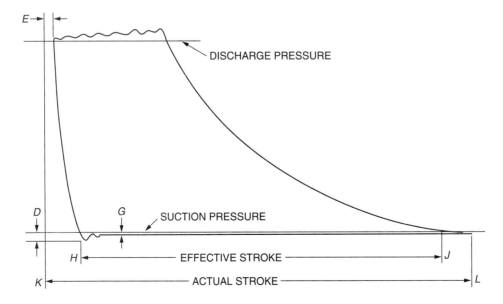

Figure 6-6 A complete indicator diagram, showing difference between effective stroke and actual stroke.

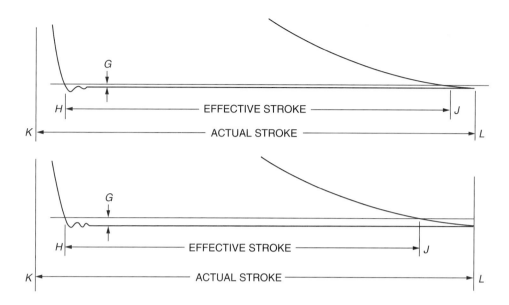

Figure 6–7 Effect of heavy suction valve spring.

Note how an increase in i shortens the distance H–J. Figure 6-8 shows how excessive clearance reduces the effective stroke. That is why the clearance between the piston head and the valve plate is made as small as possible.

Figure 6-9 compares the diagrams made by a high suction pressure and a low suction pressure.

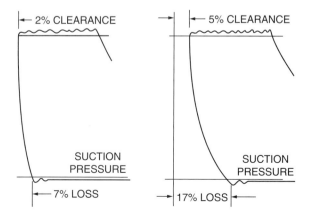

Figure 6–8 Effect of increasing clearance.

For constant discharge pressures, as shown here, the length of the discharge line B–C is an excellent comparison of the amount of vapor pumped. At high suction pressures a much greater amount of vapor is pumped, as shown by these lines. This is caused by the fact that there is actually more vapor forced into the cylinder by the higher suction pressure. It should also be noted that the volumetric efficiency is higher at the higher suction pressure.

Figure 6-10 compares the diagrams made by high and low discharge pressures. The size of the diagram is a measure of the power required to compress the vapor. The lower discharge pressure requires less power (makes a smaller diagram), and also gives better volumetric efficiency.

A thorough understanding of the indicator diagram is the best way to understand exactly what goes on inside the compressor cylinder. As mentioned above, the size of the diagram indicates the power necessary to compress the vapor. The shape of the diagram tells much about the actual operation of the compressor. It is helpful to remember that anything that increases the size of the diagram

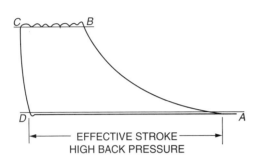

Figure 6-9 Effect of varying back pressure.

increases the horsepower required. Anything that decreases the length of the discharge line at constant discharge pressure decreases the capacity of the compressor.

It should also be remembered, as stated previously, that the compressor does not refrigerate or cause refrigeration. It is the evaporation of the liquid in the cooling coils that does this. However, the amount of vapor removed from the coils determines the speed at which the liquid can evaporate. Therefore the quantity of vapor removed from the coils indirectly determines the amount of refrigeration that can be produced at a continuous rate. So, although we sometimes speak of the amount of refrigeration produced by a given sized compressor, it should be remembered that the compressor's effect on capacity is indirect.

Historical Development: Compressor Types

Horizontal Double-Acting Compressor

The earliest refrigeration compressors were designed after the general style of steam engines built at that time, that is, in a horizontal double-acting pattern, abbreviated HDA, and shown in Figure 6-11. These were used from approximately the middle of the nineteenth century to the middle of the twentieth century. Although they had the advantage of being near the floor and accessible, they required a great deal of floor space and heavy foundations. With a large-diameter cylinder and a long stroke, a single-cylinder compressor could be made to produce large capacities of refrigeration.

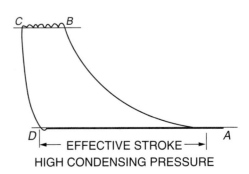

Figure 6-10 Effect of varying condensing pressure.

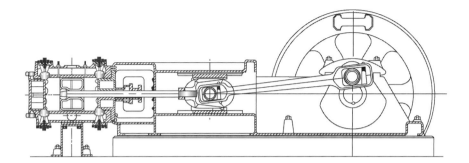

Figure 6–11 Horizontal double-acting compressor. *Courtesy of Vilter Manufacturing Co.*

Suction and discharge valves were placed in each end of the cylinder, so pumping was done on each stroke or half-revolution. Poppet valves, plate valves, and ring valves were all used, Figure 6–25.

In these units, because the pressure on the stuffing box (shaft seal area) around the shaft alternates between high and low pressure, and because the rod slides back and forth rather than rotating, the service against it is much more severe than with other types of compressors. Oil is pumped through the stuffing box, then discharged to the suction line. High-pressure vapor that has leaked into the packing is carried with the oil back to the suction of the cylinder. Oil in this line also helps cylinder lubrication. Friction on the packing generates considerable heat. Some, but not all, of this heat is removed by the oil. To offset this heat, a small amount of liquid refrigerant is piped through an expansion valve to a passage around the stuffing box, then to the suction. Thus, mechanical refrigeration is used to control the packing temperature.

Lubrication of this type of machine may be with an oil pump, lubricator, or with a combination of both. Oil is fed to the various bearings at a low pressure. Oil to the shaft packing may be at a higher pressure. Some oil is supplied directly to the cylinder walls to augment that leaking through the stuffing box.

Vertical Single-Acting Compressor

The first attempt to remedy the weaknesses of the HDA compressor was to stand it on end in a pair of A frames (Figure 6–12). Because of top heaviness, poor accessibility, and other faults, this solution was not very satisfactory.

The vertical single-acting compressor, abbreviated VSA, was the next improvement in design. Most of these compressors were built in two-cylinder designs as shown in Figure 6–13. The crankcase acted as a base, supported the cylinders and main bearings, and acted as an oil reservoir. It was originally designed for ammonia, but similar compressors were later designed for halocarbon refrigerants.

Figure 6–12 An A-frame compressor.

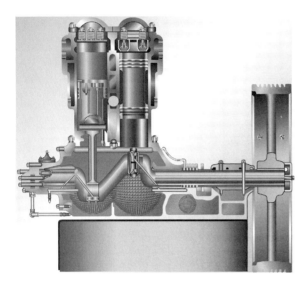

Figure 6–13 Sectional view of a two-cylinder single-acting compressor. *Courtesy of York Corp.*

In these units, the crankshaft converts the rotary motion of the flywheel to a reciprocating motion to operate the pistons. The stuffing box prevents leakage past the shaft, either of refrigerant out or of air into the system. The pistons in the cylinders draw in the refrigerant vapor on the suction stroke, and compress and discharge it on the compression stroke. Suction valves in the piston head and discharge valves in the safety head allow flow in the proper direction, but act as check valves against any reverse flow. On VSA compressors designed for ammonia, a water jacket surrounds the upper part of the cylinder and the head, to keep the cylinder walls cool.

These machines are no longer manufactured. However, because they were such a standard design for so many years, they were the most common type of compressor in industrial refrigerant plants in most areas.

V- or VW-Type Compressor

Modern industrial compressors are small, high-speed, multicylinder machines, called V- or VW-type compressors (Figure 6–14 and Figure 6–15).

Capacities are varied by using a different number of cylinders instead of building a different-sized machine. From 3 to 12 cylinders may be used. Direct drives to 1200-rpm or even 1800-rpm motors are common. With mass production methods, and

Figure 6–14 Machine room of a dairy. From left to right is a vertical intercooler between stages, six belt-driven V-type ammonia compressors, and a partial view of an insulated horizontal brine cooler. Low-stage rotary compressors are behind the intercoolers. *Courtesy of Vilter Manufacturing Corp.*

Figure 6–15 A modern VW-type compressor. *Reproduced with permission of The Trane Company.*

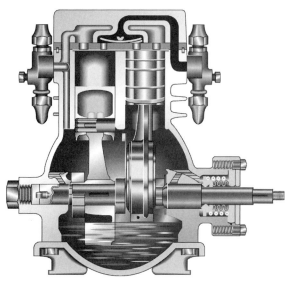

Figure 6–16 Low-pressure refrigerating compressor with suction direct to valve plate. *Courtesy of General Electric Co.*

better standardization of cylinder sizes, both the size and the cost of a given-capacity machine has been reduced. Replacement parts for repairs are also more readily available.

Open-Type Domestic and Commercial Compressor

The first experiments aimed at developing small equipment came after large equipment was well developed. Low-pressure refrigerants such as sulfur dioxide and methyl chloride were adopted. These were later replaced by Refrigerant 12 and, later, Refrigerant 22. Many different types of compressors were tried throughout the twentieth century. Before the advent of hermetically sealed equipment, most manufacturers came to produce a miniature vertical, single-acting, belt-driven compressor as shown in Figure 6–16, with from one to four cylinders. Speeds were stepped up. Later, much that was learned on these low-pressure, small compressors was applied to ammonia equipment.

Open-type compressors are no longer used in domestic refrigerators except in special cases where direct-current motors are required. Very few open compressors are used below 1 horsepower. Above 1 horsepower, both belt-driven and direct-connected open-type compressors are quite common for commercial applications (Figure 6–17).

Hermetically Sealed Unit

In attempts to build a fully automatic, trouble-free, long-lived compressor such as was needed for the maximum development of domestic and small commercial equipment, early designs had several weak points. One of the worst of these was the shaft seal. A slight oil leakage was necessary to keep the face lubricated. Over a period of years, this could drain the compressor of oil. Sooner or later the seal would pick up some dirt or grit, which would cause it to wear and eventually leak refrigerant. Another weakness was the belt drive. Over time, the belts, through normal wear, would stretch and occasionally slip. Belt tension had to be adjusted to take up stretch. Belts would wear, become frayed, and sometimes break. Oversized motors to take care of varying belt frictions as well as other variables were common.

The most successful method found to eliminate all these troubles was to direct-connect the motor and compressor, and then enclose them both in a

84　CHAPTER 6　*Compressors*

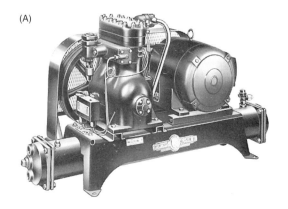

Figure 6–17 Belt-driven open-type compressors. (A) *Courtesy Teucmsch Products Company.* (B) *Photo courtesy of Noreen.*

Figure 6–18 Cross section and photo of a small, hermetically sealed compressor. *Photo by Noreen, courtesy of Articglacier Inc.*

gas-tight housing (Figure 6–18). Such a compressor is termed *hermetically sealed*. Not only does it solve the above-mentioned problems, further advantages have been found or have been developed into it. By doing all assembly work under factory conditions, much better quality control of the product is possible. The machine is made right, then sealed up so it stays right. Domestic equipment made this way may still be in operation after 15 years or more, with no repairs or service. Many of the troubles that have developed in such machines built in the last 15 years have been eliminated, so the life of the equipment built today should be even longer.

Hermetic design has been so successful that it has come to entirely dominate the domestic and self-contained commercial fields. Manufacturers are also supplying such equipment for remote installation. Some manufacturers offer sealed reciprocating units up to 100 horsepower (Figure 6–19).

Figure 6–19 Hermetically sealed compressors. *Courtesy of Trane Company 2000.*

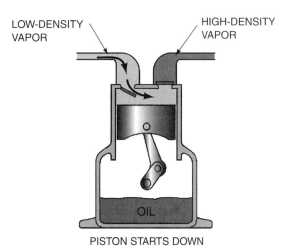

Figure 6–20 An illustration of what happens inside a reciprocating compressor while it is pumping. When the piston starts down, a low pressure is formed under the suction reed valve. When this pressure becomes less than the suction pressure and the valve spring tension, the cylinder begins to fill. Gas rushes into the cylinder through the suction reed valve.

Many different shapes and designs of hermetic compressors have been used. Most domestic and similar small sealed units are built with a vertical shaft shown in Figure 6–18. Many fractional-horsepower and nearly all sealed compressors above 1 horsepower are built on a horizontal shaft shown in Figure 6–19. They may have from 2 to 12 cylinders.

Construction Details

Valves

Domestic and small commercial compressors usually have all valves secured to a valve plate, Figure 6–16. The suction valves are on the bottom of the plate, and the discharge valves are on top. For commercial servicing, this makes valve replacement easier and faster. A new valve plate replaces all valves and valve seats. Discharge valve construction details are shown in Figure 6–24. Suction valves may be either a leaf or ring valve mounted on the bottom of the valve plate. Figures 6-20, 6-21, 6-22 and 6-23 show a reciprocating compressor as it goes through a complete suction/discharge cycle. The suction and discharge valves allow vapor to flow in one direction only.

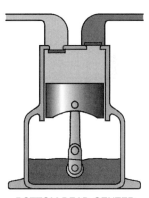

Figure 6–21 When the piston gets near the bottom of the stroke, the cylinder is nearly as full as it is going to get. There is a short time lag as the crankshaft circles through bottom dead-center, during which a small amount of gas can still flow into the cylinder.

Older horizontal or vertical ammonia compressors may use a poppet valve, feather valve, or ring valve, Figure 6–25. The poppet valve is more rugged, but the lighter valves give better volumetric efficiencies.

86 CHAPTER 6 *Compressors*

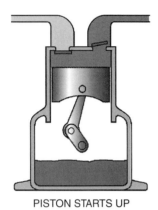

PISTON STARTS UP

Figure 6–22 When the piston starts back up and gets just off the bottom of the cylinder, the suction valve has closed, and pressure begins to build in the cylinder. When the piston gets close to the top of the cylinder, the pressure starts to approach the pressure in the discharge line. When the pressure inside the cylinder is greater than the pressure on the top side of the discharge reed valve, the valve opens and the discharge gas empties out into the high side of the system.

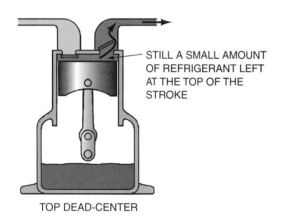

Figure 6–23 A reciprocating compressor cylinder cannot empty completely because of the clearance volume at the top of the cylinder. Manufacturers try to keep this clearance volume to a minimum but cannot do away with it completely.

New, high-speed compressors, whether for ammonia or halocarbon refrigerants, usually have a ring suction valve around the cylinder and discharge valves in a valve plate (Figure 6–26). The valve plate may

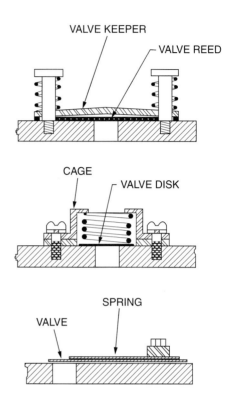

Figure 6–24 Low-pressure discharge valves.

be bolted down or may be held down with a heavy spring, also shown in Figure 6–26. A valve plate held down by such a spring is known as a *safety head*. A safety head is common on large equipment, because with more powerful driving motors as well as larger investments, liquid slugging could be disastrous with a solid head. The safety head can lift against its spring, then reseat itself, so no damage is done. Many large halocarbon compressors also use safety heads.

Gas Passages

In early designs for small refrigeration compressors, the suction vapor was introduced directly into the crankcase. However, this type of design led to considerable trouble with oil slugging. Any sudden decrease in pressure on the oil, such as happens when starting the compressor after an off cycle, will cause the oil to foam up. This foam,

Construction Details **87**

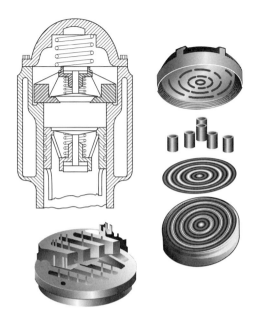

Figure 6–25 Typical ammonia compressor valves: upper left, poppet valve; upper right, ring valve; lower left, feather valve. *Feather valve photo courtesy of Worthington Pump & Machinery Corp.*

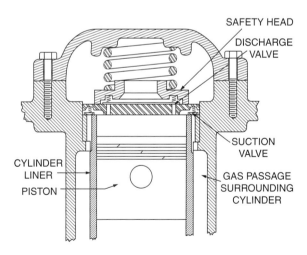

Figure 6–26 Modern valve arrangement on large-capacity compressor.

similar to the foam on beer, rises and is swept through the suction valve into the cylinder by the suction vapor (Figure 6–27). Oil is noncompressible, and it cannot change direction to flow out the

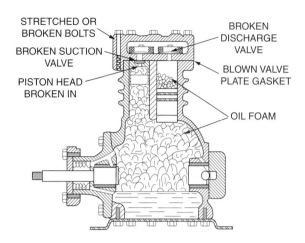

Figure 6–27 Oil foaming and slugging, and possible results.

discharge valve as rapidly as the piston moves. This causes a severe knocking, which strains all parts of the compressor.

Figure 6–27 shows some of the possible results of oil slugging. Liquid refrigerant has the same effect as oil in slugging the compressor. That is why expansion valves and evaporators must be designed and installed so that no liquid is present in the suction line. Figure 6–28 is a hypothetical indicator diagram showing the effect of a slug of oil or refrigerant. Such

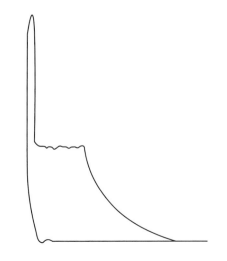

Figure 6–28 Indicator diagram of oil slug.

a diagram can only be drawn theoretically, as stops on the indicator would prevent the indicator piston and pencil rising to the peak of the slug pressure.

Present design of small, open-type halocarbon compressors brings the suction vapor in through ports cored in the cylinder wall castings, as in Figure 6-16. A breather line extends to the crankcase. This is necessary to balance the crankcase pressure, and to give oil from the suction line a means of returning to the crankcase. A check valve or screen is located in this breather line to stop or break up any foam that might rise up in the crankcase and enter this breather.

Most hermetic compressors bring the suction vapor into the hermetic case in such a way that this vapor flows over the electric motor windings. This helps provide some cooling effect to the windings, and would help evaporate liquid refrigerant that might return as a result of faulty evaporator operation. The suction into the cylinder is taken from a point high enough inside the case that the entire case would have to be filled with oil foam to get oil into the cylinder.

In modern multicylinder designs, suction vapor comes in through ports in the crankcase casting, then goes through a chamber surrounding the cylinder before going to the suction valve ports, Figure 6-26. This action helps cool the hot cylinder wall, evaporate any liquid refrigerant that might reach this point, and acts as an oil separator. Any oil that collects is drained back to the crankcase.

Stuffing Boxes; Shaft Seals

Regardless of the method used to bring the suction vapor into the compressor, it has proved next to impossible to keep refrigerant vapor out of the crankcase. There is always some leakage past the rings. The oil lubricating the cylinder walls dissolves some refrigerant, then carries it back to the crankcase. This is why an equalizer line to the crankcase is essential. The crankcase pressure is then maintained at suction pressure.

With refrigerant in the crankcase of an open-type compressor, leakage past the revolving shaft must be prevented. In older ammonia machines, this was done using some form of packing. This might be a flexible packing shown in Figure 6-29, or a metallic packing as in Figure 6-30. The flexible packing is similar to a common valve stem packing except for the oil lantern.

Lubricating oil is fed to this packing, usually under pressure, to keep the shaft and packing lubricated. The metallic packing is a series of sectional Babbitt or other bearing metal rings made to hug the shaft tightly with springs or with synthetic rubber. This, and the oil seal between the shaft and contact rings, prevents leakage past the shaft.

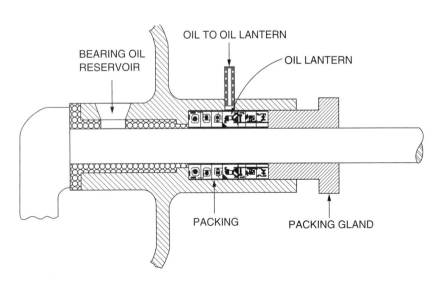

Figure 6-29 Stuffing box for ammonia.

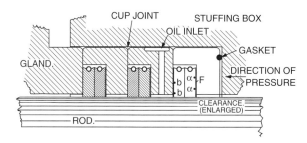

Figure 6–30 Metal rod packing. *Courtesy of Garlock Packing Co.*

Synthetic rubber packing or gaskets prevent leakage around the packing assembly.

On all modern open-type compressors, a metallic or carbon seal is used. This is a flat metal or carbon face rotating under oil against a similar flat stationary face. The accurately matched flat faces under oil seal off any leakage.

Figure 6-31 shows a bellows-type shaft seal. This usually bears against a hardened steel ring on the shaft as shown. A synthetic rubber ring forms a seal between the shaft and steel ring. The seal nose is a special alloy of soft bronze or carbon.

The steel ring revolves with the shaft, but the bronze or carbon nose is stationary. Lubricating oil makes an oil seal between these two perfectly flat surfaces, preventing leakage at this point. A spring maintains the necessary pressure between the two faces. This spring pressure is opposed by a thrust. In small compressors, this thrust is usually a steel

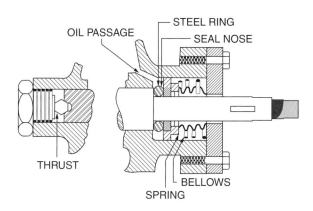

Figure 6–31 Bellows-type shaft seal.

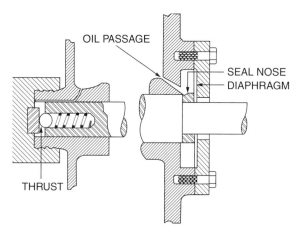

Figure 6–32 Diaphragm-type shaft seal.

ball at the opposite end of the shaft. In some larger compressors, shaft bearings are designed to oppose this thrust, Figure 6-1.

The flexible bellows allows the seal nose to move sufficiently to follow the slight variations in the shaft as it revolves. A gasket under the outside ring or collar completes the seal. Older seals pressed directly against the shaft shoulder rather than the steel ring. However, if this shoulder should become scored by long wear, or by grit or dirt, it is difficult to refinish and expensive to replace.

Figure 6-32 shows a similar type of seal using a diaphragm for flexibility instead of a bellows. With this type, the thrust spring is put on the other end of the shaft. It takes less room than a bellows seal, but it does not have as much flexibility.

Another type of shaft seal is the rotary seal, Figure 6-33. Collar A revolves with the shaft. The spring back of it serves to keep it against the stationary bronze nose piece B so there is no leakage past this face. The spring also bulges the synthetic rubber cushion C so it will ride the shaft with sufficient pressure to prevent any leakage along the shaft.

Lubrication

Older machines, both large and small, may be lubricated by splash or by force feed. Figure 6-16 shows a typical compressor that uses splash-feed lubrication. The cranks dip into the oil as they revolve. This forces

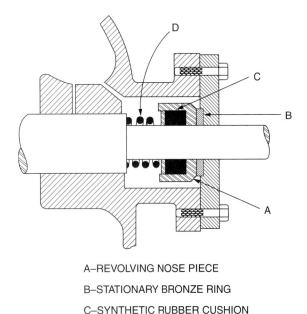

A—REVOLVING NOSE PIECE
B—STATIONARY BRONZE RING
C—SYNTHETIC RUBBER CUSHION
D—THRUST SPRING

Figure 6-33 Rotary shaft seal.

oil to the crank bearing, and also splashes a fog of oil throughout the crankcase. Some of this oil fog lubricates the cylinder walls. That which collects on the under side of the piston head drips down to the funnel-shaped opening at the top of the connecting rod. This feeds oil to the wrist pin. Oil running down the end of the crankcase collects in reservoirs, and feeds the main bearings and the shaft seal.

All lubricated bearings are designed so that some oil continually leaks from them. This makes room for a constant flow of fresh oil flowing through these points to carry away frictional heat, as well as lubricate the bearings. Splash lubrication is used on smaller units. Some sealed compressors depend on splash feed. Splash may be used on older VSA ammonia compressors up to 5 × 5 size.

Figure 6-1 shows a compressor with force-feed lubrication. The oil sump in the crank case is low enough that oil is not touched by the cranks or rods. An oil pump driven by the main shaft picks up oil from this sump and feeds it to the oil system. This oil system carries oil under pressure to all moving parts. Cylinders are lubricated by oil escaping from wrist pins.

Force feed is used on many hermetic compressors regardless of size, and on all high-speed compressors.

Cooling

As pointed out in Chapter 2, compressing the refrigerant vapor generates heat. This raises the temperature of the vapor, which in turn heats the cylinder walls, valves, and so on, that it touches. On compressors of any size, some cooling is necessary. This does not effectively cool the discharge vapor, but without cooling, the cylinder walls would become so hot they would be difficult to lubricate. There would be trouble due to oil breakdown.

Tables 3-1 and 3-2 gave the discharge temperatures for different refrigerants at standard ton conditions. Figure 14-9, Figure 14-10, and Figure 14-11 show the discharge temperatures for different operating conditions. Notice that halocarbon refrigerants do not heat up as much as ammonia. Refrigerant 21 and Refrigerant 22 have higher discharge temperatures than other halocarbons. Air cooling has proved sufficient for most halocarbon refrigerants.

The compression temperature of ammonia in most applications is such that water cooling is necessary. Therefore, a water jacket on the upper part of the cylinders, over the cylinder head, or both is used.

There are cases where air-cooled reciprocating ammonia compressors are used as booster compressors, but this is with a combination of low compression ratios and very cold entering vapor. Under such conditions, the discharge temperature does not become excessive. Lately, some booster compressors for very-low-temperature applications have been cooled by liquid refrigerant in the jacket.

Service Valves

Except for sealed domestic and small commercial equipment, most refrigeration compressors have valves at the inlet and the outlet of the compressor as shown in Figure 6-15 and Figure 6-16. The valve on the inlet or suction side is called the *suction service valve*. The valve on the outlet or discharge is the *discharge service valve*. The only difference between them is usually their size. The suction valve is larger because the low-pressure suction

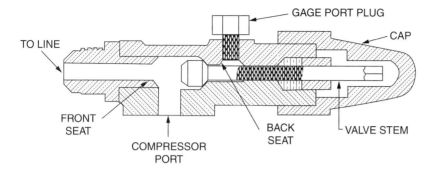

Figure 6–34 Compressor service valve.

vapor has a larger volume than the high-pressure discharge vapor.

Figure 6–34 shows the construction of valves used with halocarbon refrigerants. When the valve stem is screwed all the way forward, called *front-seated*, the "to line" port is isolated and the "compressor port and "gage port" are open to each other. If a gage is installed in place of this plug, the gage will indicate the pressure in the compressor. If the valve is screwed all the way back, or *back-seated*, the port to the "gage port" is isolated, and the main "compressor port" is open to the "to line" port. Thus, the valve may be back-seated to remove the plug and install the gage. Then, if the valve is turned one or two turns forward (cracked), the compressor will function normally, and the pressure in the system will be indicated on the gage. These ports are also useful for special operations such as charging and discharging the system. The midseat position is used for performing standard pressure tests and system evacuations. On systems that used halogenated hydrocarbon refrigerants, the gauge port was a 1/4-inch male flare fitting.

Older ammonia compressors usually have valves plus a bypass, Figure 6–35. The valves are used to close the lines when necessary. The bypass may be used for starting unloaded (with the pressures equalized) or for special pump-out operations. With the valves A and C open and valves B and D closed, the compressor can be started with no load on it—unloaded. The suction pressure is on both the suction and discharge side of the piston, so there are no forces against turning the compressor. When the compressor is running up to speed, C is closed and B is opened. The bypass can also be used to suck from the discharge line and discharge to the evapo-

rator. This may be necessary if work must be done on the condenser or receiver. The refrigerant pumped from the high side is discharged through the other bypass to the evaporator. In this case, the evaporator becomes the condenser and receiver temporarily.

Most vertical single-acting ammonia compressors combine the bypass and valves in a manifold as shown in Figure 6–36. Normal operation is with the discharge bypass valve closed, the suction valve open, the discharge valve open, and the suction bypass valve closed. With the discharge bypass and the suction valve open and the other two valves closed, the machine can be started up unloaded. The discharge vapor passes back to the suction line. After the machine is up to speed, closing the discharge bypass and opening the discharge valve puts the machine in normal operation.

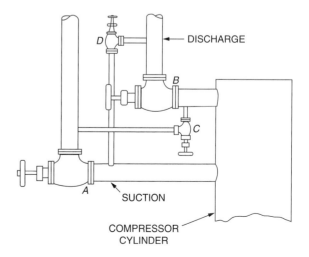

Figure 6–35 Ammonia compressor valves and bypasses.

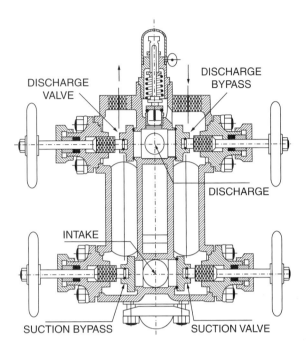

Figure 6-36 Ammonia compressor manifold. *Courtesy of Baker Refrigeration Corp.*

Most high-speed, multicylinder compressors are built with *unloaders* (see next section). Unloaders make it possible to start the compressor partially unloaded, so bypasses are not necessary.

Capacity Control

One problem in operating refrigeration compressors is balancing the compressor capacity against the refrigeration requirements. That is, the vapor should be removed from the suction line at exactly the same rate as it is produced by evaporation in the evaporator. This should hold true whether a large or small amount of liquid is evaporated.

If the compressor removes the vapor faster than it is produced, the pressure is gradually lowered. This reduces the evaporator temperature. On the other hand, if the compressor cannot remove the vapor as rapidly as it is formed, the pressure, and consequently the temperature, of the evaporator rises. With most automatic systems, particularly domestic and commercial units, a compressor that is large enough for the worst conditions likely to be encountered is usually selected.

A control is installed to shut off the compressor when it reduces a pressure below a predetermined point. Then, as the pressure rises, the compressor is turned on again. With ammonia systems, cycling may be done automatically or by hand. In large industrial systems, several compressors may be used. If one is not enough, two are put into operation; if two are not enough, three are used; and so on. If operation is not automatic, it is the duty of the operator to watch the pressure and use the proper number and size of compressors to hold this constant.

With large-sized compressors, one additional compressor makes quite a large step up in operating capacity. In the old days, with steam drives, this could be balanced by varying the speed. However, with present-day electric drives, this was not so simple until *variable-frequency drives* (VFDs) became cost-effective. VFDs permit running induction motors at variable speeds down to a certain minimum limit. Other devices have also been used to control the capacity of a single compressor. These capacity control devices increase the cost of the compressor. Whether they are to be used depends on an analysis of the savings in operating cost against the increase in first cost.

For vertical single-acting compressors, clearance pockets and bypasses have been used for capacity reduction. Figure 6-37 illustrates a clearance pocket. Part of the vapor compressed fills this pocket on the

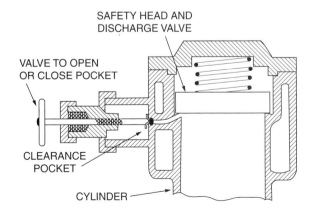

Figure 6-37 Clearance pocket.

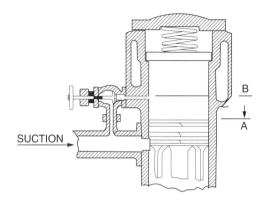

Figure 6–38 Bypass capacity control.

pressure end of the stroke. On the suction stroke, this high-pressure vapor returns to the cylinder and must be expanded until its pressure is reduced to a point that is slightly below the low pressure before more low-pressure vapor from the suction line can enter the cylinder. This actually increases the clearance, and has the effect shown in Figure 6–8. Clearance pockets can be made to give 25 or 50 percent reduction in capacity. They may be operated either by hand or automatically by a solenoid valve.

Figure 6–38 illustrates a bypass. When a valve is opened in the bypass line, the stroke A–B is ineffective. All the above capacity-reduction methods also reduce the power required to turn the compressor shaft. Therefore, they are an economical way of balancing the capacity to the requirements.

Most modern multicylinder compressors control capacity with cylinder unloaders. When a reduction in capacity is required, the pumping action of one or more cylinders is interrupted. This is accomplished by lifting the suction valve off its seat and holding it open. With the suction valve open, no compression is possible. Only some of the cylinders are capable of being unloaded, so there is always some pumping action taking place. A lever that lifts or drops the valve is operated by a small control piston. Some manufacturers use oil pressure, others use discharge gas pressure on this control piston.

A change in suction pressure operates a valve in the oil or gas line to the control piston. Design is such that the cylinders that can be unloaded remain unloaded when the compressor is idle. Thus, the compressor can be started at reduced load. As the compressor comes up to speed, a change in oil or gas pressure activates the control pistons to allow the suction valves to operate normally.

Compressor Speeds and Piston Speeds

In studying refrigeration compressors, the meaning of *compressor speed* should be well understood. Early ammonia compressors usually operated at less than 100 rpm. Many modern machines are direct-connected to 1725-rpm or to 3450-rpm motors. Reciprocating compressors built for aircraft use, where size and weight are at a premium, have been run up to 10,000 rpm. However, the actual rpm of the shaft or flywheel means very little. The rpm does indicate how fast the valves must open and close. The speed that determines cylinder wear and cylinder lubrication is *piston speed*. Piston speeds have been stepped up from speeds of 150 feet per minute in the old days to 600 feet per minute or more at the present time. Some designers today feel that 600 feet per minute should be a maximum. Others have gone considerably higher with the proper use of improved alloys, lighter parts, and better lubrication.

The distance a piston travels is once up and down or two times the stroke per revolution. Thus a machine with a 6-inch stroke turning 400 rpm has a piston speed of $(6 \times 2)/12 \times 400 = 400$ feet per minute.

A domestic machine direct-connected to a 1725-rpm motor having a 1-inch stroke has a piston speed of $(1 \times 2)/12 \times 1725 = 287$ feet per minute, or less than the above machine, which only turns 400 rpm. Compressor speeds should be reduced to piston speeds before any attempt is made to compare the speeds of any two machines.

Other Types of Compressors

All previously described compressors are piston-type or reciprocating machines. A piston draws vapor into a cylinder, compresses this vapor, then forces it out of the cylinder. Other methods of compressing vapor have been devised. For certain size refrigeration applications, some of these other methods have largely displaced piston machines.

Rotary Compressor

Rotary compressors comprise two general types. The first type is illustrated in Figure 6–39. It is referred to as a *rotating-vane rotary compressor*. The rotor, containing four or more blades, revolves off-center in the cylinder or housing. The blades are constantly held out against the outer wall, either by spring pressure or by centrifugal force. As space A moves past the suction port, the trapped volume is at its largest. As this space moves around to B and C, this trapped volume is decreased to the point that it becomes practically nothing. This compresses the vapor, then forces it out the discharge valve.

Figure 6–40 illustrates the second type of rotary compressor. It is called a *stationary-vane rotary compressor*. Here the shaft is in the center of the housing, but the rotor is on an eccentric. The blade is in the outer housing and is held against the rotor by spring pressure. Chamber A is at its maximum in Figure 6–40(1), and gradually decreases to nothing as the rotor revolves to the point shown in Figure 6–40(3).

One big difference between the action of the rotary and reciprocating compressors is that in the former, the pressure balances when the compressor stops. That is, the fit around the rotor, blades, and end plates cannot be made gas-tight

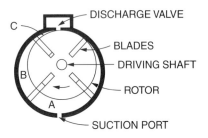

Figure 6–39 Rotary compressor blades in rotor.

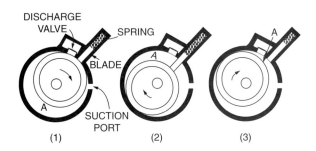

Figure 6–40 Rotary compressor blade in outer cylinder.

when not revolving in oil. Therefore, a check valve is often used in the suction line to prevent the high-pressure discharge gas from leaking through the compressor back to the evaporator. Other designs allow this leakage to balance the pressure during the off cycle so the compressor can start unloaded.

Another difference between rotary and reciprocating compressors is that the rotary draws a much better vacuum. There is no re-expansion of the high-pressure discharge vapor back to the suction side. This gives very high volumetric efficiency. All vapor is forced from this compressor. This is not the case in a reciprocating compressor. Against a closed valve, a rotary compressor will instantly pull down to within a small fraction of an inch of a perfect vacuum. A reciprocating compressor may only be able to pull down to a 25- to 28-inch vacuum. On the other hand, the rotary compressor has not been so successful in pumping against heavy discharge pressures.

Another feature of the rotary compressor is that it is easy to design to handle a large volume of refrigerant. The displacement per revolution is the difference in volume between the inner rotor and the outer cylinder. If the rotary compressor is as large as the crankcase of a reciprocating compressor, it will handle a much larger volume of vapor than the cylinders of a reciprocating compressor.

All these factors add up to the fact that the rotary compressor is well suited to the use of a low suction pressure. It easily handles the large volumes of vapor produced at low suction pressure.

Another common use of the rotary compressor is as a booster compressor for very-low-temperature jobs, both for halocarbon refrigerants and for ammonia. Again, its high-vacuum, high-volume characteristics at low discharge pressures fit perfectly to this type of application.

The rotary booster compressor is a large, multiple-blade machine (Figure 6–41). It differs from the small rotary compressor in that there is a check valve in the discharge line instead of in the cylinder. The discharge port in the cylinder is of the proper size and location that the blades uncover it at the point where compression equals required discharge pressure. For this reason, these machines are designed to give a specified compression ratio. Efficiency is reduced if they are operated at pressures that differ significantly from the design point. The machines are built with different ratios for different applications. Jacket cooling is by oil. Cold suction vapors would freeze water if it was used for cooling.

Portable rotary compressors are commonly used as service tools to evacuate all air from newly installed refrigeration systems before charging refrigerant into the system. Such portables are also used to evacuate contaminated systems before adding a fresh charge of refrigerant.

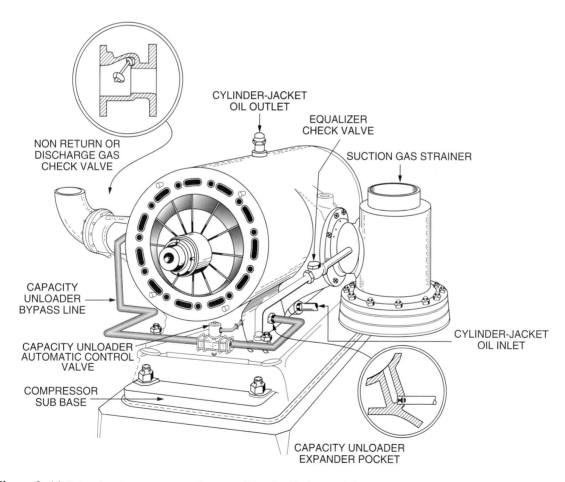

Figure 6–41 Rotary booster compressor. *Courtesy of Freezing Equipment Sales, Inc.*

Screw Compressor

Screw compressors (see Figure 6–42 and Figure 6–43) were developed after reciprocating and rotary machines but before the scroll compressors. They are applied to both ammonia and halocarbon refrigerants.

Spiral flutes in a female rotor act as cylinders, Figure 6–44. The length of the flutes in the rotor correspond to the stroke of a cylinder in a piston-type machine. Vapor flows into these flutes from a suction port on the suction end. As a flute turns beyond this port, a spiral lobe of the revolving male rotor turns into the suction end of the flute and seals it. Further rotation causes the sealing lobe to travel along the spiral toward the discharge end. This acts like the compression stroke of a reciprocating compressor. Thus, vapor flows into the flute, is trapped, then compressed and forced out a discharge port. No valves are needed or used. The suction port is open for most of the circumference of the rotors to allow vapor to flow into the flutes with a minimum of restriction. The discharge port is sized and placed to release the compressed vapor when its pressure matches the design discharge pressure. Like the rotary compressor, this machine must be designed for the compression ratio at which it is to be used.

Clearances between the male and female rotors, and between the rotors and housing, are only a few thousandths of an inch. With this clearance, there is no wear at these points. However, these small clearances are easily sealed with lubricating oil so vapor will not flow from the discharge to the suction side.

Oil is supplied by a separate oil pump. Bearings and the shaft seal are the only points requiring lubrication. However, sufficient oil is injected into the rotors to form seals between the rotors and housing, and to absorb the heat of compression. With this oil cooling, discharge temperature can be maintained at not more than 10°F to 20°F above condensing temperature.

A sliding valve is placed in the housing, which can open to allow vapor to escape back to the suction side during part of the compression cycle. Thus, compression takes place during only part of the revolution (stroke). The effect is similar to that of the bypass control of Figure 6–38.

By using a slide valve, a gradual change of capacity from 100 percent down to 10 percent is possible. This capacity-reduction valve can be built to be operated by hand or automatically. In the latter case, the valve is moved by oil pressure according to the demands on the compressor. Some screw compressors have a slide valve–limiting device. This allows the machine to effectively change its compression ratio and match its capacity to plant demands and/or ambient conditions. This provides more efficient operation.

The screw compressor has many advantages. It is easily driven at high speed and therefore is a natural mate for a steam turbine. Each flute acts like a separate cylinder. These facts make it possible to obtain large capacity from a comparatively small machine. Although they are expensive to build, they are competing in the 100- to 400-ton size, somewhat above the size of modern piston compression machines. The cost of manufacturing these machines has come down significantly in recent years with improved, automated machining methods. All motion is rotary, so there is a minimum of vibration. Volumetric efficiency is high because there is no leakage of high-pressure vapor back to the suction side. With this feature and oil cooling, it is possible to use single-stage compression on high compression ratios. Thus, a single-stage screw compressor can be used for low-temperature freezer applications whereas two-stage compressors would be necessary with reciprocating compressors. Two-stage systems are discussed in Chapter 21. A single screw compressor is cheaper than two reciprocating compressors below the 100-ton size. The screw compressor will easily pass liquid, so it is not damaged by slugging. Capacity control is easy and efficient. The cost of the manufacture of the complex rotor shapes is the biggest disadvantage of this compressor.

Figure 6-43 shows one stage of compression as refrigerant vapor moves through the screw compressor. Figure 6-44 shows a cutaway view of a screw compressor. The suction gas enters the machine at one end of the rotors and is discharged

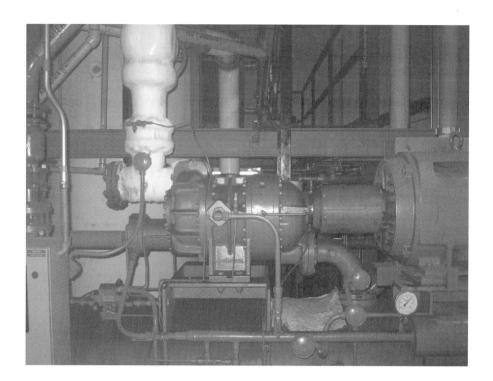

Figure 6–42 Ammonia screw compressor. *Photo by Noreen, courtesy of Arcticglacier Inc.*

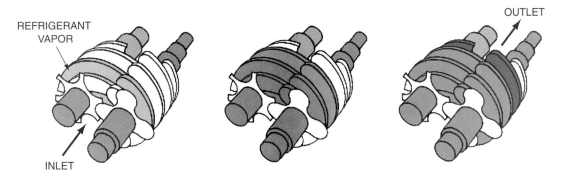

ONE STAGE OF COMPRESSION AS REFRIGERANT MOVES THROUGH THE SCREW COMPRESSOR

Figure 6-43 Tapered machined screw-type gears in a screw compressor.

at the other end. As the vapor is forced through the machine, its volume is reduced and its pressure is raised. Most screw compressors are equipped with an economizer port. This is another piping connection located between the suction and discharge connection and entering the compressor casing on the side. The vapor in the economizer port is drawn into the compressor at an intermediate pressure (approximately midway between suction and discharge pressure).

Figure 6-44 Rotors of screw compressor. *Courtesy of Lewis Refrigeration Co.*

Some systems employ a controlled pressure receiver. The controlled pressure receiver will have an intermediate suction line connecting it to the economizer port on the screw compressor. As the system's liquid refrigerant is fed from the receiver to the controlled pressure receiver, its pressure, and therefore temperature, is dropped. The resulting vapor enters the compressor's economizer port via the intermediate suction line. This entire procedure approximates to "free" cooling. There is some load increase on the compressor, but it is not in direct proportion to the amount of refrigeration capacity gained.

To summarize, we are now feeding liquid refrigerant to the evaporator at a lower temperature than is realized in a system without a controlled pressure receiver. The added cost to drive the compressor is minimal and is not in proportion to what it would cost to cool the liquid if a separate system were employed. Ask yourself this question: Would you rather feed liquid refrigerant to your evaporator at 86°F or at 50°F? If you can feed 50°F liquid with very little increase in cost, the answer is obvious. The economizer port on the screw compressor may be blanked off or plugged if you choose not to use it. The screw compressor will work as if it does not even have an economizer connection.

In recent years, screw compressors have been used in two-stage systems (see Figure 21-12,

Figure 21–14, Figure 21–15, Figure 21–16, Figure 21–17, and Figure 21–20). This works well, as screw compressors are capable of handling refrigerant vapor occupying large volumes and have high volumetric efficiencies. One (or more) large-volume screw compressor is used as the booster compressor. It takes suction from the evaporator(s) and discharges to the intercooler. The second-stage screw compressor(s) takes suction from the intercooler and discharges to the condenser(s). The intercooler removes the superheat from the first stage of compression. Two-stage systems are sometimes called *booster* or *compound systems*. These systems are capable of producing much lower temperatures than their single-stage cousins. These types of systems should at least be considered when temperatures go below 0°F or to avoid excessive compression ratios.

Finally, one manufacturer makes a screw compressor using a single screw. The single screw is vapor-sealed via rotor gates and lubricating oil. Because of its unique design, virtually all radial and suction thrust associated with twin-screw compressors is eliminated. By eliminating the lion's share of these thrust forces, bearing life in these types of machines is significantly increased. The machine is overall more efficient. It has microprocessor capacity control and safety interlocks like its twin-screw cousins.

Centrifugal Compressor

All compressors previously described are positive-displacement compressors. Vapor is drawn into a chamber; this chamber then decreases in volume to compress this vapor. After compression, the vapor is forced from the chamber by further decreasing the volume of the chamber to zero or nearly zero. A positive-displacement compressor can build up a pressure that is limited only by the volumetric efficiency and the strength of the parts to hold this pressure.

The centrifugal compressor, shown in Figure 6–45 and Figure 6–46, depends entirely on the

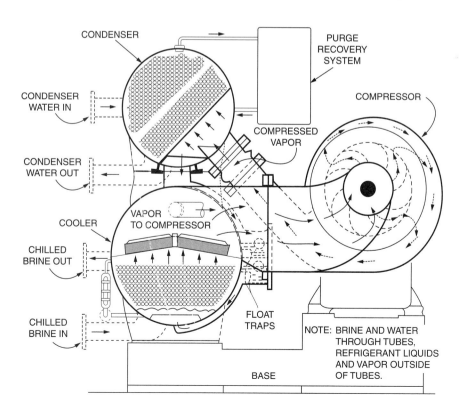

Figure 6–45 Centrifugal compression system. *Courtesy of Carrier Corporation.*

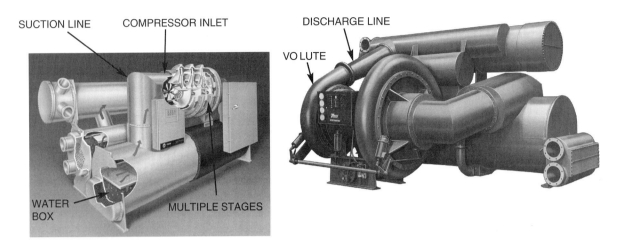

Figure 6–46 Multiple stages of compression and hot gas entering the condenser through the discharge line. *Courtesy of Trane Company 2000.*

centrifugal force of a high-speed wheel to compress a vapor passing through the wheel. There is no positive displacement; the action is called *dynamic compression*. This compressor is fundamentally a centrifugal blower designed for high-speed operation. The pressure at the discharge side of these blowers is higher than that of most other blowers. On these blowers, there is typically a 15- to 20-psi difference between the suction and the discharge. The pressure a centrifugal compressor can develop depends on the tip speed of the wheel. Tip speed is based on the diameter of the wheel and its revolutions per minute. The capacity of the compressor is determined by the size passages through the wheel. This makes the size of the compressor more dependent on the pressure required than on the capacity.

Because of its high-speed operation, the centrifugal compressor is fundamentally a high-volume, low-pressure machine. Such a machine has been built as small as 50 tons capacity, but is competitive with other types of compressors only above 100 tons. Above 300 tons capacity in one compressor, the centrifugal is the only type of machine usually considered. Centrifugal compressors of over 3000 tons capacity have been built. The custom-built machines used in the World Trade Center were much larger than this.

Many centrifugal compressors are connected inline to a 3450-rpm motor. Speed-increasing gears or direct-drive steam turbines are used to drive others at speeds up to 25,000 rpm.

Because of its low-pressure characteristics, the centrifugal compressor works best with a low-pressure refrigerant such as Refrigerant 11. With this refrigerant, suction pressure will be from 18 inches to 25 inches of vacuum, depending on the evaporator temperature required. Discharge pressure will be near atmospheric pressure. A single-stage wheel can be used with this refrigerant for air conditioning suction temperatures. Other refrigerants suitable for centrifugal compressors include R-21, R-113, R-114, and R-123. As previously mentioned, R-123 is currently dominant in centrifugal applications used for large chilled-water jobs (air conditioning and/or process). R-123 has replaced R-11, as R-11 is a chlorofluorocarbon (CFC). For water temperatures lower than about 45°F, some type of antifreeze solution must be used. Otherwise, there is danger of freezing up the evaporator and damaging the tubes. This is because the evaporator saturation temperature must be lower than the water temperature in order to absorb heat from it.

A two-stage wheel is common for standard ton conditions. In two-stage operation, the discharge of the first-stage wheel goes to the suction of the second wheel. Each stage can build up a compression ratio of about 4 to 1; that is, the absolute discharge pressure can be four times the absolute suction pressure. Centrifugal compressors have been built for high-pressure refrigerants such as ammonia, but an ammonia centrifugal machine requires at least five stages. The centrifugal compressor must be designed for the refrigerant with which it is to be used and for the temperature conditions under which it will be operating.

Because of the large volume of the low-pressure refrigerants used with a centrifugal compressor, refrigerant lines cannot be run around a plant as with high-pressure refrigerants. Therefore, the centrifugal compressor is coupled into a package unit with its evaporator and condenser (Figure 6–45 and Figure 6–46). The evaporator cools water or brine, which is then pumped to the areas requiring refrigeration.

Scroll Compressor

Scroll compressors are the newest members of the compressor family. Basically, the scroll compressor consists of two scrolls, one placed within the other. The two scrolls are identical to each other. One scroll is stationary and the other one "orbits" (not rotates) within the other (Figure 6–47). The swing link causes the orbiting action. It (the swing link) is connected to a rotating shaft on one end and the orbiting scroll on the other end. Its action is similar to that of the cranks of an internal-combustion engine. Suction is taken at

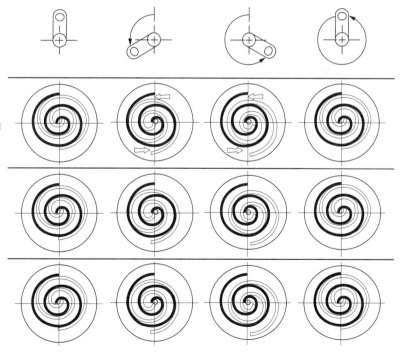

General
A 3-D compressor has two scrolls. The top scroll is fixed and the bottom scroll orbits. Each scroll has walls in a spiral shape that intermesh.

Inlet—First Orbit
As the bottom scroll orbits, two refrigerant gas pockets are formed and enclosed.

Compression—Second Orbit
The refrigerant gas is compressed as the volume is reduced closer to the center of the scroll.

Discharge—Third Orbit
The gas is compressed further and discharged through a small port in the center of the fixed scroll.

Figure 6–47 Schematic illustration of scroll compressor operation. *Courtesy of Trane Corporation.*

the periphery and the refrigerant vapor is forced to the center of the two scrolls. So, suction takes place at the outer part of the scrolls and the discharge vapor comes from the center of the scrolls. The scroll compressor is a positive-displacement compressor and reduces the volume of the refrigerant suction vapor. So, the vapor is compressed and its pressure is raised.

The necessary axial sealing is typically accomplished with "floating" seals. A thin film of oil seals the scrolls against each (similar to the way a screw compressor's rotors are sealed). However, the screw compressor's rotors are quite rigid, whereas the scroll compressor's scrolls will break their seal if necessary.

Because of the flexibility of the scrolls, its compression ratio is limited. However, this is a positive attribute. As long as the system is engineered, built, and maintained properly, the scroll compressor will perform satisfactorily. In the event of high head pressure (due to dirt, air, or other causes), the scrolls will move apart from each other as necessary and "relieve" the excess pressure internally. The operating engineers will notice reduced system performance and/or excessive head pressure and take corrective steps as necessary. An occasional shot of liquid refrigerant will not do serious damage, as the orbiting scroll will move to accommodate the liquid. However, no compressor is designed to take liquid refrigerant indefinitely. It is always necessary to find the cause of the liquid flooding and correct it.

The scroll compressors used in domestic applications are housed in hermetic enclosures, as are most scroll compressors found in light industrial jobs. The scroll compressors used for automotive air conditioning are open type with a shaft seal. This is necessary because the compressor(s) used in automotive applications are typically started and stopped via an electromagnetic clutch assembly. In some scroll compressors, the cycling of the electromagnetic clutch is reduced by the use of an internal capacity-control valve. This type is called a variable-displacement scroll compressor. The capacity control valve simply allows some of the refrigerant vapor to bypass from the discharge end of the compressor back to the suction side, effectively reducing the compressor's capacity. Scroll compressors employing this method of capacity control emit less noise because the clutch does not cycle as much. Also, the flow from the discharge is smoother and more uniform (consistent as opposed to start-and-stop).

Scroll compressors are relatively quiet, vibration-free, and efficient compared to their reciprocating cousins. There are no suction and discharge valves, thus eliminating the re-expansion of the small amount of vapor that remains trapped in the clearance of a reciprocating compressor. Most scroll compressors employ a suction check valve to prevent condenser vapor from bleeding back through the compressor to the low side. However, the compressor still equalizes internally during the off cycle, so a low-starting-torque motor may be used. The suction and discharge vapor do not mix, thus eliminating some thermal inefficiency. There is one main moving part (the orbiting scroll). Theoretically, the two scrolls never touch (under normal conditions), thus eliminating most friction heat. The refrigerant vapor is drawn in and discharged smoothly, which reduces vibration and motion in the adjacent connecting piping. One of the main advantages of the scroll compressor is that, over time, the scrolls actually wear in, making the compressor more efficient, whereas, with reciprocating compressors, the valves and other mechanical parts wear out, making them less efficient.

Steam Jets

Another device used instead of a compressor is the steam jet. It is used only with water as a refrigerant. Figure 6–48 shows the fundamentals of such a system. A steam jet blasts through the throat. In so doing it draws a high vacuum on the evaporating chamber—a sufficiently deep vacuum to bring the evaporating temperature of water down to 40°F. The jet carries the low-pressure water vapor with it to the condenser, where both are condensed by contact with condensing water. The water and condensate are drawn from the condenser by some form of

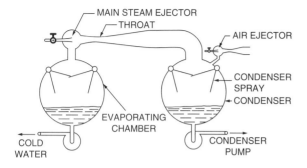

Figure 6–48 Steam jet refrigeration system.

pump. A second steam jet on the condenser is necessary to remove air that may be drawn in through possible leaks, or that is brought in dissolved in the water. This is fundamentally a purge device. Notice that the vapor condensed in the condenser is discarded with the condenser water. However, the main advantage of water as a refrigerant is that it is cheap enough that there is no need to be economical with it.

The water chilled by evaporation is pumped to where it is needed to absorb heat. It is then returned to the evaporating chamber in a spray, to give a maximum of surface from which evaporation may take place. This makes evaporative cooling possible whether the water is actually boiled or not. Water cannot be used much below 40°F without approaching the danger range of freezing. Therefore, such a system is suitable only where 40°F is the minimum temperature required.

The steam jet system has had some use in air conditioning. Because this system uses steam instead of mechanical power for the driving force, its biggest advantage is in areas where fuel is cheap.

A relatively new application of steam jet refrigeration is in cooling leafy vegetables such as lettuce. Lettuce in crates or cartons is loaded into a steel tank. Doors are closed and sealed, and the steam jet used to pull a vacuum on the tank. Lettuce is over 90 percent water. The evaporation of 4 percent of the water content will cool the lettuce about 40°F below the temperature at which it entered the chamber. Thus, the water content of the lettuce acts as the refrigerant.

The loss of 4 percent of the moisture does not hurt the lettuce, but the lettuce can be cooled completely and evenly within 20 minutes. Heads in the center of the crate cool as rapidly as the outer layers. The center of each head is cooled as thoroughly as the outer leaves.

A somewhat similar lettuce cooling system uses a combination of rotary compressors to evacuate air from the cooling chamber, and plates refrigerated by ammonia to condense out the water vapor drawn from the lettuce. Most lettuce that is precooled for shipment is now cooled by one of these vacuum systems.

Compressor Sizes

Many different methods of rating compressors are used. Small machines, such as are used on domestic or commercial boxes, may be rated in Btu per hour, or by the horsepower used to drive them. The latter is most common. Intermediate sizes, such as large commercial or small industrial machines, may be rated in horsepower or in tons of refrigeration. Most industrial compressors, and nearly all ammonia compressors, are rated in tons of refrigeration or by their bore and stroke. Thus, a 6 × 6 means a two-cylinder compressor with a 6-inch bore and a 6-inch stroke. Modern, high-speed industrial compressors may be rated in tons or in the number of cylinders.

Questions

1. What is an indicator chart?
2. Are there any disadvantages to strong compressor valve springs?
3. Do high or low backpressures have the highest volumetric efficiencies? Explain.
4. Does a large clearance increase or decrease the volumetric efficiency? Explain.

5. What are the differences between the VSA compressor and modern V-type compressors?
6. What different types of compressors are positive-displacement compressors?
7. What is the fundamental difference between positive-displacement compression and dynamic compression?
8. What harm can come from oil slugging?
9. Why are safety heads used on some compressors?
10. Name four different methods of capacity control.
11. What are some of the advantages of hermetically sealed compressors? The disadvantages?
12. Describe how a scroll compressor works. What are its advantages?
13. What factor is most important in considering compressor speeds?

Chapter 7   Condensers

Requirements

The condenser must take the superheated vapor from the compressor, cool it to its condensing temperature, then condense it. Review the condenser section of the cycle chart, Figure 2–13. The action in the condenser is just the opposite of the action in the evaporator. Whereas the evaporator absorbs heat to change a liquid to a vapor, the condenser gives off heat to change the vapor back to a liquid. However, the requirements of a condenser are similar to those of an evaporator. The condenser must have the required heat transfer surface. It must provide sufficient volume for the vapor delivered to it. It must allow the condensed liquid to separate from the uncondensed vapor.

The condenser must have the required heat transfer surface. All the heat picked up in the evaporator, and the heat of compression added in the compressor and suction line vapor, must be removed from the refrigerant by the condenser. The total heat that must be removed varies according to the conditions, but is roughly about 1.25 times the heat picked up in the evaporator (when a reciprocating compressor is used).

The condenser must provide sufficient volume for the vapor pumped to it by the compressor. Before this vapor condenses, it has a definite volume. This volume can be decreased by increasing the pressure. However, any increase in pressure increases the work done by the compressor, and therefore the cost of driving the compressor. Normally, any condenser that provides sufficient surface provides sufficient volume.

The condenser must allow the condensed liquid to separate from the uncondensed vapor. One reason for the need for sufficient volume is for the purpose of supplying room for this separation. If the liquid does not separate from the vapor and flow out, the condenser fills with liquid. This leaves insufficient volume for the additional vapor pumped to it. A second reason for adequate separation of liquid and vapor is that too much liquid between the vapor and the metal walls of the condenser can act as a partial insulator. This will retard heat transfer from the vapor to the condenser surface.

Operation

Figure 7–1 shows an air-cooled condenser in its simplest form. The hot vapor from the compressor enters the top. In the first few coils, this vapor is cooled to the condensing temperature for the pressure prevailing in the condenser—for example, 90°F at 133.8 psia for R-134a. Sensible heat is given up to the ambient air as the vapor cools from 110°F to 90°F. From here on, as heat is extracted, the vapor is condensed; latent heat is given up to the ambient air. Droplets form on the inside surface of the tube and flow down toward the outlet, similar to the way drops of water form on a window in a steam room. These droplets collect and run together until they fill the entire tube. From here down, any further

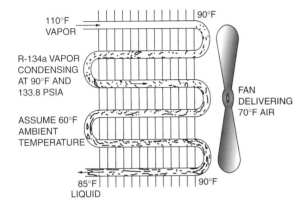

Figure 7–1 Schematic of an air cooled condenser.

extraction of heat results in subcooling the liquid—that is, cooling it below 90°F for our example. Here we see sensible heat being given up again. The pressure, however, remains at 133.8 psia because of the vapor pressure pushing down on the liquid. To keep this heat extraction continuing, a fan forces a blast of air over the outside of the condenser. This air cools, or extracts heat from, the condenser. The air in turn is heated as it does this.

When the compressor starts, the vapor pumped into the condenser causes the high-side pressure to rise. The heat of compression raises the hot discharge gas to a temperature that is warmer than that of the air flowing over the outside of the condenser. This allows the air to cool or extract heat from this vapor. The pressure is raised to the condensing pressure corresponding to the temperature to which the vapor is cooled. Finally, an equilibrium is established at a constant condition. The supply of vapor, and the heat it carries, balances the heat extracted by the air and the vapor condensed.

There are many advantages to having a cooler liquid in the condenser. Less flash gas is formed (see Chapter 20) and less liquid must be used (and pumped by the compressor) to do a given job, see Figure 22–4.

The lower the condensing temperature, the lower will be the condensing pressure. The lower the condensing pressure, the less work the compressor must do, and the less power is required to drive it. This is assuming that conditions remain constant in the evaporator.

Too low a condensing pressure is undesirable, particularly with halocarbon refrigerants, because expansion valves will then not feed properly. Sufficient head pressure is also required in systems using hot gas defrosting as well as in systems that incorporate gas pumps. In Figure 7-1, 90°F is the actual condensing temperature, although the liquid is later subcooled to 85°F. Some subcooling (perhaps as much as 15–20°F) is desirable to prevent liquid flashing to vapor in the liquid line. However, a large amount of subcooling indicates poor separation of liquid from vapor, and a higher condensing pressure than is necessary. It could also be caused by an overcharge of refrigerant. Standard efficiency systems typically subcool the liquid refrigerant 25°F or more.

Air-Cooled Condenser

Figure 7-1 is typical of small air-cooled condenser design. This design is called a single-row, single-pass condenser. Figure 7-2 shows such a condenser with a double-row, single-pass condenser. The double-row, single-pass condenser has two rows of tubes to provide greater surface, but all the refrigerant must pass through a single tube

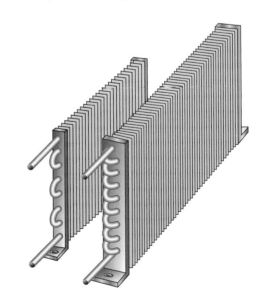

Figure 7–2 Single-row, single-pass and double-row, single-pass air-cooled condensers.

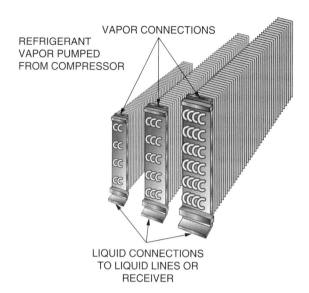

Figure 7–3 Double-row, double-pass, three-row, three-pass, and four-row, four-pass air-cooled condensers.

Most small commercial hermetic machines with air-cooled condensers include a fan, driven by a small shaded pole motor. Above 3 horsepower, it is difficult to get enough condenser surface in front of a fan of this type. Therefore, 3 horsepower is about as large a completely self-contained air-cooled condensing unit as is available.

If air cooling is wanted above 3 horsepower, a separate condenser, usually mounted on the roof, is used, as shown in Figure 7–4. This condenser is

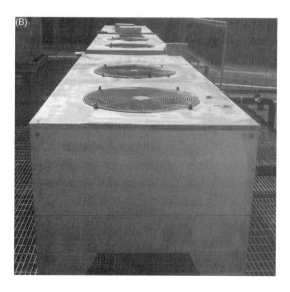

Figure 7–4 (A) Remote air-cooled condenser, usually mounted on the roof, *Courtesy Heatcraft Inc., Refrigeration Products Division*. and (B) roof-mounted remote air cooled condenser. *Photo courtesy of Noreen.*

before exiting this coil. The number of possible paths that the refrigerant can take provides the number of passes, and the number of tubes from the front of the coil to the back of the coil provides the number of rows. Double-row, single-pass condensers are suitable for domestic jobs and fractional-horsepower commercial jobs. Single-pass condensers in larger sizes do not satisfactorily separate the condensed liquid from the vapor.

Figure 7–3 shows how this weakness can be improved by headering each row so that the refrigerant is divided. In the case of the double-row, double-pass condenser, half the refrigerant goes through each coil. Because half the liquid is then condensed in each coil, the liquid does not fill the coil so quickly. Also shown in Figure 7–3 are a triple-row, triple-pass condenser and a four-row, four-pass condenser. Such designs are used for larger-sized air-cooled condensers.

Many domestic air-cooled condensers do not depend on a fan for air circulation, but rather on the fact that warm air rises and creates some air flow. Larger areas are required in such cases. Such a condenser may cover most of the bottom of a cabinet, the back of the cabinet, or both.

properly designed to give sufficient surface and the proper fan for forced convection. These remote air-cooled condensers are commonly available up to 100 tons capacity. Some manufacturers make larger sizes. For requirements above these sizes, several units can be combined to give the required capacity.

Air- vs. Water-Cooled Condensers

Air-cooled condensers are much simpler to install than water-cooled condensers. There is no need to pipe water or provide a drain from an air-cooled unit. No water-regulating valve or cooling tower is needed. There are no problems with water scale or corrosion. Restrictive legislation on the use or drainage of water is of no consequence.

However, water cooling also has advantages. Particularly in the summer, when loads are already heaviest, water is typically cooler than the air. Also, water gives better heat transfer, so it is possible to get a condensing temperature nearer to water temperature than to air temperature. And the lower the condensing temperature, the lower will be the pressure and the lower the operating costs.

In small units, the saving is not enough to pay for the added cost of the installation of a water-cooled condenser. However, as system size increases, the savings become greater, whereas the cost of piping is only a little more. The cost of installing the associated piping increases as well, but not enough to offset the savings. Air-cooled self-contained condensing units are available up to a certain limit. If air cooling is wanted above more than a few horsepower, a remote air-cooled condenser is typically used.

Condensing units with water-cooled condensers are available from about $1/3$ horsepower up. However, below 3 to 5 horsepower their use is decreasing because of greater complications and restrictions on the use of water. For larger sizes, however, the advantages of water cooling begin to outweigh their disadvantages and they are often chosen.

Water-cooled condensers are a better choice for most industrial-size plants, particularly if ammonia is the refrigerant.

Double-Pipe Water-Cooled Condenser

The double-pipe water-cooled condenser is also called the tube-in-tube and/or pipe-in-pipe condenser. A simple and once commonly used water-cooled condenser is shown in Figure 7–5. This is a double-pipe condenser created by placing one copper tube inside another, then bending the combination to the required form. The refrigerant is run in the outside space and the water in the center. This puts water on one side of the refrigerant and air on the other.

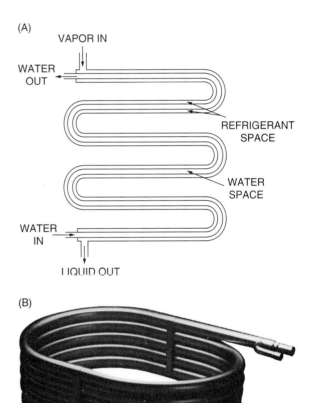

Figure 7–5 (A) Schematic of double-pipe water-cooled condenser, with (B, C) double-pipe condensers.

(C)

Figure 7–5 (Continued)

Notice that the water flows in the opposite direction to the refrigerant. This is known as counterflow and is commonly used wherever possible in all heat transfer equipment. In this way, the coolest water is used for the final cooling of the liquid refrigerant. This gets the liquid refrigerant as cool as possible, increasing the amount of condenser subcooling. The warmest water strikes the hottest refrigerant vapor. This makes it possible for the water to pick up more heat. In this way less water is used and lower condensing temperature and pressure become possible. Figure 7–6 shows comparable results that can be obtained with a given condenser with counterflow and parallel flow.

The bent-copper-tubing condenser is a simple, inexpensive form used where water cooling is wanted for small systems. However, the water side of any water-cooled condenser usually requires cleaning sooner or later. Unfortunately, the bent tubing cannot be cleaned by mechanical means such as brushes. These coils must be cleaned chemically. There is one type of double-pipe condenser on the market that has headers at the ends. These headers can be opened so that the water tubes can be cleaned mechanically (Figure 7–7). Because copper tubing cannot be used with ammonia, a double-pipe ammonia condenser must be made of iron pipe and fittings. This was a favorite type of ammonia condenser many years ago. However, the large number of joints and mechanical connections necessary to complete the installation at the end of each pipe made it difficult to prevent leaking. Therefore this condenser is not used in modern ammonia plants.

Shell-and-Tube Condenser

The shell-type condenser effectively eliminates the biggest disadvantage of the above-mentioned condensers: poor separation of liquid and vapor. Figure 7–8 shows a cross section of a horizontal shell-and-tube condenser. It could be made almost identical to the brine cooler of Figure 5-6. The water is introduced into the bottom tubes and then passes back and forth through the condenser several times. The larger liquid surface at the bottom of the condenser gives good contact to the vapor to maintain a pressure corresponding to the

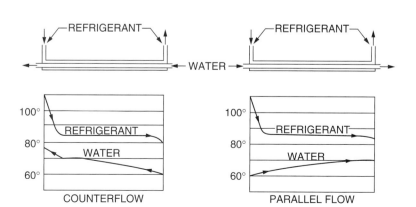

Figure 7–6 Results of counterflow and parallel flow in double-pipe condenser. (This is a visual representation of how heat transfer takes place in the condenser.)

Figure 7-7 Double-pipe water-cooled condenser, cleanable type.

liquid temperature (it significantly reduces pressure drop). If the lower tubes cool the liquid, the entire volume, including the surface, is cooled. This liquid surface is in direct contact with vapor, so it reduces the condensing pressure. The pressure drop on the refrigerant side is negligible. The heads are removable for cleaning the water sides of the tubes without disturbing any refrigerant connections.

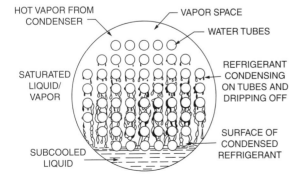

Figure 7-8 Cross section of shell-and-tube condenser.

Figure 6-17 shows how such condensers are used for commercial jobs with complete condensing units. In such installations the condenser also acts as a liquid receiver. Figure 7-9 shows an ammonia installation. Two or more such condensers may be used to provide the required capacity, to provide standby protection, or to permit one condenser to be cleaned while the other remains in service. On industrial jobs a separate liquid receiver is always used.

Figure 7-10 is a diagram of a vertical shell-and-tube condenser. Such condensers are used only with ammonia. Figure 7-11 shows the construction of the head of one of these condensers. Water is distributed over the entire head and enters each tube through a swirler. This causes the water to spin or swirl down the sides of the tubes instead of dropping through the center, where much of it would be away from the walls. Only a single pass of water is used through this type of condenser.

This type of condenser is not considered to be quite as efficient as the horizontal shell-and-tube type. Notice the lack of good counterflow. However, the vertical condenser does not foul up as rapidly as the horizontal condenser, particularly in areas that have bad water. The once-through pass of water, plus the effect of gravity, helps carry sediment and scale through the tubes instead of allowing it to accumulate on the interior surfaces of the water tubes. One benefit of this vertical variation is that the water tubes may be cleaned while the unit is in service.

Shell-and-Coil Condenser

Figure 7-12 shows a shell-and-coil condenser used in commercial units. The results obtained here would be about the same as with a shell-and-tube condenser. The only way to clean the water side of the coils is chemically. Because there are fewer joints in this type than in a shell-and-tube condenser, there is less danger of developing leaks.

Figure 7-9 Ammonia horizontal shell-and-tube condensers with receiver below.

Water Supply; Cooling Towers

Any water-cooled condenser must have a supply of water. For small commercial jobs, city water may be used. Check to see what local codes allow. Different municipalities have different codes. A water valve (see Chapter 9) is used to allow water to flow only when needed.

The large number of small water-cooled systems that depended on public water supply has introduced a serious problem in many cities. A condenser requires at least 1 gallon of water per minute per horsepower while the unit is running. For efficient cooling, two or three times this amount should be used. Cities spend large sums of money to obtain and treat water. Users run it through a coil, raise its temperature a few degrees, then dump it into a sewer. In some cities, the aggregate of many thousands of customers doing this has overloaded both the water supply system and the sewage system. It has become such a serious problem that many cities have passed legislation

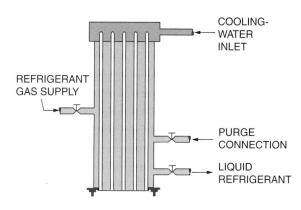

Figure 7-10 Ammonia vertical shell-and-tube condenser.

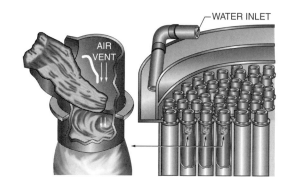

Figure 7-11 Condenser water distributor.

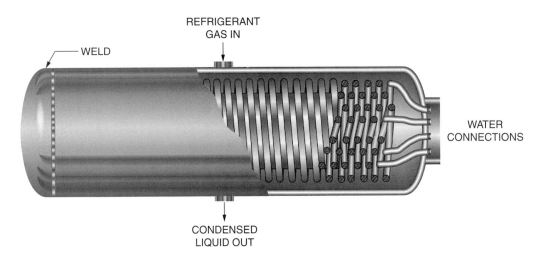

Figure 7-12 Shell-and-coil water-cooled condenser.

prohibiting such use. A cooling tower is required on any but the smallest system.

Besides the legal restrictions, the cost of water puts an economic restriction on the use of purchased water for large commercial or industrial systems.

If it is available, water from wells, rivers, lakes, or the ocean may be used. The water is used, then discarded after it has served its purpose. However, such water supplies often are not available where needed. In such cases the same water is used over and over again, with the aid of a cooling tower to remove system heat.

The latent heat of evaporation of water at 86°F is 1045 Btu, but 1000 Btu is usually taken as a round number. The evaporation of 1 pound of water will then cool 100 pounds of water by 10°F. Thus the evaporation of 1 pound per 100 pounds, or 1 percent, of water will cool the water by 10°F. There may be another 1 percent to 5 percent water loss due to windage or drift loss, that is, water blown out of the tower. However, this still keeps the water bill within reason when water must be purchased. Water may be cooled to within 5°F to 10°F above the wet-bulb temperature (see Chapter 19 for a complete discussion of wet-bulb temperature) of the air. The wet-bulb temperature reading is always less than the dry-bulb temperature reading on an ordinary dry-bulb thermometer. The drier the air, the lower the wet-bulb temperature will be. Thus, quite often, cooling tower water temperatures below that of the dry-bulb temperature of the air are possible and quite common.

Cooling towers may be natural-draft, induced-draft, or forced-draft. Natural-draft towers in turn may be divided into spray towers and deck towers, either of which must be located where prevailing winds can pass through them.

Figure 7-13 shows a typical spray tower. Water is sprayed from the top of the tower through nozzles. As it falls, air passing through the tower evaporates some of the surface of each drop. This evaporation cools the water remaining in the tower. It drops into a pan or tank at the bottom of the tower, from which it is piped to the condenser. In addition, the evaporating water absorbs heat from the air passing through the tower, increasing the heat transfer rate between the air and the water in the tower. From the condenser, it is pumped back up to the spray nozzles. Louvers on the side of the tower prevent the wind from blowing the drops away from the tower as they fall. This potential loss of tower water as a result of high wind is referred to as *drift*.

Figure 7-14 shows a deck-type tower. Here the water is distributed over the top in a trough

Water Supply; Cooling Towers 113

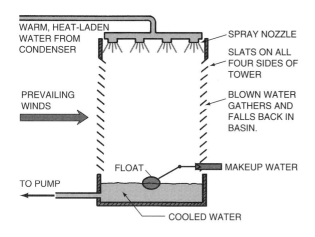

Figure 7-13 Spray-type cooling tower.

system. It runs in tiny streams from these troughs and drips to the first deck. It splashes on this deck, forming drops that fall to the deck below, where it splashes again. The added surface of the wet deck, and the interruption of the fall as it splashes, increases the cooling. Also, because there is no nozzle pressure to pump against, the pumping load is not as great as with a spray tower.

Figure 7-15 shows an induced-draft tower. This is enclosed on the sides, with louvered openings at

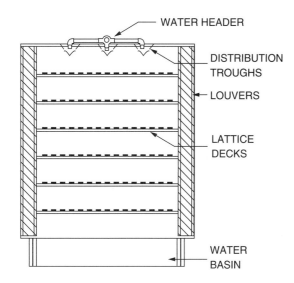

Figure 7-14 Deck-type cooling tower.

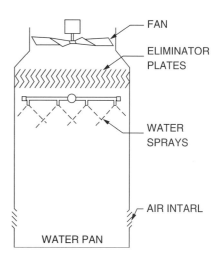

Figure 7-15 Induced-draft cooling tower.

the bottom. The fan and the heat provide a chimney effect, cooling the air from the bottom up and out the top. Eliminator plates are placed to catch the drops as they are swept up. This unit must still be located outside, but it will provide cooling on days with little or no wind.

Figure 7-16 shows a forced-draft tower. This is somewhat similar to the induced-draft tower, but the air is forced in at the bottom instead of being sucked out at the top. This type of unit may be placed out in the open, or if more convenient, located inside and the outlet carried outside through ducts.

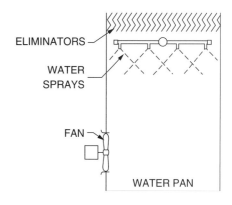

Figure 7-16 Forced-draft cooling tower.

If a cooling tower is required in a modern refrigeration system, one of the previously mentioned mechanical-draft towers is usually chosen. They take up less room than other types, and they provide better cooling on hot days with little or no wind.

Evaporative Condenser

In recent years a condenser that is a combination of an air-cooled condenser, a water-cooled condenser, and a cooling tower, all in one unit, has been developed. Wherever a cooling tower is used, air is the final cooling medium, even though a water-cooled condenser is used.

The evaporative condenser combines the condenser and cooling tower in one piece of equipment. The water provides evaporative cooling, wets the surface to give good heat transfer, and helps increase the surface area of the water drops of the spray. Figure 7–17 shows a typical evaporative condenser. The water is taken from the sump at the bottom by the self-contained pump and sprayed in at the top. It sprays over the bank of coils and down to the sump again. The fans pull the air in from the bottom openings and out of the top. Eliminator plates prevent the water from

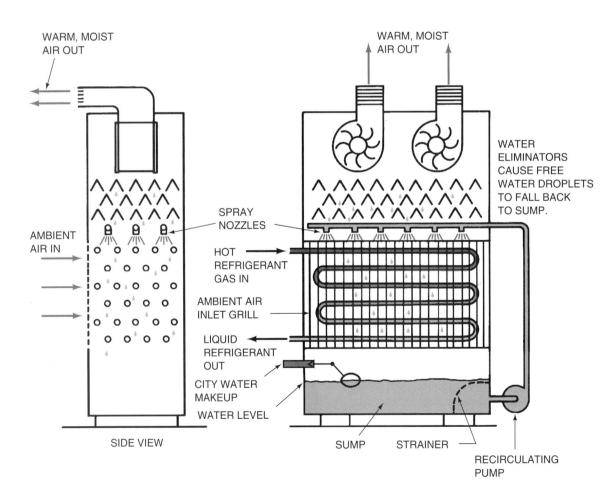

Figure 7–17 Evaporative condenser.

being blown out. The coils of the condenser must be well headered to increase to a maximum the separation of liquid and vapor. Such a condenser can be placed outside, or inside with ducts going out, whichever is convenient. Note that in a cooling tower, the cooled water is pumped from the sump to the condenser. In an evaporative condenser, the condenser coils are built right into cooling tower. This directly affects the amount of refrigerant the system holds.

An evaporative condenser is usually cheaper than the combination of a shell-and-tube condenser and a cooling tower. Therefore, it is usually the first choice for a new installation using positive-displacement compressors. A shell-and-tube condenser closely coupled to the compressor must be used with a centrifugal compressor. Therefore, an evaporative condenser cannot be applied in such an installation.

A special modification of the evaporative condenser is a single condenser shell containing several different circuits or coils. Each compressor of a multiple-unit job, such as a supermarket, has its own separate condenser circuit. This is a cheaper and better-looking solution than a multitude of small condensers all over the roof of a building.

Questions

1. What are the requirements of the condenser?
2. Which is better, a single-pass or a headered condenser? Why?
3. What sizes of machines use air-cooled condensers only? Air- or water-cooled condensers?
4. What are the advantages of water-cooled condensers? Disadvantages?
5. What is counterflow and why is it used?
6. List the common types of water-cooled condensers.
7. What are the advantages of the shell-and-tube condenser?
8. What is the purpose of a cooling tower?
9. List the different types of cooling towers.
10. Describe the evaporative condenser.

Chapter 8 Flow Equipment

Liquid Receiver

The liquid receiver is a storage chamber or tank to hold refrigerant after it is condensed but before it is needed in the evaporator. It is used to store the system refrigerant if it is necessary to pump the system down for service operations. The receiver should be large enough to hold the system's entire refrigerant charge plus approximately 10 percent.

Figure 8-1 shows two typical liquid receiver configurations: vertical and horizontal. The outlet of the receiver, which is piped into a liquid-line service valve (king valve), is usually located toward the top of the shell. A dip tube in the receiver ensures that 100 percent liquid refrigerant leaves the shell of the receiver. The dip tube is positioned far enough from the bottom so that dirt and sediment that might find its way into the system settles to the bottom and is not picked up and carried into the liquid line. Enough liquid should always be kept in the receiver to provide a liquid seal at the inlet of the dip tube. The inlet of the receiver is usually located some distance from the outlet so as not to stir dirt and sediment up into the liquid refrigerant as it leaves the device.

Receivers on commercial units equipped with air-cooled or a double-pipe water-cooled condenser are of either type shown in Figure 8-1. Where a shell-and-tube or shell-and-coil condenser is used, the lower part of the condenser is often used as the receiver (Figure 6-17).

Figure 8-2 as well as Figure 7-9 show typical industrial receivers. Orientation (horizontal or vertical) is dictated by floor space and/or ceiling height. Industrial receivers have a gage glass to indicate the refrigerant level. Commercial receivers usually do not. An ammonia-refrigerant industrial receiver should also have a drain valve in the bottom, which can be used to drain off oil, water, or sediment. The receiver may be directly under the condensers if convenient, or may be a considerable distance from them.

Most industrial systems have a vent pipe or equalizing line from the top of the receiver to the top of the condenser to balance the pressure, as shown in Figure 8-3. This is to keep an unobstructed flow of refrigerant from the condenser to the receiver. It acts similarly to the second hole in a can of condensed milk. Figure 8-4 shows what can and sometimes does happen without a vent. Some engineers feel that if the drain from the condenser to the receiver is of adequate size and is not trapped, the vent line is superfluous.

Pipes and Piping

Copper tubing is used almost exclusively with halocarbon refrigerants. For domestic and small commercial systems, soft-drawn tubing is usually used. Table A-7 in the Appendix lists data and available sizes. Notice that the size specified is the outside diameter (OD) of the tubing. Although 3/16-inch,

Pipes and Piping   117

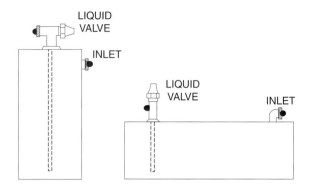

Figure 8-1 Vertical and horizontal low-pressure liquid receivers.

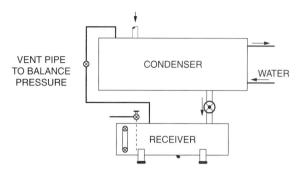

Figure 8-3 Vent pipe from condenser to receiver.

5/16-inch, and 7/16-inch sizes are available, they are not common, and fittings for them are not readily available. Therefore, the use of such sizes is not recommended. Soft-drawn tubing is easily bent to the required shape and is easily flared. For refrigeration use, this tubing is furnished in rolls. It has been thoroughly cleaned, dehydrated, filled with dry air, then the ends sealed. This tubing is classified as ACR tubing and is considerably more expensive than plumbing-grade tubing. Care should be taken to keep it in this same clean, dry condition until used.

Except in hermetically sealed units, flare connections are usually used to make joints in soft-drawn copper tubing. Figure 8-5 shows the type of flare used in refrigeration work. Notice that the flare forms a copper gasket between the two fittings. These flares are easily made in the tubing with the proper tool. Figure 8-6 shows typical flare fittings available in all sizes necessary to fit soft-drawn tubing. Flare connections are not used on copper tubing larger than 3/4 inch. It is very difficult to tighten such large flare connections enough to make a gas-tight joint.

Larger commercial equipment usually uses hard-drawn tubing, information about which is provided in Table A-7 in the Appendix. Hard-drawn tubing is also used occasionally on smaller equipment. Because rigid piping materials are more difficult to bend or form, special solder or "sweat" fittings, often abbreviated ODS, are used (Figure 8-7). Such fittings are also used occasionally with soft-drawn tubing.

Many different types of solder have been used to put piping together arrangements. Ordinary 50-50 lead-tin solder has seen limited use, but a 95-5 solder (95 percent tin, 5 percent antimony) with a melting temperature of about 450°F, is usually considered more satisfactory. Even this type of solder has been known to crack or develop leaks where there is vibration.

In many areas, codes do not permit the use of a solder having a melting point below 1000°F. In any event, it is good field practice to use higher-temperature filler materials on the high-pressure side of the system.

Silver brazing alloys have melting points above 1000°F. A properly made joint using these alloys is as strong or stronger than the tubing itself. Silver-bearing brazing materials can contain anywhere from 3 percent silver to more than 50 percent silver. Brazing alloys are copper-phosphorous combinations with or without some silver, and perhaps other elements. Those that contain silver are referred to as silver-bearing brazing alloys. Most silver brazing alloys do not make dependable bonds

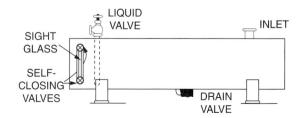

Figure 8-2 Horizontal ammonia liquid receiver.

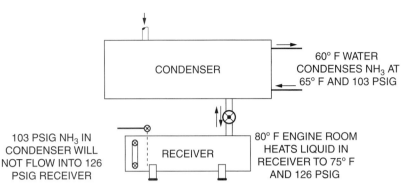

Figure 8-4 Why a vent pipe is needed.

with iron or steel, but make excellent joints with copper or copper alloys. Silver solder typically bonds well to both steel and copper as long as the appropriate flux is used.

Joints made using 95-5 or other low-temperature filler material can be made with a Prest-O-Lite torch, shown in Figure 8-8. An oxyacetylene torch can be used, but even with a small tip, great care must be taken not to overheat the joint.

Hard solder, as the silver brazing alloys are sometimes called, is best worked with an oxy-acetylene torch, although an air-acetylene torch is still the most popular choice among field technicians. Prest-O-Lite torches can be used on small-sized tubing, but they are slow to heat up and cannot get large fittings or valves hot enough to do a satisfactory job, especially when the pipe is large.

Iron pipe has occasionally been used with halocarbon refrigerants, particularly in evaporator construction. It has also been used for the refrigerant lines in locations where it is subject to more abuse than copper tubing could stand. However, using iron pipe often results in the formation of corrosion and pipe scale, so its use is usually avoided where possible. Where iron pipe is used, welded joints are recommended. If welded joints are not used, a good thread cement such as litharge and glycerine must be used on all threaded joints or a leak will result. Pipe connections made without some type of thread sealer will not hold a refrigerant in either liquid or vapor form. Make certain that the thread sealer to be used is compatible with not only the system refrigerant but also the type of refrigeration oil being used.

Because copper cannot be used where ammonia is the refrigerant, copper, iron, or steel pipe is necessary with this refrigerant. Butt-welded pipe should not be used on the high-pressure side of an ammonia system. For $1\frac{1}{2}$-inch or smaller pipe, X-heavy pipe, also called Schedule 80 pipe, should be used for the high side. Seamless wrought iron or steel pipe has been used to an increasing extent in recent years. Table A-8 in the Appendix lists iron pipe dimensions, working pressures, and other information.

Joints may be flanged or welded. Where joints are flanged, the flange fittings must be screwed

Figure 8-5 Components for a flare joint. *Photo courtesy of Bill Johnson.*

Figure 8-6 Examples of flare fittings. *Photo courtesy of Bill Johnson.*

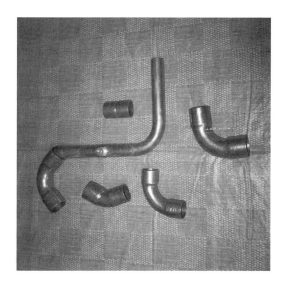

Figure 8–7 A line of sweat or solder fittings. *Photo courtesy of Noreen.*

or welded onto the pipe. Again, screw threads must be soldered or doped with litharge and glycerine to make them refrigerant-tight. Welding has been the preferred method used in almost all ammonia plants installed in recent years. A complete line of welded fittings has been developed for this, as shown in Figure 8-9.

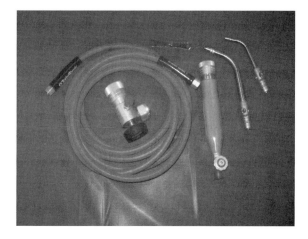

Figure 8–8 Brazing/soldering torch. *Photo by Noreen, courtesy of Johnstone Supply.*

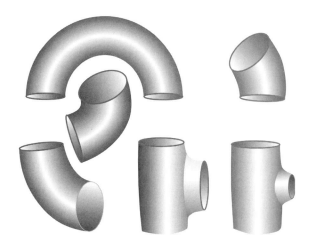

Figure 8–9 A line of weld fittings. *Courtesy of Taylor Forge and Pipe Works.*

Hand Valves

Figure 8-10 shows typical hand valves of the packed type available for halocarbon refrigerants. The valve body is of forged brass, and the stem and plug are made of steel, often of a corrosion-resistant type. The stem comes out through a packing. To take care of any possible leakage past the packing, a cap is supplied to go over the stem and fit the valve body against a copper gasket. In the larger sizes, this valve cap is so made that it can be used as a wrench to open or close the valve.

Figure 8-11 shows packless and diaphragm-type valves. Such valves are used a great deal with halocarbon refrigerants for sizes under 1 inch. They are available in larger sizes, but these are quite costly compared to packed valves. Here, the pressure of the screw stem is applied to a flexible diaphragm. A second stem and plug is on the other side of this diaphragm. Thus, turning the hand wheel in presses the valve closed through the diaphragm. When the hand wheel is turned out, a spring opens the valve. This valve should always be installed so that the flow through the valve is from under the plug. If it is installed the other way, the refrigerant pressure may hold the valve closed when the screw stem

120 CHAPTER 8 Flow Equipment

(A)

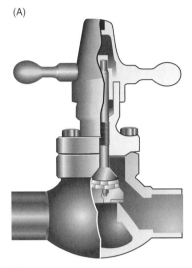

(B)

PCE NO.	DESCRIPTION	QUAN.
1	BODY	1
2	STEM	1
3	SEAT DISC ASS'Y	1
4	DISC SPRING	1
5	DISC PIN	4
6	RETAINER RING	1
7	PACKING WASHER	1
8	PACKING	2*
9	PACKING GLAND	1
10	CAP	1
11	CAP GASKET	1
12	FLANGE	1
13	ADAPTER	1
14	GASKET	1
15	CAPSCREW	4
16	PIPE PLUG	2

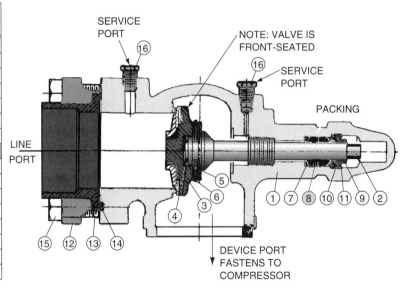

Figure 8–10 (A) A packed low-pressure valve and (B) a service valve showing packing material, which can be replaced if it leaks. *Courtesy of Henry Valve Company.*

is backed out, as shown in Figure 8–12.[1] These valves are clearly marked with directional arrows on the valve body to ensure that they are installed facing the correct direction.

Figure 8–13 shows a typical valve for used with ammonia. It is made of forged steel. The plug may be a ground metal joint, or may have a disk with a babbitt metal face. Notice the back

---

[1] Some valves are so designed that they may be safely installed either way. See the manufacturer's instructions.

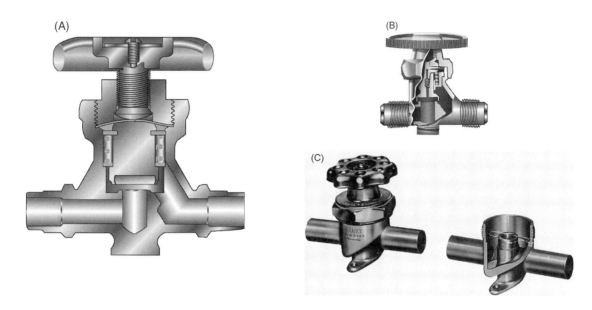

Figure 8–11 (A) Packless valve. *Courtesy of Superior Valve & Fittings Co.* (B) Packless valve. *Courtesy of Henry Valve Co.* (C) The diaphragm-type hand valve used when servicing a system. This valve is either open or closed, unlike the suction and discharge valve with a gage port. The valve has some resistance to fluid flow, called pressure drop. It can be used anywhere a valve is needed. The larger sizes of this valve are soldered in the line; care should be used that the valve is not overheated when soldering.

seat on these valves. That is, if the valve is opened wide, the stem is sealed from the open valve body so there can be no leakage out the stem. This makes it possible to repack the valve when it is open, regardless of the pressure in the line. In case of a needle-type valve such as that shown in Figure 4-2, the back seat also makes it impossible to back the stem out of the threads so the ammonia pressure in the line will blow the valve stem out of the valve and release the ammonia. All valves for use with ammonia should have this feature.

Filters and Strainers

There is always the possibility of introducing some sludge, dirt, or scale into a refrigeration system. To prevent this from plugging small valve orifices, holding valve needles open, or scoring compressor parts, filters are used. All small expansion valves have a built-in screen ahead of them, Figure 4-8 and Figure 4-11. Larger valves may have individual filters placed ahead of each one, or there may be one large filter following the liquid receiver. Most of the larger filters are designed so that the filter element may be removed and cleaned without removing the filter body from the line, Figure 8-14.

Figure 8–12 How pressure can hold a packless valve closed.

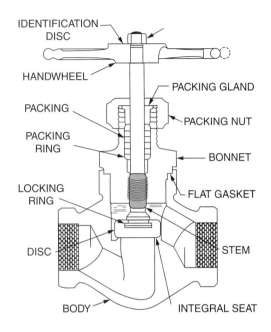

Figure 8–13 An ammonia glove valve. *Courtesy of Henry Vogt Machine Co.*

Figure 8–14 "Y"-type filter. *Courtesy of Henry Valve Co.*

Most compressors have a filter built into the suction of the compressor. If this is not the case, a suction-line drier is often installed in the line connecting the evaporator back to the compressor.

Ammonia systems do not use a filter following the receiver, but there should always be an adequate filter before each expansion valve. There is also a large filter just before the compressor, to protect it against any scale that might be picked up in the iron pipe of the evaporator or suction line.

Driers and Dehydrators

Moisture has caused so much trouble with refrigerants other than ammonia that driers or dehydrators have to be used. Filter driers do not remove moisture from the system. The moisture is simply collected in the drier, which, when completely saturated with moisture, must be physically removed from the system and replaced.

Moisture in a halocarbon system will cause corrosion. It can freeze out at the expansion valve to form an ice plug that will block refrigerant flow. In a hermetic compressor, moisture can damage the insulation on the motor winding. This damage is caused by moisture, which, when in the presence of halocarbon refrigerants and heat, facilitates the formation of acid.

A drier is a cartridge filled with a chemical that has a high affinity for moisture. Such driers may be sealed, or they may be made so they can be opened to facilitate the changing of the desiccant when it can longer absorb moisture and/or particulate matter (Figure 8-15). These driers may be put in the line at the time the system is installed or when it is opened up for servicing, then removed after absorbing any moisture that might have entered. Or they may be left in the system indefinitely. It should be remembered that a drier can hold only a given amount of moisture. If there is more moisture in the system than the dehydrator can hold, the

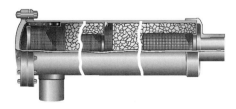

Figure 8–15 Angle-type dehydrator with cartridge, dispersion tube, and compression spring. *Courtesy of Henry Valve Co.*

excess will pass right through. Also, excessive heat against the dehydrator will release the moisture in it. In fact, the colder the dehydrator is kept, the drier it will leave the refrigerant, and the more moisture it can hold.

A drier works equally well in either the liquid or suction line if it is kept at the same temperature. Suction-line driers must be made larger because the restriction caused by a drier in the suction line can cause more reduction in capacity than a drier in the liquid line. For these reasons, a liquid-line drier is much more common than a suction-line drier.

Calcium chloride was the first moisture absorbent used. However, it dissolves in the moisture absorbed, and the resultant brine flows through the system with the refrigerant. This brine is highly corrosive and will do more damage than the moisture alone. Nonsoluble drying agents such as silica gel, activated alumina, or calcium sulfate are now available. They do a better job than calcium chloride, and they are much safer. These moisture-absorbing materials are called *desiccants*.

Ammonia holds the water in solution so it does not ice up at the expansion valve, and in these systems it is also much more difficult to remove moisture. Therefore driers are typically not used in ammonia systems.

Oil Traps

Lubricating oil is necessary on the cylinder walls of the compressor to prevent wear. It is necessary between valves and valve seats to make a valve vapor-tight. However, some of the oil vapor in the compressor will leave the compressor via the discharge line with the discharge vapor. This oil will be carried through the system with the refrigerant. Oil can contaminate the inside surface of both the condenser and the evaporator. This contamination reduces heat transfer.

Small halocarbon systems permit this oil to circulate, but the system is designed to flush the oil out of the evaporator and blow it back the suction line to the compressor. In larger systems it has been found best to install an oil trap to catch as much of the oil as possible, and return it directly to the crankcase.

Even when an oil trap is used, some oil will be circulated, and provision for its return to the compressor must be made. This is true when the evaporator is located below the compressor. Because oil is heavier than the refrigerant, traps are not needed when the compressor is located below the evaporator.

Figure 8–16 shows an oil separator. For halocarbon refrigerants it is necessary to reduce the velocity of the refrigerant, change its direction, and run it through a screen or similar device to catch the oil. The cooler the refrigerant, the easier it is to trap this oil fog. With halocarbons, however, the cooler the temperature, the more refrigerant the oil dissolves. Therefore it has proved to be best practice to place the oil trap as near the compressor as possible. A further help, particularly during the off cycle, is to insulate the oil trap. Some manufacturers put an electric heater in the oil trap. Anything that helps keep it warm helps prevent refrigerant from condensing in the trap and being returned to the compressor crankcase. The valve in the bottom of the oil trap acts exactly like a high-side float valve. Some oil must be left in the bottom of the trap, but as the oil level rises enough to lift the float, the valve is opened and the oil is forced back to the crankcase by the head pressure on it. Oil traps are located at the outlet of the evaporator coil when the coil is located below the compressor. Oil separators are located in the compressor discharge line to limit the amount of oil that travels through the system.

Oil traps have always been used as standard practice on ammonia systems. Figure 8–17 shows a typical ammonia oil trap. Here, the gas velocity must be reduced and the direction changed, but no screening device is necessary. Because ammonia will not mix well with oil, results are best if the trap is placed as far from the compressor and as near the condenser as possible. The cooler the oil refrigerant mixture, the greater the percentage of oil that will separate out.

Oil in low-temperature evaporators can be very troublesome. Plants should be arranged to keep oil in these evaporators at a minimum.

Modern high-speed ammonia systems drain oil back to the compressor crankcase, as is done in halocarbon systems. Oil separators in older ammonia systems have a drain valve as in Figure 8–17, which

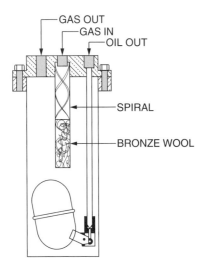

Figure 8-16 Low-pressure oil separator.

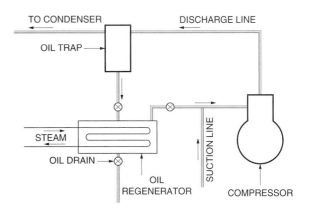

Figure 8-18 Oil regenerator.

must be drained periodically by hand. Sometimes the oil is drained directly to a bucket or container, but this oil has a small amount of ammonia absorbed in it. That makes this method of draining very smelly and unpleasant as well as wasteful. A method of eliminating this problem is to connect the oil drain to an oil purifier or regenerator, as shown in Figure 8-18. The oil is drained to the regenerator tank, then the oil line closed and a valve to the suction line opened. The suction pressure will remove most of the ammonia from the oil. To make the ammonia removal even more complete, a steam coil is sometimes put in the regenerator. The heat reduces the ammonia in the oil to a negligible amount. After the ammonia is removed from the oil, the oil may be drained from the regenerator. Both the oil trap and the regenerator usually have sight glasses so the operator can tell when draining is required.

Automatic Purgers

There is usually some noncondensable gas collecting at all times in an ammonia system. Noncondensable gas is any gas other than the refrigerant that might collect in the system. Common noncondensables that find their way into air conditioning systems include air and nitrogen. Because the compressor drives all the gas it handles over to the high side, and only liquid drains out of the liquid receiver, any gas that does not condense collects on the high side of the system. This gas adds its pressure to the normal condensing pressure, which increases the high-side pressure, and therefore the power costs associated with operating the equipment. Such noncondensable gas can be purged off by hand, but this is wasteful because some ammonia is lost with it.

Automatic purgers have been developed that separate the ammonia from the noncondensable gas and then bleed the latter out to the atmosphere. Figure 8-19 shows how this is done. Connections from the condenser or the receiver, or both, are taken to a refrigerated chamber. At the existing pressure and temperature in this chamber, any ammonia gas is condensed. The noncondensable gas gradually

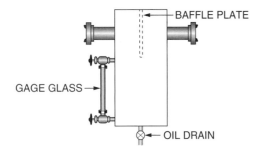

Figure 8-17 Ammonia oil separator.

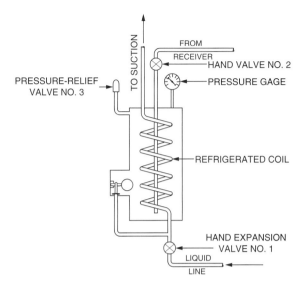

Figure 8–19 Automatic purger.

(1) To start, open hand expansion valve No. 1. This can usually be closed when sufficient high-pressure gas has been condensed to maintain operation through the float valve. (2) Hand valve No. 2 must be restricted so the pressure in the purger tank is 20 or 30 psi below the pressure in the condenser. (3) Pressure-relief valve No. 3 must be set for a pressure slightly below the condensing pressure. (4) Ammonia gas in contact with the refrigerated coil is condensed to liquid and returned to the suction line through the float valve. Air or other noncondensable gases do not condense, so will gradually collect, build up the pressure, and bleed off through the pressure-relief valve.

collects and builds up pressure to the point at which it bleeds out through a pressure-relief valve. The condensed ammonia is led to a low-pressure coil and re-evaporated to produce the chilling necessary to make the process continuous. A separate liquid line with an expansion valve must be connected to the coil to provide sufficient liquid to start operation, and occasionally to augment the condensing liquid. With a proper pressure–temperature balance, noncondensable gas can be eliminated with the loss of very little ammonia. Automatic purgers are made up in many designs, though they all work on the same principle: chilling and condensing the refrigerant, and bleeding off what will not condense. One such design uses the suction vapor from the evaporators to the compressor to do the required chilling.

Because centrifugal compressors operate at a vacuum, sometimes even in the condenser, their purgers include a small reciprocating purge compressor. This draws refrigerant from the condenser and discharges it to a purge device similar to the one described in Figure 8–19.

A water separator is also usually included in a centrifugal purger. Any accumulated water floats on top the liquid refrigerant and becomes visible in a gage glass. When water appears in the glass, it may be drained manually. See Chapter 14 for a detailed description of hand purging.

There is not so much need for an automatic purger with halocarbon refrigerants in a reciprocating compressor. Their discharge temperatures are low enough that there is no oil or refrigerant breakdown.

Sight Glasses

Small halocarbon refrigeration systems generally do not require sight glasses. On large halocarbon compressors, "bulls-eyes" have been put in the crankcase to show the oil level. Liquid indicators, Figure 8–20, are often put in the liquid line, either following the receiver or just before the expansion valve. The indicator will show if the refrigerant is low (bubbles or "milkiness" caused by many fine bubbles shows in the glass), but still will not show how much refrigerant is in the

Figure 8–20 Liquid indicator. *Photo courtesy of Noreen.*

system. Some systems will show bubbles in the sight glass even with a proper charge of refrigerant. It is best to refer to the information provided with the equipment.

Ammonia systems are generally better equipped. Sight glasses are common on compressor crankcases, liquid receivers, oil traps, purgers, and other pieces of equipment. Figure 8-21 shows the details of one of these. There is an automatic check valve in each line to the glass. This is supposed to close if the glass is broken. The hand wheels will close the valve if it is necessary to change glasses, to change the packing, or perform any other service operation. These glasses should be well protected by bars, screens, boxes, or other devices to protect them from being broken.

Heat Exchangers

Suction lines are insulated and suction vapors are carried back to the compressor with as little heat pickup as possible. This is because any heat added to the suction vapor expands its volume. The greater this volume becomes, the larger will be the compressor necessary to pump it. Also, the warmer the suction vapor entering the compressor, the hotter will be the discharge temperature. On small-diameter lines carrying a comparatively small amount of vapor, the heat leakage through an insulated line will heat the vapor somewhat in spite of the insulation. The effect of the insulation is not as pronounced as it is on larger lines, but nevertheless, it still worth having.

Part of this superheating can be made use of by using a heat exchanger shown in Figure 8-22. A heat exchanger chills the liquid entering the expansion valve as this warm liquid in turn heats the suction vapor. The chilled liquid will form much less flash gas, leaving more liquid for useful cooling. This type of system does the same amount of refrigeration with a little less refrigerant circulated. On the other hand, the warmer suction vapor from the heat exchanger will be a little nearer room temperature, so it will not pick up as much heat through the suction line. Thus, its temperature entering the compressor will be little, if any, warmer than without the heat exchanger. The colder the evaporator, the greater the savings will be. Heat exchangers are practically standard equipment on ice cream or frozen food cabinets.

As mentioned before, heat exchangers are also used in some cases to supply the superheat to operate a thermostatic expansion valve, as shown in Figure 4-16 and Figure 5-20. They will provide a more flooded evaporator, which gives better performance in this way.

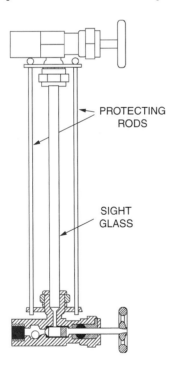

Figure 8-21 Liquid sight glass.

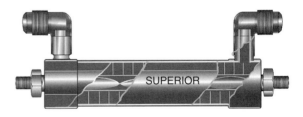

Figure 8-22 Heat exchanger. *Courtesy of Superior Valve Co.*

Questions

1. What is the purpose of the liquid receiver?
2. How large should the liquid receiver be?
3. Why are industrial receivers vented to the top of the condenser?
4. What type of pipe or tubing is most commonly used with halocarbon refrigerants?
5. What type of pipe or tubing is most commonly used with ammonia?
6. What two methods are used to make gas-tight joints in copper tubing?
7. Name two methods used to make gas-tight joints in iron pipe.
8. What prevents leakage through the packing of a halocarbon packed valve?
9. What prevents leakage past the stem of a packless valve?
10. What is the purpose of a drier?
11. What is the purpose of an oil trap?
12. How does a halocarbon oil trap differ from an ammonia oil trap?
13. What is the purpose of an automatic purger?
14. On what principle does the automatic purger work?
15. What is the purpose of a sight glass?
16. What is the purpose of a heat exchanger?

Chapter 9

Electrical Controls and Control Valves

Both controls and valves are used to control temperature as well as a number of other system conditions and parameters. However, controls and valves are entirely different devices, perform different functions, and should therefore not be confused with each other. A *control*, whether mechanical, electromechanical, or electronic, is a device used primarily to energize or de-energize an electrical circuit. A *valve* is intended to modulate (stop, start, or adjust) the flow of a fluid through a pipe. Either may be actuated by pressure or by temperature, or by another means. Electrical controls turn off the entire system or a part of the system by electrical means. They may stop the compressor, start a blower, or close a valve on part of the system, to mention just a few of the many tasks that can be accomplished. Control valves control the liquid flow in some part of the system. The greatest development and use of automatic electrical controls and valves has been in automatic small refrigeration. However, as the quality of these controls and valves has increased, they have been quickly adopted in ammonia work.

Expansion valves, discussed in Chapter 4, are one type of control valve. Their purpose is to keep the proper amount of liquid refrigerant in the evaporator. They have nothing directly to do with temperature control. The valves to be considered in this chapter are for the direct purpose of controlling pressure, which, in turn, results in temperature control, or for controlling temperature directly. For fully automatic operation, both an expansion valve and some form of temperature control are necessary.

Backpressure Control

The backpressure or low-pressure control is the most common form of electrical control used, shown in Figure 9–1. A flexible bellows is connected by tubing to the low-pressure or suction side of the system, usually directly at the compressor. The suction pressure in the system acts on the bellows inside the control. This pressure is opposed by an adjustable spring. The motion of the bellows is transmitted to an electrical switch that controls the driving motor. This device is intended to open its contacts upon a drop in pressure and close its contacts as the sensed pressure rises above the desired set point. An increase in temperature of the evaporator increases the low-side pressure, which expands the bellows against the spring. This closes the contacts on the circuit to the compressor motor. The operation of the compressor reduces the pressure in the low side and in the bellows. When the pressure (and temperature) is sufficiently reduced, the spring pressure collapses the bellows, which opens the switch.

For the control to operate satisfactorily, there must be a difference in pressure between the cut-in point (the pressure at which the contacts on the low-pressure switch close) and the cut-out point (the pressure at which the contacts open). This pressure difference is called the *differential*. The average temperature or pressure at which the control operates is the *range setting*. In Figure 9–2, the difference A–B, 20 psi, is the differential. The range is set to average 25 psig.

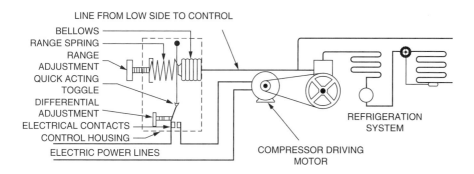

Figure 9-1 Schematic of pressure control.

The range setting may be changed by changing the pressure of the adjustable spring. The adjustment on this spring may be a screw available only to service personnel, or it may be a knob placed where the refrigerator owner may change it to suit. Any change in the range does not change the differential.

The differential may be changed by an adjustment that changes the gap between the pressures at which the contacts open and close, or that limits the movement of the arm operating these contacts. Changing the differential moves either the upper or lower limit, depending on the type or construction of the control. Such a change will obviously change the average of the range setting somewhat. This adjustment is nearly always inside the control. Therefore the control must be opened by a service technician to reach it. Figure 9-3 shows a typical backpressure control with the adjustments. The control contacts may be used directly up to 3/4 or 1 horsepower. A magnetic controller should be used on anything larger than this, and is also used on all sizes of three-phase power.

The principal advantage of the backpressure control is that it can be placed right at the compressor. With the control near the compressor, transmission lines between the system and the control bellows are short, and electrical lines from the control to the motor are short as well. There are, however, two disadvantages to the low-pressure control:

1. A cracked bellows, which, if not noticed immediately, could drain the entire refrigerant charge from the system.
2. Short cycling at certain irregular operating conditions. Short cycling means turning on, then off, every few seconds. This is very hard on electric motors and can be caused by leaking compressor valves.

Short cycling results when the high-side pressure leaks back through the valves during the off cycle

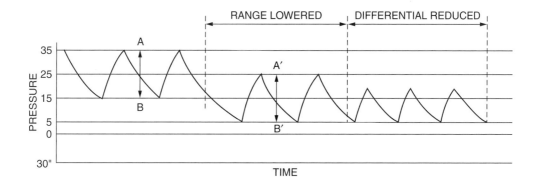

Figure 9-2 Effect of changing adjustments on control operation.

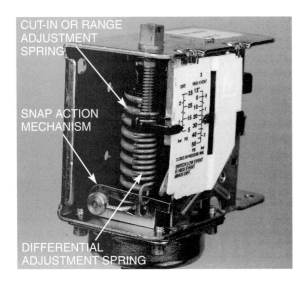

Figure 9-3 Typical pressure control.

high-pressure vapor that leaked into it is evacuated, the control shuts off the compressor. In such cases, the extra pumping on the evaporator sometimes causes too cold a temperature under light load conditions.

Temperature Control

Figure 9-4 shows a schematic diagram of a temperature control and Figure 9-5 shows the device. This is identical to the pressure control except for the method of applying the pressure to the bellows. A sensing bulb filled, at least part of the way, with a charge of saturated refrigerant is clamped to the evaporator, or hung in the room to be controlled. A change in sensed temperature of the coil or refrigerated box results in a change in the temperature of the refrigerant in the bulb. This then changes the pressure within the bellows, causing it to expand or contract. This bellows movement, in turn, causes the electrical contacts in the control to open or close. Many companies supply temperature controls and pressure controls that are identical except for the type of bellows. Naturally, adjustments are the same. In the case of a control to be clamped to the evaporator coil, a differential of 10°F to 25°F is used. If the thermostat must operate from room temperature, a much closer differential is needed, 2°F to 4°F. This device is typically an automatic reset.

Temperature controls are the type used on domestic machines. This includes not only the temperature element, but also a hand switch.

and raises the back pressure. This expands the bellows and immediately starts the machine. However, the machine first shut off because the evaporator was cold enough. So as soon as the machine starts, it pumps out the high pressure vapor that has leaked back to the low side, and then shuts off again.

A low-side float valve evaporator with a low refrigerant charge will do the same. The liquid level in the evaporator does not hold the float valve closed, and a mixture of high-pressure liquid and vapor enters the evaporator and raises the pressure. The control reacts to this and starts the compressor. The evaporator is cold enough, so as soon as the

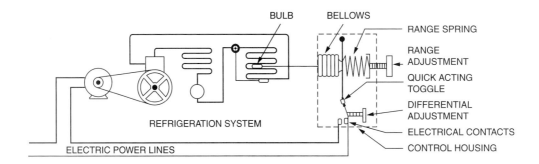

Figure 9-4 Schematic of temperature control.

High-Pressure and High-Temperature Cut-outs

A high-pressure cut-out or pressure-limiting device is similar in operation to a backpressure control. This device, however, is located on the high-pressure side of the system, shuts off (opens its contacts) as the pressure rises, and is often a manually reset device, as opposed to the automatic reset commonly found on the backpressure switch. This device is classified as an "open on pressure rise" device. It is used to shut off the compressor if the head pressure rises abnormally. Backpressure or temperature controls are available either with or without a high-pressure cut-out built in. A high-pressure cut-out is recommended on all water-cooled condensing units, and on air-cooled condensing units containing more than 20 pounds of refrigerant. A separate high-pressure cut-out (not combined with any other control) is nearly always used on ammonia installations.

A high-temperature cut-out or temperature-limiting device is a temperature control adapted to safety applications. It also opens the circuit upon rising temperature. Some ammonia systems are beginning to be operated fully automatically. In such cases the bulb of the temperature-limiting device is usually strapped to the side of the cylinder wall, or inserted in a location to check discharge gas temperatures. This will stop the compressor in case of a broken discharge valve, faulty lubrication, or other condition that could cause excessive overheating. This device may reset automatically.

Low-Oil Cut-out

Another safety device is a low-oil cut-out. This is a form of pressure control put into the oil line of a force-feed lubrication system. In case of no or low oil, the circuit opens and stops the compressor.

For many compressors, this must be a differential control. That is, it is made with two bellows, and the difference between the pressures in the two bellows operates the contacts. One bellows is connected to the suction pressure and the other to the oil pressure. The oil pressure must be a set to an amount above the crankcase's vapor pressure to permit the compressor to operate. With this setup, an abnormally high suction pressure cannot offset a low oil pressure and permit the compressor to operate.

Compressors with a low-oil cut-out must have some sort of time-delay feature to allow the compressor to start. Obviously, there is no oil pressure

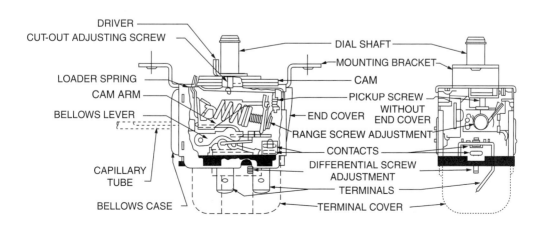

Figure 9-5 Domestic refrigerator control. *Courtesy of Ranco Inc.*

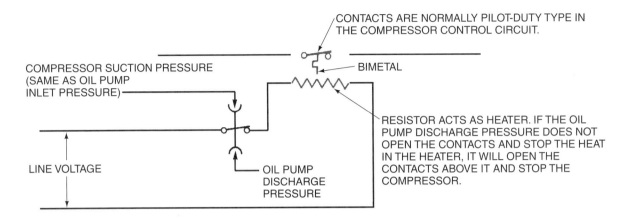

Figure 9-6 Oil pressure control.

before the compressor is started. It would be impossible to start the compressor if this control could not be bypassed. See Figure 9-6.

Low-Water Cut-out

Most large-sized systems, particularly ammonia systems, that use a water-cooled condenser have a low-water cut-out as a safety feature. This is a control very similar to a backpressure control. (Backpressure controls have in fact been used as low-water controls.) The pressure element is connected to the water system. If the water supply fails, the pressure in the water lines drops and the control opens the contacts, thus stopping the compressor. This makes it impossible to operate a system when the condenser water has failed.

Float Switch

As mentioned in Chapter 4, a float valve is a common device used to control the liquid level in evaporators or other liquid headers. Another method, which often makes more flexibility in piping possible, is the use of a float switch. A float mechanism opens and closes a set of contacts. These can operate a magnetic valve placed at the most convenient point in the refrigerant circuit, not necessarily near the float switch.

Float switches may be used to start liquor pumps to pump out liquor traps when they fill. Liquor traps are used in suction lines and provide a place to trap liquid so it does not enter the suction side of the compressor. Float switches may also be used to stop compressors if the refrigerant level in evaporators or accumulators reach a dangerously high level. This prevents floodback, which could damage the compressor.

Control Applications

Normally, either a pressure control or a temperature control can be used on any type of fixture or evaporator coil, but there are two exceptions. A temperature control must be used in conjunction with an automatic expansion valve or a capillary tube. It takes a changing pressure to operate the pressure control, but the automatic expansion valve is a constant-pressure device. Recall that the automatic expansion valve is intended to maintain a constant pressure in the evaporator, regardless of the system conditions.

Even a temperature control would be nearly as bad if the bulb was clamped to the evaporator near the valve. In practice, the bulb is clamped to the last coil of the evaporator. In Chapter 4 it was pointed out that the automatic expansion valve will gradually refrigerate farther and farther through the evaporator, until finally, if nothing stops it, it will refrigerate the suction line to the compressor. With the control bulb on the last coil of the evaporator, when liquid refrigerant enters this last coil, the control shuts off the motor.

The temperature control must not be set too cold for the expansion valve. For instance, if the expansion valve was set for 12 psig with R-134a, this gives a 10°F evaporator coil. If a colder box was wanted, and the control cutout point set to 5°F, the coil would never get cold enough to shut off the control, and the machine would run continuously. Thus, the automatic expansion valve is an exception to the rule that the expansion valve is not used to control temperature. It must always have a setting that gives a colder coil than the control setting.

Notice the similarity between the automatic expansion valve plus a temperature control and the thermostatic expansion valve. The thermostatic expansion valve has a bulb that clamps on the outlet of the coil. This bulb controls the flow of refrigerant to the coil. In this case, the bulb from the thermostatic control is better set anywhere on the evaporator except near the end. The control will then be affected by the refrigerant temperature in the evaporator, and not by any varying superheat at a point where there is no liquid refrigerant. The automatic expansion valve depends on a temperature control with the bulb clamped to the last coil. It takes considerable time before liquid refrigerant reaches the end of the coil. This chills the control bulb and turns off all refrigeration. The pressure control is also unsuitable for a system that use a capillary tube. Because the pressure balances in this system during the off cycle, the rising low pressure would immediately start the compressor again. Thus, it would run nearly all the time.

The thermostatic expansion valve or a low-side float valve will operate equally well when used in conjunction with either a temperature or a pressure control. The high-side float valve with a bleeder tube cannot be used with a pressure control because the balancing pressure would immediately restart the compressor. This short cycling can have negative effects that reduce the operational life of the compressor.

A temperature control operated by the evaporator temperature is used with domestic equipment. Commercial equipment may use a temperature or a pressure control. If there is more than one evaporator or fixture, a pressure control is a simple method of operating the compressor for the best average condition. If more accurate control of cabinet temperatures is desired, a temperature control operated by the cabinet air temperature is preferred.

Solenoid Valve

Figure 9-7 illustrates a direct-acting and a pilot-operated solenoid valve. Such valves are suitable for liquid lines, suction lines, water lines, or for any other fluid requiring control. The direct-acting valve is used for small-size lines. Solenoids can be either normally open or normally closed devices and are selected depending on their desired function. The normally closed valve is in the closed position when the coil is not energized. The plunger is lifted off the seat, and the valve opens, when the coil is energized.

For larger valves, a pilot-operated piston is used. Line pressure bleeds past the piston or through a bleed hole to the top of the piston and holds it closed. When the coil is energized, the small pilot valve is opened. This allows the pressure above the piston to escape to the low side. Line pressure then pushes the piston up to open the main valve port. When the magnetic coil circuit is broken, the pilot plunger drops and closes. Enough pressure then builds up behind the main valve disk so this pressure plus the weight of the disk closes the valve.

In refrigeration, these valves are usually operated by a thermostatic control, but they can be operated by float switches, backpressure controls, or any other device for making or breaking an electric circuit. Such a valve is sometimes used in a single-evaporator system, but is more commonly used where more than one evaporator is operated by the same compressor. Any number of boxes can be paralleled to one compressor (if big enough), and each box individually controlled as in Figure 9-8. They can all be operated at the same temperature, or at different temperatures, depending on the setting of each thermostatic control. A backpressure control can be used to shut off the compressor

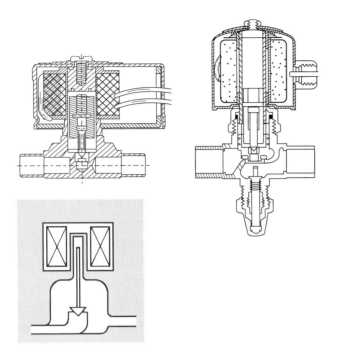

Figure 9-7 Solenoid valves. *Courtesy of Sparlan Valve Co.*

when all the solenoids are closed and the suction pressure drops. This control strategy is referred to as an automatic pump-down cycle. It typically occurs after the system is off on temperature or, in the case of a low-temperature application, after defrost. The space thermostat in the box is wired in series with the liquid-line solenoid. When the box reaches the desired temperature, the thermostat contacts open and the liquid-line solenoid valve is de-energized. This causes the system to pump down.

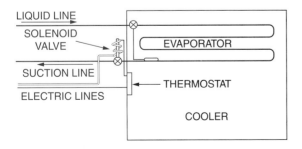

Figure 9-8 Solenoid valve operated by room thermostat.

When the suction pressure drops below the predetermined set point, the low-pressure control opens its contacts and de-energizes the compressor.

For a warm and a cold box operating from the same compressor, the warm box may be controlled by a solenoid valve placed as in Figure 9-9. The backpressure control on the compressor is set to give whatever conditions are required in the cold evaporator, in this case, a cut-in of 20 psig and a cut-out of 10 psig. If the solenoid is open, the warm evaporator maintains a high enough pressure to keep the control contacts closed and the compressor in operation. When the solenoid closes, all the refrigerant is pumped from the warm coil so it no longer refrigerates, and the compressor pumps only on the cold coil. When the cold coil pulls down to its cut-out point, the backpressure control opens and the compressor stops.

In Figure 9-8 the solenoid is shown in the suction line. It is often placed in the liquid line. A smaller (which means cheaper) valve may be used. Also, because the evaporator is pumped dry after

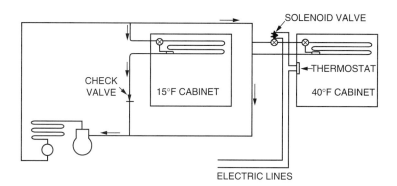

Figure 9-9 Solenoid valve in liquid line of two-temperature system.

the solenoid closes as there is no danger of flooding the compressor at the beginning of the cycle. Its one disadvantage is that it sometimes allows a coil or box to get too cold while pumping down after the solenoid closes.

The solenoid in the suction line gives instant control, as refrigeration starts or stops with the action of the thermostatic control. However, when a warm evaporator full of liquid refrigerant is suddenly opened to a low-pressure suction line, it will begin to boil so violently it is apt to "boil over" and flood back to the compressor. Also, if the expansion valve leaks at all during the off cycle, the evaporator coil fills to a point at which it will surely flood back at the beginning of the next cycle.

Evaporator Pressure-Regulating Valve

The evaporator pressure-regulating (EPR) valve is sometimes called an inline pressure regulator (IPR). It has also been called an evaporator-responsive regulating valve. In the past this valve was also called a backpressure-regulating valve, a constant-pressure valve, or a suction-pressure valve. However, these last terms could also describe a crankcase pressure-regulating valve as discussed in the next section, so these terms have become ambiguous. Figure 9-10 shows an evaporator pressure-regulating valve. Fundamentally, this valve prevents the pressure in the evaporator from going below a set value.

The EPR valve is placed in the suction line following the evaporator. The pressure entering the valve is applied to a bellows or diaphragm that is opposed by a spring. The valve is designed to open upon rising inlet pressure and close upon a dropping pressure. If the pressure in the evaporator rises, the valve opens to release this pressure to the suction line. If the pressure drops, the valve closes to maintain the pressure in the evaporator. This is a throttling-type valve. That is, it holds the evaporator pressure constant by opening exactly as required to balance a constant pressure. Therefore, there is no differential. Its range,

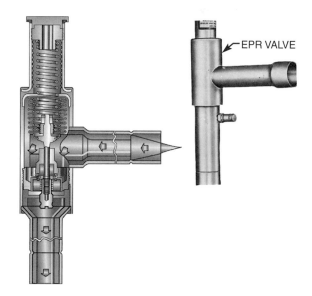

Figure 9-10 Evaporator pressure-regulating valve. *Courtesy of Controls Co. of America.*

or operating pressure, is adjusted by a screw that changes the spring pressure against the bellows or diaphragm. Turning the adjustment screw into the valve increases the spring pressure and, as a result, increases the evaporator pressure.

Figure 9-11 shows how this valve is used to maintain two coil temperatures. Coil 1 opens directly to the suction line. Its operating pressure is controlled by the backpressure control on the compressor. Coil 2 has an evaporator pressure-regulating valve between it and the compressor. The evaporator pressure-regulating valve is set for 25 psig to give a 25°F temperature in this coil. The backpressure control is set for a cut-out pressure of 10 psig. If it was to be set for Coil 1 only, its cut-in point would be about 20 psig. However, with the evaporator pressure-regulating valve in the circuit, the cut-in point must be above the valve setting. Otherwise, the pressure bleeding through this valve from refrigerant evaporating in Coil 2 will raise the suction-line pressure and short-cycle the compressor. So the cut-in point is set for about 30 psig. This will operate Coil 1 at an average of about 10°F. Thus, a 25°F and a 10°F coil are being operated by the same compressor.

This type of valve is sometimes used in conjunction with a solenoid valve, as shown in Figure 25-6. If a solenoid valve in a warm evaporator opened to a low-pressure suction line, the compressor could pull the coil down to such a low temperature that goods near the coil might be damaged by freezing before the rest of the box got cold enough. Or, if freezing did not take place, humidity could be lowered more than is desirable. This double valve arrangement is commonly used on warm boxes where a single large compressor operates on several boxes of different temperatures.

An evaporator pressure-regulating valve is sometimes used on a single coil to limit its temperature. Such an application is common on water coolers. The evaporator pressure-regulating valve is set for a pressure that gives a temperature above the freezing point of water. Then, regardless of the type of control on the compressor, or compressor running time, it is impossible to freeze and damage the cooling coil.

Crankcase Pressure-Regulating Valve

The crankcase pressure-regulating valve is similar to the evaporator pressure-regulating valve except that the outlet pressure operates the diaphragm. The crankcase pressure-regulating valve closes upon rising pressure and opens upon dropping pressure. It is a crankcase-responsive or downstream-responsive valve. It is used in applications where a start-up with a warm evaporator could overload the compressor, see Figure 14-7 and Figure 14-8. If this valve is set for 50 psig and the pressure in the evaporator is more than 50 psig, the valve will throttle down so the crankcase pressure does not go above 50 psig. Some suction vapor passes through the valve,

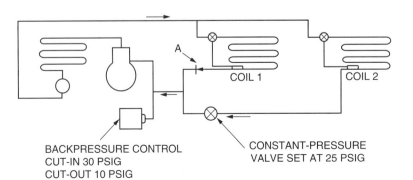

Figure 9-11 Application of constant-pressure valve.

gradually bringing down the evaporator pressure and temperature. When the evaporator pressure falls below 50 psig, the valve opens wide. Evaporator cooling from a high temperature is slower with this valve, but the compressor and motor are protected against overload.

Crankcase Heater

Compressors mounted outside (in the ambient) or in any area where the temperature surrounding the compressor may be lower than that of the refrigerated space must have crankcase heaters. Refrigerant vapors, if left unchecked, will always migrate to the coolest part of the system, where they can condense into a liquid. There may be a walk-in cooler (or any box maintained between approximately 38°F and 45°F) somewhere inside the building and its condensing unit located outside of the building. In the summer, the temperature inside the walk-in cooler will be lower than that of the ambient (in most cases). In the winter, however, the temperature of the ambient will be below that of the cooler most of the time. Also, it is likely that the load on the cooler will be less in the winter than it will be in the summer. During the compressor's off cycles, the refrigerant vapor in the system will condense in the compressor crankcase if the pressure/temperature conditions will allow this to happen. If the compressor's crankcase condenses enough liquid refrigerant, the compressor will not be able to start the next time the cooler's controls call for cooling. The compressor will burn out if this situation is allowed to continue. The proper way to avoid this is to install a thermostatically controlled heater in the compressor's crankcase. The temperature of the heater is set 10–15°F above that of the cooler. Under these conditions, the refrigerant vapor cannot condense in the compressor's crankcase and the compressor is protected. Some systems are designed so that the crankcase heaters are energized all the time, regardless of the ambient conditions. By wiring the heaters directly to power, there are no switches or contacts that can become defective and prevent the crankcase heater from doing its job.

Check Valve

A check valve should always be used at the outlet on the cold coil in a two-temperature system. In Figure 9–11, during the off cycle, Coil 1 will be somewhere between 10 psig and 20 psig. If Coil 2 gets warm enough to open the evaporator pressure-regulating valve, the vapor at 25 psig in this coil will flow back to Coil 1, because the latter is at a lower pressure. This vapor will come in contact with the colder liquid in Coil 1 and recondense. This will continue to produce refrigeration on Coil 2, but will not allow the pressure in the suction line to rise to the point that it would close the backpressure control. This will warm up Coil 1 to nearly the temperature of Coil 2 before the compressor does start. Worse, Coil 1 is now filled with refrigerant from Coil 2, which will flood back to the compressor at the beginning of the on cycle. A check valve placed at *A* will prevent this. Any refrigerant from Coil 1 cannot get to Coil 2, so it will continue to build up the backpressure until the control starts the compressor. A system with solenoids acts similarly.

Figure 25-2 and Figure 25–6 also show typical applications of all the above-mentioned valves.

Water-Regulating Valve

In Chapter 7 we noted that municipal water is sometimes used for condenser cooling. Where the water is not recirculated over a cooling tower, it is not economical or satisfactory to allow water to flow through the condenser at all times. A water-regulating valve, shown in Figure 9–12, is used to control this flow. A diaphragm or bellows opposed by an adjustable spring is connected to the high-pressure side of the system. This operates the valve mechanism so that water is supplied when needed. The spring pressure pushes against the high-side pressure of the system As the pressure rises, the valve is opened to allow a greater flow of cooling water through the condenser. As the pressure drops, the valve closes. The valve should be adjusted so that it shuts the water off completely

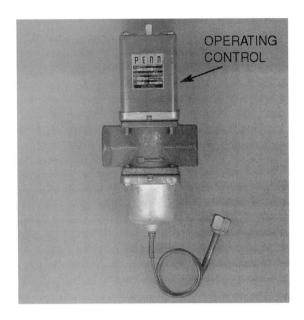

Figure 9–12 Water-regulating valve.

One of the simplest of these devices is the fusible plug. This is commonly used on the liquid receivers of most commercial equipment. It is also used on all refrigerant drums except the smallest sizes used for service operations. The heat of a fire will melt the fusible plug and release the refrigerant before excessive pressures develop. Naturally, however, this gives no protection against excessive pressures developed other than by heat.

A device to protect systems from excessively high pressures caused by any means is the rupture disk or rupture member, as in Figure 9–13. This is a thin sheet of metal clamped across an opening. The thin metal will break or tear out if excessive pressure is applied to it. This device may be used on liquid receivers, flooded brine coolers, compressor discharge lines, and other equipment.

Larger systems, particularly ammonia equipment, usually use pressure-relief valves, Figure 9–14. These are spring-loaded valves that open at a predetermined pressure, but that also close when the pressure returns to normal. On refrigerant-containing vessels such as receivers and brine coolers, they discharge to the air, or to a line to the roof. On compressors they are so connected that the discharge pressure is discharged back to the low-pressure side of the system, as shown in Figure 6–32. Therefore, if this valve opens, the refrigerant is not wasted. On chiller systems, for example, the rupture disk, which is a type of pressure-relief valve that opens when the sensed pressure rises above 15 psig, will vent the refrigerant to the atmosphere outside the structure.

during the off cycle. If there is much variation in water pressure, a pressure-reducing valve should be put in the water supply line. This should be set for a pressure a little lower than the lowest point reached by the varying water pressure. This keeps a steady flow of water through the condenser for a given valve setting, regardless of the water pressure.

Pressure-Relief Devices

In case of a fire in a building that has refrigerating equipment, the heat of the fire will raise the refrigerant pressure just like steam pressure in a steam boiler without an outlet. The result could be a severe explosion if nothing is done to prevent it. Also, excessive pressures can build up in parts of the system by closing the wrong valves, by excessive air in the high side, by reduced water flow through a water-cooled condenser, or occasionally by a part of the system trapping liquid refrigerant or oil, which warms and expands. Several different types of devices have been designed to release the refrigerant from the system in such cases before an explosion occurs.

Figure 9–13 Rupture disk. *Courtesy of Black, Sivalls and Bryson Inc.*

 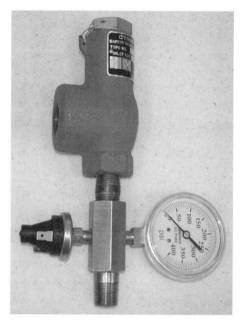

Figure 9–14 Pressure-relief valve: two-way valve assembly shows the only type of stop valve permitted ahead of a relief valve. *Courtesy of The Cyrus Shank Corporation.*

Computers and Refrigeration

Electronic controls are being used on more and more equipment and systems. Refrigeration systems are no exception. The subject of computers, programmable logic controllers (PLCs), and microprocessors could fill an entire separate volume. We only attempt to give a very brief introduction here. The microprocessor is essentially a small computer that is typically an integral part of a piece of equipment (compressor, condenser, recirculator, etc.). The PLC is a close cousin of the microprocessor but is usually not integral. Both the PLC and the microprocessor can interface with a central computer that monitors not only the building's refrigeration system(s), but virtually all mechanical building systems. The sophistication, control, and capabilities of such an arrangement are limited only by the availability of funds. The entire system taken together is described as *direct digital control* (DDC).

Such a system can be set up to monitor virtually every activity and condition in the system. This might include process temperatures and humidity, oil pressure and temperature, suction pressure and temperature, discharge pressure and temperature, motor winding temperature, capacity, flows, and on and on. Equally as important, if any parameters are exceeded that may damage equipment, the DDC system can be programmed to shut the equipment down and alert operating engineers via an alarm system. This may even include the use of beepers, cell phones, and home phones—all done automatically and electronically. Some systems may allow the operating engineer to take corrective steps from a remote location, or the system may even take some corrective actions on its own. Not too long ago, this would have all been considered science fiction, but today it is all quite real. The younger operating engineers and technicians have grown up in the computer age and have a knack for both mechanical systems and computer controls.

Energy Management

DDC systems are a natural choice to use to help manage energy and control costs associated with operating refrigeration systems. Electricity is commonly used as the power source to drive compressors, pumps, and other equipment. It is generally reliable, and electric motors require very little maintenance. However, the electric tariff(s) for most commercial and virtually all industrial operations are quite a bit more complex than those associated with private homes. Commercial and industrial tariff(s) are broken down into energy and demand. (Demand is also mentioned in Chapter 12.) The energy portion of the electric bill is tied to the energy that the utility has to purchase in order to generate the electricity. This could be in the form of oil, natural gas, coal, or other forms of fuel. The demand portion of the bill is tied to the capital costs of the electricity-generating equipment. Most of us use air conditioning equipment in the summer months. Refrigeration loads are heavier in the summer. So, the demand for electricity is greater in the summer than in the winter. Electric companies have capital tied up in equipment whether they are actually generating electricity or not.

Most electric utilities break the usage periods down into "peak" and "off-peak." Some even have an "intermediate" period. On-peak may be from 8:00 AM to 8:00 PM Monday through Friday (the times when many businesses are operating and need electricity). Off-peak usually includes all other hours. On-peak energy and demand rates are typically the highest. Off-peak energy and demand costs are generally the lowest. Intermediate rates (if available) fall between those of on-peak and off-peak. So it is obvious that electric utilities provide incentive to use electricity at night and avoid using it in the middle of the day. This is all very fascinating and offers the opportunity to reduce costs associated with operating refrigeration plants. *Demand metering* is sometimes used, typically when the power company uses the highest 15-minute demand period that occurs during on-peak hours and multiplies that demand by a fixed dollar amount. This dollar amount is then added onto the consumer's monthly electric bill.

Again, this is the subject of another book. However, we can give a brief introduction here. How can we turn equipment off in the daytime when we need to keep food, beverages, and other items refrigerated? In large refrigerated warehouses, there are often such large masses of goods that we can turn the system off or at least run the system at partial capacity during on-peak hours. The temperature of the product may rise by a degree or two, but this is acceptable as long as it does not rise above an established ceiling temperature. Manufacturing and quality control/quality assurance people can usually provide guidelines. When the off-peak period begins, we can add refrigeration compressors and/or lower the temperature set point and make up for the lost refrigeration time. The temperature of the product is maintained within acceptable parameters, but the total electric bill will be lower. Other strategies may include turning off electrically driven machines altogether during on-peak hours. Partial capacity can be maintained with gas- and/or steam-powered compressors during the on-peak hours. Large volumes of water can be chilled at night using the electrically driven equipment. This chilled water can then be pumped through the building's air handlers during the day to provide air conditioning. The serious energy miser may even opt to drive compressors by falling water and a water turbine (if falling water is nearby). This possibility might even prompt one to choose a site very carefully. There are as many money-saving strategies as there are operating engineers.

Questions

1. What is the difference between a control and a control valve?
2. Is an expansion valve a control or a control valve?
3. Why are both an expansion valve and another control device necessary on the simplest automatic system?
4. What is the difference between a temperature control and a pressure control?

5. Is an increase of pressure in the bellows of a control caused by an increase or a decrease of temperature?
6. What is the difference between range and differential?
7. What is a high-pressure cut-out, and what is its purpose?
8. With what expansion valves can a pressure control be used?
9. What is a constant-pressure valve, and what is it for?
10. What is the purpose of a crankcase heater?
11. Why is a check valve sometimes needed?
12. What is the purpose of a water-regulating valve?
13. What is the purpose of a low-water cut-out?
14. List the safety devices used on refrigeration equipment.

Chapter 10 Lubrication

Oil Types

Lubricating oils may be derived from animal, vegetable, or mineral sources. There is also a relatively new group of synthetic oils. Animal and vegetable oils do not perform very well in refrigeration applications because the oils can gum up relatively easily when introduced to impurities. Therefore, the overwhelming majority of refrigeration systems use either mineral oil or some type of synthetic oil.

The mineral oil family includes aromatic, paraffinic, and napthenic oils. Paraffinic oils contain wax, which can separate out from the oil and cause problems such as the formation of sludge and plugging of strainers. Most refrigeration oils are napthenic. The synthetic oils include glycols, esters, and alkylbenzenes. They all have their appropriate application(s).

Lubricating oils used in refrigeration systems may also contain additives to help improve the properties and lubricating effects of the oil. Some additives help control foaming. Others add stability to the oil and help prevent it from breaking down. Still others help the oil to maintain its viscosity (within limits) over a wide range of conditions.

The subject of lubricating oil (and lubricants in general) would fill an entire text. We offer some generic information here as it applies to refrigeration systems. Always consult with the compressor and systems vendors to help select the proper oil for a particular application.

Polyalkaline glycol (PAG) oil, as its name suggests, is a glycol-based synthetic oil. PAGs work very well in automobile air conditioning applications. PAG oils are hygroscopic. This means they have a high affinity for water and, once they absorb moisture, they tend to hold it under the conditions found in refrigeration systems.

Ester-based oils, unlike their mineral-based cousins, are wax-free. Wax is undesirable in a refrigeration system. At extremely low temperatures, wax may separate from the oil and cause problems. The overwhelming majority of applications involving HFC refrigerants and their azeotropes calls for some type of polyol ester oil (POE). Ammonia will also work with some POEs. New-generation refrigerants such as R-134a, R-410a, and R-407c, which are HFCs or HFC blends, rely on ester-based lubricants.

Alkabenzene oils have gained popularity in refrigeration applications in recent years. Alkabenzenes seem to perform well with most of the popular HCFC refrigerants and their azeotropic and ternary blends.

Most of the more popular CFC refrigerants perform well with mineral oils or alkabenzene. One notable exception is R-11, which works best with a mineral-based oil.

We have warned previously against mixing refrigerants. The same warning applies to lubricants used in refrigeration systems. Whenever an existing system is being retrofitted, follow the vendor's instructions carefully. The system may have to be cleaned and flushed several times. Failure to follow instructions may cause irreparable equipment damage.

Requirements

As with any moving piece of machinery, lubrication is an important consideration in refrigeration. Crankshafts, connecting rods, and pistons must be lubricated. This requires oil in the compressor. Compressor valves and all automatic valves in the system operate best with some lubrication. This means that some oil should circulate in the system.

Therefore, the problem is to lubricate all compressor parts and to circulate enough oil through the system to supply an oil film on all working valve parts. This circulated oil must be taken care of in some way. With ammonia, most of the excess discharged by the compressor is caught in an oil trap and periodically drained, as shown in Figure 8-17. That which goes on to the evaporator settles to the bottom, and must also be periodically drained.

Oil traps are also usually used in larger halocarbon systems. With or without an oil trap, some oil is carried by the liquid refrigerant to the evaporator. From the evaporator the oil is expected to be swept back through the suction line to the compressor. Here it is separated from the suction vapor by a drain, and allowed to flow back to the crankcase.

Regardless of the refrigerant used, oil in the evaporator introduces certain problems. It tends to thicken and coat the inside evaporator walls, reducing heat transfer. The colder the evaporator, the worse this condition becomes. At very cold temperatures, wax may separate out of the oil. Oil mixes in varying amounts with the refrigerant and alters the boiling temperature. This means that for a given evaporator temperature, a lower evaporator pressure is necessary because more oil is dissolved in the refrigerant. This penalizes the compressor. Thus, the oil circulated should be kept at a minimum. That is why oil separators are recommended for most halocarbon systems, although for reasons of first cost and incomplete oil separation, they are sometimes omitted. The oil separator is located in the compressor discharge line, thereby limiting the amount of oil that is circulating through the system at any given time.

Splash lubricating systems are used in many small open-type commercial compressors. The amount of oil circulated can be fairly well controlled by the oil level in the crankcase. The higher the oil level, the more oil is splashed up to the cylinders and pistons. This allows more oil to be carried through the compressor valves, and the refrigerant vapor carries it into the system. Removing the cylinder head allows one to see whether the oil is at the proper level in the crankcase. If the discharge valves are perfectly dry, the oil level is too low. If they are just moist with oil, the oil level is proper. If there is considerable oil standing on top of the pistons, the oil level is too high. Such a test should not be made after a quick pull-down. When the pressure in the crankcase is reduced rapidly, the oil foams up, and more of it is pumped up to the high side.

Force-feed lubrication provides more accurate control of oil distribution. There is less oil foaming on the starting cycle. The amount of oil delivered to the high side is designed for, and is not subject to, the variations possible with splash systems. Force-fed lubrication systems utilize oil pumps that force oil into the spaces between the mating surfaces inside the compressor.

Oil Foaming

Oil foaming has been previously mentioned several times. Figure 10-1 shows a solubility curve for a refrigerant in a particular oil. Values vary for different refrigerants in different oils, but these conditions are typical. In this case, if the refrigerant is at 20 psig pressure on this oil, about 50 percent by volume of refrigerant vapor will be dissolved in the oil. If the pressure is reduced to 10 psig, only 16 percent of the refrigerant will be soluble. The other 34 percent tries to escape immediately. It separates from the oil as bubbles of vapor. Thousands of tiny bubbles, or foam, form throughout the entire mass of oil. Naturally, this foam takes more volume than the solid oil, so it rises as in Figure 6-23. This is exactly the same action that occurs in beer when it is released from the pressure of the keg or bottle.

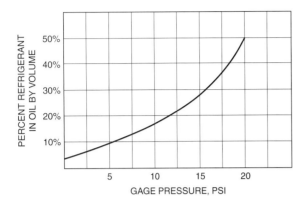

Figure 10–1 Solubility of a typical halocarbon refrigerant in oil.

Because a great deal more oil is soluble in the halocarbons than in ammonia, the former will foam much more than the latter. With the halocarbons, there will be a certain amount of foaming on the beginning of each on cycle, as the crankcase pressure is reduced. If all the gas is pumped out of the crankcase to do some work on the compressor, there will be foaming. Modern design has pretty effectively separated the suction gas from this oil foam by the use of check valves, baffles, or screen separators. This prevents the foam from reaching the cylinders and slugging the compressor.

Oil Selection

Selection of the proper oil for a particular refrigeration system is very important. Unlike the oil used in automobiles, the oil in a refrigeration system is not changed periodically. A proper oil does not have to be changed. Many thousands of hermetically sealed machines have been running for as long as 15 years on the same oil, and are good for many more years. So a proper oil can be chosen.

Some of the most important terms used to describe the properties of oil should be introduced. The *viscosity* of an oil is a measure of how thick or heavy it is. This is measured by the length of time it takes a measured quantity of the oil at a given temperature to flow through a standard orifice. A viscosity of 120 Saybolt seconds at 100°F temperature means that it takes 120 seconds for a given sample of the oil at 100°F to flow through the orifice. A viscosity of 200 Saybolt seconds means a heavier oil, because it takes longer for the measured sample to go through the orifice.

The *pour point* of an oil is the temperature at which a given amount of oil in a standardized vessel thickens to the point where it will sag but not pour out when the vessel is turned on its side.

The *flash point* of an oil is the temperature at which a flammable vapor is distilled from the oil. A flash point of 225°F means that a flammable vapor is distilled off at 225°F. It does not mean the oil will ignite at this temperature unless a blaze is provided to light it. Although this does not necessarily create a hazard in the system, the flash point is an indication that the oil will start to break down at this temperature.

The *cloud point* is the temperature at which wax will begin to separate from the oil. This wax separates as separate crystals thicken the oil. As these crystals form, their opaque white color gives the oil a cloudy appearance.

A properly refined oil for refrigeration is almost clear, usually with a yellowish tinge. Its viscosity is determined by the refrigerant used and by the evaporator temperature. All halocarbon refrigerants thin the oil considerably, so a heavier oil is used with them to give the required weight after it has been thinned; see Figure 10–2. Usually, viscosities and pour points are closely related. So, as evaporator temperatures go down, the viscosity must be decreased to get the required pour point. The thinning of the oil by the halocarbon refrigerants will also reduce the pour point and the cloud point. Therefore it is possible in some cases to accept an oil with a pour point or cloud point nearer to the evaporator temperature than would normally be considered safe. Ammonia does not have this thinning action on oil.

The flash point of the oil should always be above the compression temperature. Any breakdown of

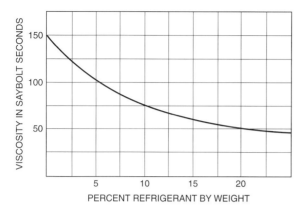

Figure 10-2 Viscosity of oil mixed with a halocarbon refrigerant.

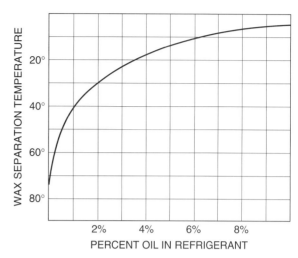

Figure 10-3 Effect of different mixtures of one sample of oil with refrigerant on wax separation temperatures.

the oil has several harmful effects. Noncondensable gases are formed that, if not purged out, will raise the head pressure. The remaining oil will have a higher viscosity, and sometimes a higher pour point. Some oils will begin to sludge in the presence of ammonia after the lighter constituents have been separated by heat. This sludge may or may not show in the compressor. If it does, it is apt to form on the cylinder walls and piston rings as a "varnish." If sludge forms, some of it will carry over to the low side. Here, as soon as it hits the cold coil surface, it thickens to a coal-tar consistency. This insulates the coils from the cold refrigerant. It has been known to stop up coils.

There is one very contradictory set of conditions to consider when choosing an oil. The colder the evaporator, the lower the pour point of the oil should be. However, a low pour point usually means a low viscosity, and a low viscosity is hard to achieve without getting a low flash point or breakdown point. The lower the evaporator temperature, the greater is the compression ratio between the low side and the high side. However, the higher the compression ratio, the higher will be the discharge temperatures, Figure 14-9 and Figure 14-11. Thus, unless two-stage compression is used, cold temperatures require an oil with a low viscosity and a high flash point, which is very difficult to get in the same oil.

Most refrigeration oils are sufficiently dewaxed for normal evaporator temperatures, but some of them are troublesome at subzero temperatures. Figure 10-3 shows how a typical oil reacts with regard to dewaxing at low temperatures. Notice, again, that the less the oil in proportion to the refrigerant, the lower will be the permissible temperatures. This is still another argument for an oil trap on the high side to keep the oil circulation down as much as possible.

Any oil for refrigeration purposes must contain no water. Every effort is made to keep water out of the refrigeration system. An oil with moisture present will undo all other precautions. The moisture in the oil is related to the dielectric strength, which is much easier to measure. A moist oil will break down in a high-voltage test sooner than a dry oil will. Not only must the oil be purchased dry, it must be kept dry. It will absorb moisture from the air if left in an open container.

Figure 10-4 gives some general recommendations for the properties of oil for refrigeration applications. However, these specifications alone are not enough. The oil supplier should be given all pertinent information, particularly evaporator temperatures, head temperatures, and the refrigerant to be used. Once a satisfactory oil is found, it should be used exclusively.

Viscosity	150–320 Saybolt universal seconds at 100°F
Pour point	−10°F (or lower to suit application)
Moisture	Not over 0.01% passes 25 to 30-kV test
Acidity, ASTM	Not over 0.01 mg KOH/g
Wax	Depends on evaporator temperature
Flash point	320–400°F
Sligh oxidation number	10 or less
Saponifiable matter	None
Sulfur	0.15% or less
Specific gravity	0.87 to 0.89
Color	White or pale
Carbon residue (Conradson)	0.005–0.01%

Figure 10-4 Oil specifications.

Questions

1. Is it desirable to keep all oil out of all parts of the refrigeration system except the compressor?
2. What are the objections to oil in the evaporator?
3. Compare splash-feed with force-feed compressor lubrication.
4. What causes oil foaming?
5. What is the meaning of pour point, and how is it measured?
6. What is the meaning of viscosity, and how is it measured?
7. What is the meaning of flash point, and how is it measured?
8. What is the importance of the cloud point?
9. How does the compressor discharge temperature affect lubrication?
10. Why should refrigeration oil be kept perfectly dry?

Chapter 11 Defrosting

An evaporator surface operating below 32°F will, unless prevented by special means, collect frost. Where frost is formed, some provision must be made to get rid of it. If this is not done, the frost increases in thickness and acts as an insulator between the air passing through the coil and refrigerant within the coil. Where frost is present, a colder refrigerant temperature must be carried to maintain the room at the required temperature. This colder refrigerant temperature requires a lower backpressure, which penalizes the compressor. This lower temperature, however, increases the rate of ice formation.

The exact amount of the insulating effect of frost is difficult to calculate. Pure ice is about 20 percent as effective as cork to reduce heat flow. However, frost on a coil will contain varying amounts of air, which increases the insulating value. Frost having an insulating value up to 50 percent of that of cork has been seen. That is, 1 inch of such frost will hold back the heat flow to the evaporator as much as $1/2$ inch of cork would. Frost found at only a few degrees below freezing (32°F) will be nearly solid ice. It forms more slowly at this temperature, but it is so solid that it is slower and more difficult to melt off. At colder temperatures, the frost becomes more like snow. At low subzero temperatures, the frost becomes very light and fluffy. It forms much more quickly, but it will melt faster because it is not so solid. Means of solving the frost problem can be divided into two methods: prevention, and defrosting or periodic removal. In all, defrosting is a necessary evil, a messy job, but one that must be done. To neglect it is to reduce the refrigeration available.

Defrost Cycle

For rooms that are kept above freezing temperatures, coils operating on a defrost cycle are common. Except for relatively higher room temperatures, a blower coil, as shown in Figure 5–17, is necessary for this. During as shown in the running cycle, the coil drops below freezing and collects frost. The fan blows the room air through the coil and cools it. During the off cycle, room air that is above 32°F is blown through the coil. This melts the frost. Some of it is re-evaporated into the room air, which helps maintain a high humidity. The rest of the melted frost flows out a drain to the sewer. This is called random or off-cycle defrost. This method of defrosting is very commonly used in commercial applications. This type of evaporator will fail to do any room cooling if it is allowed to fill with frost. Air circulation is necessary to get the necessary cooling. However, there is no air circulation if the evaporator fills with frost. Defrost cycles may be initiated at predetermined times via a timer or may be initiated by demand (when the coils become frosted).

Hot-Gas Defrost

A common defrosting method used with commercial and industrial systems is a hot-gas defrosting system. Here the hot gas directly off the compressor is fed to the evaporator. Figure 11–1 shows the elements of this system. By closing valve A

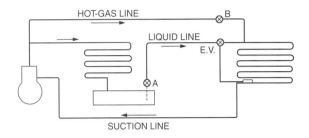

Figure 11–1 Simple hot-gas defrost system with one evaporator.

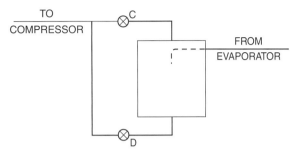

Figure 11–2 Liquid trap in suction line.

and opening valve B, the hot discharge gas from the compressor is fed directly to the evaporator. This method of defrost is typically found on systems whose evaporator and box temperatures are both below freezing.

This hot gas not only carries sensible heat to the evaporator; as it strikes the cold evaporator surface, it will condense and give up its latent heat. It pushes the cold liquid on ahead of it toward the compressor. This system has been used, but valves must be opened very slowly, carefully, and not too far. Otherwise the liquid pushed from the evaporator will slug the compressor badly.

If a tank such as shown in Figure 11–2 is put in the suction line to trap the liquid, slugging can be eliminated. During defrosting, valve C is opened and valve D closed. The tank will trap the liquid but return it to the compressor as it evaporates. After defrosting, valve C must be closed and valve D opened. Otherwise the trap will catch oil, which should be returned to the compressor.

One coil manufacturer has put a combination liquid trap and heat exchanger on the outlet of the evaporator. The coil can be rapidly drained of cold refrigerant without danger of slugging the compressor. Heat is applied to evaporate this liquid.

Figure 11–3 shows a common hookup when there is more than one evaporator in the system. Normally the liquid valve A and the suction hand valves C are open; and the hot gas valves B and the expansion valve bypasses D are closed. To defrost Evaporator 1, the liquid valve A and the suction hand valve C-1 are closed, and valves B-1 and D-1 are opened. The hot, high-pressure gas flows into the evaporator through valve B-1 and pushes the liquid out backwards through valve D-1 into the liquid line. Thus the liquid required for Evaporator 2 is supplied from Evaporator 1 instead of from the receiver. Evaporator

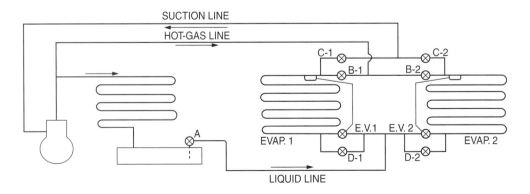

Figure 11–3 Hot-gas system with two evaporators arranged so that liquid from one is forced into the other upon defrosting.

1 acts as the condenser to supply this liquid. When Evaporator 1 is completely defrosted, the valves may be reversed and Evaporator 2 defrosted. Such a combination can be worked out for any group of two or more evaporators.

This method is very commonly used on large systems that are in the charge of an operating engineer. Defrosting can also be initiated by automatic defrost controls. The hot-gas system has been simplified for commercial installations by putting check valves in place of valves D. These are placed so as to prevent flow from the liquid line to the evaporator. Thus they do not interfere with normal operation, but they do allow flow from the evaporator back to the liquid line when the evaporator pressure rises due to the compressor, and the liquid-line pressure drops as a result of draining the line into the second evaporator. Thus the operator only has to manipulate the main liquid valve and two evaporator valves. The two evaporator valves are usually placed conveniently close together, which makes operating the wrong valve less likely. Or solenoid valves can be used that are opened by a defrost timer. Most modern systems operate in this fashion.

Thermobank System

The thermobank, shown in Figure 11–4, is another system developed to overcome some of the disadvantages of a simple hot-gas system. It consists of a water tank, re-evaporator coil, and hot-gas heating coil. During normal operation hot, high-pressure vapor from the compressor goes through the heating coil and then on to the condenser. Thus, the water in the tank is heated and stored until needed. In small systems, the suction vapor goes through the re-evaporator coil. In large systems, the re-evaporator coil is bypassed during normal operation.

The only valve that has to be manipulated is the one valve in the hot-gas line. It may be either a hand valve or a magnetic valve controlled by a time clock. The latter is recommended. To defrost the evaporator, this valve is opened. The hot, high-pressure vapor flowing into the evaporator condenses and defrosts the coil. The condensed liquid flows to the re-evaporator to be re-evaporated by the warm water. The holdback valve in this line acts like an automatic expansion valve to prevent overloading or slugging the compressor. Such a system can be installed so as to work on several evaporators multiplexed to one compressor. The defrost valve is placed in the branch line to the evaporator to be defrosted. Only one evaporator is defrosted at a time, but each evaporator can be defrosted in turn.

Water-Spray Defrost

Figure 11–5 shows a typical water-spray defrost system. A spray header fed by a special three-way valve is installed over the coil. The compressor and fan are stopped and this three-way water valve turned on. The water sprays over the frosted coils. As long as the water velocity is kept high, even cold water will not freeze, and it washes or melts the frost off the coils to the pan below. Melting must be sufficient that no chunks of frost or ice remain in the drain to plug it up. The valve is then turned to the drain position so that all the water drains out of the spray line and header so that these lines will not freeze later. If this kind of system is properly installed and intelligently operated, it is very satisfactory, even on freezer work.

Electric Heater Defrost

Electric heaters are very commonly used for defrosting domestic and commercial evaporators. Although the cost of electricity to operate heaters is appreciable, defrosting is sure and rapid. Evaporators can be defrosted quickly enough so that little or no heat escapes to the refrigerated space or cabinet.

Domestic no-frost refrigerators are defrosted electrically. The air-circulating fan is turned off when the heat is turned on. The water drains to a pan under the refrigerator, where it is evaporated by another heater. With this arrangement, the pan does not have to be emptied.

150 CHAPTER 11 Defrosting

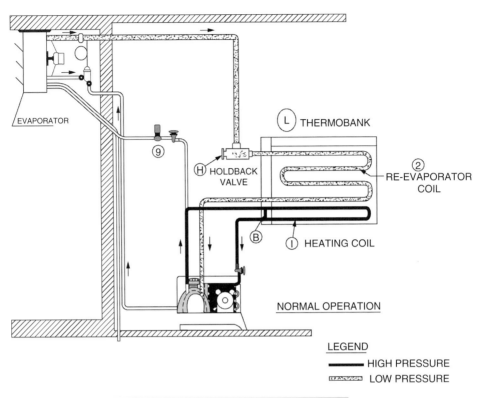

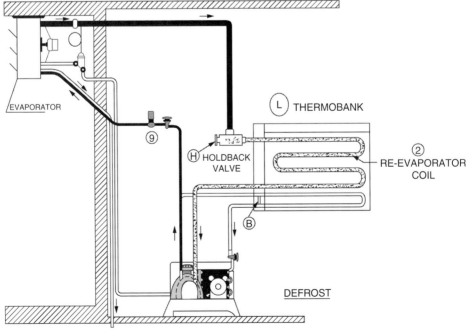

Figure 11–4 Thermobank system. *Courtesy of Kramer Trenton Co.*

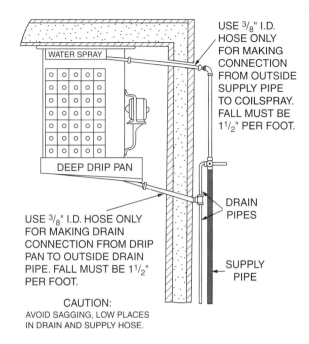

Figure 11-5 A water-spray defrost system. *Courtesy of Recold Corp.*

Open freezer cabinets are also commonly defrosted by built-in electric heaters. Drains must be supplied in these installations to carry away the water.

Defrost Controls

Defrosting may be done by hand or automatically. However, more and more automatic equipment is being commissioned and built. Many large industrial and some commercial evaporators are defrosted by hand. That is, refrigeration and fans, if used, are turned off. Valves are then opened in a hot-gas line or a water-defrost line, or a heater is turned on. When defrosting is completed, the valves or switches are returned to the proper settings for normal operation and the refrigeration and fans are turned back on.

Most commercial and domestic equipment now use some form of automatic control to operate the defrost cycle. Various systems are used. One of the simplest is a time clock that both starts and terminates defrosting. In domestic systems, defrosting once a day is usually sufficient. Commercial open display cases may need defrosting every few hours. The length of the defrost cycle must be long enough so that all frost melts off under the worst conditions. This usually results in a longer cycle than necessary under average or low-frost conditions.

Another defrost system uses a suction pressure control. As frost insulates the coil, suction pressure decreases. At a preset point, the suction pressure control initiates defrosting. This control is also set to restart refrigeration when the suction pressure rises to a value indicating a temperature above 32°F. This point will be reached only after all the frost has melted. Thus, this control adjusts the defrost cycle according to defrosting needs.

A third automatic defrost control starts defrosting by a time clock, but stops defrosting when suction pressure reaches a point indicating that no frost is left. This also matches the length of the defrost cycle to the need.

Any defrost control must make several changes in the system. These changes can be made through relays or by multipoint switches. First, refrigeration must be turned off. On a single-evaporator system, this means stopping the compressor. If more than one evaporator is on the same compressor, a solenoid valve, usually in the liquid line to the evaporator to be defrosted, is closed. The fan on the evaporator must be stopped. Then defrosting is started by opening a hot-gas line or a water line, or by closing a switch on an electric heater. At the end of the defrost cycle, all the actions must be reversed to return to normal refrigeration. See Figure 11-6 for defrost control wiring diagrams.

Brine Spray

Brine spray can be classified as a method of frost prevention. By spraying a brine or glycol solution over the coils, frost formation can be prevented. Table 23-4 gives concentrations and freezing points of sodium chloride (ordinary salt) brine, and Table 23-5, of calcium chloride. The proper brine concentration must be chosen to prevent the brine from freezing.

152　CHAPTER 11　*Defrosting*

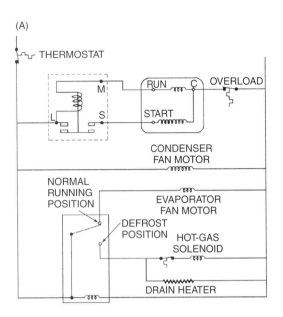

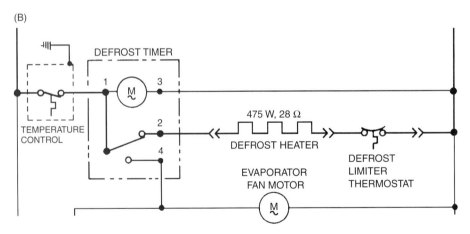

Figure 11-6 Wiring diagram for (A) hot-gas defrost and (B) electric heat defrost. *Courtesy of White Consolidated Industries, Inc.*

As the moisture that ordinarily causes frosting is absorbed in the brine, the brine is diluted. If nothing is done to correct this, it will become diluted to the point at which the solution itself freezes on the coils. Two methods are used to prevent this. The first is to add salt to the solution continually to maintain its strength and allow the excess solution that collects to drain to the sewer. This method, however, wastes salt.

The second method of maintaining the solution at the proper strength is to keep boiling off part of the solution (Figure 11-7). The solution is boiled down, or reconcentrated, in a separate chamber that may be heated using steam, gas, electricity, or any other convenient heat source. A heat exchanger is used to help heat the cold liquid from the cooler and to help cool the hot liquid from the concentrator. This method saves the salt or glycol,

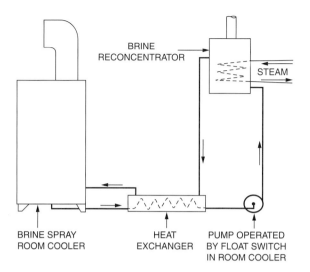

Figure 11–7 A brine reconcentrator.

but requires more equipment, requires a heat source, and requires more refrigeration to recool the liquid back to refrigerating temperatures. It is seldom used with salt or calcium brine, but has been used with glycols, which are too expensive to let drain to the sewer.

Questions

1. Why must frost be removed from evaporator coils?
2. How does a defrost cycle work?
3. Can a defrost cycle be used in a freezer box?
4. Briefly describe hot-gas defrosting.
5. What are some of the possible problems of hot-gas defrosting?
6. How does the thermobank system work?
7. Briefly describe water defrosting.
8. Can a cold-water defrost be used on a freezer box?
9. Briefly describe brine-spray defrosting.
10. Describe two methods of maintaining the brine strength in a brine spray.

Chapter 12 Compressor Drives

Requirements

The motor drive to supply power to the compressor equipment should be chosen with all requirements in mind. The primary requirement in a refrigerating plant is reliability. Where refrigeration is depended on to keep food in storage, a single breakdown or power failure at an inopportune time could cause the spoilage of food worth several times the cost of the equipment. This is true whether one is designing a large cold storage warehouse or a small commercial cabinet. Therefore it is of utmost importance that everything else be subordinated to completely *dependable* operation.

The second requirement is *economy*. This is of great importance in the installation of any equipment. Refrigeration equipment is bought as a business proposition. If its cost is greater than the financial return from it, it will not be used for long. The lower the cost, the more service can be rendered for a given price, or the more profit can be earned. It is important to remember that total cost includes many factors besides first cost.

First cost is an important part of total cost, but operating costs as well as maintenance and repair costs are also part of total cost. These latter can be more important than first cost if they become excessive, because first cost must be paid only once, but operating and maintenance costs continue for the life of the installation. It is very poor economy to jeopardize operating and maintenance costs to save a little on first cost. Consider the old adage, "You get what you pay for." Another part of economy is space economy. The more space is required for the driving equipment, the larger the required engine room, and naturally, the greater the cost. If the available space is limited, a larger engine room reduces the usable or pay space.

The third requirement for driving equipment is *simplicity*. The drive should be simple to understand, simple to operate, simple to maintain, and simple to repair. The easier and simpler it is to take care of equipment properly, the better the care that is usually taken of it.

A refrigeration compressor without unloaders usually requires a drive with a high starting torque. *Torque* is the actual twisting power of a motor, whether it is stationary or turning. A compressor that is not unloaded on the off cycle must start under load. That is, it starts to compress and build up a resistance to turning as soon as it starts to revolve. This requires a motor that has lots of twisting power, or torque, before it gets up to speed. Some types of drive have good starting torque. Others have almost none, and must be brought to speed before a load is applied or they will stall. Motors of this latter type may be used where the compressor is unloaded and can be brought up to speed before it starts pumping.

Electric Drive

Electric motors have proven to be the most satisfactory drive for average conditions. An electric motor is only as dependable as the power supplied to it, but power systems have become so interconnected that power failures occur extremely infrequently in most locations.

Electric motors run longer with less care than any other type of drive. Therefore they are one of the most dependable types of drive available. They are very simple. Their principal need is to be kept clean and dry, with sufficient oil in the bearings. Power cost can be higher than with other types of equipment, but this is usually offset by lower maintenance and repair costs. Space requirements are less than with any other drive. High-torque motors are available where they are required. These all point to the electric motor as being dependable, economical, and simple. Its application to all domestic, to practically all the commercial, and to well over half of all industrial drives prove it to be so.

Fractional-horsepower (less than 1 horsepower) motors are nearly all single-phase motors; that is, they can be run from the common two-wire system used for lights or small power requirements. The most common type of single-phase motor used at the present time is some form of induction motor. An *induction motor* has a set of field coils and a rotating armature, as shown in Figure 12–1. The armature may be made of copper or aluminum bars, sometimes called a squirrel cage, shorted so that each bar makes a coil or electric circuit with the next one. Or they may be wound coils that are shorted on each other. In either case, they are set in a laminated iron armature to provide a path for the magnetic flux or forces. The bars or windings in the armature act as the secondary of a transformer, while the stator or field coils act as a primary. Current is induced in the armature and no brushes are needed, hence the name induction motor.

A single-phase induction motor with one set of windings can be compared to a single-cylinder gasoline or steam engine that always stops on dead center. Therefore, it will not start by itself. A resistance split-phase induction motor is one of the most common methods used to obtain starting. Such a motor has two sets of field coils, a running winding and a starting winding. The running winding has a high inductance or choke coil effect and a low resistance. The starting winding has a low inductance and a high resistance. The greater

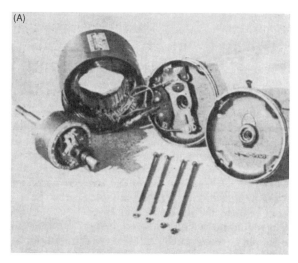

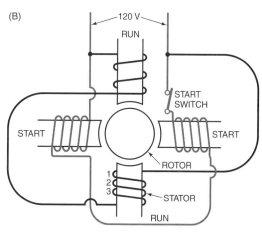

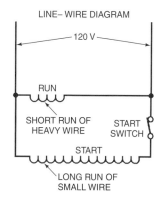

Figure 12–1 (A) Dismantled single-phase (split-phase) induction motor and (B) split-phase motor wiring diagram.

choke coil effect in one set of windings than in the other holds back, or causes the current to lag sufficiently, that the alternating current impulses do not act at exactly the same time. This gives an effect that can be compared to a two-cylinder gasoline engine (Figure 12-2). The angle between the cylinders is small, but it is sufficient to get the motor started. When the motor is started, both coils are in the circuit. When it gets up to speed, the starting winding is cut out of the circuit, usually by a centrifugal switch. The motor then runs on the running winding only.

Common, general-purpose fractional-horsepower motors are of this type. However, because of the small difference in electrical effect between the high-inductance and low-inductance coils (such as the small angle between the cylinders of Figure 12-2), their starting torque is poor. For this reason, a general-purpose or resistance split-phase induction motor is not satisfactory for refrigeration if the compressor does not unload.

A capacitor split-phase motor is the most common single-phase motor used for refrigeration systems that must start under load. This kind of motor works very similarly to the resistance split-phase motor, except that the capacitor acts just the opposite to the inductance or choke coil effect. This gives a condition similar to the gasoline engine of Figure 12-3. There is sufficient difference in the angle between the electrical effects of the two coils to give excellent starting torque (see Figure 12-3A). As with the resistance split-phase motor, the starting winding and the capacitor are cut out of the

Figure 12-2 Split-phase induction motor: gasoline engine analogy.

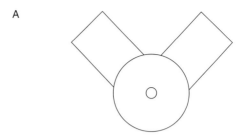

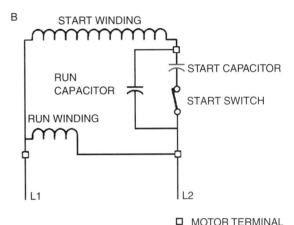

Figure 12-3 Capacitor motor: gasoline engine analogy.

circuit when the motor gets up to speed. This is an excellent single-phase motor for refrigeration use where there is no unloading. The start capacitor is wired in series with the start winding.

Potential Relay

Potential relays, also known as voltage relays, are used on single-phase motors that require large starting torque. During the off cycle, the contacts are closed to the start windings. This prevents arcing and possible burning of the contacts. When the thermostat contacts close, electricity is applied to the start and run windings in the motor. As the motor picks up speed, voltage is actually generated in the windings and causes a small amount of current to flow. As the motor approaches operating speed, this counter electromotive force becomes strong enough to open the start winding relay and power is no longer applied to the start windings. Power remains on the run winding and the motor continues to run.

Contactor and Motor Starter

Contactors are very similar to relays. Contactors are generally larger and can be rebuilt, whereas relays are usually replaced. Motor starters are basically contactors with overload protection built in. This overload protection supplements the protection offered by breakers. Breakers protect the entire circuit, while the protection built into the starter protects a specific motor.

Single-phase motors are available in either 110- or 220-volt varieties. In many single-phase motors the winding is in two sets of 110 volts each, shown in Figure 12–4. These can be hooked up in parallel for 110-volt operation, or in series for 220-volt operation. For $3/4$-horsepower or smaller motors, either 110 volts or 220 volts may be used, depending on the power available at the site. Smaller wire can carry the required power to a 220-volt motor without excessive line losses or voltage drops. Therefore 220 volts is preferred if it does not require installing additional electrical service.

For 1-horsepower or larger motors, 110 volts should not be used. Line voltage drops, particularly in starting, are excessive. With either 110 or 220 volts, it is best to run separate lines from the switch box to the refrigeration unit. There is always a surge of heavy starting current when a motor starts and comes up to speed. This causes a voltage drop in the feed lines, which will affect anything else on the same circuit.

Small, hermetically sealed motors are usually of the resistance split-phase type. They are unloaded by the capillary tube as it balances pressures during the off cycle. Commercial units to be used with an automatic or thermostatic expansion valve require a capacitor split-phase motor. Automatic and thermostatic expansion valves do not allow the system pressures to equalize during the off cycle, so when the compressor is started, there is a significant imbalance between the pressure on the suction side and the pressure on the high side.

Centrifugal devices, as used in open motors, cannot be used for starting motors contained within hermetically sealed compressors. This is because the arcing of the switch will burn the oil and break down the refrigerant sealed in with the compressor and motor. This can result in the formation of acid within the system. Therefore a starting relay that is operated by the starting surge is used, as shown in Figure 12–5. This device is referred to as a current relay, current magnetic relay, or simply a CMR.

Locked rotor amperage (LRA) occurs instantly, as the rotor is stationary when the motor is first started. It is typically five to eight times full-load amperage (FLA). As soon as the rotor begins to

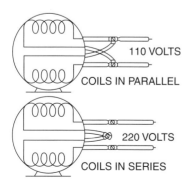

Figure 12–4 110–220-volt winding with proper hook-ups for both voltages.

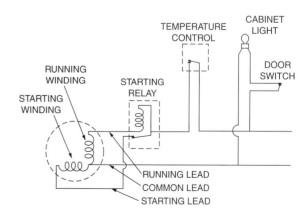

Figure 12–5 Domestic system showing starting relay.

move, the current drops quickly to well below LRA. FLA is the amount of current the motor should draw at its rated horsepower output. Running load amperage (RLA) is the amount of current the motor draws as it drives the compressor. It may be equal to FLA but is typically less than FLA, because refrigeration systems don't always run at their full-load rating. Refrigeration systems are usually oversized to accommodate temperature extremes and start-up requirements.

Upon starting, the surge of current (LRA) is sufficient to pull the magnetic contactor closed, which connects the starting winding to the active electric circuit. This facilitates the starting of the motor. As the motor comes up to speed, the current drops to a value too low to hold the relay contacts closed (FLA or RLA), so it opens the starting winding circuit. This relay is mounted outside the compressor case, and the start, common, and run leads are brought from the case through insulated, pressure-tight bushings.

As mentioned above, a single-phase motor can be compared to a one-cylinder gasoline engine. Like a one-cylinder gas engine, the power is in pulsations, which gives rise to considerable vibration. Therefore, single-phase motors are nearly always rubber- or spring-mounted so that this vibration is cushioned from the foundation.

Four-pole single-phase induction motors run at speeds of 1725 to 1740 rpm. The magnetic pull rotates at 1800 rpm. However, there must be some "slip" to create power. *Slip* is the speed at which the motor spins under load. The slip speed is always less than the speed at which the magnetic field rotates. Some motors run as fast as 1790 rpm under no load. Few drop below 1725 rpm unless they are overloaded. They are therefore practically constant-speed motors—that is, there is very little variation in speed, regardless of the load.

Two-pole induction motors have a full-load speed of about 3450 rpm. Small open-type motors are usually four-pole. Hermetically sealed motors may be four-pole or two-pole. There is a trend toward two-pole motors to get more refrigeration capacity in a smaller package. The formula for calculating the synchronous speed of an induction motor is

$$S = Hz \times 120/\text{no. of poles}$$

where

S = speed in rpm
Hz = frequency (in the United States, it is 60 cycles per second; in Europe, it is 50)
no. of poles = number of poles in the motor

From this formula, we can see that as the number of poles increases, the speed of the motor drops.

All hermetically sealed compressors, whether domestic or commercial, are direct-connected so that they run at motor speed. Open-type compressors may be direct-connected, or they may be driven by pulleys and V-belts at something less than motor speed.

Wherever possible, motors should be three-phase. Three-phase motors are available in fractional-horsepower sizes if needed. Naturally, a three-phase electric circuit is necessary to run any three-phase motor. The three-phase motor in a given power rating is smaller and usually cheaper than a single-phase motor. Three-phase power is generally more efficient and provides smoother power pulsations. Figure 12–6 shows a three-phase power diagram.

The autotransformer starter starts the motor at reduced voltage. Start-up current is applied to the motor windings through coils that act like transformers. This allows the motor to start spinning, but with very little power. This method may be used when very little starting torque is required. When the motor is spinning on the reduced voltage, the power is interrupted to the coils and the motor receives full line voltage.

The part-winding start system involves the three-phase motors with nine wiring leads. These motors consist of two motors in one and may be wired at different voltages, depending on what is available. First, one motor is started. Then, after a second or so, current is applied to the second motor's windings. This has the effect of starting two motors that each have half of the motor's full rating.

Up to about 50 horsepower, the three-phase induction motor is the type commonly used. This is similar to the single-phase induction motor except that there are three sets of windings in the stator. Because electric impulses come at different times in each of the three phases, no special starting

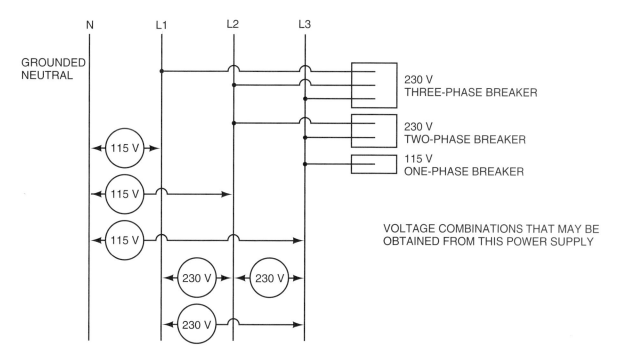

Figure 12–6 Diagram of a three-phase power supply.

devices or centrifugal switches are needed. The three phases are in the circuit at all times, which makes a simpler motor that is self-starting and runs more smoothly. Vibration is negligible. Its power pulsations can be compared to those of a three-cylinder gasoline engine.

As sizes go up, speeds usually go down. Some drives are belted, using V-belts. As motor speeds go down, direct-connected motors become more common. Motors are available that run on 220 volts, 440 volts, or 550 volts. As sizes go up, particularly for older types of compressors, manual operation with manual unloading on the start becomes more common. This makes standard rather than high-starting-torque motors usable. Modern high-speed industrial compressors with unloaders can also be started with standard motors. Where there is no unloading, high-starting-torque motors are still required.

Above 50 horsepower, either synchronous or induction motors can be used. A synchronous motor is usually more expensive than an induction motor.

The original synchronous motors required direct current for the field coils. This was typically supplied by a motor generator set or by a built-in exciter (a small generator on the motor shaft). The newer synchronous motors generate the necessary field current on the rotor itself. This is made possible by the silicon thyristor. Now we have synchronous motors without brushes. These motors can be used in applications where their "brushed" cousins cannot. The synchronous motor offers better efficiency and a better power factor than an induction motor. In fact, where other induction motors are on the line (induction motors have notoriously poor power factors), the synchronous motor can be operated so as to improve the line power factor.

A synchronous motor is something like a single-phase induction motor in that special starting means are necessary. The synchronous motor is usually started as a three-phase induction motor and has special built-in windings for this purpose. When it gets up to as near synchronous speed as it can on the induction windings, switches throw it into

Synchronous motors	3600	1800	1200	900	600
Induction motors		3450	1740	1160	875

Figure 12–7 Standard motor speeds on 60-cycle power.

synchronous operation. It then pulls into step at synchronous speed. With this type of system, however, the starting torque is very low. This makes unloading during starting essential. Thus, as sizes (and power bills) go up, so does the greater first cost of a synchronous motor, and more manual operation becomes justified to offset the power savings possible.

Synchronous motors run exactly at synchronous speed, with no variation due to changing loads. They can be selected to run at any speed shown in Figure 12–7. The speed is obtained by dividing 3600 (60 cycles per second for 1 minute) by the number of pairs of poles in the motor. As sizes go up, slower speeds are possible without exorbitant increases in cost. Therefore the proper speed for direct connection to the compressor is usually selected. Where electric motors are used to drive centrifugal compressors, the required speeds can only be obtained through step-up gears. Synchronous motors rely on the frequency of the power supply more than on the voltage.

Voltages available for synchronous motors are the same as for three-phase induction motors. Higher voltages are sometimes used, but as a rule they are not recommended because of the hazard involved.

Variable-Frequency Drive

Variable-frequency drives (VFDs) are relatively new compared to other motor drives. Prior to their development, it was generally accepted that induction motors had to spin at a fixed speed. VFDs changed this. When they were first introduced, they were quite expensive and generally cost-effective only for very large motors. The cost of VFDs has since come way down to the point that they are now found on even the smallest induction motors. VFDs allow induction motors to spin at reduced rpm. This results in considerable savings in energy and more convenient capacity control.

Where direct current is the power supplied, direct-current motors must be used. These are more costly than alternating-current motors. Brushes and commutators increase the maintenance required. A shunt-wound or compound-wound motor is usually used to obtain nearly a constant speed. The operating speed of these motors can easily be changed, so they do provide a simple method of capacity control. Direct current is standard on most shipboard work.

When considering electric drives, some thought should be given to the different power ratings available. Not only must the actual cost per kilowatt be considered, but also such factors as demand rate, power factor, and so on. *Demand* is the greatest amount of power required in a given time compared to the total power consumed. Large amounts of power used for only short periods can be more costly than more total power used at a steadier rate. Thus, although one large motor may be more efficient than two small ones, the power cost of one large motor cycled may be greater than that for two smaller motors, one cycled and the other running steadily.

Most power companies charge penalties for low-power-factor loads or offer bonuses for high-power-factor loads. The *power factor* is the amount of power available in a given circuit compared to the circuit's full potential. If current and voltage peak at the same time, the power factor is 100 percent (unity), but this rarely happens in practice. The current peak usually lags behind the voltage peak. This distance divided by the full distance between the current or voltage peaks is the power factor and is expressed as a percentage. Increasing power factor provides more cost-effective use of the electric distribution system, so the use of synchronous motors to raise the power factor may more than pay for the extra cost of this type of motor. If the horsepower requirement exceeds the speed, a synchronous motor should at least be considered. Or, in sizes for which a synchronous motor is hardly justified, an investment in static condensers or capacitors may be worthwhile. These are all factors that vary with different power rate schedules, but they should not be overlooked when selecting a motor. Static condensers have been applied to the compressors used on window air conditioners to reduce the total current needed. This makes it possible to put larger-capacity units on ordinary house circuits.

Steam Drive

Before electric drives were so well developed or power lines so dependable, steam was the time-honored power drive for refrigeration as well as other uses. In some cases it is used today, even installed in new plants. With proper equipment and proper care it can be as dependable as electricity. Fuel costs are usually lower than electric power costs. However, the larger amount of equipment necessary—boilers, steam engines, and auxiliaries—make first cost and space requirements much greater than with electric drives. The increased amount of equipment, machinery, and so on, increases maintenance costs and the potential for problems.

Where fuel is cheap and electric rates high, steam power may be justified. Steam power is particularly well suited where steam heat is used for processing. Boilers must be installed to generate the process steam needed, and the extra cost of generating steam at a higher pressure for power is small. The steam is first used in the steam engine, then the exhaust steam is used to do the required heating.

One problem inherent in steam-drive systems is the fact that the steam has to be condensed back to water before it can be used again in the boiler. This gives rise to a considerable amount of "waste heat" because of water's high latent heat. In some cases, the exhaust steam from the turbines or engines can be used in an absorption machine (or machines). In this way, two objectives are met: The steam is condensed; and the heat can also be used in the generator section of the absorber. Such an arrangement makes for more efficient operation. Wherever latent heat can be used, there is an opportunity to increase plant efficiency. Dairies, breweries, and such are excellent examples of situations in which such an installation can work out well. One other advantage of a steam drive is the ease with which the refrigeration capacity can be controlled by controlling the speed of the engine.

Many older steam systems using Corliss engines are still in use. Where new steam driven-refrigeration systems are installed, either a poppet-valve engine or a uniflow engine is usually selected.

Direct-connected steam turbines are a natural drive for centrifugal compressors because they are both high-speed machines. Turbine-driven electric generators that in turn drive electric motors on the compressors have been installed in some plants. However, these installations are very costly, and other considerations usually dictate the choice of equipment in such cases.

Diesel or Gasoline Engine

In some plants, diesel or gasoline engines have proven to be economical. Somewhat like steam plants, they are as dependable as the care taken of them. Their first cost, operating costs, and repair costs are usually much higher than with electric power, but fuel costs may be low enough, depending on the location, to offset these costs. This is particularly true where oil or gas is cheap and electric power expensive. Such engines may be direct-connected, but more often drive through a belt. The latter makes it possible to operate the engines at higher and more economical speeds.

There has been an increase in the use of internal combustion engine drives in what is called a "total energy system." The new buzzword used to describe many of these systems is *cogen* (short for cogeneration). Oil or gas but no electricity is purchased. Engine or gas turbine-driven electric generators supply the electricity needed. Electric motors from this electric system supply small power requirements. Large mechanical drives such as refrigeration use other internal combustion engines or gas turbines. Exhaust heat from these engines is used in exhaust boilers to heat hot water or to generate low-pressure steam to supply heating requirements for processes or to heat a building. The coolant pumped through the jackets of the engines may also be used as a heat source for an absorption machine. Steam turbines may also be used to drive refrigeration compressors, and the associated exhaust steam may be used as described previously.

Water-cooled air compressors typically operate at cooling-tower temperatures (85°F to 105°F). These temperatures are not high enough to use for direct heating, but this heat may be used as the source in a heat pump system (see Chapter 25). Instead of throwing the heat away, it can be used to heat a building. In large commercial buildings and skyscrapers, interior heat (present all year) may be used to heat the perimeter of the building via an air conditioning system. The possibilities are endless. In today's economic climate, we mustn't think in terms of "waste heat." We should always find a useful home for our excess heat.

Questions

1. What is the most important requirement for a refrigeration power drive?
2. Name two other important requirements of a power drive.
3. Why are electric drives so common in refrigeration systems?
4. What is the most common type of fractional-horsepower motor used for refrigeration at the present time?
5. How are hermetically sealed motors started?
6. What is the operating speed of most single-phase refrigeration motors?
7. How do three-phase induction motors compare with three-phase synchronous motors?
8. What is the purpose of static condensers?
9. Where, in modern plants, have steam drives proved very suitable and economical?
10. What is a total energy system, and how does it fit into refrigeration drives?

Chapter 13 Food Preservation

Refrigeration technicians sometimes becomes so engrossed in the technical aspects of the business that they forget refrigeration is only a means to an end. In most cases, that end is the preservation of foods. It is important that anyone who is responsible for the operation of a refrigeration system understand what the refrigeration is for and how to control it to get best results.

When short-time storage is required (average domestic and commercial applications), conditions are not critical and some variation from best conditions is allowable. For industrial storage, however, an understanding of food technology is necessary. Some products cannot be kept for more than a week, even under refrigeration. Others can be kept for months or, with proper freezing, for years. Some varieties of fruits and vegetables keep in storage better than others. Sometimes the same variety grown under two different climatic conditions shows marked differences in keeping properties. There is great variation in the temperatures and humidities at which different foods keep best. Sometimes other factors besides temperature play an important role in proper storage. Wrapping with treated papers, an atmosphere with an excess of carbon dioxide, and ultraviolet radiation are common aids to storage.

This book cannot provide an entire course in food technology, but a few guiding principles will be provided about the use of refrigeration to aid in the storage of food products. Most important of all, though too often overlooked, is that no food is improved by refrigeration. No food ever comes out of storage better than it went in, except for the one exception of the aging of meat. And in that case the aging is only controlled, not produced by the refrigeration. In the early days of refrigeration, the best food products were sold fresh and only what could not sell in competition with the highest quality products was put into cold storage.

Such food brought from storage and offered to the public gave a bad name to all cold storage products, yet cold storage has made fresh food products available in all communities all year around. Our present economy, with millions of people living in cities, some of which are long distances from food-growing regions, would be impossible without our present cold storage systems. Any good food products, properly stored, will leave the storage plant in good condition. Often they are in much better condition than so-called fresh foods that have taken a week or two to get from the farm, through commission houses to retail stores, and finally to the consumer, without refrigeration.

Natural Ripening and Enzyme Action

In any growing thing, several factors tend to promote spoilage. First is the natural ripening process. Any food picked green will gradually ripen. Anything picked ripe will become overripe. Fruits will become so soft that they are easily bruised or mashed. Vegetables become hard, woody, or pithy. Warm temperatures speed up this effect, whereas cold temperatures slow it down. Meats, however, are not affected in this way.

One result of this ripening process that is sometimes overlooked is that it generates a certain amount of heat, usually called *vital heat*. The amount of vital heat produced varies considerably among different food products. Also, because ripening occurs more quickly at warm temperatures, more heat is generated at warmer temperatures than cold ones. This heat is usually not enough to increase the load on the refrigeration plant significantly, but it does explain why the center of a closely packed stack of food in a cooler will remain at very nearly the same temperature for weeks, rather than dropping to room temperature. The vital heat produced by the food offsets the heat that escapes from the center of the pile.

All plant and animal products contain chemicals that promote deterioration or breakdown of the entity. These chemicals are called *enzymes*. During life, their actions are controlled by growth and metabolic processes. After harvesting or slaughtering, growth and life cease, but the enzyme action continues. This action turns green vegetables yellow and gives them "off" flavors. It is slow enough that it is not an important factor in short-term storage, but in frozen foods, which may be kept up to a year or sometimes more, it is of great importance.

Enzyme action can be controlled by heat, which destroys the enzymes. The enzymes in fruits, in most cases, cause no trouble in the frozen product. In meats, the enzyme action tends to break down or dissolve the meat tissue. The process is much more rapid than in vegetables. A certain amount of this action is desirable, and is used to tenderize the meat by holding it long enough for this action to take place. Enzyme activity, like all chemical processes, is slowed down by lower temperatures.

Bacteria and Molds

Bacteria causes rotting of food products. Bacterial spoilage takes place inside the food, under the surface. Molds cause spoilage as they form over the surface of foods. Molds attack the surface and may taint the food a short distance below the surface.

Bacteria and mold spores are always in the air. Most food products already contain a great deal of bacteria and mold spores. These microorganisms cannot be eliminated entirely. They can be destroyed by high temperatures—considerably above the boiling point—or by other germicidal action, but such actions modify or change the flavor of fresh food products.

The activity or growth of both molds and bacteria can be slowed down by lowering the temperature. At sufficiently low temperatures, usually considerably below freezing, their growth is stopped entirely. The temperature at which this happens varies, both for different food products and for different varieties of bacteria or molds. Although the growth of the molds or bacteria has stopped, they are not killed. They are still there and will immediately become active when the food is brought up to warm temperatures again.

Other Changes

If food is allowed to dry out or dehydrate, other undesirable changes take place. Fruits and vegetables wilt, causing them to look and taste unappetizing. Meats lose their juicy, fresh, red look upon drying. Both meat and vegetable products lose weight, which penalizes the commercial seller. If the vendor has to buy 100 pounds and it dries to 90 pounds, she has 10 pounds less to sell. The amount of drying or shrinkage depends on whether the food is stored in a dry or humid storage room.

Certain foods require an atmosphere that is not too humid. Dried foods of all kinds, nuts, onions, and cheese are some of these. High humidities decrease keeping qualities and increase surface molds.

The vitamin content of many foods deteriorates rapidly after picking. Temperature has a very material effect on this change. At cold-storage temperatures, most foods will retain from 40 percent to 70 percent of their vitamins. Some foods, such as corn and peas, also lose much of their sugar content if kept at ambient temperatures. This means a loss of sweetness or flavor. Cold temperatures reduce this loss.

Cold Storage

Because so many undesirable changes take place at ambient temperatures, storage at reduced temperatures is desirable for any perishable food product. Experience has shown that ripening and bacterial spoilage is cut approximately in half for each 18°F reduction in temperature. This means that for each 18°F reduction in temperature, the possible storage period can be doubled. Thus, the colder the temperature, the better the results, unless the foods involved have a low temperature tolerance. Certain foods, particularly tropical foods such as citrus fruits, bananas, tomatoes, avocados, melons, and pineapple, have such temperature tolerances, which means that other undesirable changes occur if temperatures are brought too low. Table A–2 in the Appendix outlines generally accepted storage conditions.

Mold growths are reduced by a reduction in temperature. Molds are also much less active in a dry atmosphere than in a moist one, but a dry atmosphere causes more dehydration than a moist one. So a dry atmosphere is desirable to hold down molds, but undesirable because of the resulting dehydration. The compromise that must be made depends on the type of food to be stored and its exact requirements. Sometimes, where high humidities are required and molds give trouble, ultraviolet light will eliminate nearly all molds. That is, high humidities are possible without molds by combining other food preservation methods with refrigeration.

To make cold storage most effective, food products should be chilled as soon as possible after picking and not allowed to warm up again until ready for use. In some warm climates, this fact is beginning to be accepted for all perishable produce. It could be applied with worthwhile results to practically all perishable products for all markets in warm and even in temperate climates.

Freezing

Many changes take place in food products when they are frozen. Some of these changes are beneficial, some are not. The biggest changes are the formation of ice crystals and the solidification of all other parts of the product. These do not happen simultaneously. Ice crystals begin to form around 30°F to 28°F. The freezing points of the other parts of the food are lower, but vary. The colder the temperature, the more is frozen solidly. Some food constituents are not completely frozen even at temperatures below 0°F.

This progressive freezing, first of internal water, then of other products, tends to separate the moisture from the other parts of the food. Also, the slower the rate of freezing, the more the water combines to form a few large crystals of ice. At rapid freezing rates, the water forms a myriad of microscopic ice crystals. Such large crystals have two undesirable effects on the food. They puncture or damage the individual cell walls, and they cause more complete separation of moisture from the other food constituents. Then, when the product thaws, the moisture, including soluble flavors, runs out of the punctured cells as juice. Fruits and vegetables become bruised, wilted, and usually discolored. Juices also run out of meat products, leaving them dry and tasteless. These results differ only in degree when stored products are frozen accidentally, or when fresh produce is sharp-frozen or quick-frozen improperly. Some popular producers of frozen food products utilize quick-freezing to improve the quality of the finished product.

When the freezing is done very rapidly, the ice crystals that result are so small they break through comparatively few cell walls and there is very little leakage upon thawing. Also, because the ice crystals are so small, each one thaws to such a tiny bit of water that it can be easily reabsorbed by the surrounding food matter. This leaves the product in a condition similar to when it was frozen.

Freezing has various effects on enzyme action. For meats, it is practically stopped. Therefore, meats must be aged or tenderized before freezing: They will not age while frozen. For most frozen fruits, enzyme action has no outward effect on the food's properties and can usually be neglected. For frozen vegetables, the enzymes present will in time destroy the chlorophyll (green coloring matter). This turns the product yellow and gives it a haylike flavor. The enzymes that produce this effect can be destroyed by heat. Vegetables are therefore blanched (heated) with scalding water or steam before freezing.

The lower the temperature of the refrigerated space, the slower is the evaporation rate. Therefore, the evaporation or sublimation (from a solid directly to a gas) from frozen products will be slower than from products kept above freezing temperature. The freezing is only incidental as far as the evaporation is concerned, and has no direct effect on its rate. It is the reduced temperature that makes the difference. Although drying is a slow process, frozen goods are usually held for long-term storage. A great deal of dehydration takes place over this extended period of time if no means of prevention is used.

The method that is most commonly used to reduce the rate of dehydration is to keep the food in a moisture-proof container or in a vapor-proof wrapping. The latter may be parchment paper, wax paper, or metal foil. A method that is commonly used for sharp-frozen fish is glazing. After freezing, the fish is dipped in water. The frozen fish is cold enough for a shell of ice to form around it. Any evaporation then takes place from the outside ice instead of from the product. Drying can also take place during the freezing process. Such drying causes a characteristic effect known as "freezer burn."

The great increase in production, popularity, and use of quick-frozen foods in recent years has been well justified by the quality of the product produced. However, freezing is only one link in a chain. This chain consists of the selection of proper food varieties for freezing, picking at the proper stage of ripeness, proper grading, proper processing, proper freezing within a few hours of picking time, and finally, proper storage.

Although "quick-frozen" foods have caught the public fancy, some of the above-mentioned points are of more importance than the speed of freezing. And for best-quality results, no link of the chain can be overlooked.

To repeat a statement made at the beginning of the chapter, food taken from storage (frozen or otherwise) will be no better than the food put in. Only if first-quality food is frozen will first-quality results be available. If it is done properly, however, the freezing process plus the other above-mentioned quality points make it possible to market food products that are far superior to the average fresh produce reaching our retail markets.

Questions

1. What is vital heat? Why is it important?
2. How do enzymes affect meat?
3. How do enzymes affect vegetables?
4. What is bacterial spoilage?
5. What is mold spoilage?
6. How does humidity affect foods?
7. Does humidity affect frozen foods?
8. How does reducing the temperature affect the keeping properties of foods?
9. Will the process of freezing a food make it keep indefinitely?
10. Is quick-freezing alone sufficient to give high-quality food?

Chapter 14 Operating

There are four principal requirements for operation of a refrigeration system: (1) to maintain the required temperature, (2) to obtain the required results economically, (3) to obtain the required results safely, and (4) to maintain the plant so that it will remain able to provide these results year after year.

Maintaining Temperature

A refrigeration plant must be operated so that required temperatures are maintained. On domestic and commercial systems, after the equipment is started and controls are adjusted, this is done more or less automatically. Occasionally some attention is required to adjust temperatures slightly, defrost coils, or to start and stop units as required. The balancing of the load to the system is taken care of by controls that cycle the unit as needed.

On a few large commercial systems, and on most industrial systems, more or less manual operation is necessary. The proper number of compressors must be kept running to balance the load. Constant backpressure must be maintained to maintain required temperatures. As loads increase and the backpressure begins to rise, another compressor must be started to hold the pressure and the temperature constant. If backpressures go down, one or more compressors must be shut off. Where only one compressor is available, it can be cycled. If clearance pockets or other capacity controls are available, they can be manipulated to balance the load.

When heavy loads occur suddenly, such as when milk starts over a milk cooler, a carload of warm produce is loaded in, or any similar event, refrigeration equipment to handle this load should be "on line" (running and ready) when the load hits. Otherwise, temperatures may rise too high while the equipment is getting started. It is much harder to reduce high temperatures than to prevent the temperature from rising above a point already established. An example may be found in your own home. If the air conditioning is running and the house is cool (assuming it's warm and humid outside), you feel better as soon as you come indoors. If you come in under the same circumstances with the air conditioning off and have to turn it on, it may be several hours before you feel really comfortable.

One thing that helps the operator during all load fluctuations is a condition that can be considered a self-regulating characteristic of refrigeration equipment. This self-regulation is apparent at two points, in the evaporator and in the compressor.

The heat transfer to the evaporator is proportional to the temperature difference between the cooler temperature and the refrigerant temperature (see Chapter 18 for a complete analysis). Thus, if the refrigerant is maintained at 20°F in a 30°F room, that is a 10°F temperature difference. Warm produce loaded into the room will raise the air temperature. Let's assume that the temperature in the room rises to 40°F. The temperature difference is now 20°F. This is twice the former 10°F temperature difference, so twice as much heat is transferred through the coils to the refrigerant. In this case, twice as many compressors will be needed to maintain the refrigerant at 20°F. Normally, of course, the room temperature is not allowed to rise this much, but this serves as an illustration. If there is a 10°F temperature difference between the refrigerant and the room, each one-degree rise in room temperature will increase the heat flow to the evaporator by $1/10$, or 10 percent. The opposite is also true. As the temperature difference becomes smaller, less heat will flow to the coil. If the refrigerant is at the same temperature as the room, no heat will flow to the coil. Thus, the room will get no colder. Thus, the more heat is on the evaporator, the more heat it absorbs. This is what is meant by the self-regulating characteristics of the evaporator.

The above example assumed that the evaporator pressure was held constant by using just the right amount of compression equipment to withdraw the vapor at exactly the same rate it was formed. If this pressure is allowed to vary, the compressor also has some self-regulating characteristics. A given amount of heat in the evaporator will vaporize a given amount of refrigerant. However, the volume of the evaporated vapor depends on its pressure. If the absolute pressure doubles, the density of the vapor doubles, and the volume of the vapor is cut in half. If the vapor has only half the volume, a given compressor will be able to pump twice as much by weight. The number of cubic feet per pound of vapor is cut in half. It is the weight of the refrigerant pumped that determines the amount of refrigeration produced (see Chapter 20 for a complete analysis). Therefore, a given compressor can produce twice as much refrigeration if the absolute pressure of the suction vapor is doubled. With only one compressor running, an increased load increases the backpressure, and the compressor pumps a larger amount of refrigerant. Thus, more liquid can evaporate to vapor in the evaporator. As it is usually stated, the higher the backpressure, the greater is the refrigeration produced. Figure 14-1 shows this in graphic form.

This condition can also work in reverse. With a decreased load the pressure will drop, and the volume of the vapor increase. Therefore the compressor will not remove as much refrigerant by weight as previously, and will not produce as much refrigeration. Thus, instead of getting colder and colder, each degree the evaporator drops requires more work from the compressor and makes it harder to do the job. Thus, a point will be reached at which the extra work required balances the drop in temperature, and the temperature will go no lower.

To depend entirely on these self-regulating characteristics would cause too much temperature variation in cold rooms, but they do help to smooth out small irregularities in the load without continual attention from the operator. And they also help correct extreme load variations when they do occur.

Economy

For any refrigeration equipment, required conditions need to be maintained as economically as possible. Part of the cost of operating such equipment depends on the nature of the original installation, but power and maintenance costs can vary considerably with operating conditions. Figure 14-1 shows the variation in refrigeration capacity and the variation in power for a given compressor operating at different suction temperatures. This graph is for ammonia, but the trends are similar for any refrigerant.[1] Notice that although it takes more power to operate a compressor at a higher suction temperature, the power curve does not rise as rapidly as the capacity curve. This means that more refrigeration is produced per horsepower at higher suction temperatures.

Figure 14-2 shows the same idea a little differently. This graph shows the horsepower required to produce the same amount of refrigeration at different

---

[1] The plot in Figure 14-1 is idealized in that it neglects variations in compressor efficiencies. Such variations may cause even larger changes than those shown on the curve. See Chapter 21 for a complete analysis.

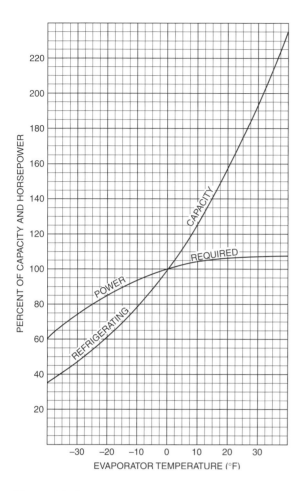

Figure 14–1 Capacity and power for a given compressor for varying evaporator temperatures: ammonia system, −80°F condenser, 100% based on a 0°F evaporator.

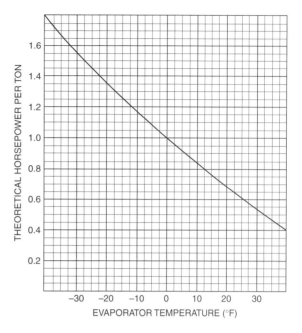

Figure 14–2 Effect of varying evaporator temperature on the theoretical horsepower per ton, calculated for ammonia with an 80°F condenser.

suction temperatures. Notice that the higher the suction temperature is, the less power is required, which means the less the cost to produce the power. The higher the suction temperature can be maintained and still maintain the required temperatures in the rooms, the cheaper will be the operation. The more coil there is, the higher the suction temperature that can be used to hold the room. Operators usually have no control over the amount of coil surface in the room, but they can at least be sure that all the evaporator coil is properly filled with refrigerant and active. A coil with liquid refrigerant in only two-thirds of its length is no more effective than two-thirds as much coil that is completely filled. Careful adjustment of the expansion valve is necessary.

It might seem backwards that opening the expansion valve, which allows the coil temperature and pressure to rise, actually increases the refrigeration done by the coil on the room, but such is the case. A review of exactly what happens shows why. An open expansion valve allows more refrigerant into the coil, making more of the coil surface effective in evaporating the liquid. This larger amount of evaporated vapor raises the backpressure, which allows the compressor to pump more vapor. The expansion valve should be kept as wide open as possible without running liquid back along the suction line. All these actions have more effect on lowering the room temperature than the slight lowering of the coil temperature caused by pinching down the expansion valve could possibly have.

Figure 14–3 shows the effect of varying condensing temperatures on the capacity and power. As the condensing temperature and head pressure rises,

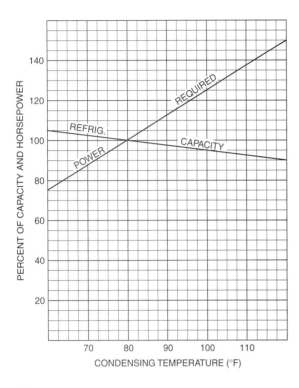

Figure 14-3 Effect of varying condensing temperature on capacity and power requirements of a given compressor, calculated for ammonia with a 0°F evaporator, 100% based on an 80° condenser.

the capacity is slightly decreased but the power is rapidly increased. Thus, the lower the head pressure, the more cheaply refrigeration can be produced. Under extreme conditions the condensing pressure could become so low that it is not sufficient to force the required liquid around the system and through the expansion valves, but such conditions are unusual in a plant with properly balanced equipment. It might happen in a poorly engineered plant, or under winter conditions with excessively cold condensing water. In general, the lower the condensing temperature is, the cheaper the operation. As the head pressure is lowered and/or the suction pressure is raised, the compression ratio is reduced. Lower compression ratios produce more economical operation.

In some plants, condensing water is purchased or is obtained only at high pumping costs. Beyond a certain point, the cost of excess water may exceed

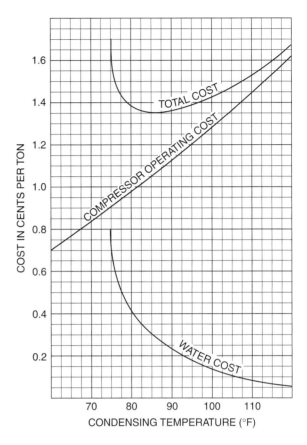

Figure 14-4 Operating costs for varying amounts of condensing water, calculated at ¾ cents per kilowatt and for 65°F water at 20 cents per 1000 cubic feet.

than the savings in power. Figure 14-4 shows how this can be worked out for an individual situation. The cost of the compressor operating at different condensing temperatures is plotted. The amount of condensing water needed to produce these condensing temperatures with existing equipment is then found by test or calculations (see Chapter 18), and its cost is plotted.

A third curve, which is the sum of these two curves, is then plotted. The compressor power cost decreases with lower head pressure, but the cost of water increases. The curve showing the sum of these two costs has a low spot that gives the proper operating condition for least expensive operation. Notice that this low point in Figure 14-4 is rather shallow, so anywhere from 82°F to 95°F condensing in this case could be taken as the proper operating condition. The steep rise below the 80°F condensing temperature occurs because, for any given condenser with a given condensing water temperature, there is a point at which additional water will pick up very little additional heat. Such a curve would have to be worked out for each individual case, using the individual costs involved.

One of the most common causes of too high a condensing pressure is noncondensable gas. Air may get into the system, either through leaks in a suction side operating at a vacuum or allowed to enter while part of the system was open for repairs. The breakdown of oil or refrigerant by compression heat, acids from oil, or moisture reactions may cause "foul gases." These will be sucked through the compressor with the refrigerant gases to the high side. Here the refrigerant condenses but the "foul" gases do not, hence their name of noncondensables. These noncondensable gases are therefore trapped between the compressor discharge valves and the liquid in the receiver.

In Chapter 1 it was pointed out that each gas acts independently of the others but creates a pressure equal to the sum of the gases' individual pressures. Thus, if a system using ammonia that normally condenses at 150 psig (164.7 psia) has air introduced that alone would be at atmospheric pressure (14.7 psia), the total pressure is 179.4 psia or 164.7 psig. The compressor then has to pump this mixture up to 164.7 psig to get the conditions that normally take place at 150 psig. The ammonia vapor will not condense until it reaches a vapor pressure of 150 psig, regardless of what other gases at what other pressures are present. Such excess pressures due to non-condensables often get much higher than the above example.

Two methods may be used to remove the air. The most economical method is to use an automatic purger, as shown in Figure 8–19. The time-honored method used before purgers were developed was to purge or let out this mixture of air and refrigerant by hand. To do this most effectively, the machine should be run long enough to ensure that all the air or noncondensable gas has been pumped to the high side. Then the compressor is stopped. (To continue to run it would just add fresh refrigerant to the mixed gases.) Water is allowed to run over the condenser until it is as cool as it will get. This condenses all the refrigerant possible. Then a valve or joint at the highest point of the system is opened slightly, or "cracked." This allows the mixture of refrigerant and noncondensable gas to escape. Fresh refrigerant re-evaporated in the receiver pushes this impure mixture toward the opening, as shown in Figure 14–5. The purging should be slow enough that the refrigerant from the receiver does not churn up the impure mixture it pushes out.

This method is still legal for ammonia. However, if the plant is located in a densely populated area, the foul gases and ammonia vapor need to be diffused into water. The water will hold the ammonia in solution and the foul gases will be allowed to escape. The current policy on releasing CFC, HCFC or HFC refrigerants to the atmosphere is *"de minimus"*—in other words, release none or as little as possible. The reader is encouraged to visit the website of the U.S. Environmental Protection Agency and to check local laws as well. Systems that use halocarbon refrigerants must be purged with a high-efficiency automatic purger. This purger must be maintained in peak operating condition.

At the beginning of the purging, the high-pressure gage will show a steadily decreasing pressure. This decrease becomes continually slower as the impure mixture escapes and the proportion of pure refrigerant to the total amount of fluid in the system increases. As soon as the pressure stops dropping, the valve should be closed. Do not purge so rapidly that the evaporation of the liquid in the receiver cools it sufficiently to bring the pressure of the pure refrigerant down. To do so and to wait for the pressure to stop dropping will eventually bring the pressure down to atmospheric. If there is a lot of air or noncondensable gas in the system, it may have to be purged several times to get rid of all of it. The compressor should be operated long enough between purgings to warm the condenser to a normal condensing temperature, then stopped and the water allowed to cool it off.

Another factor that may cause too high a head pressure is too large a refrigerant charge, shown in Figure 14–6. If the liquid covers part of the condensing surface, this does not leave sufficient

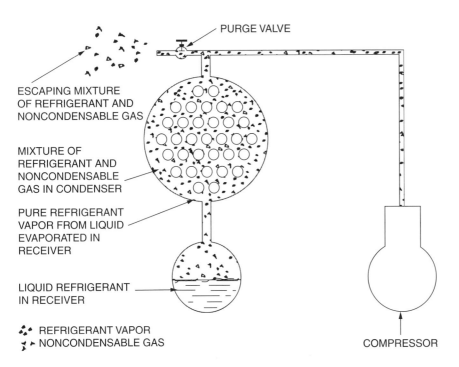

Figure 14–5 What happens when purging.

surface exposed to the vapor to extract the heat from it. It acts exactly like too small a condenser for the compressor. The only effective cure for this is to remove some of the refrigerant from the system. Refer to EPA requirements. Excess CFC, HFC, or HCFC refrigerants have to be transferred into a recovery system.

Figure 14–7 combines in one chart the amount of refrigeration produced by a given compressor under varying suction and condensing temperatures. Standard ton conditions, 5°F evaporator, and 86°F condenser temperatures are taken as 100 percent capacity, and the increase or decrease from that point is given in percent. Figure 14–8 shows the variation in horsepower of a given compressor operating under different conditions. Again, 100 percent is taken at standard ton conditions, and variations are calculated from there.

Safety

All the above required conditions must be maintained safely. This means that all operation and maintenance must be performed in a way that is not hazardous to the people working in the plant, to the equipment, to the plant, or to any inhabitants of the area. The first necessity is to be sure of the positions of all valves. Always be aware of the consequences of opening a closed valve or closing an open valve. In particular, never close a discharge valve on an operating compressor or start a compressor with this valve closed, unless a bypass is open. For water-cooled equipment,

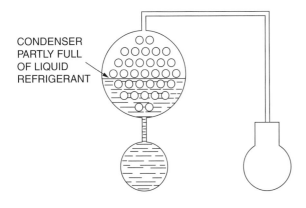

Figure 14–6 What happens with too much refrigerant.

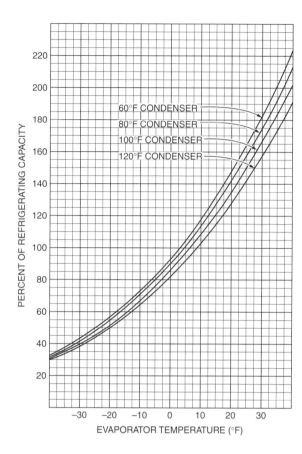

Figure 14–7 Refrigerating capacities at various suction and discharge temperatures: ammonia system, 100% based on standard ton conditions.

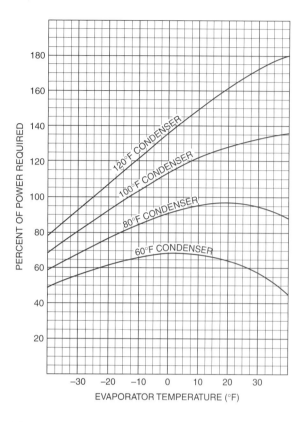

Figure 14–8 Variation in power required for different suction and discharge temperatures: ammonia system, 100% based on standard ton conditions (approximately 1 horsepower per ton of refrigerant).

always ensure that water is circulating in the condenser before starting.

Small machines may be easily started with all valves open. For intermediate-sized machines (5 to 25 horsepower), if the motor seems to labor on starting, closing the suction service valve makes starting easier. This is often necessary when starting a system with a warm evaporator. If the compressor need not pump any vapor, it will start much more easily. When it gets up to speed, the suction service valve can be cracked, then gradually opened. Be careful that the backpressure does not drop fast enough to start oil slugging. If that begins to happen, open the suction valve more quickly to let enough pressure into the compressor to increase the pressure on the oil.

Large machines without unloaders are started with the bypass open. Care must be taken that both high and low-side valves are not open at the same time. (This would allow all the high-pressure gas in the system to blow directly to the low side.) These machines are usually started with the discharge bypass valve and the suction valve open, and the suction bypass and the discharge valve closed, Figure 6–31 or Figure 6–32. When the machine is running at speed, the discharge bypass valve is closed. Just as this valve begins to seat, the discharge valve must begin to open. Otherwise, the compressor will be pumping against a closed valve.

As mentioned in Chapter 6, the clearance on a compressor head is so small that it will not safely handle any liquid. Therefore liquid refrigerant or excess oil must not be allowed to enter the suction. Proper manipulation or adjustment of the expansion valve is the first line of defense against liquid refrigerant returning. Surges or "boilovers" at sudden loads are best caught by liquid traps in the suction line. If these are not provided, the only safe procedure is to keep expansion valves pinched down.

This is not efficient operation, of course, but if the plant is so designed that it will "slop over," efficiency must be sacrificed to safety. In an emergency, to stop slugging already started, the only option for the operator is to throttle (partially close) the suction service valve. Unless it is absolutely essential, this valve should not be completely closed. To do so will allow the suction line to fill completely with liquid. Then, unless another compressor is available to pump it out, it is particularly difficult to get rid of, because to open the valve later feeds raw liquid directly to the compressor.

Two very important indications can tell the operator the condition of the returning vapor— whether it is wet, saturated, or superheated. The first indication is the frost line. On small, low-pressure machines, the suction line usually is not insulated. It will not frost out of the cabinet except on freezer jobs unless the vapor is wet. On larger machines, whether low-pressure or ammonia, particularly with insulated suction lines, usually sufficient vapor at a temperature below freezing is present to cause frost. This frost should end at the compressor and should not spread over a large part of the compressor castings. The latter condition indicates that there is liquid with the suction vapor.

A more exacting check of the condition of the compressor suction is the compressor discharge temperature. This is varied by three separate factors. First, different refrigerants have different discharge temperatures for the same operating conditions. Table 3-1 lists the discharge temperatures of the common refrigerants at standard ton conditions. Note that ammonia has the highest and R-500 and R-12 the lowest discharge temperatures. R-134a has about the same as R-12, and R-123 has a much lower discharge temperature than R-11. Note also that R-22 has a higher discharge temperature than any other halocarbon refrigerant. Second, the discharge temperature varies for the same refrigerant operating at different pressures. The more the vapor entering the compressor has to be compressed, the more work must be done on it, and the more heat is added to it.

Figure 14-9 shows the effect for ammonia at various evaporator temperatures but a constant 80°F condensing temperature. It is calculated for dry saturated vapor entering the compressor. That is, no unevaporated liquid is present, but the evaporated vapor has picked up no superheat.

The third factor is that the wet, saturated, or superheated condition of the suction vapor varies the above conditions. Figure 14-10 is calculated for ammonia with a constant 0°F evaporator and a constant 80°F condensing temperature, but for different suction vapor conditions. The saturation point shows that the temperature of the discharge gas is 205°F when the suction vapor is dry saturated. (This information can also be obtained from Figure 14-9.)

If the discharge temperature is lower than 205°F, unevaporated liquid is entering the compressor. This liquid evaporates as the vapor carrying it is heated by compression. The evaporation of

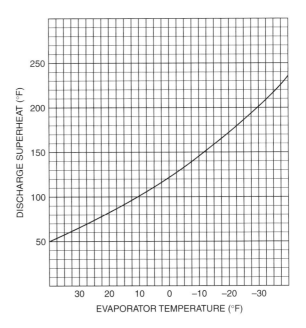

Figure 14-9 Effect of varying evaporator temperature on discharge superheat temperature, calculated for dry saturated suction and an 80°F condenser.

this cools the vapor so it drops to a lower temperature. If enough liquid is present, its evaporation brings the temperature of the discharge gas down to 80°F. In this case, an 89-percent quality vapor, which means 89 percent vapor and 11 percent liquid, would do this. If more than 11 percent liquid is present, the additional will not be evaporated, and it will slug. From the operator's point of view, knowing the percentage of liquid present is not important. When the discharge temperature drops and begins to approach the condensing temperature, the danger point is approaching.

The other side of the saturation point, superheat condition, Figure 14–10, is a measure of the superheat in the suction line. The warmer the vapor entering the evaporator at a given suction pressure, the higher will be the discharge temperature. For every 10°F the discharge temperature is above 205°F, the suction vapor has nearly 10°F of superheat. Any plant will usually have a small amount of superheat in the suction vapor, but this superheat should be kept to a minimum. Superheat increases the unit volume of the suction vapor, which decreases the capacity of the compressor and the refrigeration plant. Also, any added superheat, when added to the discharge vapor, increases temperatures to the point at which lubrication can begin to be a problem.[2]

The three factors mentioned above—refrigerant, operating temperatures, and condition of the suction vapor—all of which change the discharge temperature, may seem to create a confusing problem. However, with the help of a few curves, determining discharge temperature is actually quite simple. Figure 14–11 shows the theoretical superheat, or temperature rise above condensing temperature, plotted against compression ratio, for a saturated suction vapor entering the compressor. The various curves are for several different common refrigerants. These curves make it easy to find the discharge temperature for these refrigerants operating at different saturated suction conditions. The absolute discharge pressure divided by the absolute suction pressure gives the compression ratio. Using this value with the proper refrigerant curve in Figure 14–11, the theoretical discharge superheat above the condensing temperature can be found. Each degree of superheat in the suction line adds slightly more than one degree of superheat to the discharge temperature.

EXAMPLE 1 An ammonia system is operating at a suction pressure of 30 psig and a condensing pressure of 140 psig. The discharge temperature is 143°F. What is the condition of the suction vapor?

$$30 + 14.7 = 44.7 \text{ psia suction}$$
$$140 + 14.7 = 154.7 \text{ psia condensing}$$
$$\frac{154.7}{44.7} = 3.45 \text{ compression ratio}$$

From the 3.45 compression ratio on the ammonia curve, the discharge temperature is 90°F above the

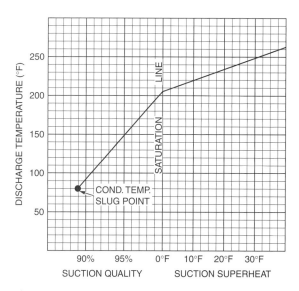

Figure 14–10 Discharge temperatures for various superheats in the suction vapor, calculated for ammonia at 15.7 psig suction and an 80°F condenser.

---

[2] There is some error in determining discharge temperatures by this method, but this error is small enough that it makes no practical difference in the results.

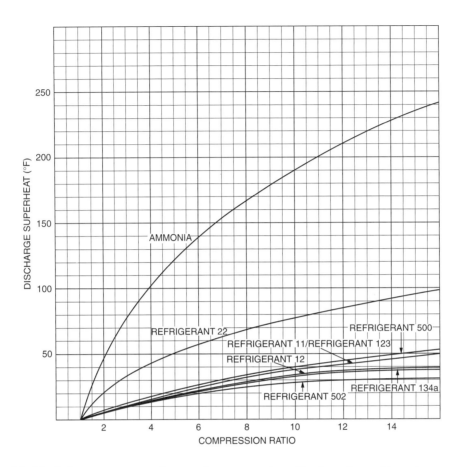

Figure 14–11 Discharge temperature rise at various compression ratios for various refrigerants.

condensing temperature if the suction vapor is saturated. Given the 140 psig pressure, the condensing temperature is 81°F. Therefore the theoretical discharge temperature is

$$90 + 81 = 171°F$$

Because a discharge temperature of 143°F is given, the suction vapor is slightly wet. This small amount of liquid returning is not enough to be hazardous if the conditions remain constant. However, if the discharge temperature is dropping, this is a sign that there are worse things to come.

EXAMPLE 2 If the discharge temperature of the machine working at the pressures of Example 1 rises to 205°F, what is the condition of the suction vapor?

In this case the discharge temperature is

$$205 - 171 = 34°F$$

above the theoretical temperature. Therefore the suction vapor must have nearly 34°F of superheat. This value is higher than it should be, and indicates a starved expansion valve, uninsulated suction lines, or a faulty compressor.

EXAMPLE 3 If an R-134a system is operating at the same suction and condensing temperatures as Example 1 and has a discharge temperature of 112°F, what is the condition of the suction vapor?

The 30-psig suction pressure of ammonia yields a 17°F evaporator temperature. The 140-psig condenser yields 81°F. An R-134a system has to be working at 16.3 psig and 88.3 psig for these temperatures.

$$16.3 + 14.7 = 31 \text{ psia}$$
$$88.3 + 14.7 = 103 \text{ psia}$$
$$\frac{103}{31} = 3.32 \text{ compression ratio}$$

This compression ratio and the R-134a curve shows a theoretical discharge superheat of 11°F. The theoretical discharge temperature is then

$$81 + 11 = 92°F$$

The discharge temperature given is 112°F.

$$112 - 92 = 20°F$$

If the discharge temperature is 20°F above the theoretical superheat point, the suction gas must have about 20°F of superheat. This is about normal for an R-134a system if the suction lines have at least some insulation on them.

EXAMPLE 4 A system using R-134a is running at an evaporator temperature of −20°F and a condensing temperature of 96°F (similar temperatures may be found in a domestic refrigerator). The discharge temperature is 145°F. What is the condition of the suction vapor?

From the thermodynamic tables, we find that −20°F yields 12.9-psia suction pressure. Now, 96°F corresponds to a discharge pressure of 130.8 psia. Then, $^{130.8}/_{12.9}$ yields a compression ratio of 10.14. From Figure 14–11, a 10.14 compression ratio corresponds to a discharge superheat of 34°F.

Now, we add 34°F to 96°F to get a theoretical discharge temperature of 130°F if the suction vapor is saturated. The problem statement gave a discharge temperature of 145°F. Thus, 145°F − 130°F shows that the discharge temperature is 15°F above theoretical, and therefore the suction vapor contains about 15°F of superheat. This is a good number for a system using R-134a.

A properly designed and operated system will not slug oil, but some older halocarbon compressors may slug when starting, as a result of the reduction of crankcase pressure. If this is a problem, lower crankcase oil levels will help. If this does not remedy the situation, stop the compressor as soon as it starts to slug. Give the oil a few minutes to subside, then restart the compressor. It may slug again but not so quickly or so violently. Stop for a few minutes more, then restart. Each start after a short rest period will remove more refrigerant from the oil, and soon make normal operation possible.

Compressors that are particularly prone to slugging oil on automatic starting cycles have been helped by keeping them warm. Manufacturers have inserted heaters in the crankcase to keep the oil warm. The crankcase heaters are wired to ensure that heat is being applied to the compressor. A small electric lamp burning directly under the crankcase has been known to help. The warm oil will not absorb as much refrigerant.

Part of the operator's duty with respect to plant safety is to keep an eye on the lubrication system. Low oil, plugged oil lines, broken oil lines, and so on, are all possible. If problems such as these are caught in time, the compressor can be stopped and the condition remedied. However, allowing a compressor to operate for long without oil will damage it, sometimes beyond repair.

Another duty of the operator is to see that coils are defrosted. Sometimes this must be done in cooperation with the plant manager to clear rooms or otherwise prevent damage to stored produce. Economy dictates that coils should be defrosted before the frost forms too much insulation on the coils. Another reason is that if frosting is allowed to build up indefinitely, it can reach a point at which its weight may pull down the coils. A simple method of defrosting bare pipe coils is to melt the frost from the coils using hot gas or other methods.

One more very important safety precaution is never to apply heat to a coil with refrigerant in it when it is closed off so the refrigerant cannot escape. Some disastrous accidents have happened as a result of closing both the inlet and outlet of coils such as in milk coolers, then turning a steam hose on them to sterilize them.

Automatic Refrigeration

The trend has been toward more and larger automatic plants. Automatic controls and valves were first applied to domestic and small commercial systems. As the reliability of automatic devices was proved on this equipment, they were applied to ever larger systems. All sizes of systems, both with halocarbon refrigerants and with ammonia, are operating automatically today.

This does not mean, however, that such large systems will run automatically with no more attention than is given domestic refrigerators. They must be checked regularly to forestall any condition that might cause trouble. If trouble does develop, automatic controls shut down the compressor and indicate trouble on a pilot light. Often, when this happens, a trouble signal is sent to a central office, and a maintenance technician is dispatched to the faulty installation to make necessary repairs.

The operators or maintenance people who maintain such plants must be better qualified than operators of days gone by. Formerly, operators maintained the mechanical equipment to the best of their ability. When trouble did occur, knocking or odors usually indicated a problem, and often indicated the source of the trouble. Today, with so much automatic and semiautomatic operation, it is critically important to find to eliminate conditions that might trouble before a problem develops. If a plant is down, the operator must be able to trace and check the automatic equipment to find why it is down, and do so without too much lost time. This requires more knowledge of the principles of refrigeration operation, and particularly of the automatic controls and valves. These automatic devices both keep the system balanced to its load and shut it down in case of trouble.

General Maintenance

A refrigeration plant must be properly maintained so it can continue to produce the required conditions both economically and safely over a period of years. Hermetically sealed domestic units need nothing more than to be kept clean and free of frost, and the condenser kept free of dirt and lint. Small commercial machines with open-type compressors should, in addition to the above, have the motor oiled and belts checked at least once a year. Twice a year is better.

Larger automatic systems should be given periodic inspections. All of the following should be done at least every six months:

1. Check the condition of the compressor valves.
2. Check the crankcase oil level.
3. If the oil level is low, is the evaporator full of oil?
4. Check all expansion valve settings.
5. Clean the condenser.
6. Check the flow of condenser air or water.
7. If the system is water-cooled, check the condition of the water drain.
8. Thoroughly check the entire system for leaks.
9. Clean the motor.
10. If there are motor brushes, check their length.
11. Check that motor lead insulation is not cracked or oil-soaked.
12. Take motor amperage readings and record them. Compare them to previous readings.
13. Check the motor bearing oil.
14. Clean control contacts.
15. Check control settings.
16. Check that insulation on control wires is not cracked or oil-soaked.
17. Check motor and belt alignment.
18. Check belt tension.
19. Check for frayed, worn, or cracked belts.
20. If there are any fans, check air circulation, fan oil, leads.
21. If there are any pumps, check water flow, lubrication, motor leads, packing glands.

In addition to the above thorough six-months' check, a quicker check should be made monthly. This need not cover all the above items, but it should include a check of all points of lubrication and a thorough check for leaks. For a system of any size, if one leak is caught before it becomes bad, it will save more than the cost of several years of monthly checking. The monthly check should also include a check of all cabinet or fixture temperatures.

If something is not right about these, then further checking should be done.

Automatic equipment, if given proper maintenance, should not need a complete overhaul for many years. Systems that have proper care can run for 10 or 15 years with nothing more than replacement of worn valve plates and shaft seals. Careful checks can allow most problems to be "nipped in the bud."

In large plants, these checks may be made by qualified operators. In small plants or commercial systems where there is not enough work to justify a permanent, adequately trained employee, a contract service may be used. Such service is supplied by many reputable refrigeration service organizations. For a monthly fee, the service organization contracts to check the system periodically and make necessary minor repairs and adjustments. The owner is notified of the need for any major repairs. Or the service contract may include any and all work necessary at no additional cost to the owner (for a higher fee, of course). Such contracts can give any owner of refrigeration equipment the benefit of proper inspection and maintenance by trained personnel.

At one time, large industrial machines that were operated 24 hours a day for months at a time were partially disassembled on an annual basis. The wearing parts were gaged or measured accurately. A record of the wear on these parts was kept so that it was possible to estimate whether they would last through another season. If the parts were significantly worn, the machine would be rebuilt. These machines could then be depended on to work through a busy season under 24-hour operation without fear of a breakdown. Most modern, high-speed equipment has hour meters. A great many maintenance functions are performed after a predetermined amount of running time(s). Most of the necessary maintenance procedures are recommended by the manufacturers/vendors.

In addition to checking the equipment itself, the condition of the refrigerated space must be watched and kept in good repair. Insulation of both rooms and lines must be kept in good condition. Any damaged places should be properly patched before frost can get under the insulation and cause serious damage. Insulation must be kept dry and protected from damage. Doors must fit properly, and any wear in the hinges or latches must be corrected. Door gaskets must be kept in good condition or replaced. Most freezer door seals have heaters. These must be checked and maintained in good working order.

Questions

1. Why is an operating engineer employed in most industrial refrigeration plants, but not in commercial operations?
2. Why should backpressures be maintained as high as possible?
3. Does closing down on the expansion valve with the same compressor capacity raise or lower the backpressure?
4. Does closing down on the expansion valve with the same compressor capacity raise or lower the room temperature?
5. Does the cost of condensing water have any effect on the economics of operation?
6. What keeps the temperature from continually getting colder and colder when the room load diminishes but the compressor capacity is not reduced?
7. Why should condensing temperatures be kept as low as possible?
8. Why should noncondensable gas be eliminated?
9. How can noncondensable gas be eliminated?
10. What would happen if the discharge service valve of an operating compressor was closed?
11. What would happen if a water-cooled system was started with no water flowing through the condenser?
12. How can bypasses be used to aid in starting?
13. Why should liquid slugging be guarded against?
14. An ammonia plant is operating at 0-psig suction and 155-psig discharge pressure. The

discharge temperature is 290°F. What is the condition of the suction vapor?

15. If the discharge temperature of Question 14 drops to 140°F, what is the condition of the suction vapor?

16. If an R-134a plant is working at 30°F suction temperature and 90-psig discharge pressure, and the discharge temperature is 84°F, what is the condition of the suction vapor?

17. If the discharge temperature of Question 16 rises to 140°F, what does this indicate?

18. A package water chiller that uses R-134a is running with a 35°F evaporator and a 96°F condensing temperature. The discharge temperature is 130°F. What is the condition of the suction vapor?

19. Briefly describe a desirable commercial maintenance service.

20. When is minor maintenance done on industrial machines?

21. When are industrial machines usually completely overhauled?

Chapter 15 Servicing

Most industrial refrigeration plants are kept in proper repair by the operating crew or maintenance personnel. Domestic and commercial equipment will work automatically if nothing is wrong, but when something does go wrong, a refrigeration service technician must be called in to locate and correct the difficulty. There are two parts to the service person's job: first, to find the problem; and second, to correct it. To find the problem requires extensive knowledge of the exact requirements of each part of the refrigeration plant. Finding the trouble is usually the most difficult and trickiest part. Then it is usually possible to find the missing requirement. Once the problem has been identified, its correction should be evident to one who knows the equipment. The technician must decide whether the faulty part needs adjusting, repairing, or replacing. Service manuals printed by the company that made the equipment are invaluable in listing adjustments, replacement parts, assembly procedures, and other information. However, it is important to determine not only which component is defective, but also why that component failed. Simply replacing a defective component without determining the underlying cause may result in the new component failing as well.

Variations from Refrigeration Requirements

Figure 2–13 gave pressure and temperature conditions for an entire ammonia refrigeration system. The service technician must know how such conditions vary for different refrigerants and for different operating conditions. Then the service person must trace through the system to find where conditions are not as they should be.

Figure 15–1 shows various system faults and the conditions that indicate them. First, for reference, approximate normal operating conditions are given. Of course, there will be some variation in these conditions for light or heavy loads and other operating variables. Where variations are given, they are for situations that vary sufficiently to make noticeably greater changes than with normal operating variables. Many of the conditions listed will still provide some refrigeration, but with more or less excessive operating costs. However, these are usually not readily apparent to the service technician who only sees the equipment for a short time, so these are not listed.

If controls are out of adjustment, backpressures and temperatures are the only things that are significantly affected. If everything including the cycling time appears normal except the required temperatures, control adjustments should be checked first.

An expansion valve open too wide will flood liquid refrigerant into the suction line and back to the compressor. This actively refrigerates the suction line and the crankcase. As explained in Chapter 14, this condition results in a low discharge temperature. If enough refrigerant floods back, the evaporation in the suction line and crankcase uses up the compressor capacity and the backpressure rises. This raises the evaporator temperature. If this happens, the running time will

Figure 15-1 Faults and conditions indicating them

	Pressure		Temperature						Cycle Time		
	Low	High	Evaporator	Suction Line	Crankcase	Discharge Line	Condenser	Reciever	Temperature Control	Pressure Control	
Normal condition	From evaporator temperature	From condensing temperature	As required	Cool or cold	Room temp. or warm	Hot	About 30°F above cooling medium	Condensing temperature	Run 2/3 time	Run 2/3 time	
Control setting											
Too high	High	Normal	High	Normal	Normal	Normal	Normal	Normal	Normal	Normal	
Too low	Low	Normal	Low	Normal	Normal	Normal	Normal	Normal	Normal	Normal	
Expansion valve setting											
Open too wide	Normal or high	Normal or high	Normal	Low	Low	Low	Low	Normal	Normal or excessive running	Normal or excessive running	
Closed too much	Normal or low	Normal or low	Normal latter part high	High	Normal	High	Normal or low	Normal	Normal	Normal or short cycles	
Plugged-up system	Low	Low	High	High	Normal	Low	Low	Normal or low	Runs continuously	Short-Cycles or remains off	
Air in system	Normal	High	Normal	Normal	Normal	Normal or high	Normal	Normal	Normal	Normal or excessive	
Dirty condenser	Normal	High	Normal	Normal	Normal	High	High	Normal or high	Normal	Normal	
Cabinet overloaded	High	High	High	Normal or high	Normal or high	High	High	Normal	Runs excessively	Runs excessively	
Faulty compressor valves	High	Low	High	Normal or high	Normal or high	Normal or low	Normal or low	Normal	Runs excessively	Runs excessively or short-cycles	

continued

Figure 15-1 Faults and conditions indicating them (*continued*)

	1	2	3	4	5	6	7	8	9	10
Too much refrigerant										
High-side float or capillary tube	Normal or high	Normal or high	Normal or high	Low, frosted	Low	Low	Normal	Normal	Normal or runs excessively	Normal or runs excessively
Other systems	Normal	High	Normal	Normal	Normal	High	High	Normal or low	Normal or runs excessively	Normal or runs excessively
Not enough refrigerant										
High-side float or capillary tube	Normal or low	Normal or low	Top high, bottom normal	Normal or high	Normal	Normal	Normal or low	Normal	Normal	Normal
Low-side float	Normal	Normal	Normal or high	High	Normal	Normal or high	Normal or high	Normal or high	Normal or runs excessively	Short-cycles
Dry expansion evaporator	Normal	Normal or low	Normal or high	High	Normal	Normal or high	Normal or low	Normal or high	Normal or runs excessively	Normal or short-cycles

increase. The unit may even run continuously. This may cause the compressor to fail. Simply replacing the compressor without determining that the cause of failure was an overfeeding metering device will cause the new compressor to fail as well.

A starved expansion valve does not let enough refrigerant through to actively refrigerate the entire evaporator. Therefore the latter part of the evaporator will be warmer than it should be. This causes higher-than-normal suction-line temperature and discharge temperature. Sometimes, enough refrigerant is not evaporated to supply the full compressor capacity, and the backpressure and temperature drop. As explained in Chapter 14, this does not mean better refrigeration. A problem that starved evaporators are apt to cause that is not readily apparent from outside the evaporator is oil trapping. Enough refrigerant is not flushed through the coil to carry the oil through to the suction line. This allows the oil that should be returned to the compressor crankcase to collect in the evaporator. An evaporator that is too warm for its suction pressure will indicate this.

A plug or stoppage in the system may occur in the liquid line or on the low side of the system. Strainers or driers may fill and plug up. Expansion valves may stick closed or stop up with dirt. If water is present in a halocarbon refrigerant, it will freeze at the expansion valve and plug it. This stops refrigeration, which usually allows the ice to melt. Intermittent refrigeration, proper freezing, then defrosting are the usual results. Occasionally, the ice only partially plugs the expansion valve, which allows enough refrigerant to pass to prevent it from melting. This situation will not show the intermittent operation characteristic of an iced expansion valve. Water in a refrigeration system can also facilitate the formation of acid in the system, which can lead to component failure.

With a plugged system, the backpressure will be low. If there is a temperature control, the system will run continuously and produce a high vacuum. If there is a pressure control, the system will pull down to the cut-out point and cut off. If the stoppage is complete, the unit will remain off. More often, there is enough leakage through the stoppage to allow the pressure to build up, and the machine short-cycles. Naturally, the evaporator will be warm. Because no refrigerant is being pumped even when the unit is running, the head pressure, head temperature, and discharge temperature are low. The receiver temperature may be low because the receiver is filled with liquid pumped from the evaporator. The receiver usually feels a little cooler to the hand where liquid is present.

If there is air in the system, it raises the head pressure, sometimes excessively. This usually increases the discharge temperature because the compressor must work harder against the higher pressure. If the head pressure becomes excessive, high-pressure cut-outs or motor overloads may stop the unit. The biggest change from normal will be in power consumption, but the refrigeration service technician cannot check this without a record of normal power consumption for comparison.

A dirty condenser will show high head pressure, high condensing temperature, and high discharge temperature. The condensed liquid will usually be so warm that the receiver is quite warm to the touch. This problem also increases power costs. Depending on the metering device being used, it may also result in high suction pressure.

An overloaded cabinet will show high evaporator temperature and pressure, high condensing temperature and pressure, and a high discharge temperature. The unit runs continuously or nearly so. The entire compressor, including the crankcase, overheats.

Faulty compressor valves cause high backpressure because the compressor does not remove vapor from the evaporator rapidly enough. Because less vapor is pumped, head pressures and temperatures are usually low. High backpressure and low head pressure produce low discharge temperatures and low compression ratios. If there is a temperature control on the system, the system runs continuously or nearly so. A pressure control system may run continuously or may short-cycle. The latter happens when the backpressure becomes low enough to shut off the control, then the head pressure leaks back through the faulty valves to raise the low pressure and start the unit again. When the unit runs continuously, the crankcase often overheats. A broken discharge valve will cause the cylinder and compressor head to overheat.

Too much refrigerant—that is, an overcharge—reacts differently with different refrigerant controls. If there is a high-side float valve or capillary tube, most of the refrigerant charge is stored in the evaporator. Therefore, too much refrigerant will fill the evaporator to overflowing and refrigerant will flood back through the suction line. This reduces the amount of superheat. The result is the same as when an expansion valve is open too wide. The suction-line temperature, crankcase temperature, and discharge temperatures are low, perhaps with higher-than-normal evaporator temperatures.

If there is too much refrigerant and a low-side float valve, automatic expansion valve, or thermostatic expansion valve, the valve does not let any more liquid through to the evaporator than is required. Then any excess backs up into the receiver. If there is so much refrigerant that the receiver is filled, and liquid backs into the condenser as shown in Figure 14–6, then insufficient condenser surface is exposed to the vapor to give adequate condensation. This results in high head pressure and temperature and increased subcooling of the liquid.

Insufficient refrigerant in a system with a high-side float valve or a capillary tube starves the evaporator, with about the same results as starving it with an expansion valve closed too much. The lower part of the flooded evaporator will be down to temperature, but the top will be warm. The suction-line temperature and the discharge temperature will be high. Suction-vapor superheat is increased.

With a low-side float valve, insufficient refrigerant will not raise the float enough to close the needle. This allows high-pressure gas to leak to the evaporator. High suction-line and high discharge temperatures result. Running time will be longer with a temperature control, although not always enough to be readily apparent. If there is a back-pressure control, the high-pressure gas leaking to the low side causes short cycling. Also, when the refrigerant quantity is low, a distinct hiss can be heard at the valve needle as gas instead of liquid rushes through.

Too little refrigerant with an automatic expansion valve or thermostatic expansion valve is not always readily apparent, unless the shortage is severe. The action is similar to that of a starved expansion valve, with high suction-line temperatures and high discharge temperatures. However, the expansion valve will hiss just as with a low-side float valve.

Figure 15–1 can be used to help familiarize yourself with the different results of common faulty operating conditions. It does not list all possible faults, or all possible checks. To be able to find the source of a problem without loss of time, a systematic method of checking the entire system should be followed. Figure 15–2 provides a suggested procedure. Checks do not have to be made in this exact order, but it is important to establish a routine system of checks that covers the entire list of possible faults and eliminates them one by one. Some checks can be made at a glance. Others take a little more time.

First, a glance will tell whether the compressor runs or not. If it does not, a check should be made as to whether the compressor is stalled or the motor does not run. Drive belt(s) or coupling(s) may be removed to determine if either the motor or the compressor will spin independently of the other. If the motor does not run, the electric power supply should be checked. Check that switches are not off. Then check fuses, breakers, and/or permissives. Fuses fail more often than anything else in the electric system. If these are O.K., the control should be checked. Of course, if the control setting is higher than the box temperature, whether because of an improper control setting or too cold a box, the control switch will remain open. A temperature control that has lost its charge (no pressure in the bellows) acts exactly like a control that is cold enough to reduce the bellows pressure. Or sometimes control points burn or become so worn that they do not come together. If there is electric power at all these points, check for power at the motor. If there is, and the motor still fails to run, then the motor itself must be faulty.

Very loose or broken belts allow the motor to run without driving the compressor. Flywheel keys may have sheared. Some small compressors use a taper fit between the flywheel and shaft, without a key. These may come loose. In such cases, the compressor appears to be running because the flywheel is

Figure 15-2 Operating Problems and Their Causes

Problem	Preliminary Check	Cause
	Mechanical Problems	
Compressor does not run	Motor runs if belts are removed	Stalled compressor
	Motor will not run	Switch off
		Fuse burned out
		Control off due to:
		Cabinet cold enough
		Improper control setting
		Faulty control
		Faulty or broken wiring
		Faulty motor
	Motor runs, compressor does not	Broken or slipping belts
		Sheared key on flywheel
		Loose flywheel
Compressor runs but faulty refrigeration	Low suction pressure and low discharge pressure	Low-side valves closed
		Refrigerant lines plugged
		Strainers plugged
		Expansion valve plugged with dirt, ice, or wax
		Expansion valve stuck closed
		High-side float valve air-bound
	High suction pressure and low discharge pressure	Faulty compressor valves
		No refrigerant
	Normal or high suction pressure and high discharge pressure	Overloaded system
		Low refrigerant
		Too little water or air over condenser
		Too warm water or air over condenser
		Dirty condenser
		Dirty or plugged discharge line
		Valves in discharge line closed
		Air in the system
		Too much refrigerant
	Compressor short-cycles	Leaky compressor valves.
		Low refrigerant, particularly with a low-side float valve
		Expansion valve starving coil
		Partially plugged expansion valve
		Improper multiplexing
		High-pressure cut-out cycling
	Erratic refrigeration; too cold, then too warm	Sticky control
		Oil-logged float valve or evaporator
	Frost backs	Improperly set expansion valve
		Thermostatic expansion valve bulb loose
		Thermostatic expansion valve bulb in warm draft or other heat source
		Expansion valve stuck open
		Dirt in valve seat
	Leaks	All joints
		All gaskets
		Service valves, gage port plugs

Figure 15-2 Operating Problems and Their Causes (*Continued*)

Problem	Preliminary Check	Cause
	Leaks (*continued*)	Expansion valves, bellows, or diaphragms Control bellows or diaphragms Oil plugs Compressor cap screws
Noises	In compressor	Worn parts Squeaking shaft seal Slugging oil or refrigerant Loose flywheel
	Elsewhere	Poor foundation Chattering water valve Loose parts Ratting refrigerant lines Squeaking belts Shipping bolts
	Food Storage Problems	
Food freezes	Cabinet too cold	Improper control setting Faulty control; stuck closed Unbalanced multiplex system Leaky expansion valve Low refrigerant charge Oil-logged evaporator
Food does not keep satisfactorily	Cabinet too warm	Any of the points under faulty refrigeration Improper control setting Faulty control; bellows lost charge Poor air circulation in cabinet Overloaded cabinet; too much warm food or too high surrounding temperature Faulty insulation Faulty door hardware Coils need defrosting
Foods dry out	Humidity too low	Poor air circulation Cooling coil too small Starved cooling coil Blower not running on defrost cycle (if unit has blower)
Foods slime	Humidity too high	Excessively large cooling coil Light cabinet loads Excessively warm, damp food
Odor, taste		Temperature too high Humidity too high Improper location of foods Dirty cabinets Wrong motor oil in blower motor Improper insulation Improperly sealed cabinet getting musty Clogged drain pipes

turning, but the shaft is stationary. The locknut on the end of the shaft is the easiest thing to see to check this.

If the compressor runs but produces little or no refrigeration, pressures should be checked. Large machines usually have permanently installed gages. If there are no gages, service gages must be connected. Figure 15–2 lists operating problems that occur if pressures are not normal. Figure 15–1 pretty well covers the reasons for abnormal conditions.

A quick, sure check for faulty compressor valves is to front-seat the suction service valve to pump the compressor down. By front-seating the suction service valve, the suction line is sealed off from the compressor, but the service port (where the low-side hose from the gage manifold is connected) will be open to the suction side of the compressor. It should be able to pull a 25-inch vacuum or better. Be careful that it does not slug oil during the pull-down. If it does, stop it long enough for the oil to subside, then restart it. After all the refrigerant has been pumped from the oil, shut off the compressor. The vacuum on the low side should hold or, if it rises, do so very slowly. If the compressor can pull a vacuum and hold it, there is nothing wrong with it and you need to look elsewhere for the fault.

Occasionally, a refrigeration system will work very erratically. It will be too warm for a while, then suddenly get too cold. This may be caused by a faulty control, or an oil-logged evaporator. A sticky control that requires considerable force to turn on or off will allow the pressure, and therefore the temperature, to vary excessively. An oil-logged evaporator can reduce boiling by coating the liquid refrigerant with a layer of oil, which reduces its boiling action. Under this condition, the cold control is set colder to get proper refrigeration. Then a larger heat load will at some time start a violent boiling that will force the oil back to the compressor. Then the colder control setting with the normally boiling refrigerant causes too cold an evaporator.

Any time a service technician finds insufficient refrigerant, he or she should check the entire system for leaks. The refrigerant in the system does not wear out, and if it is not lost it is will last indefinitely.[1] Therefore low refrigerant indicates a leak. And finding one leak does not prove that was the only one in the system. All joints should be checked, as well as all mechanical parts containing refrigerant, such as the compressor, all valves, controls, and so on. Oil seepage around any joint indicates a leak, which otherwise may be difficult to find. Many modern plants have refrigerant detectors permanently mounted in the engine room and throughout the facility. You can usually read refrigerant vapor concentration off of these directly. Many of these systems have detectors hard-wired to a direct digital control (DDC) or other system for alarming purposes. Ammonia plants may even have detectors hard-wired to the building's fire alarm system. In the case of an excessive refrigerant leak, the building's personnel will immediately leave the building.

Noisy equipment is usually caused by worn parts. If the parts are so badly worn that they are noisy, they need to be replaced. In such cases the entire compressor must be torn down for a complete check. The most common knocking noise that is not caused by worn parts is the slugging of liquid refrigerant or oil. This can seriously damage the compressor, so it should be corrected as soon as possible. One knocking noise that is sometimes hard to trace is caused by a loose flywheel. The flywheel may become loose enough that it knocks against the key on each compressor stroke, without being loose enough to be easily spotted. Squeaks or squeals may come from a shaft seal or from belts. A squealing seal may need polishing, may need more oil than it is getting, or may need replacing. If a belt is squeaking, try cleaning it with a wire brush, then apply a little soap. There are also aerosol agents available to help quiet noisy belts.

Practically all domestic and many small commercial machines are mounted on rubber or spring

---

[1] The ternary blends will fractionate (separate), as they are near-azeotropic or zeotropic. System(s) that use these refrigerants may have to be evacuated and recharged if excessive refrigerant is lost due to leaks.

cushions to deaden vibration and noise. To prevent damage during shipment, these units are fastened solidly with shipping bolts. These bolts must be removed to allow the unit to float on its cushion mounting before putting it into service. Occasionally, someone fails to do this. If a new unit is objectionably noisy, this is one of the first things to check.

Food Storage Problems

As mentioned in Chapter 13, one should not lose sight of the fact that the prime purpose of refrigeration is to improve the preservation of materials and products that require refrigeration. The condition of the food kept under refrigeration is the best indication of whether the system is operating properly and is properly adjusted. In fact, the refrigeration system may be operating perfectly as a mechanical device, and still food preservation is unsatisfactory. So food storage problems must be considered as well as mechanical problems.

If food products or materials are sometimes freezing, the cabinet is too cold, or the food/materials may be too close to coils or cold air blasts. The service technician must determine whether rearranging the product or a relocating the control is necessary. A faulty control could also cause unwanted freezing. Sometimes freezing is caused by faulty multiplexing. Consider two cabinets operating from the same compressor without auxiliary valves. If there is a heavy load in one cabinet and a light load in the other, the one with the heavy load will keep the compressor running. But the continuously running compressor will continue to cool the other cabinet, and its temperature may get too low. This situation may be corrected with evaporator pressure regulators (see Chapter 9).

Too cold a cabinet has in some cases been caused by a slight shortage of refrigerant in a system with a backpressure control. Enough refrigerant is supplied to refrigerate the coil, but with just enough high-pressure vapor to keep the control on continuously.

If food does not keep satisfactorily, this points to too warm a temperature. Faulty or poor refrigeration will cause this. A faulty control can do the same. A leaky and collapsed bellows in a temperature control will have the same effect as a sufficiently cold evaporator which would keep the control bellows collapsed.

Sometimes the coil is cold enough, but there is poor or insufficient air circulation. This will not carry the heat from the different parts of the cabinet to the coil. Forced-draft fans must not have the air flow blocked with stored products or any other material. Cabinets with convection coils must not be too crowded.

An overloaded cabinet, whether the overload is due to too much warm product in the cabinet, too high an outside temperature, or leakage due to faulty insulation, will naturally be warmer than it should be. Yet the compressor may be doing all that is possible, so the only cure is to remove the cause of the overload, or sell the customer a larger system.

Frost on the coils will partially insulate them and prevent their proper absorption of heat. Cabinet temperatures will often drop as much as 5°F with the same coil temperature or control setting after coils are defrosted.

If foods dry out excessively, the cabinet air is too dry. The cabinet humidity is closely tied to the design of the system (see Chapter 19), but, in general, anything that reduces the temperature difference between the coil and the cabinet air increases the humidity. The easiest way to do this is to raise the cabinet temperature. This reduces the load on it. If this cannot be done, a larger coil is about the only answer. Excessive dehydration can also be caused by improper storage/wrapping of the product (see Chapter 13).

A forced-draft system designed to cycle must do so to keep the humidity up. It is during the off cycle that moisture removed from the air in the form of frost is melted off the coils and blown back into the room.

If there is trouble with foods sliming, it is due to too high a humidity. Less coil or a colder coil will reduce the humidity. Some systems that are properly engineered for summer conditions are oversized for winter conditions. Oversize equipment is one cause of high humidity. In such a case, starving the expansion valve sometimes helps. A slower speed fan, or a two-speed fan that operates at slow speed during the off cycle, is sometimes used. This problem can be very troublesome in cases where there is a large difference between summer and winter loads.

Sometimes there are complaints of odors or "off" flavors in the food. Foods with strong odors are apt to flavor other foods. If foods with strong odors must be placed in the same cabinet with other foods, they should be covered. Onions or garlic, or foods that contain them, cheese, and cantaloupes are some of the foods that may be troublesome. Dairy products, particularly butter and cream, will absorb other odors. They are best kept in cabinets by themselves. If this is impossible, they should be covered. It helps to place them directly under the coils, where air from the frosted coil will flow over them.

Questions

1. On what class or size of equipment does the refrigeration service technician usually work?
2. What are the two principal parts of any service problem?
3. How many parts of a system will be affected by a control that is set too warm?
4. How many parts of a system will be affected by an expansion valve that is open too wide?
5. How will a liquid line strainer that is filled with dirt affect the system?
6. How does an expansion valve plugged with ice affect the system?
7. What is apt to be the difference between the effects of Question 5 and those of Question 6?
8. Name two entirely different problems that can be created by a leak in the low side of a system operating on a vacuum.
9. What are the results of a dirty condenser?
10. What are the results of an overloaded cabinet?
11. What are the symptoms of insufficient refrigerant?
12. On what types of systems is too much refrigerant apt to give the most trouble? What is this trouble?
13. If a compressor will not run, what should be looked for?
14. If there is no refrigeration, high backpressure, and low head pressure, what should be looked for?
15. How would you check to see if a compressor is pumping properly?
16. If there is no refrigeration, low backpressure, and low head pressure, what should be looked for?
17. What are some of the causes of short cycling?
18. What are some of the causes of high head pressure?

Chapter 16 Refrigerated Enclosures

Any refrigerated enclosure such as a cabinet or a room must meet certain requirements. It must be well insulated against heat. It must provide a vapor barrier against the infiltration of moisture. It must have doors to bring food in and out, but these doors must close tight and form a continuation of the insulated wall. And it must be protected against excess heat, such as from stoves, boilers, and, where possible, direct sunshine.

Insulation

A good insulator should meet the following requirements:

1. It must provide maximum resistance to heat flow.
2. It should not rot or otherwise decompose.
3. It should be nonflammable.
4. It should not settle.
5. It should not absorb moisture.
6. It should be nonodorous.
7. It should be rodent- and vermin-proof.
8. It should be self-supporting.
9. It should be reasonable in price.
10. It should not be difficult to handle or install.
11. It should be light in weight.

1. A good insulator must provide maximum resistance to heat flow. The first purpose of an insulator is to hold out heat. Its insulation value is measured by the number of Btu or heat units that will leak through 1 square foot of the substance 1 inch thick (see Chapter 18). This value is called the *k factor*. For the time being, it is sufficient to remember that the smaller the k factor, the less heat flow will get through the surface. Therefore, for maximum effectiveness, the k factor should be as small as possible. Most insulation materials used for refrigeration have a k factor of about 0.25. Some types of insulation have k factors much less than this.

2. A good insulator should not rot or otherwise decompose. Otherwise, in time the insulation will no longer be effective and will have to be replaced.

3. A good insulator should be nonflammable. Nonflammable insulation not only reduces the fire hazard, but in case of a fire will act as a fire stop.

4. A good insulator should not settle after being put in place. Settling will leave uninsulated spaces around the tops of the walls.

5. A good insulator should not absorb moisture. Water is a very poor insulator. If the insulation becomes water-soaked, heat will travel through the water very rapidly. A hot iron rod can be held with a dry rag, but using a wet rag it will soon become too hot to hold.

6. A good insulator should be free of odors. If it is not, it will taint the food stored with these odors.

7. A good insulator should be rodent- and vermin-proof. An insulator that makes good rat nests, or that ants or other such insects can tunnel through or make nests in, becomes a definite sanitary hazard.

8. A good insulator should be self-supporting. It is then not necessary to build walls to support it.

9. A good insulator should be reasonable in price. Naturally, the lower its price, the more desirable the

insulator will be from the customer's point of view. If the needed insulation is very expensive, it may be cheaper to buy a larger machine or operate the machine longer than to pay for a well-insulated wall.

10. A good insulator should not be difficult to handle or use. Anything that adds to the difficulty of installation adds to the installation cost. Just as the material cost cannot be excessive, neither can the installation cost.

11. A good insulator should be light in weight. When insulation is installed in cabinets, excessive weight adds to the difficulty of moving and the cost of shipping. When insulation is installed in buildings, its weight affects the cost of foundations and other supporting members.

As in most situations when a list is made of desirable properties, nothing available meets all these requirements. So the requirements of the individual job must be studied and an insulation chosen that provides the best compromise.

As mentioned in Chapter 1, heat may be transferred by conduction, convection, or radiation. Conduction occurs by physical contact. It can be reduced by using a minimum mass of material to conduct the heat, and by using a material that is a poor conductor. Nonmetals are poorer conductors than metals.

Convection is the movement of air or other medium, which carries heat with it. The reduction of the amount of air that flows, or the distance it flows, will reduce convection heat losses.

Radiation is a heat effect that shines or is radiated. Light-colored or shiny surfaces are the poorest radiators of heat and the poorest absorbers of radiant heat. Polished metal, glass, or certain plastics can be used to keep radiant heat transfer at a minimum.

All these methods of heat transfer contribute to heat leakage in most insulations, but the importance of each transfer method will vary with different forms of insulation.

The perfect insulator would be a vacuum. If there is nothing to conduct or convect heat through a space, and if radiation is guarded against, no heat will flow. Of course, outside such a small device as a thermos bottle, the cost and technical difficulties of maintaining an air-tight wall space necessary to hold a vacuum make it prohibitive.

Air and other gases would have very good insulating values if they could be kept still. Halocarbon gases have a k factor of less than 0.10; air has a k factor of 0.16. However, air or gas in an open wall space will not remain still. It will carry or convect heat from the warm side to the cool side, as shown in Figure 16–1. The only way air can be held still is to divide it up into so many small spaces that the movement of any part of it is negligible, shown in Figure 16–2. That is all many insulations do: trap or

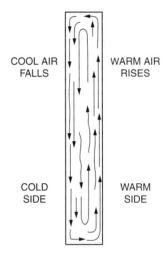

Figure 16–1 Air circulation in a "dead air space."

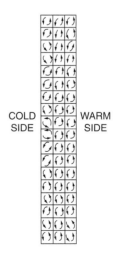

Figure 16–2 How air circulation is reduced by reducing size of space.

Figure 16–3 Microscopic view of foamed plastic insulation, *Courtesy of Armstrong Co.*

hold air so it cannot circulate. Some heat will be conducted through the solid parts of the insulating material, but not as much as would be carried by the air if it was left free to circulate.

Two different methods are used to trap and hold the air: enclosed bubbles and packed fibers. In the first method, bubbles are small enough that the convection within each bubble is negligible. And the walls are thin enough that there is a minimum mass to conduct heat across the material. Cork is a natural form of insulation of this type. Synthetic insulations of this form are made of foamed glass, plastic, or rubber (Figure 16–3).

The second method of trapping and holding air is with felted fibers. The tangle of fibers causes enough restriction that air flow is negligible. The denser the felted mass, the less the convection, but the more the conduction. Much study has been given to determining the proper size fiber and the proper packing density to yield a minimum overall heat flow. Natural fibers such as shredded wood or bark, or synthetic fibers such as glass or mineral wool, are of this type.

Another method of classifying insulations is into board, bat, loose fill, or foamed plastic insulation. Any insulating material can be mixed with a binder and pressed together into board form. Foamed plastic can also be molded into board forms.

Board insulations have the advantage of being self-supporting. They can be attached to a supporting masonry wall with a suitable adhesive or to a wooden wall with nails, as shown in Figure 16–4. If more than one layer of insulating board is used, they are held together with wooden or plastic skewers. Adhesives seal the joints. No other support is needed. Inside partitions have been made entirely of board-type insulation. If a harder, more durable surface is required, the surface of board insulation can be plastered.

In bat form, felted materials are put in the proper size and shape to fill the required space. Some cling together like cotton batting, and can be shipped, handled, and placed in position with no further protection. Others are glued between sheets of paper as a protection and an aid in holding their size and

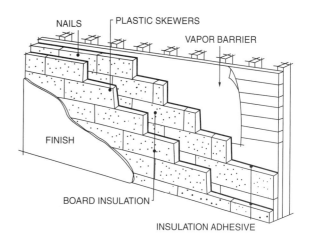

Figure 16–4 Method of laying up board insulation, *Courtesy of Armstrong Co.*

shape. Some bats can be compressed for shipping; many cannot. Bat-type insulation requires a rigid wall on each side of it. This type of insulation is more commonly used in refrigerated cabinets, showcases, or other fixtures than in large buildings.

Loose fill insulation may be supplied in bales or sacks. Some can be poured or blown directly into the space to be insulated. Other types must be stuffed in by hand to get the proper packing. Loose fill insulation has less use in refrigeration than other types of insulation. The biggest use of loose fill is to insulate uninsulated existing buildings for air conditioning.

Foamed plastic can be applied in two ways. The first method is to build outside and inside walls with a hollow space between. Uncured plastic is mixed with a curing agent and a foaming agent. A measured amount of this mix is poured into the hollow wall space. The foaming agent causes it to foam up and fill the space. The curing agent causes a reaction that hardens the newly formed plastic foam. This method of insulation is used both in large buildings and in refrigerated cabinets and fixtures. In the latter case, the foam makes a solid bond between the inner and outer walls of the fixture to strengthen the structure. The foam flows easily around any corners, brackets, tubing, or other shapes in the wall space. Thus, no cutting and fitting of insulation material is necessary.

Foamed insulation is also used around valves, pumps, and other complex shapes. A form of plywood or corrugated cardboard is made around the object to be insulated, and the mixed plastic is added. The foam fills the space between the object and the form. Regardless of how intricate the shape is, the foam fills in around it perfectly.

A second method of using foamed plastic is to spray it on the wall to be insulated. A special spray gun is used. This gun combines the uncured plastic, a foaming agent, and a curing agent delivered to it under pressure. The mixture is sprayed onto the wall to be insulated in several coats or layers until the required thickness is built up. It is almost impossible to build up a wall to exactly the same thickness throughout, so high spots are cut down after spraying is complete. The surface is then usually finished with a plaster.

In many cases a single insulating material can be supplied in any of the above forms. Any of the fibers can be cemented together into board form, although not all suppliers of insulating fibers do this.

The board form is usually best from a self-supporting or ease-of-handling point of view. Where a binding material must be used, however, it adds to the density of the insulating material, which increases its conductivity. It is therefore usually cheaper to handle or install, but it provides less insulating value. Bats or loose fill require support and more labor to install; but the k factor is usually better.

There is an increasing trend toward foamed-in-place insulation.

Organic Insulation

In early refrigeration systems, cork was the best-known high-quality insulation. However, cork is all imported to the United States from places such as Spain, Portugal, or North Africa. Shipping costs plus processing costs make it expensive compared to most other insulations. As a result, as synthetic insulations have improved in quality, they have almost entirely replaced cork. Corkboard was the standard insulation for so long that many specifications were written as "cork or equivalent." Today, we use either R or k factors to define insulation performance. R here stands for resistance and k for conductance. However, they are not inverse functions of one another, because the thickness of the insulation is included in the R factor number. The k factor of corkboard is 0.25 to 0.30, depending on the weight. The lighter the corkboard, the better the k factor but the less the mechanical strength. See Table A–9 in the Appendix.

Many other older insulations were also of organic origin. Sawdust and wood shavings were commonly used as fill insulation. In lumbering areas, they could be had for hauling away. Modern fire codes, however, do not allow sawdust or any flammable material to be used as an insulator.

Shredded bark, exploded or shredded wood fiber, kapok, and other similar materials have all been used as insulation. They are available in loose wool form, in blanket form, and in pressed board form.

Most of these organic materials had very good insulating values and, except for cork and kapok, cost less than other insulations. However, unless they were specially treated, they were flammable, tended to absorb moisture, and were not vermin-proof. Many are still found in older structures, but few if any are used in new construction.

Mineral and Glass Insulation

Molten limestone or molten glass can be formed into threads that when cool resemble wool. This wool (fiberglass) can be made into bats, blankets, or boards. Its insulation value is good, but it also varies according to the form used. Different sized fibers and different packing densities give k factors from over 0.3 down to about 0.24. Such insulation is fireproof, nonodorous, and rodent-proof. It is moisture-proof in that it will not absorb moisture, but it should not be overlooked that it is possible for the air spaces in a bat of this material to fill with water. Its weight is also low, except for the board form.

A glass foam has also been developed and is available in board form. Its structure is similar to that of cork, being made of a multitude of air or gas bubbles. However, the walls of these bubbles are completely inert; fireproof, waterproof, odor-proof, and rodent- and vermin-proof. Its biggest disadvantage is that it is brittle, so it should not be used in any application where there is vibration. Also, its k factor is not as low as those of some other insulating materials.

Plastic Insulation

Plastic foams consist mainly of two types, polystyrene and polyurethane. Both are made in board form or may be foamed in place. Polystyrene foam has a k factor of 0.25, polyurethane foam a k factor of 0.17. Plastic foams are moisture-proof and inert. Vermin will not attack or nest in the material. Some are flammable, some are not. All can be softened and the foam collapsed by heat. Some adhesives and sealers will dissolve them. Because of these restrictions, installation methods must follow the manufacturer's instructions.

Reflective Insulation

Crushed metal foil, curtains of cloth or paper painted with aluminum or coated with aluminum foil, and layers of polished aluminum or steel plates have all been used as insulation. The theory is to reduce radiant heat transfer to almost nothing, and do it in such a way that conduction and convection are also minimized. Such insulations have the advantage of being light, but they have not seen much acceptance for refrigeration work.

Other Types of Insulation

Many other materials have been used occasionally as insulators, or may be used in certain areas because of local availability or for other special reasons.

Hair felt is commonly used in certain industries, both as an insulation and as a sound deadener, but it has not seen much acceptance in the refrigeration industry. It is expensive, it is not fireproof, and it is not vermin-proof or moisture-proof.

The question is sometimes asked why other common heat insulators are not used in refrigeration. And what is the difference between insulating for heat and insulating for cold? Theoretically, there is no difference. In either case we must retard the flow of heat. However, there are two very important practical considerations. A heat insulator must be more than fire-resistant. It must be fireproof. Some previously mentioned insulating materials such as mineral wool or glass wool can be used to insulate steam or hot water lines and tanks, but even these insulations would melt if they were used to insulate a furnace or firebox. Plastic foams will melt and organic materials will char at high temperatures.

Those materials that are fireproof do not have as good a k factor as the insulations used in refrigeration.

A second practical consideration is that a cold insulator must be moisture-proof. There is no concern about this with heat insulators because the heat keeps them dry.

Moisture Proofing

When a glass of ice water is allowed to stand in a room at the room temperature, moisture or dew collects on the outside surface. Similarly, if a refrigerator cabinet was so built that air could reach the inside surface, moisture would collect at this point, as shown in Figure 16-5. The least this moisture would do would be to soak up the insulation. As stated earlier, wet insulation is a very poor heat retarder. Even with those fiber insulators that will not absorb moisture, such as glass wool, if the spaces between the fibers are filled with water, there are no trapped air spaces left to form the required insulation. With such insulation, water has been known to fill the space in the bottom of the cabinet and several inches up the walls (Figure 16-5). If the insulation absorbs moisture, it will soak up this water like a blotter. This keeps all the insulation wet. Foamed plastic insulation is not as troublesome because moisture cannot penetrate the plastic. Each bubble is a moisture-proof cell.

At below freezing temperatures, such moisture would freeze. Freezing water expands. This will crush and destroy the insulation. If allowed to continue long enough, it may even damage the cabinet walls. For these reasons, moisture must be kept out of the insulation. Three general methods are used

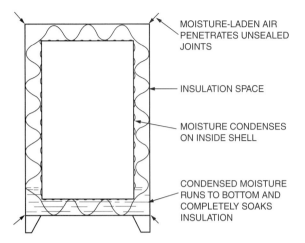

Figure 16-5 Moisture in cabinet insulation.

to do this. The first is to make both the inside and the outside walls moisture-proof and practically airtight. If no moisture or moist air can enter the insulated space, it cannot cause trouble. This sealing is done by using metal walls.

The second method is to seal only the outside wall and leave the inside wall unsealed or porous. If a breaker strip is used on a cabinet, it is not sealed, but passage is allowed to the interior of the cabinet as in Figure 16-6. With this method, all the moisture possible is kept out of the insulation. The

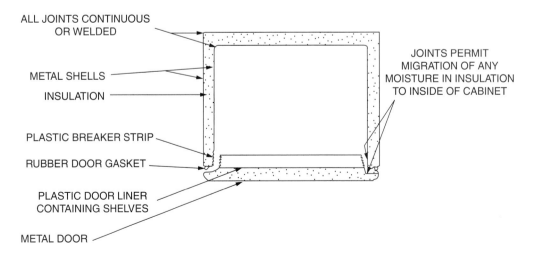

Figure 16-6 Construction of domestic refrigerator cabinet.

porous inside wall allows any moisture that might get through to go on through the inside wall. It eventually reaches the coils, where it collects as frost. Here it can cause no damage to the insulation. This method is used for most refrigerated cabinets and much cold-storage construction.

The third method of sealing out the moisture is to build a completely moisture-proof wall. A moisture-proof membrane is first applied to the wall to be sealed. Then a moisture-proof board-type insulation is installed with a moisture-proof sealer at all joints. A foamed-in-place moisture-proof plastic also forms an impervious wall if floor and ceiling joints are properly designed. This type of moisture proofing is used mainly in cold-storage construction when the insulated walls are expected to last indefinitely. All joints are welded except the breaker strip around the door. This breaker strip is sealed with a moisture-proof sealer. This completely sealed system is common in freezer cabinets, particularly those with coils on the insulation side of the inside liner.

Openings and Doors

The doors into a cabinet or cooler have the potential to cause many problems. To be effective, the door must be as well insulated as the cabinet walls. The thicker the insulation, the heavier this makes the door. These heavy doors must be strongly built and hung from a strongly built frame. The door must have a gasket that fits tight when the door is closed. The hardware must be heavy enough to hold the door, yet adjustable so that it pulls the gasket up tight.

Glass windows, glass doors, and glass-fronted showcases present special problems. Glass itself is a poor insulator. To improve its insulating effect, several thicknesses of glass are used with air spaces between them. Even this makes a poor insulator compared to a standard insulated wall. The spaces between the glass must be dehydrated and filled with dry air or nitrogen. For exterior applications, the space may be filled with argon or another noble gas. These gases help reduce the negative effects of ultraviolet rays. Otherwise, dew forms on the glass as a fog that interferes with vision. Then the glass must be so sealed in its frame that no air or moisture can leak in. Some manufacturers place a chemical drier in each air space to absorb any moisture that might leak in during the life of the case. This all makes glass expensive to install as well as expensive to refrigerate. It is therefore used only when the display through the glass has a definite sales value.

Cabinet Construction

Modern domestic and commercial cabinets are made of sheet steel with plastic breaker strips. Domestic cabinets usually have plastic door liners. Outer and inner shells are formed and welded into the required shape. The inner shell is given an inside finish of porcelain enamel. The sides of these shells next to the insulation may be painted or may be given a special anticorrosion treatment. The outside finish is usually a synthetic enamel paint, but on a few higher-priced models it may be porcelain enamel.

Practically all types of insulation have been tried by various manufacturers. Mineral wool, glass wool, and foamed plastics are all being used. The present trend is to a high-k-factor foamed plastic. A 1.5-inch-thick wall of the best insulators will have little or no more heat leakage than a 3-inch-thick wall having a k factor of 0.25 to 0.30. Yet the thinner insulation makes it possible to increase the usable space within the cabinet by a quarter to a third more with no increase in outside dimensions. Modern refrigerator cabinets have an inner and outer shell made of sheet metal and/or plastic. The space between the inner and outer shells may be filled with some type of poly foam.

Design is usually such that the inner shell will go in through the door opening of the outer shell. The door jamb or breaker strip between the inner and the outer shell is made of plastic. This dresses off the finish and also breaks the metal connections so there is no metal-to-metal conductivity from the outer to the inner shell. The door gasket seals against a metal lip turned in on the outer metal shell, as shown in Figure16-6. Leakage is permitted from the inside edge of the breaker strip to the interior of the cabinet. This leakage permits any moisture that might have entered the insulation to migrate to the evaporator.

The inner door wall may be porcelain on steel with plastic breaker strips, or the entire inner face may be of one-piece molded plastic. Extra steel reinforcements are added to the shells at the necessary points to hold the screws of the hardware.

In the past, many commercial walk-in coolers were built up with wood walls on the customer's property. Now such coolers can be purchased knocked down so that they may be moved into the building through doors and assembled in place. Or they may be built up of standardized metal-clad, insulated panels.

Several types of construction are used on large industrial boxes or entire cold-storage buildings, depending on the type of insulation used and the type of building construction. Figure 16-4 shows the recommended assembly method for a board-type insulation. The material may be corkboard, mineral or glass wool board, fiber board, or a foamed plastic board. It may be laid up against a concrete, brick, or wood wall. Insulated partitions are sometimes built up without any other support.

If the room is to operate below 20°F, it is desirable to have it above the first floor or have an air space between it and the ground. A freezer floor on the ground will chill the earth under the floor. Given enough time, this chilling takes place regardless of how much insulation is used in the floor. Subsoil water vapor is attracted to the coolest point because of a lower vapor pressure. This water vapor in earth below 32°F will condense and freeze. The freezing causes expansion, which can buckle and lift the floor. Floors and even walls have been ruined in this way.

If a freezer floor must be put on the ground, air ducts, pipes, or electric heaters plus remote-reading thermometers or thermocouples should be installed under the floor. If the earth approaches freezing temperature, heat applied through the ducts, pipes, or heaters will prevent freezing and buckling.

Cold-storage doors are usually purchased from builders who specialize in their construction. The prefabricated door, mounted in its door frame with all hardware properly set, is shipped complete.

Tank and Pipe Insulation

In addition to walls, most cold surfaces in plants of any size should be insulated. Not only does this prevent losses of refrigeration, it also prevents sweating and dripping of cold surfaces.

Tanks can be insulated with bats, with boards molded to the proper curvature, or with flexible boards bent to the required shape.

Pipes are usually insulated with a molded form made in halves. The halves are put together over the pipe and then cemented, wired, or taped in place. Small pipes may be covered with a molded foam rubber tube of the proper dimensions. This rubber insulation can be slipped over the pipe before it is assembled. Or a split insulation is available that can be slipped over the pipe after installation.

Excessive Heat

As far as protecting a cold-storage cabinet against excessive heat, it is only common sense not to place it too near a range or boiler. If it must be placed in the sun, it is sometimes desirable to build a shading roof over the building with an air space between it and the main roof. If this cannot be done, white or aluminum paint helps to reflect the heat away.

Questions

1. What are the requirements of a refrigerated room?
2. What are the requirements of a good insulator?
3. What are some of the difficulties caused by moisture penetration into insulation?

4. What methods are used to protect insulation against moisture?
5. Discuss the use of sawdust or any other flammable material as an insulator.
6. Why is a vacuum not used commercially as an insulator?
7. How do most insulators hold back the flow of heat?
8. How have metals been used as insulators?
9. What are the objections to windows in refrigerators?
10. Why is it best not to build freezer rooms directly on the ground?
11. What precautions should be taken if freezer rooms must be built directly on the ground?

Chapter 17 Instruments and Meters

Nearly everything that happens in a refrigeration system occurs inside pipes and is thus invisible. What conditions are detectable from the outside, such as the temperature of the pipe surfaces, can only be estimated if there is no accurate means of measuring them. Therefore, to know what is happening in a refrigeration system, means of measuring temperatures and pressures are essential. Liquid indicators to show liquid levels are desirable. If any check is to be made of electric driving equipment, electric meters and instruments are the only way to know the current flow or power consumed. Therefore, refrigeration technicians need to know something about the indicating and recording devices that show temperatures, pressures, liquid levels, and electric quantities. Thermocouple heat sensors with digital displays are fast becoming the temperature-reading method of choice. These devices may be read locally or remotely. They also interface easily with direct digital control (DDC) systems. As with any piece of equipment, these devices need to be checked occasionally for proper operation.

Stem Thermometer

Because the control of temperature is the primary purpose of a refrigeration system, means of measuring this temperature accurately must be available. This is done with a thermometer. In the past, the most common form was the stem thermometer with mercury or spirit in a glass tube, shown in Figure 17-1. The bulb is filled with a liquid that expands as it is heated and contracts as it is cooled. The change in volume of the liquid causes it to rise or fall in the tube or stem. The volume of the bulb is large compared to the volume of the stem, so the temperature of the stem makes very little difference in the volume compared to the temperature of the bulb. In laboratory work, corrections are sometimes made for the small error introduced by the stem temperature, but in practical work the error is not large enough to be significant.

The temperature indications may be engraved directly on the glass stem of the thermometer, Figure 17-1, or may be on the background or support of the thermometer, Figure 17-2. The latter is easier to read but not quite as accurate. There are two reasons for the better accuracy of having the scale on the stem. There is always a chance of the stem shifting or slipping slightly in its clamps. This would introduce an error in the entire reading if the figures were on the background. Also, if the reading is not taken at right angles to the thermometer, the sighting error (parallax) is larger when the engraving is on the background, as shown in Figure 17-2.

The fluid in the thermometer may be mercury or a spirit such as alcohol or turpentine with a colored dye to make it visible. The mercury-filled thermometer is accurate only above −38°F. Because mercury freezes at this temperature, a mercury thermometer will not work below −38°F. On the other hand, the purity of mercury is easily controlled, so it is just about the same for all thermometers. Mercury does not change in any way

Figure 17-1 Stem thermometer, engraving on stem.

with age. Its expansion is directly proportional to temperature. When the proper scale is found for the bore size of the tube used, this is standard for all thermometers using the same bulb size. Readings at the top of the scale have the same spacing as readings at the bottom of the scale.

The big disadvantage of a mercury thermometer is that it is difficult to read. The mercury is approximately the same color as a reflection on the glass, so if the light is not right, the height of the mercury column may not be visible.

A spirit that has been dyed red, blue, or green is much more easily seen. For this reason it is preferred by some, although its accuracy is not considered to be as dependable. It may change slightly with age. Its expansion is not proportional, so the scale is not evenly divided. Under some conditions, some of the fluid will vaporize and recondense at the top of the stem (Figure 17-3). Because this is a distillation process, the color is left behind, and what is condensed is colorless. It can be seen if looked for, but when one is looking at a colored column, the clear fluid at the top of the stem may be easily overlooked. Naturally, an error due to the length of the distilled fluid will be introduced into the thermometer reading. Spirit thermometers are also slower in reacting to a change of temperature than are mercury thermometers.

Special designs of mercury thermometers have been introduced that provide the accuracy of mercury with improved visibility. Figure 17-4 shows a cross section of one so designed that a red ribbon is invisible from the front of the thermometer, but reflects on the mercury column as on a mirror. This red reflection is seen up the height of the mercury column, but there is no reflection above it. Therefore, it looks exactly like a red spirit thermometer from the front. It has one disadvantage: It must be viewed from directly in front. If the eye of the observer is a little to one side, the diffraction of the glass makes both the mercury column and the reflection invisible.

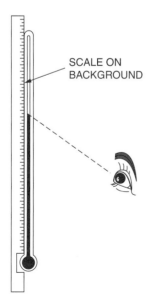

Figure 17-2 Parallax in reading.

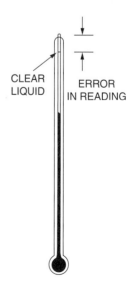

Figure 17-3 Spirit thermometer with vapor condensed at tip.

202 CHAPTER 17 *Instruments and Meters*

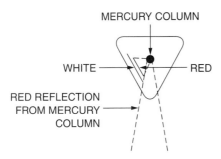

Figure 17–4 Reflecting-type mercury thermometer.

Figure 17–5 shows another method of improving the visibility of the mercury thermometer. The center round section, A, magnifies the mercury column similar to the way a round stem does. The two side bulges, B, B, collect and concentrate light on the mercury column. This makes the mercury show up much better. This type of thermometer is usually called a binoc thermometer. It cannot be viewed from as extreme an angle as is possible with a round stem, but it can be seen through a much greater angle than the reflecting type mentioned above.

Thermometer wells, either stubs of small pipe welded in, as in Figure 17–6, or threaded wells shown in Figure 17–7, should be installed in industrial plants where temperature measurements may be required. Or an insertion-type thermometer may be used. The thermometer well should be filled with oil or copper paste to provide good conductivity from the well to the thermometer bulb. Insertion-type thermometers are available in a variety of shapes to fit different requirements. Figure 17–8 shows the points in a system where thermometers or thermometer wells are needed to provide a complete record of the operation of the equipment. This is, of course, in addition to thermometers in the rooms or products being cooled.

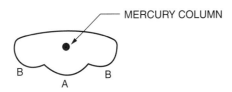

Figure 17–5 Binoc thermometer.

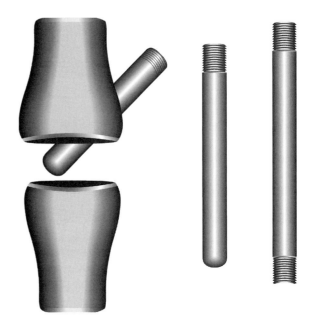

Figure 17–6 Construction details of a welded thermometer well.

Temporary wells that clamp onto the outside of the pipe are available. One type is a set of brass blocks made in sizes to fit onto the more common sizes of copper tubing. A hole drilled in this block holds the thermometer bulb. Another type is a ribbon of silver (silver is an excellent conductors) that is wrapped around the pipe and held with a spring-and-chain arrangement. One end of the silver is curled up to hold the thermometer bulb. Temporary wells can also be fashioned from wooden

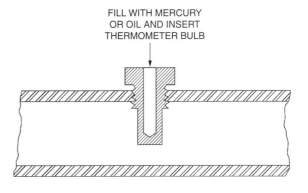

Figure 17–7 Thermometer well, screw type.

Stem Thermometer   203

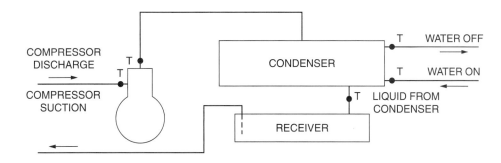

Figure 17-8 Points in refrigeration system that require thermometers.

blocks, Figure 17-9, or from a piece of sheet metal cut and bent to fit, and then soldered to the pipe. For a quick check, a wad of putty or similar material has been used. Grease sometimes works well on cold lines, where the low temperature thickens it. Figure 17-10 illustrates a set of temporary thermometer wells.

It should be remembered that any check on the outside of the pipe measures the pipe temperature, not the temperature of the fluid in the pipe. These two temperatures may or may not be the same. The thermometer well or insertion thermometer does a much better job of measuring the fluid temperature in the center of the pipe.

The limits of error to be expected in these thermometers depend on how well they are made and on their location. Test thermometers are available with graduations to 0.5°F. The reading may be estimated to 0.1°F. Such thermometers are available with an accuracy of 0.1°F. That is, the true temperature at the bulb will not be more than 0.1°F different from the reading on the stem. In general, unless otherwise guaranteed, the error of

Figure 17-9 Temporary thermometer wells cut from wood or strapped on.

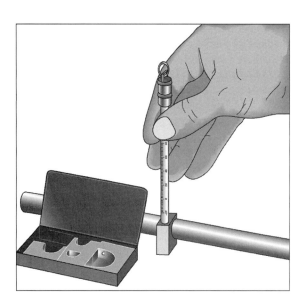

Figure 17-10 A set of clip-on thermometer wells.

an industrial thermometer can be depended on to be no more than the graduations. Thus, if the thermometer is graduated to 0.5°F, its error will be 0.5°F or less.

The second possible error in a thermometer may result from location. The reading of a thermometer on the wall does not always mean that the products in the room are at that temperature. Temperatures may differ by several degrees in different parts of a cabinet. Goods have frozen in one part of a cabinet when the temperature at another point in the same cabinet is unsatisfactorily high. Naturally, the thermometer will only record the temperature at the bulb location. It is too often overlooked that the bulb location may not always be where it will give a good average of the temperature in the cabinet or room.

The accuracy of a thermometer can easily be checked by placing it in a mixture of ice and water. A mixture of ice and pure water will be at 32°F, and the thermometer should show this. If the thermometer reads as high as 212°F, this end of the scale can be checked with boiling water (at sea level). Thermometers that measure temperatures this high are typically used to measure compressor discharge temperature.

Checking for the best location for a thermometer in a room is not so easy. Probably the best procedure is to place several thermometers in different parts of the room, but doing this is expensive, and some of the thermometers may be obstructed by product. If there are inside posts, these make good locations. Thermometers are often put on walls near doors for convenient reading, but this is often a poor place for them because readings will fluctuate every time the door is opened.

Dial Thermometer

Another type of thermometer is the dial thermometer, as shown in Figure 17–11. This type has the advantage that it can be easily seen and read from a distance. Its accuracy is not as high as that of a stem thermometer, but it can be easily recalibrated if a check shows it to be off. Dial thermometers may

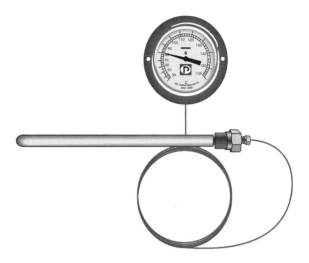

Figure 17–11 Dial thermometer with remote bulb.

be either self-contained or remote-reading. The self-contained type is usually activated by a bimetallic coil made of two different metals that have different rates of expansion. As a result, the coil winds up or unwinds with a change of temperature. This movement turns a needle that travels over a suitable scale.

A remote dial thermometer may utilize an expanding liquid or the vapor pressure of a liquid. In either case, the liquid is contained in a bulb that is connected to the thermometer mechanism by a capillary tube. The operating part is fundamentally a pressure gage, a closed circular tube. An increase in the pressure in the tube forces it to straighten. This motion is suitably transferred to a needle. The quantity of liquid in the bulb is so large compared to the rest of the instrument that unless the connecting tube is too long, the temperature of the liquid any place but in the bulb has a negligible effect on the needle. The bulb may be placed in an air stream, in a pipe, a tank, or any other place a temperature reading is required. If excessively long tubes are necessary to connect the bulb to the instrument, special means have been devised to correct for the error introduced by the temperature of the liquid in the tube.

Recording Thermometer

A recording thermometer is a dial thermometer with a pen on the end of the needle. A clockwork mechanism turns a calibrated paper chart under this pen point to produce a graph or picture of the temperature conditions, as shown in Figure 17–12. Some of these instruments are self-contained. They must be placed in the box or other location where a temperature record is required. They are valuable for checking conditions over a 24-hour (or other) period. The remote type is commonly used in large plants, where it may be permanently mounted on an instrument board to provide a continuous record or log of temperatures at required points.

The DDC systems mentioned in Chapter 9 are capable of recording all conditions in today's building systems, including, of course, refrigeration systems. A knowledgeable technician can look at the computer monitor and track the history of the system's conditions. Some information technology (IT) professionals describe the memory capability of these systems as "virtual infinity."

It should also be noted that a thermostat is nothing more than a thermometer with electric contacts instead of a needle or recording pen. It may be self-contained or of the remote type.

Pressure Gage

A pressure gage is a device for measuring the gas or liquid pressure in a closed system. Figure 17–13 shows the essentials of a common pressure gage. It employs a circular tube of flattened section. Such a tube will tend to straighten out as the pressure inside the tube increases. The action is the same as in a coil of garden hose, which tends to straighten out when the water is turned on, because increased internal pressure makes the tube more nearly round. Such a tube is called a Bourdon tube, after its inventor. The movement of the end of the tube is linked to a rack-and-gear mechanism that moves a needle. A suitable dial is placed under the needle, as shown in Figure 17–14. A pipe connection with a standard pipe thread is supplied to make a connection to the system where a pressure measurement is required.

Figure 17–12 Recording thermometer. *Courtesy of Bacharach Industrial Instruments.*

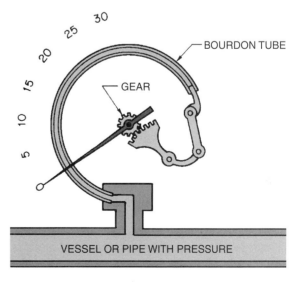

Figure 17–13 Interior of a pressure gage.

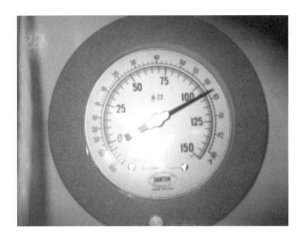

Figure 17–14 Pressure gage, 0 psi to 150 psi. *Photo by Noreen.*

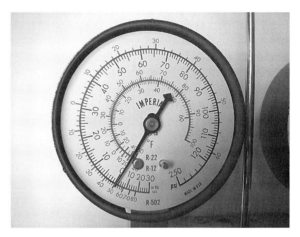

Figure 17–15 Compound gage scale.

Any reduction in pressure inside the tube causes the opposite action. The atmospheric pressure outside the tube flattens it further, and it bends into a smaller circle. This moves the needle in the opposite direction. Such a gage can be used to measure vacuums. Gages are made with bronze or steel Bourdon tubes, depending on whether the pressure of ammonia or other substances are to be measured. They are available in different sizes. The larger the size, the larger the scale, and the better the accuracy. They are also available in different pressure ranges. A 0- to 300-psi gage will measure any pressure between 0 psig and 300 psig pressure. Notice that the difference between 40 psi and 45 psi would be difficult to read on such a scale (Figure 17–14). A 0- to 60-psi scale would have more spread, so such a reading could be determined with more accuracy. However, if much over 60 psi were applied to this gage, it would damage it.

Such gages are called pressure gages. A gage showing 0 to 30 inches of vacuum is called a vacuum gage. Gages are available showing vacuums on one side of the scale and pressures on the other. These are known as compound gages (Figure 17–15).

Most gages used in refrigeration are calibrated in pounds per square inch for pressures and inches of vacuum for vacuums. To measure extremely deep vacuums such as are associated with system evacuation, a micron gauge is used. The micron gauge is an electronic device. It measures pressures so small that they cannot be detected with a pressure gage or mercury column. 1000 microns = 1 millimeter of mercury. One millimeter of mercury = 1/760th of an atmosphere. Another way to look at this is that 1 micron = 0.0000013157 psia. That is an extremely small amount of absolute pressure. Figure 17–19 shows an analog micron gage.

Because evaporating pressures and condensing pressures are often used to determine evaporating and condensing temperatures, some gages have a temperature scale as well as a pressure scale printed on them. If the gage is permanently mounted on a system, the temperature scale is for the refrigerant in the system. Some service gages have several common scales printed on them, Figure 17–16. Although this is valuable in providing a temperature scale in convenient form, the figures are usually so small they are hard to read. Even the pressure figures must be reduced in size to make room for the additional scales. For this reason, some service technicians do not like these temperature-scale gages.

It should always be remembered a gage measures only gage pressure. That is, it measures the difference between the pressure imposed on the inside of the Bourdon tube and the atmospheric pressure on the outside of the tube. If the gage is being used to check the safety of the system, the difference between these two pressures is what is needed.

Pressure Gage **207**

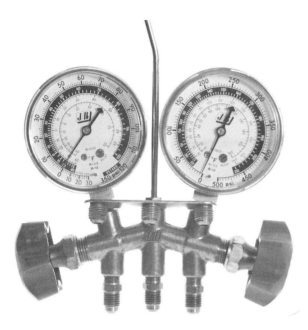

Figure 17–16 Gages with temperature scales mounted on service manifold.

However, if the pressure is being checked as a guide to evaporating or condensing temperature, a correction must be made if the gage 0 is not at an atmospheric pressure of 14.7 psi. (See Chapter 27 for a complete discussion.)

To maintain the accuracy of a gage, it should be checked regularly. A gage on a board should be checked at least twice a year. A gage used by a service technician should not get banged around in a tool box. It is subjected to all kinds of pressures, pressure fluctuations, temperatures, and so on, and should be checked once a month. A rough check can be made by opening the gage to atmospheric pressure and seeing whether it registers zero (compensate for elevation if necessary). Some gages have a screw that can be used to adjust the needle and bring it to zero if it is found to be off. To be sure the gage is accurate over the entire scale, it should be checked against a gage that is known to be accurate. Or better, have it checked on a dead-weight gage tester and adjusted if it is off. Instrument companies are available in any large city that can do this job.

It is further recommended that, to keep a gage accurate, it should be protected from vibration, either internal or external, as much as possible. By internal vibration is meant a pressure that fluctuates so much the needle continually vibrates. A valve or restriction in the gage line is the proper remedy for this. Vibration dampeners are also available to attenuate the transfer of vibration from the system to the gage.

Recording gages are also available. These are similar to standard gages but have a pen, clockwork, and chart similar to a recording thermometer.

To measure pressures at or near atmospheric and to measure vacuums with the greatest accuracy, a mercury column is sometimes used (Figure 17–17). This instrument is approximately the same as a barometer. The difference in height between the mercury in the two legs is the pressure or vacuum in inches of mercury. If the open end is higher than the connected leg, the pressure is greater than atmospheric. If the connected leg is higher, there is a vacuum in the system being checked. Mercury tubes are more common in testing than for measuring conditions in operating systems. They are valuable in operating

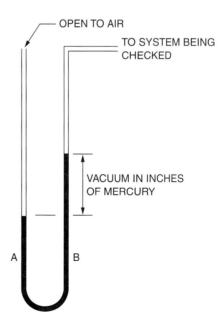

Figure 17–17 Mercury column.

systems to show certain conditions. One is where it is desired to carry as low a pressure as possible without going into a vacuum. The other is to accurately determine temperatures from pressures. Looking at refrigerant tables, it can be seen that the higher the vacuum, the greater will be the error in temperature if the pressure reading is missed by 1 inch.

Transducer

A transducer is an electronic device that mates very well with DDC systems. Essentially, the transducer responds to a change in the refrigeration system and causes a change in electrical resistance that is measured by a potentiometer. The signal (usually 4 to 20 milliamperes) is read and processed on a microprocessor or PLC. This signal is then amplified and further processed within its electronic interties. This information may be used to make adjustments to the system and/or interface with a central computer.

Transducers can be used to monitor any of the conditions in a system, including pressure(s), temperature(s), flow(s), vibration, and others. They can also be used to measure power (electric) draw and therefore, via the DDC system, to monitor and manage energy usage. This makes for an overall optimally efficient operation.

Thermocouple

Thermocouples are used to measure temperature(s) and/or establish the presence of a flame. We may be interested in the conditions in an absorption machine used for air conditioning or refrigeration. A thermocouple consists of two dissimilar metals bonded together. When heat is applied to this arrangement, a small electrical potential is generated. The output will vary depending on the intensity and on the nature of the metals. This signal is sent to some type of electronic intertie for processing (just like a transducer). A group of thermocouples (a thermopile) may be used to provide direct control. These are typically found on some gas-burning equipment.

Thermistor

A thermistor is an electronic device whose resistance changes with temperature. It may be direct-acting in that if its temperature rises, its resistance will increase. This is described as having a positive temperature coefficient. Some thermistors are reverse-acting. These thermistors' resistances decrease when temperature rises, and they are described as having a negative temperature coefficient. Current is sent to the thermistor as previously described for transducers. The return current (usually 4 to 20 milliamperes) is processed and usually displayed digitally. Thermistors may be used wherever temperature needs to be monitored. They may also be used as safety devices. Figure 17–18 shows an electronic thermometer.

Electric Instruments

Because so many refrigeration systems are driven by electricity, it is often necessary to measure electrical quantities. Electric measuring instruments are best divided into portable and switchboard-mounted

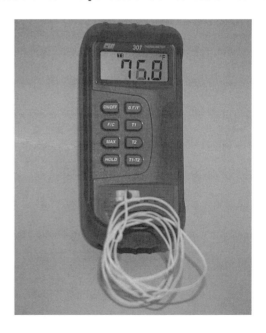

Figure 17–18 Electronic thermometer.

Figure 17-19 Analog micron gage.

types. Portable instruments are sometimes necessary to find trouble in commercial or industrial systems that have no switchboard instruments of their own. Voltmeters are necessary to check for improper voltages or widely fluctuating voltages. Watt meters are the surest check as to whether systems may be overloaded. Ammeters are of use to determine reasons for thermal overloads going out.

For domestic or commercial units, self-starting electric clocks are sometimes paralleled with the motor to check running time.

Large industrial plants usually have voltmeters, ammeters, and watt-hour meters mounted on an instrument panel. These again are part of the electric circuitry and are installed with the electrical equipment. The operating engineer must take readings from these meters and enter them on log sheets. Such data provide an accurate record of power costs. They will sometimes show the gradual decrease of efficiency of the mechanical equipment, or a sudden occurrence of something wrong, if the power supplied is correlated with the refrigeration requirements.

In many modern plants, this information is read directly on a human/machine interface (HMI) panel and/or on a remote computer screen (DDC). It may be stored and used to track maintenance activities as well as the aforementioned items.

The serious student of refrigeration technology is encouraged to take courses in and study electricity and electric instrumentation.

Questions

1. Why is the refrigeration engineer (service or operating) so dependent on instruments and meters?
2. Why is a stem thermometer more commonly used than a dial thermometer?
3. What is the advantage of a dial thermometer?
4. Why is a mercury thermometer more commonly used than a spirit thermometer in the refrigeration industry?
5. What is the advantage of a spirit thermometer?
6. What is the purpose of a thermometer well?
7. How can a thermometer be checked on the job?
8. What is the difference between a thermostat and a thermometer?
9. How does a pressure gage work?
10. What is the difference between a compound gage and a pressure gage?
11. How can a gage be checked?
12. What pressures are mercury tubes used to check?
13. What conditions in the system might a transducer help to monitor?
14. What is a thermocouple? Describe some applications.
15. What instrument should be used when an electric overload release continues to kick out?
16. What should be used to check power consumed by a motor?

Chapter 18 Heat Calculations

The previous 17 chapters have been devoted to describing the mechanical and practical aspects of a refrigeration system. Beginning with this chapter, we see show how calculations are made to choose or check the proper size of equipment to do a given refrigeration job. Basic, fundamental methods are stressed. For smaller-sized equipment, guesses based on past experience are often depended on. Short-cut factors, or short-cut tables, are often used to choose commercial equipment. However, even when using shortcuts, it is of great value to know something of the data from which the shortcuts arise. And if the factors that affect the results are known and understood, even guesses can be better modified to fit existing conditions. Finally, a beginner with a knowledge of a few fundamentals can do a better job of selecting equipment than some experienced technicians who depend only on guesses.

Design Temperature

One of the first factors that needs to be well understood is design temperature. The *design temperature* is the basic temperature from which we calculate heat loads. It is sometimes called the *ambient temperature,* meaning the temperature surrounding the equipment. There is sometimes some variation in the methods used to choose the design temperature, which sometimes yields a few degrees difference in the results obtained. That is, all tables do not show exactly the same design temperatures for the same place. In general, however, the design temperature is assumed to be the average of the hottest conditions under which the unit has to work.

Sometimes the design temperature is considered to be a temperature that is not exceeded for more than a certain percentage of the time. For instance, the highest temperature ever recorded in a certain city is 104°F, but a temperature of 100°F is reached on only an average of every other year. A temperature of 90°F is reached or exceeded, on average, for three days per month during the three summer months. A temperature of 87°F is reached or exceeded on an average of five days per month. Temperatures of more than 87°F are never experienced for more than five hours per day, and usually for not more than three hours. The average of the hottest temperature reached each day for July (the hottest month of the year) is 80°F. Different tables show design temperatures from 87°F to 90°F for this city. In general, these differences are less important to the final results of a problem than other variables that are encountered.

The question might be asked: Why not choose a machine that will operate at average conditions? The reason is that on the hottest days, when refrigeration is most needed, equipment selected on the basis of average temperatures would be too small. That answer might raise another question: Why not choose a system based on the maximum temperature, rather than on some theoretical temperature less than this? Any insulated box or building will have considerable cold-holding capacity. If the outside temperature goes above the design temperature, it will take some time for this higher temperature to leak through the walls. Before the extra temperature actually increases the load on the equipment, the outside temperature will probably drop again. This, plus the fact that any equipment should be chosen with a factor of safety makes it possible to do calculations from a temperature that is less than the worst condition that may be encountered. It works out best to choose a unit that will be approximately fully loaded at average operating conditions, rather than one that is too big most of the time. These are the reasons for working from a design temperature that has been developed to give an economical choice of equipment.

The design temperature of a given locality can usually be obtained from the local United States Weather Bureau office, the local newspaper, or refrigeration or air conditioning handbooks. Or it can be taken from Figure 18-1.

After selecting a design temperature, it must be modified to fit individual conditions. The heat load on a certain wall surface will certainly not be the same regardless of whether it is in the shade or in the sun, in a cool basement or in a hot kitchen. Mistakes have been made in sizing equipment just because such factors as these have been overlooked.

The k Factor

Chapter 16 described how heat leaks or is conducted through insulated walls. Test data have been collected on practically all substances used for wall construction or insulation. These data are given in the form of k factors, shown in Figure 18-2, or more completely in Table A-9 in the Appendix. Figure 18-3 shows the condition this k factor is to represent. The k factor is the number of Btu that will

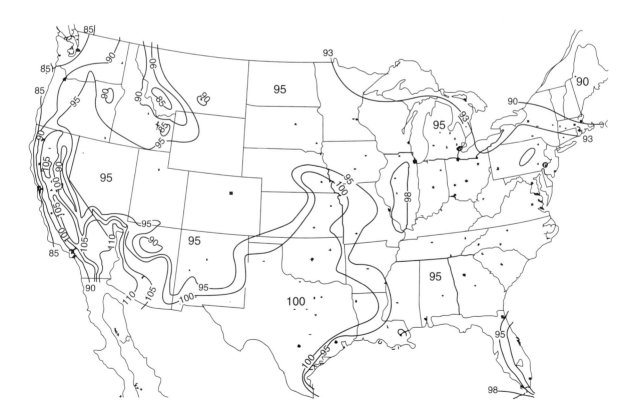

Figure 18-1 Map of outside design dry-bulb temperatures for summer cooling estimates.

Material	k Factor
Most refrigeration insulation	0.3 or less
Most soft woods	0.8
Most hardwoods	1.1
Most heat insulation	0.5
Concrete	12.0
Brick	5.0

Figure 18-2 Conductivities of some common materials (see Appendix Table A-9 for a complete table).

leak through 1 square foot of surface 1 inch thick in 1 hour's time for a 1°F temperature difference between the two sides of the surface.

The k factor is used in the following formula:

$$Q_1 = \frac{Ak(t_2 - t_1)}{x}$$

where

Q_1 = total heat in Btu per hour leaking through the surface
A = total area in square feet
k = leakage factor from Figure 18-2 or Table A-9
t_2 = temperature on the warm side
t_1 = temperature on the cold side
x = thickness of the surface in inches

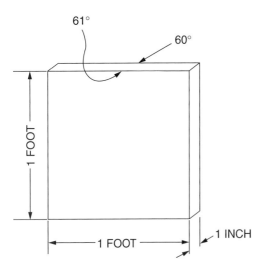

Figure 18-3 Diagram of k-factor conditions.

($t_2 - t_1$ is sometimes taken together and referred to as (t or delta t. Both of the latter terms mean "temperature difference.")

R Value

Virtually all modern building materials and types of insulation are assigned an R value. The R value for a particular material has an inverse relationship to the k factor. The difference is that with the thickness of the material is already factored into the R value. The inverse of the R value is the U factor ($1/R = U$). The following formula can be used to give the transmission (heat leakage) through a given building component (wall, door, window, etc.):

$$Q_1 = U \times A \times (t_2 - t_1)$$

where

Q_1 = total heat in Btu per hour leaking through the surface
U = inverse of the R value ($1/R$)
A = total area in square feet
t_2 = temperature on the warm side
t_1 = temperature on the cold side

We can further relate the k factor to the R value by setting their respective equations equal to each other and rearranging them algebraically as follows:

If

$$Q_1 = U \times A \times (t_2 - t_1) \quad \text{and} \quad Q_1 = \frac{Ak(t_2 - t_1)}{x}$$

then

$$U \times A \times (t_2 - t_1) = \frac{Ak(t_2 - t_1)}{x}$$

Rearranging, we get

$$\frac{1}{R} \times A \times (t_2 - t_1) = \frac{Ak(t_2 - t_1)}{x}$$

$$\frac{1}{R} \times \frac{k}{x}$$

$$\frac{R}{1} \times \frac{x}{k}$$

and finally,

$$R = \frac{x}{k}$$

Here we can see that insulation thickness is already factored into the R value but has to be specified when using k factors.

R values are easy to use when computing the transmission gain (or loss) through a composite wall. Take the sum total of all the R values of the various materials in the wall. Take the inverse ($1/R_1 + R_2 + R_3 + R_N$). The resulting number can be substituted into the formula as the U factor. The remainder of this chapter uses k factors to calculate heat flow(s) through given materials.

EXAMPLE 1 How much heat will leak into a household refrigerator 5 feet high, 3 feet wide, 2 feet 6 inches deep, with 1.5-inch polystyrene foam, if the inside temperature is 40°F and the outside temperature is 75°F, as in Figure 18–4?

The wall areas are

Front and back: $2 \times 5 \times 3 = 30$ sq ft
Both sides: $2 \times 5 \, x \times 2.5 = 25$ sq ft
Top and bottom: $2 \times 3 \times 2.5 = 15$ sq ft $\times 2$
$\qquad\qquad\qquad\qquad\quad = 30$ sq ft for both
Total area: $30 + 25 + 15 = 70$ sq ft

The formula is

$$Q_1 = \frac{Ak(t_2 - t_1)}{x}$$

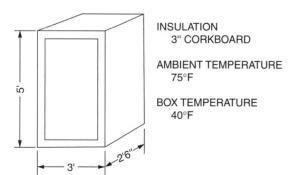

Figure 18–4 Reach-in refrigerator (Example 1).

Substituting,

$$\begin{aligned} Q_1 &= 70 \times 0.17(75 - 40)/1.5 \\ &= 70 \times 0.17 \times 35/1.5 \\ &= 278 \text{ Btu per hour heat leakage} \end{aligned}$$

EXAMPLE 2 How much ice would be required to maintain the refrigeration of Example 1?

In Chapter 1 we found that the latent heat of ice is 144 Btu per pound. Therefore we need $278/144 = 1.93$ pounds of ice per hour. (Here we take only latent heat into account.)

If a check is made of the insulating value of materials usually used for refrigeration insulation, it will be found that, except for expanded polyurethane, their k factors vary from 0.24 to 0.30 Btu per square foot, with an average value nearer the latter figure. Therefore, for general calculations, 0.30 can be used as the k factor for most types of insulation. The small error involved provides a small factor of safety, which could help offset the fact that the insulation used might not be exactly like the test sample, or it may not be installed under the same conditions as the test sample.

However, do not use 0.30 as the k factor of sawdust, shavings, or any insulation that has been allowed to become wet (remember our caveat about these materials from Chapter 16).

A factor of 0.17 can be used for expanded polyurethane.

If an insulated wall is made up of two or more types of material, the insulation usually has such a large insulating value compared to the other materials in the wall that everything but the insulation can be neglected for all practical purposes. If it is desired to check the value of materials other than the insulation, the insulation equivalent of other materials can be calculated.[1] To do this, divide the k factor of the insulation by the k factor of the material considered.

EXAMPLE 3 A wall is made up of 3 inches of glass wool, $1/2$-inch fir plywood inside, and $3/4$-inch plywood outside. A layer of building paper is between the outside plywood and the glass wool,

---

[1] See page 399 in the Appendix for the mathematical method of doing this.

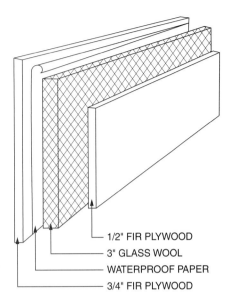

Figure 18–5 Wall construction, glass wool and waterproof plywood (Example 3).

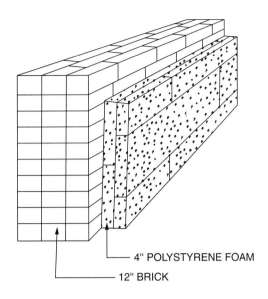

Figure 18–6 Wall construction, polystyrene foam and brick (Example 4).

Figure 18–5. What is the insulation equivalent of the entire wall?

The building paper is so thin that it can be neglected as far as insulation is concerned (although it is essential for waterproofing). The k factor of glass wool is 0.24 Btu per square foot and that of fir plywood is 0.80. The insulation equivalent of the plywood is

$$\frac{0.24}{0.80} = 0.30 \text{ inch}$$

This means that 1 inch of this plywood has the same insulation value as 0.30 inch of glass wool. The wall has a total of 1.25 inches of plywood. This 1.25 inches of plywood is equivalent to

$$1.25 \times 0.30 = 0.375 \text{ inch of glass wool}$$

The decimal can be shortened to 0.38 for this type of calculation.

Thus, the entire wall of Figure 18–5 has the same insulating effect as a glass wool wall with no wood of the following thickness:

$$3 + 0.38 = 3.38 \text{ inches of glass wool}$$

EXAMPLE 4 A 4-inch polystyrene foam wall is set against a 12-inch brick wall, as shown in Figure 18–6. What is the insulation equivalent of the entire wall?

The k factor of the polystyrene foam is 0.25; of the brick, it is 5.0.

$0.25/5.0 = 0.050$ inch of polystyrene is equivalent to 1 inch of brick.
$12 \times 0.050 = 0.60$ inch of polystyrene is equivalent to 12 inches of brick.
$4 + 0.60 = 4.60$ inches of polystyrene is equivalent to total wall thickness.

Usage Factor

Examples 1 and 2 included only leakage loads, with no provision for cooling the food or other products inside the cabinets, the air that fills the cabinet at each door opening, or other factors. For small-sized cabinets, the easiest way to calculate the *usage*, as the food loads are called, is on a volumetric basis. This is done with the help of volumetric factors, as shown in Table A–11 in the Appendix. The factors given are averages that have been worked out from experience. They give the Btu load per cubic foot per degree per hour. They are given for low, medium, and heavy usage. That is, there is considerable variation in the use that different establishments will give a certain sized cabinet. In general, heavier loads are applied in busy downtown markets or in chain stores. The lightest loads are for small community stores.

Two factors that determine the size of the load are (1) the temperature of the goods as they are loaded into the cabinet and (2) the frequency of loading or the amount loaded at a time. The warmer the product that is loaded, naturally, the heavier is the load. If everything loaded is already precooled, there will be little load on the system. The more product is loaded at a time, or the more that is loaded per hour, the heavier is the load. By more product is meant more weight. Bulky foods such as leafy vegetables will not create as great a load per crate as solid products. More weight of meat is often handled in a given sized cabinet than most other products. Some judgment is required surveying a job and obtaining from the owner enough information to make an intelligent guess as to the usage factor to use. However, these usage factor guides have been found to give a better measure of the loads in average cabinets than attempting to measure the actual weight of food to be loaded in and refrigerated.

Notice that the factors for a walk-in cooler are lower than those for a reach-in cabinet. The larger the cooler, the less in proportion it can be loaded. Smaller coolers are filled with shelves, and the shelves may be "loaded to the limit." However, the larger the cabinet, the greater the variation possible between a light load and a heavy load—and the greater the possible error if one is made. Above about 1500 cubic feet (a 15 foot by 10 foot by 10 foot cabinet), this method of calculation should not be used.

These usage factors are used with the following formula:

$$Q_u = VF(t_2 - t_1)$$

where

Q_u = heat load due to usage, Btu per hour
V = net volume in cubic feet
F = usage factor from Table A–11

EXAMPLE 5 What is the total heat load for Example 1 using average usage?

The net inside dimensions of the cabinet are the outside dimensions minus the thickness of the two insulated walls. In this case the inside dimensions are 4.75 feet by 2.75 feet by 2.25 feet. The net volume is thus

$$4.75 \times 2.75 \times 2.25 = 29.4 \text{ cubic feet}$$

The formula is

$$Q_u = VF(t_2 - t_1)$$

Substituting,

$$Q_u = 29.4 \times 0.22(75 - 40) = 226 \text{ Btu per hour}$$

In Example 1 we found the leakage load to be 278 Btu per hour. Therefore the total load is

$$278 + 226 = 504 \text{ Btu per hour}$$

Condensing Unit Sizing

Commercial-size condensing units should not be chosen to operate continuously. They should have rest periods to cool off if they are to give a maximum of trouble-free service. Running times of from 14 hours to 22 hours per 24 hours have been used. A 16-hour running time is a conservative figure to use; that is, if all the time per day the unit was running was added up, it would amount to not over 16 hours. To choose a condensing unit under this condition, use the following formula:

$$Q_c = Q_t \times \frac{24}{16} = 1.5 Q_t$$

where

Q_c = Btu per hour capacity of the required condensing unit
Q_t = total Btu capacity of load

The K Factor

The heat leakage through some surfaces is better given for the average thickness in which such surfaces are made rather than for 1 inch. Glass is one of these surfaces. Heat leakage is given per square foot per degree per hour, but not per inch thick. It is given for the average thickness or average construction used. These data are also given in Table A–9 in the Appendix. Notice that capital K is used for these values, rather than the lowercase k we have used before. The capital K means that no thickness is included. The only difference in the use of K is that the thickness factor x is eliminated from the heat leakage formula. Where different walls of a box are made of different materials, each must be calculated separately.

EXAMPLE 6 What is the Btu capacity of the required condensing unit for the cabinet in Figure 18–7? A 40°F cabinet is required in Sacramento, California. The usage is estimated to be average. The cabinet has 3 inches of a standard insulation. The doors are triple glass.

Because this cabinet will be used in a retail store, the ambient temperature around the cabinet will probably be no higher than the design temperature. This is given as 105°F for Sacramento.

The gross area of the cabinet is calculated as

Front and back: $2 \times 6 \times 6 = 72$ square feet
Side: $2 \times 3 \times 6 = 36$ square feet
Top and bottom: $2 \times 3 \times 6 = 36$ square feet

Thus the total gross area is $72 + 36 + 36 = 144$ square feet.

There are five glass doors, each 1.5 feet square. Therefore the area of the glass is

$$5 \times 1.5 \times 1.5 = 11.25 \text{ square feet}$$

Thus the net insulated wall area = 133 square feet to the nearest whole number.

Heat leakage through the insulated walls is calculated as

$$Q_1 = \frac{AK(t_2 - t_1)}{x}$$

$$= 133 \times 0.3(105 - 40) / x = 864 \text{ Btu per hour}$$

Heat leakage through the glass doors is

$$Q_{1g} = AK(t_2 - t_1)$$
$$= 11.25 \times 0.29(105 - 40)$$
$$= 212 \text{ Btu per hour}$$

The net volume is

$$5.5 \times 5.5 \times 2.5 = 75.6 \text{ cubic feet}$$

The usage load is

$$Q_u = VF(t_2 - t_1)$$
$$= 75.6 \times 0.22 \times (105 - 40)$$
$$= 1081 \text{ Btu per hour}$$

The total heat load is 2157 Btu per hour.
The required size of condensing unit corrected for running time is

$$Q_c = 1.5 Q_t = 1.5 \times 2157 = 3236 \text{ Btu per hour}$$

Evaporator Sizing

The same principles of heat transfer used to calculate the heat flow through a wall can also be used to calculate the size of evaporator needed. Table A–12 in the Appendix includes K factors for various types of evaporator surfaces. The formula $Q = AK(t_2 - t_1)$ must be rearranged to solve for A.

$$A = \frac{Q}{K(t_2 - t_1)}$$

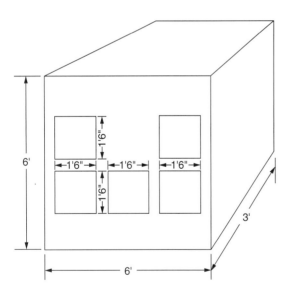

INSULATION: 3"
DOORS: TRIPLE GLASS
AMBIENT TEMPERATURE: 105°F
BOX TEMPERATURE: 40°F

Figure 18–7 Reach-in box with glass doors (Example 6).

In this case,

- A = surface area of required evaporator in square feet
- Q = total Btu load on cabinet corrected for running time
- K = heat transfer of the evaporator under conditions of operation
- t_2 = temperature of the air in the cabinet
- t_1 = temperature of the refrigerant in the evaporator

EXAMPLE 7 How much ⅝-inch bare copper tubing is needed to make an evaporator for the cabinet in Example 6? Assume 20°F temperature difference between the cabinet and the refrigerant, and a dry expansion evaporator. The K factor for the copper tube can be assumed to be 2.5 Btu per square foot per degree temperature difference.

$$A = \frac{Q}{K(t_2 - t_1)}$$

$$= \frac{3236}{2.5 \times 20} = 64.7 \text{ square feet}$$

From Table A–7 in the Appendix we find that it takes 6.1 lineal feet of ⅝-inch tubing to make 1 square foot. Therefore this evaporator requires

$$6.1 \times 64.7 = 395 \text{ lineal feet of 5/8-inch copper tubing}$$

A bare-pipe evaporator would not be chosen to refrigerate a cabinet of this type. Such an evaporator would be too costly, too heavy, and would take up too much room. By using fins on the tubing to increase heat transfer surface, the same amount of cooling can be obtained with less tubing and a smaller, lighter evaporator. A fan or blower blowing air through the evaporator further increases heat transfer. Thus, an even smaller evaporator can be used.

The size of fins in relation to the size tubing, the number of fins per inch, the type of contact between the fins and the tubing, and the velocity of air flow all affect heat transfer. Because of these variables, it is almost impossible to calculate accurately the heat transfer from a finned coil. Manufacturers' tables, which are based on test data, should be relied on to tell what the equipment will do. These tables are set up in different ways by different manufacturers, but any variation in these coils follows the formula just given for evaporator areas. A given coil can have a variety of capacities, depending on the temperature difference between the refrigerant and the cabinet air. Figure 18–8 shows how one manufacturer set up its rating table. Notice how a correction is made for different temperature differences.

EXAMPLE 8 What forced-draft coil in Figure 18–8 would be suitable for the cabinet in Example 6 if 20°F temperature difference is permissible?

A C3C195 coil has a capacity of 3900 Btu per hour. The next size smaller does not have the required capacity of 3236 Btu at 20°F temperature difference, so the C3C195 coil must be chosen.

Some tables only give the coil ratings for one temperature difference, usually 15°F or 20°F. Sometimes it is necessary to find the capacity of these coils at some other temperature difference to fit a particular job. Even when ratings are given for several temperature differences as in Figure 18–8, it is sometimes desirable to compute at some temperature difference that is not given. The easiest way to do this is to find the Btu capacity per degree from the conditions given, then calculate it for the required condition.

EXAMPLE 9 What is the Btu capacity of the C3C195 coil in Figure 18–8 at 12°F temperature difference?

If the Btu capacity at a temperature difference of 10°F is 1950 Btu, the capacity per degree is

$$\frac{1950}{10} = 195 \text{ Btu per degree}$$

The capacity for 12°F is then

$$12 \times 195 = 2340 \text{ Btu}$$

EXAMPLE 10 If coil C3B130 of Figure 18–8 is chosen for Example 6, what temperature difference between the cabinet air temperature and the refrigerant temperature is necessary?

The capacity per degree is

$$\frac{1300}{10} = 130 \text{ Btu}$$

Figure 18–8 Typical ratings and data for unit coolers. *Courtesy Fedders Manufacturing Co.*

Pounds Unit R-12 Model No. Charge. (30% Vol)	BTU Per Hr. at Specified TD Between Ent. Air and Refrig.*			CFM	Outlet Air Vel. FPM	Motor		Fan Dia.
	20° TD	15° TD	10°TD			RPM	HP	
C3A90 0.85	1,800	1,360	900	285	528	1,500	1/30	8 in.
C3B130 1.23	2,600	1,950	1,300	451	543	1,500	1/30	10 in.
C3C195 1.46	3,900	2,925	1,950	738	590	1,500	1/30	12 in.
C3D304 2.12	6,080	4,560	3,040	1040	518	1,140	1/20	14 in.
C3E428 3.02	8,560	6,420	4,280	1,300	518	1,140	1/20	16 in.
**C3E600 4.03	12,000	9,000	6,000	1,437	573	1,140	1/12	18 in.
**C3F780 5.30	15,600	11,700	7,800	1,880	595	1,140	1/8	20 in.
**C3G900 6.15	18,000	13,500	9,000	2,065	570	1,140	1/9	20 in.

*Capacity based on continuous running of compressor and unit at test conditions of 40° Ent. DB and 85% RH. To select unit, it will be necessary to take into consideration the off period of the condensing unit.
**Capacities based on units equipped by the user with valves having an external equalizer.
NOTE: On some applications, it may be desirable to operate at 25° TD. In such cases, multiply the 20° TD Btu ratings by 1.25.

General Unit Cooler Recommendations:

As a general practice we do not recommend these Unit Coolers for temperatures less than 38 degrees F., average (36 to 40 degrees F.) An attempt to operate these Unit Coolers to maintain lower than 38°F. temperatures may create a tendency to build up an excessive amount of frost on the unit cooler core. We recommend fans on all Unit Coolers to operate continuously. An intermittent control switch may be used provided no difficulty is experienced in obtaining the proper setting in synchronization with Condensing Unit. Motors furnished with these Unit Coolers are satisfactory for continuous operation and only require attention for oil occasionally.

The required load is 3236 Btu. Therefore the following temperature difference is required:

$$\frac{3236}{130} = 24.9°F \text{ or } 25°F \text{ as near as it is practical to read temperature}$$

Because the required cabinet temperature is 40°F, the required evaporator temperature for the job is

$$40 - 25 = 15°F$$

EXAMPLE 11 A walk-in cooler, Figure 18–9, is to be maintained at 38°F. It is in Denver, Colorado. The walls are 3½-inches of glass wool. The floor is 4 inches of corkboard. The floor is on a concrete slab that is laid directly on the ground. The cooler is to be used for meat in a busy market. What is the Btu rating of the required condensing unit and coil, if a blower coil with a 10°F temperature difference is to be used?

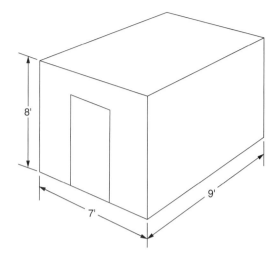

INSULATION: WALLS AND CEILING
3½" GLASS WOOL
INSULATION: FLOOR 4" CORKBOARD
AMBIENT TEMPERATURE: 90°F
BOX TEMPERATURE: 38°F
USAGE: HEAVY

Figure 18–9 Walk-in cooler (Example 11).

The cabinet wall and ceiling areas are

Front and back: $2 \times 8 \times 7 = 112$ square feet
Sides: $2 \times 8 \times 9 = 144$ square feet
Ceiling: $7 \times 9 = 63$ square feet

The total area of glass wool is thus $112 + 144 + 63 = 319$ square feet.
The floor area is $7 \times 9 = 63$ square feet.
The heat leakage figures are

$$Q_1 = \frac{AK(t_2 - t_1)}{x} \quad \text{(Walls)}$$
$$= 319 \times 0.29 \times (90 - 38)/3.5$$
$$= 1380 \text{ Btu per hour}$$

Because of the poor insulating value of the concrete, it will be neglected. Where a floor is laid directly on the ground, it is not necessary to calculate the heat transfer from the design temperature. Probably the best method is to assume that the ground is at the mean (day and night) summer temperature. Such mean temperatures are also available from United States Weather Bureau data. If this mean temperature is not available for the job being calculated, an average between the design temperature and the required temperature can be used. If this average is below 50°F (which only happens on freezer jobs), 50°F should be used instead of the average. In this case the average temperature is

$$90 + \frac{28}{2} = 64°F$$

Now,

$$Q_1 = \frac{AK(t_2 - t_1)}{x} \quad \text{(Floor leakage)}$$
$$= 63 \times 0.28 \times (64 - 38)$$
$$= 115 \text{ Btu per hour}$$

The net volume is

$$6'5'' \times 7'4\tfrac{1}{2}'' \times 8'5''$$
$$= 6.42 \times 7.38 \times 8.42$$
$$= 400 \text{ cubic feet}$$

A meat load in a busy market will certainly have heavy usage. The usage load is

$$Q_u = VF(t_2 - t_1) = 400 \times 0.12 \times (90 - 38)$$
$$= 2500 \text{ Btu per hour}$$

The total heat load is

$$1380 + 115 + 2500 = 3995 \text{ Btu per hour}$$

The Btu rating of the required condensing unit and coil is

$$Q_c = 1.5Q_t = 1.5 \times 3995 = 5993 = \text{Btu per hour}$$

The coil will be chosen for a 10°F temperature difference, and the condensing unit will be chosen for a 28°F suction temperature.

Boxes Larger than 1500 Cubic Feet

As mentioned earlier, usage factors are apt to give misleading results if they are used for boxes or cabinets larger than 1500 cubic feet. In such cases, the amount of product loaded into the cabinet, its temperature, and the time allowed to cool it should be determined accurately. The heat load is then figured by the method given in Chapter 1. Also, all cases in which a liquid or foods are frozen should be calculated. A large freezer load can be put in a small space, and average volumetric factors are not large enough to cover them.

EXAMPLE 12 A cooler, 20 feet × 20 feet × 8 feet high is to be built. It will sit out in the sun in a location where the design temperature is 90°F. The walls are insulated with 12 inches of sawdust. It should have capacity to cool 15 tons of peaches per 24 hours from a field heat of 90°F to a temperature of 35°F. What is the refrigeration load?

Because the building sits out in the sun, a correction to the design temperature must be made for sun effect. Corrections for sun effect are given in Table A-10 in the Appendix. Here we find that even the color of the wall makes a difference. Dark colors absorb more heat than light colors. Aluminum paint or polished metal surfaces absorb the least. In this case, we learn that the building will be painted a light color. Light-colored paint does not reflect as much heat as aluminum paint, so average figures are used. Thus we need to add 20°F to the temperature of the roof, 10°F to the east or west wall, and 5°F to the south wall. Because the sun cannot shine on the east and west walls at the same time, either can be chosen on which to make the correction, but not both.

Roof area: 20 × 20 = 400 square feet
Area of one wall: 20 × 8 = 160 square feet
Heat transmission (roof):

$$Q_1 = \frac{AK(t_2 - t_1)}{x}$$

$$= \frac{400 \times 0.45 \times (90 + 20 - 35)}{12} = 1125 \text{ Btu per hour}$$

Heat transmission (north and east wall):

$$Q_1 = \frac{AK(t_2 - t_1)}{x}$$

$$= \frac{(160 + 160) \times 0.45(90 - 35)}{12} = 660 \text{ Btu per hour}$$

Heat transmission (south wall):

$$Q_1 = \frac{AK(t_2 - t_1)}{x}$$

$$= \frac{160 \times 0.45 \times (90 + 5 - 35)}{12} = 360 \text{ Btu per hour}$$

Heat transmission (west wall):

$$Q_1 = \frac{AK(t_2 - t_1)}{x}$$

$$= \frac{160 \times 0.45 \times (90 + 10 - 35)}{12} = 390 \text{ Btu per hour}$$

Heat transmission (floor):

$$Q_1 = \frac{AK(t_2 - t_1)}{x}$$

$$= \frac{400 \times 0.45 \times \{[(90 + 35)/2] - 35\}}{12}$$

$$= 412 \text{ Btu per hour}$$

Total heat leakage is thus 2947 Btu per hour.

In addition to the actual leakage through the walls, there will be some heat loss every time the door is opened. This loss is caused not only by the warm air that enters and must be cooled, but also by the latent heat of the moisture in the air as it condenses and freezes on the coils. The usage factors given in the previous method of calculation include this. Here, however, a separate allowance must be made for these losses.

These door losses vary widely. During the day, when product is being loaded in or out, they can be as much as four times the heat transfer leakage. During a storage period when a door is only opened occasionally for checking, door losses may be 25 to 50 percent of heat transfer leakage. During the night, when doors remain closed, door losses may be negligible.

For averages over a 24-hour period, losses can be assumed to be twice the heat transfer leakage during loading periods and 25 percent of the heat transfer leakage during long-term storage periods. In this problem, the "two times" figure should be used during the loading and precooling times.

Door leakage: 2 × 2947 = 5894 Btu per hour

Usage load to cool 15 tons of peaches:

$$Q_u = WS(t_2 - t_1) = 15 \times 2000 \times 0.92 \times (90 - 35) = 1{,}520{,}000 \text{ Btu per day}$$

This usage can can occur over 24 hours, so

$$\frac{1{,}520{,}000}{24} = 63{,}200 \text{ Btu per hour}$$

The total heat load is thus

Leakage:	2,947 Btu per hour
Door losses:	5,894 Btu per hour
Usage:	63,200 Btu per hour
Total:	72,041 Btu per hour

If a machine using a halocarbon refrigerant is chosen for this job, we still have to correct for running time:

$$Q_c = 1.5 Q_t = 1.5 \times 72{,}041$$
$$= 108{,}060 \text{ Btu per hour}$$

Ammonia equipment is usually made to withstand continuous operation. Therefore, if an ammonia machine is chosen for this job, it should be selected on a different basis. However, it would not be safe to choose a machine with a capacity of 72,041 Btu per hour, because this allows no safety factor. An increase of 25 to 50 percent should be made to cover variables such as higher temperatures, extra loads, and so on. Therefore a machine as small as

$$1.25 \times 72{,}041 = 90{,}050 \text{ Btu per hour}$$

might be chosen.

If it might be necessary to cool a product load such as this without affecting other goods already in storage, it is not uncommon to double the estimated loads. Also, ammonia machines are usually rated in tons of refrigeration rather than Btu per hour. In this case, then, the actual load to use is

$$\frac{90{,}050}{12{,}000} = 7.5 \text{ tons of refrigeration}$$

To provide a safety factor, equipment capable of producing 7.5 to 12 tons of refrigeration should be chosen. Whether the lower figure or the higher should be chosen depends on the dependability and accuracy of the data, on the possible consequences of an overload, and on allowable costs.

In practice, a halocarbon refrigerant would be chosen for this size job, or for a job on a farm where maintenance would probably be minimal. The data for ammonia is given merely to illustrate the method of calculation.

EXAMPLE 13 An ice maker is to be designed to make 150 pounds of ice per 24 hours. Incoming water is to be 70°F. Ice freezing surfaces will be 15°F. Neglecting all cabinet losses, how much refrigeration is required to freeze the ice?

The product load per pound of water is

To cool water to 32°F: (70 − 32) = 8 Btu
To freeze water (latent heat): 144 Btu
To cool ice from 32°F to 15°F: 0.5(32 − 15)
= 8.5 Btu
Total load per pound of water: 190.5 Btu

Therefore the load to freeze 150 pounds of water in 24 hours is

$$\frac{190.5 \times 150}{24} = 1190 \text{ Btu per hour}$$

EXAMPLE 14 A combination cooler-freezer, Figure 18–10, is to hold a temperature of 35°F in the cooler and a temperature of −10°F in the freezer. The insulation is polystyrene foam. The design temperature

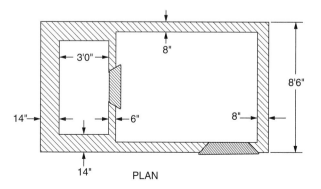
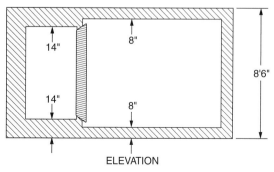

Figure 18–10 Cooler-freezer (Example 14).

222 CHAPTER 18 Heat Calculations

is 90°F. There is a fan in the freezer compartment that draws 50 watts of power. It is specified that the freezer must be able to freeze up to 25 pounds of miscellaneous food products per hour. What is the capacity in Btu of the required compressor?

1. Areas to cool from 90°F to 35°F:

   Sides: $2 \times 8.5 \times 10 =$ 170 square feet
   End: $8.5 \times 8.5 =$ 72.2 square feet
   Ceiling $8.5 \times 10 =$ 85 square feet
   Total: $=$ 327.2 square feet

2. Area of cooler floor: $8.5 \times 10 = 85$ square feet
3. Areas to cool from 90°F to −10°F:

   Sides: $2 \times 4.5 \times 8.5 =$ 76.5 square feet
   End: $8.5 \times 8.5 =$ 72.2 square feet
   Ceiling: $8.5 \times 4.5 =$ 38.2 square feet
   Total: 186.9 square feet

4. Area of freezer floor: $4.5 \times 8.5 = 38.2$ square feet
5. Area to cool from 35°F to −10°F: $8.5 \times 8.5 = 72.2$ square feet

Heat transfer, Area 1:

$$Q_1 = \frac{AK(t_2 - t_1)}{x}$$

$$= \frac{327.2 \times 0.25 \times (90 - 35)}{8} = 562 \text{ Btu per hour}$$

Heat transfer, Area 2:

$$Q_1 = \frac{AK(t_2 - t_1)}{x}$$

$$= \frac{8.5 \times 0.25 \times (55/2)}{8} = 73 \text{ Btu per hour}$$

Total leakage to cooler $= 635$ Btu per hour
Assume medium usage.
Volume $= 9 \times 7.2 \times 7.2 = 466$ cubic feet

$$Q_u = VF(t_2 - t_1) = 466 \times .08 \times 55$$
$$= 2050 \text{ Btu per hour}$$

Total load on cooler $= 2685$ Btu per hour
Correction for running time:

$$Q_1 = 1.5Q_t = 1.5 \times 2685 = 4027 \text{ Btu per hour}$$

Heat transfer, Area 3:

$$Q_1 = \frac{186.9 \times [90 - (-10)]}{14} = 334 \text{ Btu per hour}$$

Heat transfer, Area 4:

$$Q_1 = \frac{38.2 \times 0.25(100/2)}{14} = 34 \text{ Btu per hour}$$

Heat transfer, Area 5:

$$Q_1 = \frac{72.2 \times 0.25[35 - (-10)]}{6} = 136 \text{ Btu per hour}$$

Total leakage to freezer $= 504$ Btu per hour
The freezer load is to be made up of a variety of products. A check of the tables shows that the following figures will give a fair average.

Specific heat before freezing: 0.8 Btu per pound
Specific heat after freezing: 0.4 Btu per pound
Freezing point: 28°F
Latent heat of freezing: 110 Btu per pound

The load per pound of food is

To cool to 28F :
   $0.8 \times (90 - 28) = 49.6$ Btu per pound
To freeze: $= 110$ Btu per pound
To cool to −10°F:
   $0.4 \times [28 - (-10)] = 15.2$ Btu per pound
Total: $= 174.8$ Btu per pound

The load for 25 pounds per hour is

$$25 \times 174.8 = 4370 \text{ Btu per hour}$$

The heat from the fan is

$$50 \times 3.41 = 170 \text{ Btu per hour}$$

The total load on the freezer is thus

   Leakage: $= 504$ Btu per hour
   Product: $= 4370$ Btu per hour
   Fan: $= 170$ Btu per hour
   Total: $= 5044$ Btu per hour

The correction for running time is

$$Q_c = 1.5Q_t = 1.5 \times 5044 = 7566 \text{ Btu per hour}$$

The total load can be handled by two condensing units. The one to the cooler requires a capacity of 4027 Btu per hour at 15°F suction temperature. The second, to the freezer, requires a capacity of

7566 Btu per hour at −20°F suction. Alternatively, the load can be handled by one condensing unit of 4027 + 7566 = 11,593 Btu capacity at −20°F. Discussion of the methods of two-temperature multiple hookups is given in Chapter 25.

Insulation Thickness

The cost of refrigeration is made up of two parts, the cost of cooling the product and the cost of holding the temperature. The cost of cooling the product cannot be changed, other than through savings resulting from economical operation. The cost of holding the temperature is directly proportional to the heat transmission, which is in turn proportional to the thickness of the insulation. Economical thicknesses of insulation have been worked out for different temperature differences, as shown in Figure 18–11.

The thicker the insulation, the more it costs, but the less operating costs will be. The thicknesses of Figure 18–11 have been established by practice for all types of insulation that have good k factors, although they were first worked out for corkboard. For sawdust, shavings, or an insulation of similar k factor, a greater thickness is used to make up for the poorer insulating value. However, some designers feel that in many cases where overall economy is important, it is good practice to use a greater thickness of one of the cheaper forms of insulation, as long as it has a good k factor. If 8 inches of one type of insulation can be installed for the same price as 4 inches of another having the same k factor per inch, the annual savings in refrigeration can be considerable.

Domestic and commercial freezer boxes are an exception to the above rules. An additional factor is that a domestic box must be able to go through a 32-inch door and a commercial box through a 36-inch door. If they do not, they cannot be moved into the building where they are to be used. To install the conventional thickness of insulation in such cabinets with these limiting exterior dimensions would make the usable space too small. Therefore compromises must be made between space limitations and desirable insulation standards. This makes a wonderful application for some of the synthetic, low-k-factor types of insulation.

Condenser Calculation

The heat transfer in a water-cooled condenser has several variables. One of the first of these is the fact that as the water absorbs heat, its temperature rises. Therefore the temperature difference is not constant. Figure 7–6 shows graphically the temperature changes that actually take place. Because there is a much higher temperature difference at the beginning because of the highly superheated discharge gas, it might seem that the heat transfer would be greatest at this point. However, a wet surface is a better heat conductor than a dry surface. Therefore the k factor of the metal is lower until sufficient condensation starts to wet the condensing surface. So these two factors more or less cancel each other.

In practice, superheating and subcooling are usually neglected in calculations. Conditions for a typical problem are assumed as in Figure 18–12. Even here, notice that there is variation in the temperature differences. The temperature difference that must be used in the heat transfer formula is a logarithmic mean temperature difference. This can be calculated from the following formula:

$$\text{MTD} = \frac{D_2 - D_1}{2.3 \log D_2/D_1}$$

Figure 18–11 Recommended thicknesses of insulation based on cork or equivalent.

Cooler Temperature (°F)	Thickness (inches)
> 45	2
35 to 45	3
20 to 35	4
5 to 20	5
−5 to 5	6
−20 to −5	8
−35 to −20	10
−50 to −35	12

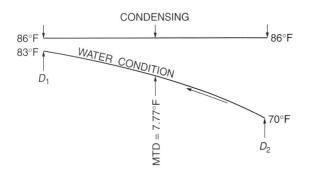

Figure 18-12 Mean temperature difference conditions.

where

MTD = log mean temperature difference
D_2 = greater temperature difference
D_1 = least temperature difference

EXAMPLE 15 What is the log mean temperature difference of Figure 18-12?

$$\text{MTD} = \frac{D_2 - D_1}{2.3 \log D_2/D_1}$$

$$\text{MTD} = \frac{16 - 3}{2.3 \log 16/3} = 7.77°F$$

To save time and the need to use logarithm tables to solve this formula, tables of answers have been made up (see Table A-14 in the Appendix). From these tables, the answer to the above or any other similar problem can be read directly.

This formula or table will work whether the temperature of one or both sides of the heat transfer surfaces vary. For instance, if one fluid is used to cool another fluid, the temperature of both will vary, but this formula or tables can still be used with accurate results.

A second variable in the performance of a condenser is the total quantity of water circulated. The less water is used, the higher will be its temperature rise to carry away the required heat. However, the higher the temperature rise, the higher will be the condensing temperature. And the higher the condensing temperature, the more heat of compression is generated by the compressor, which means more heat to be absorbed by the water. The compression ratio will also be higher.

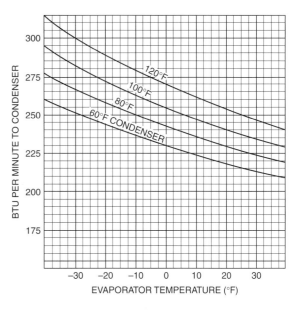

Figure 18-13 Btu per minute to condenser per ton refrigeration.

Figure 18-13 shows the total heat per ton per minute to be removed by the condenser for different operating conditions. This chart is for ammonia, but the chart for any common refrigerant will be similar.

A third variable that ties in very closely with the quantity of water is the effect of the velocity of the water. The faster the flow, the higher will be the heat transfer or k factor of the surface. Naturally, the more water passes through a condenser, the faster it must flow. The effect of velocity on heat transfer is shown in Figure 18-14. Average heat transfer values are also given in Table A-2 in the Appendix.

Still another variable in condenser operation is the condition of the heat transfer surface. This is also shown for a particular condenser and a particular water condition in Figure 18-14. Its exact condition for any condenser will depend on the water hardness, water treatment, and the length of time between cleanings. The designer must take this into account if the condenser is to be adequate under average cleaning schedules. The operating engineer must recognize the significance of this and maintain proper cleaning schedules.

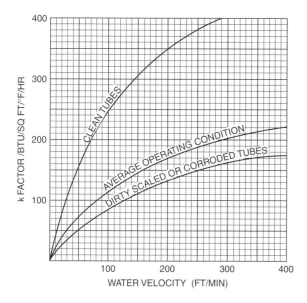

Figure 18-14 Heat transfer in a horizontal shell-and-tube condenser.

To combine all these factors, complete data on the condenser to be used must be obtained. Different manufacturers set their rating tables up differently, but they should include all the above variables. It is usually possible, using these manufacturers' tables, to choose two or three sizes of condensers to fit a particular job. However, it is important to recognize that to save a little money on the first cost of a condenser usually results in higher operating costs for the life of the equipment. It should also be remembered that the only disadvantage of an adequate size of condenser is the first cost, but it will keep condensing pressures at a minimum, which means reduced power costs. An oversized condenser is cheaper than an oversize compressor, yet the oversize condenser will aid the compressor in handling an overload by keeping the head pressure down. However, an overload on even an oversized compressor will overload everything in the system if the condenser is inadequate.

EXAMPLE 16 A horizontal shell-and-tube condenser is to be chosen for an ammonia plant rated at 50 tons with a 5°F evaporator and an 86°F condenser. If 70°F water is available, what size condenser will be suitable? Choose from Figure 18-14 at average operating conditions.

From Figure 18-13, the heat to the condenser will be 242 Btu per minute per ton at these conditions.

$$50 \times 242 = 12{,}100 \text{ Btu per minute total load}$$

High water velocities are desirable for best heat transfer values, but they are undesirable in that they increase pumping costs. If a velocity of 250 feet per minute is chosen, from Figure 18-14 we find a k factor of 190 Btu per hour per degree temperature difference.

If the same water is recirculated over a cooling tower, a 5°F rise is permissible. To absorb 12,100 Btu with a 5°F rise requires

$$\frac{12{,}100}{5} = 2420 \text{ pounds of water per minute}$$

There are 8.34 pounds of water per gallon, so this is

$$\frac{2420}{8.34} = 290 \text{ gallons of water per minute}$$

The heat transfer formula uses heat in terms of Btu per hour. Therefore the load of 12,100 Btu per minute is

$$12{,}100 \times 60 = 726{,}000 \text{ Btu per hour}$$

If a mean temperature difference of 7°F is assumed, the required heat transfer area is

$$A = H/K \, td$$

where $td = (t_2 - t_1)$, which in this case is the mean temperature difference, so

$$A = 726{,}000/190 \times 7 = 546 \text{ square feet of pipe surface}$$

If 2-inch pipe is used, from Table A-8 we find that it takes 1.61 lineal feet of pipe to make 1 square foot of surface. Therefore the condenser must have a total of

$$546 \times 1.61 = 880 \text{ lineal feet of 2-inch pipe}$$

If a condenser 10 feet long is chosen, this is

$$880/10 = 88 \text{ tubes}$$

The volume flowing through a pipe is found by multiplying the pipe area by the velocity of flow.

From Table A-8, the internal area of a 2-inch pipe is 0.0233 square feet. Then the volume flowing through one pipe is

$$250 \times 0.0233 = 5.82 \text{ cubic feet per minute}$$

There are 7.48 gallons per cubic foot. Therefore,

$$5.82 \times 7.48 = 43.6 \text{ gallons per minute per pipe}$$

Therefore, to carry 290 gallons per minute of water, the condenser must have the following number of parallel pipes:

$$290/43.6 = 6.65 \quad \text{or} \quad 7 \text{ tubes}$$

To parallel 7 tubes with a total of 88 tubes requires

$$88/7 = 12.6 \text{ passes}$$

The total number of tubes, the number of tubes per pass, and the total number of passes must all be rearranged to make them come out to be whole numbers. Usually the inlet and outlet water headers are on the same end of the condenser for convenience in piping. This requires an even number of passes in the condenser. In this case, 12 passes of 7 tubes per pass will not give the required surface. Therefore 12 passes of 8 tubes each are required.

Because a 70°F water temperature was included in the data given, and we assumed a 5°F rise in water temperature and a 7°F mean temperature difference, the condensing temperature will be less than 86°F. However, the figures given follow more nearly the conditions found in practice where plenty of water is available. As mentioned before, a 5°F evaporator and an 86°F condenser is a standard at which refrigeration equipment is rated. Actual operating conditions are usually at different temperatures.

EXAMPLE 17 What is the condensing temperature for the condenser chosen under the conditions of Example 16?

The data given tells us that there is a mean temperature difference of 7°F, entering water is at 70°F, and leaving water is at 75°F. Figure 18–15 shows graphically what is known of the problem. The beginning and final temperature differences are 5°F apart. So, by trial and error from Table A-14, temperatures 5°F apart are checked until a mean temperature of 7°F is found. A start can be made with temperature differences of 4°F and 9°F.

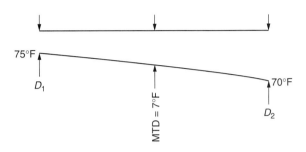

Figure 18–15 Mean temperature difference conditions (Example 17).

This gives a mean temperature of 6.17°F, which is too small. Temperatures of 5° F and 10°F give a mean of 7.21°F. This is slightly more than the 7°F required, but it is the nearest even temperature. This will give a condensing temperature of

$$70 + 10 = 80°F \quad \text{or} \quad 75 + 5 = 80°F$$

EXAMPLE 18 What is the condensing temperature if the water flow over the condenser of Example 16 is reduced to 4 gallons per minute per ton?

The total water over the condenser is now

$$4 \times 50 = 200 \text{ gallons per minute}$$
$$200 \times 8.34 = 1668 \text{ pounds of water per minute}$$

Absorbing 12,100 Btu with 1668 pounds of water causes the following temperature rise:

$$12{,}100/1{,}668 = 7.25°F$$

The velocity of the water is the volume divided by the area of pipe through which it flows. The area will be the area of eight 2-inch pipes. The volume of water in cubic feet is thus

$$200/7.48 = 26.8 \text{ cubic feet}$$

The velocity is then

$$26.8/8 \times 0.0233 = 143.5 \text{ feet per minute}$$

From this velocity and Figure 18–14, we find the K factor is 140 Btu per square foot per degree per hour. To find the required temperature difference, the heat transfer formula must be rearranged:

$$H = AK \, td$$
$$td = H/AK$$
$$td = 12{,}100 \times 60/546 \times 140 = 9.5°F$$

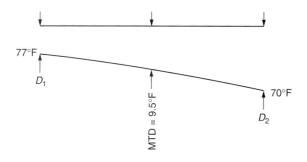

Figure 18–16 Mean temperature difference conditions (Example 18).

Figure 18-16 shows graphically what is now known about the problem. Now, by trial and error we must find a D_2 and D_1 that are 7°F apart (7.25°F if we can) and that give a mean temperature difference of 9.5°F. Temperatures of 6°F and 13°F give a mean of 9.08°F; temperatures of 7°F and 14°F give a mean of 10.1°F. The 6°F and 13°F figures give the closest to the required answer, and if the 13°F is increased to 13.25°F, which is the accurate figure, this will raise the 9.08°F to a point even closer to the required 9.5°F. Using the 13.25°F figure, the condensing temperature is

$$70 + 13.25 = 83.25°F$$

Questions

1. A reach-in cabinet is 7 feet high, 8 feet long, and 3 feet deep. The insulation is 3 inches of polystyrene foam. The cabinet is to be maintained at 40°F in New Orleans, Louisiana (design temperature 94°F).
   a. What is the heat leakage load?
   b. If average usage is expected, what is the usage load?
   c. What is the Btu rating of the required evaporator coil and condensing unit?

2. A combined cooler and freezer, Figure 18–17, is to be built for farm use near Buffalo, New York (design temperature 83°F). The

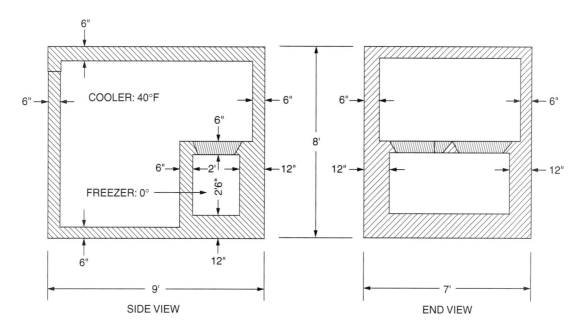

Figure 18–17 Combination cooler and freezer box (Question 2).

insulation is to be packed mineral wool. The walk-in space is to be held at 40°F, the small compartment at 0°F.

a. What is the heat leakage in the 40°F section?

b. The usage is estimated as light. What is the usage load in the 40°F section?

c. What Btu coil capacity should be chosen for the 40°F section?

d. What is the heat leakage on the freezer section?

e. If the freezer section is to be used only for storage, no freezing (light usage), what will be the usage load?

f. If the freezer has to be able to freeze 300 pounds of mixed food products per 24 hours and cool them to 0°F, what will be the usage load?

g. What Btu capacity condensing unit should be chosen for the entire job if the freezer is to be used only for storage?

h. What Btu capacity condensing unit should be chosen for the entire job if the freezer is to be used as in f?

3. A small steak box the size of Figure 18–3 is to go beside the stove in a restaurant in Atlanta, Georgia (design temperature 91°F). It has 3 inches of average insulation. It is estimated that the kitchen will be 20°F warmer than the design temperature. One of the 5-feet × 29-feet sides faces the stove. It is estimated this side will get 20°F warmer than the kitchen.

a. What is the heat leakage load?

b. What is the usage load? (Use heavy usage.)

c. What is the required Btu capacity of the condensing unit?

4. The following data are known about a water-cooled condenser:

Temperature of the water on is 66°F.

Temperature of the water off is 72°F

Quantity of water circulated is 412 gallons per minute.

Condensing temperature is 78°F.

Total surface in the condenser is 960 square feet.

a. How much heat is the condenser absorbing?

b. What is the mean temperature difference?

c. At what K is the condenser surface operating? (Rearrange the formula to $K = H/A \, td$.)

d. If the quantity of water is reduced to 280 gallons per minute, what will be the temperature of the water off? (Neglect the effect of a rising condensing temperature on the total heat to be absorbed.)

e. If the change of K due to reduced velocity in d is neglected, the mean temperature difference will not change. In this case, what will the new condensing temperature be?

5. Explain what is meant by R value. Relate it to k factor.

Chapter 19 Humidity in Refrigeration

Relative and Absolute Humidities

Humidity is the amount of moisture in the air. This moisture is in the form of water vapor, or low pressure steam. *Absolute humidity* is the actual weight of water vapor in a given amount of air, usually in grains[1] of moisture per pound or grains per cubic foot. *Relative humidity* is the actual amount of moisture in an air sample compared to the maximum amount of moisture the air sample can hold at the same conditions. That is, a relative humidity of 60 percent means that the air has absorbed, or is holding, 60 percent as much moisture as it is capable of holding with no change in temperature. Air at 100 percent humidity can absorb no more moisture. Air at a relative humidity of 100 percent is called *saturated air*.

Although absolute humidities are a measure of the actual moisture in the air, relative humidities are a better measure of how the air will affect other things. An air sample with 100 percent relative humidity can hold no more moisture, so it will not cause any drying. Air at anything less than saturated will absorb moisture from any wet product. The lower the relative humidity, the faster is this moisture absorption or the greater is the drying action. Air at 50 percent relative humidity is very drying to food products. As mentioned in Chapter 13, high humidities prevent drying, but also increase molds and sliming. Therefore the best humidity at which to keep food products is something of a compromise. Table A-2 in the Appendix lists the recommended humidities for different products for long-term storage. Figure 19-1 lists this information in a simpler form. This latter is usually used as a guide for commercial applications.

The thing that complicates the problem of relative humidity is the fact that air can hold different amounts of moisture at different temperatures. For every 20°F rise in temperature, air can hold approximately twice as much moisture. Air at 60°F can hold 5.8 grains per cubic foot. If it has this much moisture, it is saturated and can pick up no more moisture. If this same air is heated to 80°F, it will hold 11.5 grains per cubic foot. Because the air contains only 5.8 grains per cubic foot, it has a relative humidity of 5.8/11.5 = 0.50 or 50 percent.

Air at 50 percent humidity is very drying. If the temperature of the saturated 60°F air is dropped to 40°F, it will hold only 2.9 grains per cubic foot. It will spill out or condense the extra 2.9 grains and remain saturated. The moisture that is condensed will form as dew on the cooling coils or whatever else is used to cool this air.

Figure 19-2 shows the moisture solubility in air for different temperatures. Actually, this is nothing more than the lower end of Figure 2-2 drawn to a larger scale. The light lines in Figure 19-2 show various relative humidity points. That is, the 50 percent relative humidity line is halfway from the line corresponding to no moisture to the saturated line.

---
[1] A grain is 1/7000 of a pound.

Temperature Difference	Forced Air	6°F	10°F	16°F	20°F Up
	Gravity Coil	15°F	20°F	25°F	30°F Up
Relative Humidity		Over 90%	80% to 90%	70% to 80%	55% to 70%
Effect on Food Products		No drying or shrinkage. Surfaces usually moist. Molds and slimes hard to control.	Shrinkage noticeable on long-term storage. Molds and slimes easier to control, but not eliminated	Drying, shrinkage, wilting are Noticeable; very objectionable on long-term storage	Drying and "case hardening" excessive
Foods Suitable For Packaged		Fish and shellfish. Long term storage of frozen foods.	Meats and poultry. Average fruits and vegetables.	Nuts, Bulbs Dried food products.	Packaged goods only.

Figure 19–1 Effect of coil temperature differences and humidities on food products.

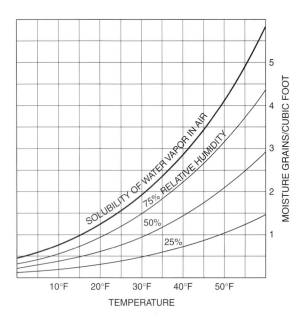

Figure 19–2 Moisture solubility in air.

Humidity Measurement

Throughout our refrigeration work we have constantly stressed the fact that evaporation absorbs heat. The evaporation of water is no exception; and the more water is evaporated, the more heat is absorbed. Without the application of additional heat, this heat absorption causes cooling. This fact is made use of to measure the relative humidity of air. Two thermometers are used, and a wet wick is wrapped around the bulb of one shown in Figure 19–3. The evaporation of water on the wet wick cools the bulb of this thermometer below the temperature of the other. The drier the air, the greater is the evaporation and cooling. The thermometer with the wet wick is called the *wet-bulb thermometer*. The thermometer without a wick is called the *dry-bulb thermometer*. The complete device is called a *sling psychrometer*.

From the readings taken from the dry-bulb and wet-bulb thermometers, psychrometric tables or

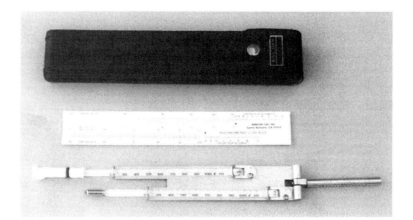

Figure 19-3 Sling psychrometer. *Photo by Bill Johnson, reprinted with permission of Abbeon Cal Inc.*

charts give the relative humidity of the air. Most newer sling psychrometers include a slide rule-type device right on the instrument to determine the relative humidity. Figure 19-4 is a standard psychrometric chart used for air at ordinary temperatures. Figure 19-5 is a low-temperature psychrometric chart that covers refrigeration temperatures. These charts are similar to the moisture solubility chart of Figure 19-2 with other information added.

Figure 19-6 shows an outline of the standard chart with one set of readings pointed out. The upper left-hand curved portion of the chart represents air that has a relative humidity of 100 percent. Dry-bulb temperatures are listed from the bottom of the chart. Wet-bulb temperatures are laid out on the saturation curve. If the air is at 70°F, the dry-bulb thermometer will show 70°F. If this air is saturated, there will be no evaporation from the wick. Therefore the wet-bulb will remain at 70°F. So the 70°F wet-bulb point is directly above the 70°F dry-bulb. If the air is 50 percent saturated, there is evaporation, and the wet-bulb temperature will cool down to 58.5°F. By picking out 58.5°F on the wet-bulb or saturated line, and following it along the diagonal wet-bulb lines to the intersection of the 70°F dry-bulb line, 50 percent relative humidity is found.

Notice, for a 70°F dry-bulb reading, a 43.5°F wet-bulb reading is the lowest that can possibly be obtained. This means an air with no moisture, which actually is almost impossible to obtain.

From the intersection of the 70°F dry-bulb point and the 58.5°F wet-bulb point, one can go horizontally to the left to intersect the saturated line and find the dew point. The *dew point* is the temperature to which this 70°F air with 50 percent relative humidity can be cooled to make it saturated. That is, if this air is cooled below 50°F, dew will condense or freeze out. At the outside edge of the chart, the absolute humidity in grains per pound can be found.

Air quantities are commonly measured in cubic feet, but the air volume changes as the temperature changes. Therefore the *weight* of air is more commonly used in problems in which air temperature changes are expected. A pound of air remains a pound of air, regardless of what happens to the temperature. The volume of air necessary to make a pound is given in the steep slanting lines of the chart. Air in the above case has a volume of 13.52 cubic feet per pound.

Figure 19-5 is a low-temperature chart that shows temperatures below 20°F, which are often used in refrigeration work. Notice that the freezing point, 32°F, does not change the overall shape or continuity of the chart, although there is a slight change in the slope of the wet-bulb lines. The lower the temperature, the less moisture air is able to hold. However, cold air still

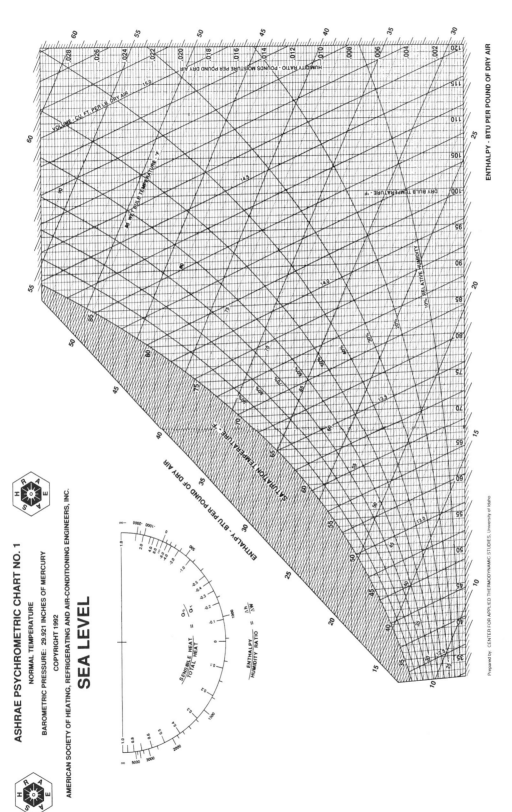

Figure 19–4 High-temperature psychrometric chart. *Reprinted with permission of ASHRAE.*

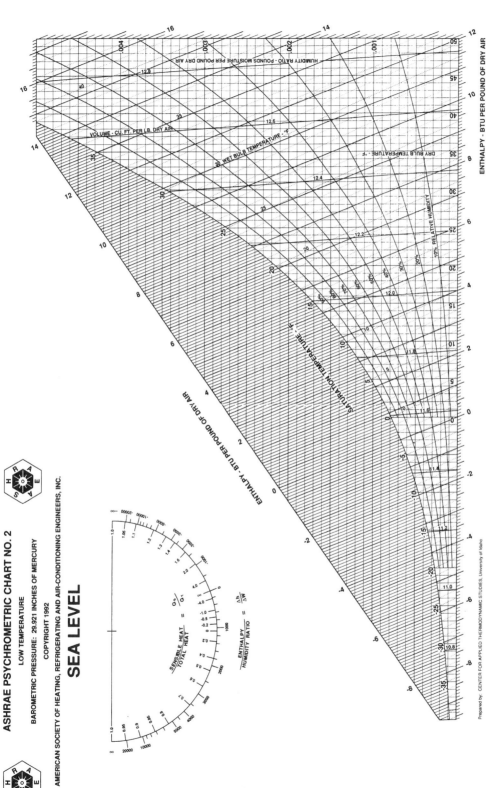

Figure 19-5 Low-temperature psychrometric chart. *Reprinted with permission of ASHRAE.*

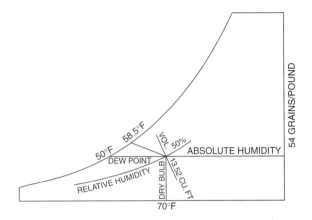

Figure 19–6 Outline of psychrometric chart with problem.

contains some moisture below 32°F. Vapor will be present in air even though any moisture condensed will form a solid (ice) instead of a liquid. The fact that water freezes somewhere along this curve has no direct effect on the moisture vapor still held by the air.

EXAMPLE 1 Psychrometric readings give a dry-bulb temperature of 46°F and a wet-bulb temperature of 39°F. What are the relative humidity, the dew point, and the absolute humidity?

These data can be found on either chart, but are easier to read on Figure 19-5.

Relative humidity = 53%
Dew point = 29°F
Absolute humidity = 24 grains per pound

EXAMPLE 2 Air at 70°F and 60 percent relative humidity (RH) is cooled to 40°F. What happens?

This example is best worked from Figure 19-4. The 70°F, 60 percent RH air contains 66 grains of moisture per pound. As the air cools, it will first come to 56°F dew point. At this point, the absolute humidity has not changed, but the relative humidity has increased to 100 percent. As the air cools further, it remains saturated, dew condenses out, and the absolute humidity decreases. At 40°F, the air is still saturated, and it now contains 36 grains

of moisture. Notice that when the air is saturated, the dry-bulb, wet-bulb, and dew points are all equal. In this case, the air has lost:

$$66 - 36 = 30 \text{ grains per pound}$$

EXAMPLE 3 Air at 70°F dry-bulb temperature and 40 percent relative humidity enters a refrigerator. It first passes over a coil where it is chilled to 20°F. Then, in passing through the cabinet, it is heated to 38°F by the cabinet walls, food, and other elements. What is its final condition?

Figure 19-7 illustrates on a skeleton psychrometric chart exactly what happens. First the air is cooled to its dew point, 45°F. Then moisture condenses as it cools down to 20°F. As it reheats from 20°F to 38°F, it changes from a saturated condition at 20°F to a relative humidity of 50 percent at 38°F. The exact conditions that occur when 20°F saturated air is heated to 38°F can be read more accurately on Figure 19-5.

Effect of Evaporator Temperature

As with many other things, the desired temperature difference between an evaporator and the cabinet air is something of a compromise. If a large temperature difference is chosen, a small evaporator will be sufficient, but this requires a lower backpressure,

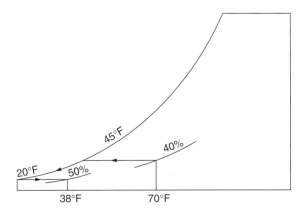

Figure 19–7 Outline of Example 3.

which penalizes the condensing unit. It may, in extreme cases, require a larger compressor and a larger motor. It also results in very dry air in the cabinet. The smaller the temperature difference is, the larger the coil required. Backpressures can be brought up to approach a pressure corresponding to the box temperature, but can never reach it. Also, the lower the temperature difference, the less moisture is taken out of the air as it passes over the coil. This means higher cabinet humidity.

Figure 19-1 shows temperature differences that have been worked out from practice for average commercial applications. These figures assume approximately 16 hours running time in 24 hours. Much variation from this running time will give different results. What actually happens on a typical job with a blower coil is shown by Figure 19-8. During the running cycle, the air flowing over the coils is cooled down to very nearly coil temperature. This brings its dew point down to the temperature to which it is cooled, 26°F at the end of the running cycle. As this air rises to the cabinet temperature of 38°F, its relative humidity is reduced to about 65 percent. During the off cycle, the air is not actively refrigerated, and it soon begins to warm the coil. Thus the dew point of the air leaving the coil rises, and with it the cabinet humidity. Toward the end of the cycle, the cabinet humidity approaches 100 percent. Thus, the 80 to 90 percent given in Figure 19-1 is an average over a period of time.

A colder coil will lower the dew point and relative humidity on the running cycle, thus lowering the average relative humidity. A longer running cycle will increase the length of time at low humidity and decrease the time at high humidity. This will lower the average relative humidity. Thus, where a certain job is chosen to control humidity as well as temperature, it must be computed accurately. If only temperatures are involved, an adjustment of the temperature difference will correct any variations from design conditions. However, every change of the temperature difference also changes humidities. So where humidities are involved, one item cannot be changed without affecting the other.

One difficulty encountered is in the seasonal variations of refrigeration loads. When equipment is properly sized to give 16 hours running time in the summer, this running time will decrease in the winter. This decrease may be enough that the humidity rises to the point where cabinet walls sweat and food begins to slime. Such conditions have been reduced in many ways.

"Starving" the coil by closing down on the expansion valve is probably the simplest solution. This makes less of the coil surface active, so it requires a greater temperature difference. This in turn lowers the backpressure, which increases the running time. Special types of controls have been used. Some are started by coil temperature and stopped by backpressure. Some are started by coil temperature and stopped by cabinet temperature. They all attempt to do a better job by balancing coil temperatures to cabinet temperatures than can be done by either a common pressure control or a common temperature control. Two-speed fans have been used, which cut down to a slow speed on the off cycle. This returns less moisture to the cabinet on the off cycle. The direct digital control (DDC) systems mentioned previously have contributed tremendously to automatically maintaining the proper humidity for virtually any application. This is done by constantly monitoring the ambient environment as well as the conditions in the controlled space and making the necessary adjustments.

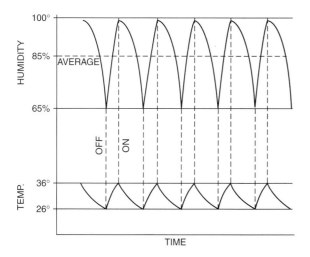

Figure 19-8 Chart from recording humidistat.

Electronics and Humidity Measurement

DDC systems require the use of electronic devices to measure humidity and transmit the information. Many of these devices involve the use of hygroscopic materials such as wood, human hair, and other materials. The most accurate measurement of humidity is done with hair. The length of the hair changes as the moisture content of the air increases or decreases. This length change is measured and transmitted to a computer, which processes the raw data into meaningful relative humidity readings. The computer then responds by adjusting humidification equipment, compressor capacity, and/or fluid flow rates.

One device uses an electronic element whose electrical resistance changes as the moisture content of the air changes. A small potential goes through the element. The amount of current returning back to the processor is converted electronically to a humidity reading. The DDC then uses this reading to make any necessary adjustments in the system.

Figure 19-9 shows a hand-held device that displays relative humidity and other readings at the push of a button.

Figure 19-9 A hand-held electronic device; models measure CO, CO_2, barometric pressure, temperature, and humidity. *Reprinted with permission of Bacharach.*

Questions

1. What is the difference between relative humidity and absolute humidity?
2. Why is relative humidity important in connection with food storage?
3. How is the relative humidity of a space found?
4. If the dry-bulb reading is 52°F and the wet-bulb reading is 41°F, what are the relative humidity, the dew point, and the moisture content?
5. If the dry-bulb reading is 50°F and the relative humidity is 40 percent, what are the wet-bulb temperature, the dew point, and the moisture content?
6. If the dry-bulb reading is 54°F and the dew point is 40°F, what are the wet-bulb temperature, the relative humidity, and the moisture content?
7. Show exactly what happens when 70°F air at 80 percent relative humidity enters a refrigeration box where it is first cooled to 25°F as it passes over the coil, then heats to 35°F in the box.
8. Show why, in Question 7, if the air is cooled to 15°F instead of 25°F by the coil and then reheated to 35°F, the cabinet humidity will be lower.
9. What temperature difference should be chosen for a blower coil to get an 85 percent cabinet humidity?
10. How will an increase in the percentage of running time decrease the relative humidity?

Chapter 20 Compressor Calculations I

It has previously been shown how liquid is evaporated to vapor in the evaporator. To absorb a given amount of heat in the evaporator, it is necessary to evaporate a certain amount of liquid. This evaporated volume of vapor must be removed, pumped out of the evaporator by the compressor. This in turn determines the size of compressor needed. Compressors are classified as constant-volume pumps, so the required rate of vapor removal from the evaporator is provided in cubic feet per minute, which is the capacity at which the compressor is rated.

Refrigerant Tables

The weight of liquid to be evaporated to absorb a given amount of heat varies with the refrigerant and with the temperature of that refrigerant. The size of each pound varies with the refrigerant and with the pressure. All these variables must be found from tables. These tables list properties of each refrigerant that have been determined by laboratory tests.

Figure 20–1 illustrates one line from a saturation properties table for R-134a. The first column gives the temperature for which these conditions hold. The second and third columns give the absolute and gage pressures at which R-134a will evaporate at 5°F. This is the same information as is given in graphic form in Figure 3–1. The next two columns give the volume of the liquid and vapor. This is the size in cubic feet of 1 pound at the pressure given.

Following this is the density of the liquid and vapor. The density is the weight of 1 cubic foot of the substance at these conditions. It could be calculated by dividing the volume into 1, but here it is calculated for us. Actually, the volume is more often used in calculations than the density.

The next three columns give heats, also called *enthalpies*. The first of these is the heat of the liquid. This is the heat necessary to heat the liquid from some point established as zero (−40°F for common refrigerant tables) up to the point under consideration. The second heat column is latent heat. As mentioned previously, latent heat is the heat necessary to evaporate 1 pound of fluid without a change of temperature. The third heat column is the total heat (enthalpy) of the vapor, which is the sum of the heat of the liquid and the latent heat.

The last columns are entropies. *Entropy* is a useful relationship between heat and temperature. The entropy of the liquid and the entropy of the vapor are given. For most work, the entropy of the vapor is the only one needed. The complete table gives this information for all temperatures required, from low evaporator temperatures to high condensing temperatures. Saturation tables for ammonia, R-134a are tables of superheated vapor, R-123, and R-22 are included in the Appendix. Following the saturation tables for ammonia and R-134a are tables of superheated vapor. For each saturation temperature, if the vapor is superheated after leaving the liquid, its volume, heat, and entropy are increased. It would

237

SATURATION PROPERTIES-TEMPERATURE TABLE

TEMP. F°	PRESSURE		VOLUME cu ft/lb		DENSITY lb/cu ft		ENTHALPY Btu/lb			ENTROPY Btu/(lb)(°R)	
t	Psia	Psig	Liquid V_f	Vapor V_g	Liquid $1/V_f$	Vapor $1/V_g$	Liquid h_f	Latent h_{fg}	Vapor h_g	Liquid S_f	Vapor S_g
5	23.772	9.072	0.0119	1.9327	63.81	0.6174	13.7	90.2	103.9	0.0308	0.2250

Figure 20–1 Line from table (5°F) Refrigerant 134a.

take a whole book to list all the possible combinations of superheats and pressures, so these data are given for each 10°F change of superheat for various pressures.

Some tables list the data in a little different order than that described, but if the column heading is always checked, there should be no reason to make an error.

Compressor Size

Figure 20-2 represents a schematic compression system operating at standard ton conditions of 5°F evaporator and 86°F condenser. The refrigerant is R-134a. A liquid at 86°F enters the expansion valve and evaporator. Leaving the evaporator is a 5°F vapor. From the R-134a tables, the heat of the liquid at 86°F is 40.2 Btu per pound, and the heat of the vapor at 5°F is 103.9 Btu per pound. Thus with 40.2 Btu per pound brought into the evaporator, and 103.9 Btu per pound taken out, the difference, or 103.9 − 40.2 = 63.7 Btu, is absorbed in the evaporator for every pound of liquid circulated. This is called the *refrigerating effect* or the *net refrigeration effect* (NRE).

As pointed out in Chapter 1, a ton of refrigeration is 200 Btu per minute. Therefore, if 1 pound of refrigerant gives 63.7 Btu, to get 1 ton of refrigeration requires the circulation of 200/63.7 = 3.14 pounds per minute of refrigerant.

From the tables, the size of 1 pound of vapor at 5°F is 1.9327 cubic feet. Therefore, to get 1 ton of refrigeration requires 3.14 × 1.9327 = 6.07 cubic feet of vapor per minute to be withdrawn from the evaporator. This must be pumped out by the compressor, so a compressor with a theoretical displacement of 6.07 cubic feet/min per ton of refrigeration must be chosen. This must be further corrected for losses in the compressor. If the compressor is 70 percent efficient, the actual displacement required is 6.07/0.70 = 8.67 cubic feet per minute per ton.

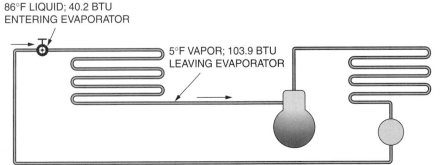

Figure 20–2 Schematic refrigeration system, with conditions entering and leaving the evaporator.

As a guide to these calculations, formulas can be used:

$$RE = h_g - h_f$$
$$w = 200/RE = 200/h_g - h_f$$
$$D_t = wV$$
$$D_a = D_t/E_v$$

where

RE = refrigerating effect in Btu per pound
h_g = heat of vapor at evaporator temperature in Btu per pound (from table)
h_f = heat of liquid at condensing temperature in Btu per pound (from table)
w = weight of refrigerant circulated in pound per minute per ton
V = volume of vapor (from table)
D_t = theoretical displacement in cubic feet per minute per ton
D_a = actual displacement in cubic feet per minute per ton
E_v = volumetric efficiency of the compressor

The volumetric efficiency factor, E_v, is a variable that is difficult to pick exactly. As pointed out in Chapter 6, the volumetric efficiency varies with different head pressures, backpressures, compressor clearances, and valve actions. It also varies for different refrigerants, being lower for dense refrigerants such as halocarbons. It is less for a worn compressor, in which valves and rings leak, than for one in good condition. For some compressors it varies with the speed. At high speeds, some valve actions are not as fast as they should be. With other types that depend on inertia to open and close the suction valves, efficiency becomes very low below certain speeds. Usually, the smaller the machine, the lower will be the efficiency. Figure 20–3 shows variations in efficiencies for different sized compressors made by one manufacturer.

The effects of head pressure and backpressure can best be combined under compression ratio, although even here, at low pressures, vapors may be so light that they affect valve action. Figure 20–4 gives the range of efficiencies for ammonia, and

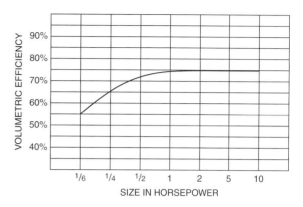

Figure 20–3 Efficiency vs. horsepower.

Figure 20–5, for two halocarbon refrigerants. (*Compression ratio* is the absolute discharge pressure divided by the absolute suction pressure.) Some manufacturers claim higher efficiencies for their equipment than is shown, and worn compressors may show lower values than the lowest ones shown. Therefore, to be safe, the average value is nearer the lower limit than the upper one. If it is suspected that a compressor is better or worse than average, the efficiency can be varied within the limits shown.

EXAMPLE 1 What are the theoretical and actual displacements for an ammonia compressor at 5°F evaporator and 86°F condenser conditions?

It saves time to pick as much data as possible from the tables before starting calculations. Then one is not interrupted repeatedly by having to go back to the tables for more information. From the ammonia tables, the following is found:

The heat of the liquid at 86°F = 138.9 Btu per pound.
The heat of the vapor at 5°F = 613.3 Btu per pound.
The volume of the vapor at 5°F = 8.15 cubic feet per pound.
The absolute pressure at 86°F = 169.2 psia.
The absolute pressure at 5°F = 34.27 psia.

240 **CHAPTER 20** *Compressor Calculations I*

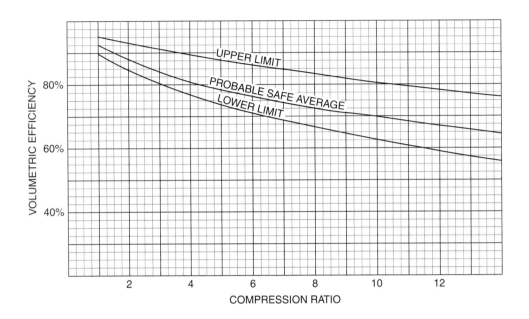

Figure 20-4 Volumetric efficiencies of ammonia compressors.

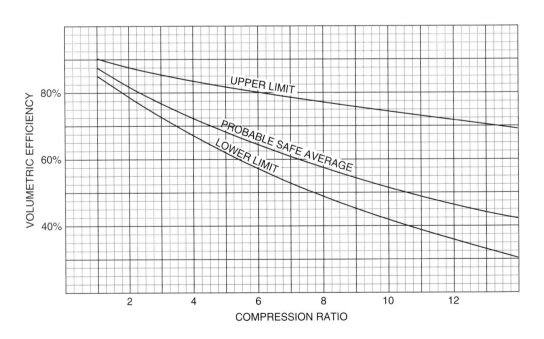

Figure 20-5 Volumetric efficiencies of R-134a and R-22 compressors.

From these data, the following values can be calculated.

$$RE = h_g - h_f = 613.3 - 138.9$$
$$= 474.4 \text{ Btu per pound}$$
$$w = 200/RE = 200/474.4$$
$$= 0.422 \text{ pounds per minute per ton}$$
$$D_t = wV = 0.422 \times 8.15$$
$$= 3.44 \text{ cubic feet per minute per ton}$$
$$\text{Compression ratio} = 169.2/34.27 = 4.93$$

From Figure 20-4, the average volumetric efficiency is 79 percent.

$$D_a = D_t/E_v = 3.44/0.79$$
$$= 4.36 \text{ cubic feet per minute per ton}$$

Compressor Horsepower

The amount of power necessary to drive the compressor can also be calculated from data obtained from the tables. In Chapter 2 it was stated that all the work done on the refrigerant vapor by the compressor ends up as heat. So the difference in heat between the vapor entering and that leaving the compressor is a measure of the power necessary to compress that vapor. This is where the useful relationship, entropy, is used. If there is no heat gain or loss to the compressor from outside sources, there is no change in entropy through the compressor. By "no heat gain or loss from outside sources" we mean that the vapor is neither heated nor cooled by contact with the compressor. Although this is a theoretical ideal, the variations from ideal are usually small. Therefore the real horsepower is very close to the theoretical figure.

In the case of the R-134a compressor at 5°F evaporator and 86°F condenser conditions, the following information can be taken from the tables.

Entering the compressor:
  Heat of vapor at 5°F = 103.9 Btu per pound
  Entropy of the vapor at 5°F = 0.2250

Leaving the compressor:
  Entropy of the vapor = 0.2250 (same as entering)
  Absolute condensing pressure = 112.0 psia

To find the remaining conditions leaving the compressor, turn to the R-134a superheat tables. First find the section headed as near 112.0 psia as possible. In the thermodynamic tables, we find 110 psia and 120 psia. The first is the closest. The third of these columns under this heading, labeled S, is entropy. Go down this entropy column until you find a figure as near 0.2250 as possible. The nearest is 0.2264. From this we find a heat of 118.3 Btu. Under the 120 psia column, the heat is 117.7 Btu. Interpolation between these two heats gives 118.18 Btu per pound leaving the compressor.

On the left column of the page, we find a temperature of 100°F. This 100°F temperature is the discharge temperature, and is of interest to check probable head temperatures.

If the heat of vapor leaving the compressor is 118.18 Btu per pound and that entering is 103.9 Btu per pound, the difference is

$$118.18 - 103.9 = 14.28 \text{ Btu per pound}$$

That is, for every pound of R-134a circulated, mechanical energy enough to create 14.28 Btu of heat must be used. In this case, 3.14 pounds per minute per ton is circulated. Therefore, this uses

$$3.14 \times 14.28 = 44.84 \text{ Btu per minute}$$

From Table A-4, we find that 42.42 Btu per minute is equivalent to 1 horsepower. Therefore, this will require

$$44.84/42.42 = 1.057 \text{ horsepower per ton}$$

This 1.057 horsepower is only the power necessary to compress the vapor; sufficient additional power must be added to make up for the mechanical losses in the compressor. The relation between the theoretical power and the actual power is the *mechanical efficiency*. This also varies according to compressor design, speed, size, and other factors. About 85 percent is a fair estimate for large equipment, and 75 percent is about right for small or fractional-horsepower equipment. When these data are being used to choose a motor for a small compressor, the motor size is usually made 50 percent larger than the theoretical horsepower. This provides some factor of safety. If a mechanical efficiency of 85 percent is assumed here, the actual driving power will be

$$1.057/0.85 = 1.24 \text{ horsepower}$$

The formulas for these horsepower calculations are:

$$\text{Theoretical horsepower} = thp = w(H_s - h_g)/42.42$$
$$\text{Actual horsepower} = ahp = thp/E_m$$
$$= w(H_s - h_g)/42.42 E_m$$

where

thp = theoretical horsepower
ahp = actual horsepower
H_s = heat of superheated vapor leaving compressor
h_g = heat of vapor leaving evaporator (entering compressor)
E_m = mechanical efficiency

Actually, there is some error in the above problem because we used the nearest figure rather than interpolating to get an exact figure. However, a four-way interpolation is usually required, which is somewhat tedious. An easier method, taking these figures from a chart, will be explained later.

EXAMPLE 2 What is the horsepower per ton in the case of Example 1?

The required data worked out in Example 1 are

$w = 0.422$ pounds per minute per ton
$h_g = 613.3$ Btu per pound

From the saturated ammonia tables we can find the following:

Entropy of the vapor at 5°F = 1.3253
Condensing pressure at 86°F = 169.2 psia

Then, in the superheat ammonia tables, the nearest to 169.2 psia to be found is 170 psia. The entropy nearest 1.3253 found under this pressure is 1.3249. From this the heat of superheat leaving the compressor, H_s, is 713.0 Btu per pound, and the discharge temperature is 210°F. The 210°F discharge temperature explains why water-jacketed cylinders are used with ammonia to reduce lubrication problems on the cylinder walls.

$$thp = w(H_s - H_g)/42.42 = 0.422(713.0 - 613.3)/42.42$$
$$= 0.99 \text{ horsepower, theoretical}$$
$$ahp = 0.99/0.85 = 1.16 \text{ horsepower, actual}$$

Complete Analysis of Expansion Valve and Evaporator

The above method of considering only the heat entering and leaving the evaporator gives the quickest method of finding the conditions required. However, a more exact analysis brings out some interesting points. For the R-134a example, the heat of liquid at 86°F entering the expansion valve was found to be 40.2 Btu per pound. As this liquid passes through this expansion valve, at the reduced pressure it cools to 5°F. From the table we find the additional data:

Heat of liquid at 5°F = 13.70 Btu per pound
Latent heat at 5°F = 90.2 Btu per pound

In passing through the expansion valve, the heat of liquid must be reduced from 40.2 Btu per pound to 13.70 Btu per pound.

$$40.2 - 13.70 = 26.5 \text{ Btu per pound}$$

This 26.5 Btu is removed from this liquid by evaporation of part of the liquid. To get 26.5 Btu of cooling requires the evaporation of

$$26.5/90.20 = 0.294 \text{ pound}$$

This 0.294 pound evaporated per pound of liquid circulated is 29.4 percent. So, at a 5°F evaporator and an 86°F condenser, 29.4 percent of the R-134a liquid must be evaporated to cool the rest of it. The remaining 70.6 percent of the liquid enters the evaporator, where it can do useful cooling. Because the latent heat was 90.20 Btu and 26.50 Btu were used up in getting to 5°F, we have

$$90.20 - 26.50 = 63.70 \text{ Btu per pound}$$

remaining. This is the same refrigerating effect found previously by the shorter method.

EXAMPLE 3 What percent of ammonia is evaporated in the expansion valve of Example 1?

Heat of liquid at 86°F = 138.9 Btu per pound
Heat of liquid at 5°F = 48.3 Btu per pound
Heat reduction = 138.9 − 48.3
= 90.6 Btu per pound
Latent heat at 5°F = 565 Btu per pound

Therefore,

90.6/565 = 0.16 or 16% of liquid evaporated in the expansion valve
565 − 90.6 = 474.4 Btu per pound refrigerating effect

Pressure–Enthalpy Diagram

A pressure–enthalpy diagram called a Mollier diagram provides the same information as the tables. The properties of saturated vapor can be read more accurately from the tables but, as mentioned previously, superheat conditions leaving the compressor are more easily found from the diagram. Also, problems involving wet or superheated suction vapor entering the compressor, as well as special problems such as two-staging, are much more easily worked out with the help of a chart.

Figure 20–6 shows an outline pressure–enthalpy diagram for R-134a. Heats in Btu per pound are given along the bottom and top, and absolute pressures on the side. The bold curved line on the left side of the diagram represents saturated liquid for different pressures. The area between the left side of the curve and the left-hand border corresponds to a subcooled liquid, that is, a liquid cooled below its boiling temperature. Note the temperature lines. The right-hand side of the bold curve represents saturated vapor. The distance between the left-hand side of the curve and the right-hand side of the curve represents the latent heat of vaporization. In this area, there is some liquid and some vapor. The percent of liquid that has evaporated is also shown. That is, one-tenth of the distance from the right to the left represents an evaporation of 10 percent of the liquid to vapor, leaving 90 percent liquid still unevaporated. This is called 10 percent *quality*. Temperature lines are horizontal in this area. This is because the addition of heat to an evaporating liquid does not change its temperature.

The space to the right of the bold curve is superheated vapor, that is, vapor heated above its corresponding condensing temperature. Here the temperature lines turn down again, but they are not straight because the specific heat of the vapor is different at different temperatures and pressures. Here also are constant entropy lines. The vapor, when compressed, will rise in pressure parallel to these constant entropy lines. Also in this region are plotted the volume lines, giving the volume of the vapor.

Figure 20–7 shows the pressure–enthalpy diagram for R-134a with the 5°F evaporator and 86°F condenser cycle drawn in on it. Point *A* represents 86°F saturated liquid entering the expansion valve. It has a pressure of 112 psia and a heat of 40 Btu (as near as can be read from the scale on the chart). In passing through the expansion valve the pressure drops, but there is no change in total heat. This is referred to as an *adiabatic process*. This is shown by the line *A–B*. The line *F–B–C* is drawn at constant pressure from the 5°F point on the saturated liquid line. It is at 23.8 psia pressure. Point *B* represents the condition of the refrigerant as it leaves the expansion valve and enters the evaporator. Its heat is still 23.8 Btu, but its quality is 30 percent.

Line *B–C* represents the change in the evaporator. The refrigerant evaporates at constant pressure and constant temperature, increasing in heat content as it does so. At *C* it leaves the evaporator to enter the compressor with 78 Btu of heat. The difference, 104 − 40 = 64 Btu per pound, is the refrigerating effect.

The length of the line *B–C* represents this refrigerating effect. The line *C–D* represents the vapor through the compressor. This rises from *C* at 23.8 psia to *D* at 112 psia pressure. It follows the entropy lines up to the condensing pressure. At *D* the heat, 118 Btu, and the temperature, 102°F, can

244 CHAPTER 20 Compressor Calculations I

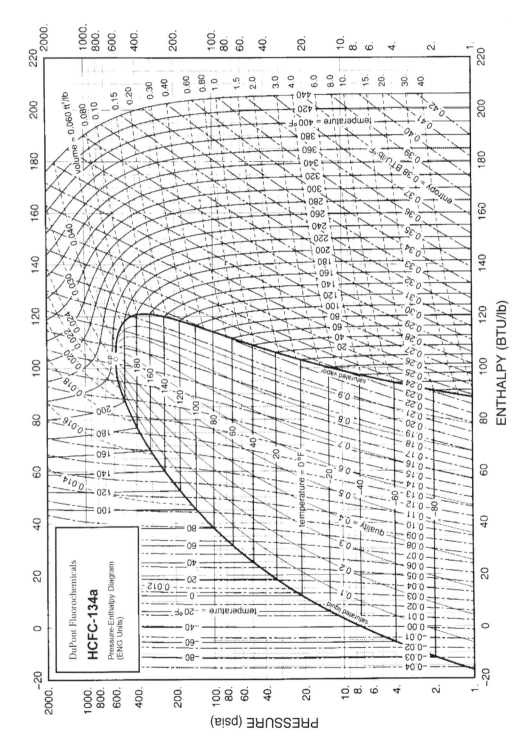

Figure 20–6 Outline pressure–enthalpy diagram for R-134a. Reprinted with permission of DuPont.

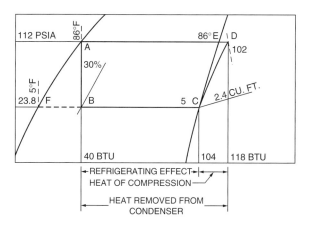

Figure 20-7 The compression cycle on the pressure-enthalpy diagram pressure.

be read directly. The difference, 118 − 104 = 14 Btu, is the heat of compression of the compressor, so it is the amount of energy in heat units necessary to compress the vapor.

The line D–A represents the condition in the condenser. From D to E, the superheat is removed from the vapor. From E to A the vapor is being condensed back to a liquid. The total heat removed in the condenser is 118 − 40 = 78 Btu per pound.

Notice, as mentioned in Chapter 2, that this heat removed in the condenser equals the heat absorbed in the evaporator plus the heat of compression, as shown in Figure 20-8. Also notice that the heat of compression, which is a measure of the power involved, is small compared to the heat absorbed in the evaporator. If we want to know the efficiency of such as system in which 64 Btu of refrigeration are produced with 78 Btu handled, the efficiency is

Figure 20-8 Heat from condenser equals heat to evaporator plus heat of compression.

64/78 = 0.82 or 82%

However, this is of less interest than the relationship between the refrigeration produced and the power required. In this case, that is

64/14 = 4.57

Because this is more than 1.00 or 100 percent, it cannot be considered an efficiency. This value is called the *coefficient of performance*. In this case, 4.57 times as much refrigeration is produced as power is consumed. This is not a form of perpetual motion, because the heat is merely transferred from the evaporator to the condensing water. The compressor is acting essentially as a heat pump to move the heat to the condenser.

Sizing Compressors

When available data have been used to calculate the displacement needed per ton, and the required tonnage is known, a machine with the required displacement can be chosen. Actually, most of these calculations have been done and are available in tables or charts. Figure 20-9 and Figure 20-10 show the required theoretical displacements per ton for ammonia, R-134a, and R-22 for various operating conditions.

The displacement required for R-502 is very near that for R-22. These figures must still be corrected for actual displacement by using the volumetric efficiency data of Figure 20-4 or Figure 20-5.

Figure 20-11 shows the theoretical horse power per ton for ammonia. Because the variation in power among most refrigerants is so small, this curve is suitable for the majority of refrigerants.

EXAMPLE 4 An ammonia compressor is required to produce 25 tons of refrigeration with a 20°F evaporator and an 80°F condenser. What size compressor and what size driving motor are required?

From Figure 20-9 we find that 2.4 cubic feet per minute are theoretically required. From the tables, the 20°F evaporator is at 48.21 psia and the 80°F condenser is at 153.0 psia. The compression ratio is

153.0/48.21 = 3.17

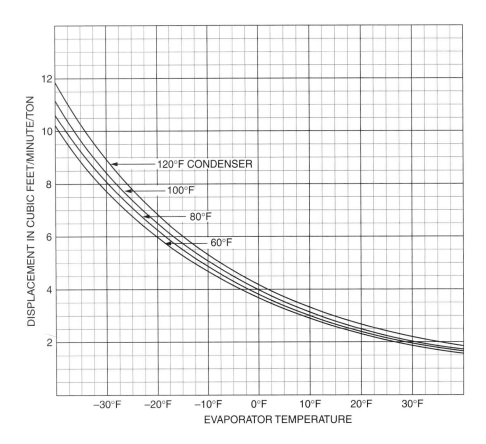

Figure 20-9
Displacement per ton, ammonia.

From Figure 20-4, a compression ratio of 3.17 gives a volumetric efficiency of 84 percent. Therefore, the actual required displacement per ton is

$$2.4/0.84 = 2.86 \text{ cubic feet per minute}$$

For a 25-ton compressor, the total displacement must be

$$25 \times 2.86 = 71.5 \text{ cubic feet per minute}$$

From Figure 20-11 we find that the theoretical horsepower per ton is 0.67 horsepower. We will assume that the motor is 85 percent efficient. Therefore, the actual power needed is

$$25 \times 0.67/0.85 = 19.7 \text{ horsepower}$$

This is very close to a 20-horsepower motor, which is a standard size. However, before choosing a 20-horsepower motor for this job, conditions should be checked that neither head pressures or backpressures will ever exceed design conditions. If they might, a 25-horsepower motor should be chosen so it will not be overloaded even at an abnormal condition. In this case, many designers would choose a 25-horsepower standard motor rather than a 20-horsepower high-starting-torque motor.

The actual displacement of a compressor can be calculated by multiplying the volume of each cylinder by the number of cylinders and by the speed. In formula form this is

$$D = 0.7854 B^2 L N \text{ rpm}/1728$$

where

D = displacement in feet per minute
B = diameter of bore in inches
L = length of stroke in inches
N = number of cylinders
rpm = revolutions per minute
1728 = cubic inches per cubic foot

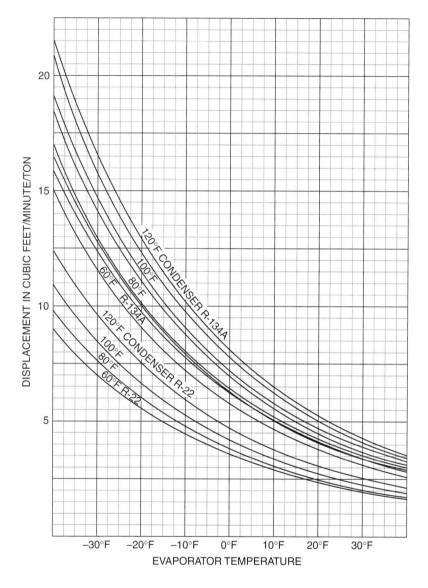

Figure 20–10 Displacement per ton, R-134a and R-22.

EXAMPLE 5 A manufacturer supplies a two-cylinder compressor with a bore of 4½ inches and a stroke of 3½ inches. The compressor can be connected directly to an 1140-rpm motor. Would this compressor be satisfactory for the conditions of Example 4?

$$D = 0.7854B^2LN \text{ rpm}/1728$$
$$= 0.7854(4.5)^2 \times 3.5 \times 2 \times 1140/1728$$
$$= 73 \text{ cubic feet/minute}$$

This compressor is satisfactory.

Tables are available that can be used instead of performing most of the above calculations. Figure 20–12 lists the displacement per inch of stroke per cylinder per 100 rpm for several small machines. The total displacement can be calculated by multiplying the appropriate factor in the table by the length of the stroke, the number of cylinders, and rpm/100, or

$$D = FLN \text{ rpm}/100$$

where F is the volume factor from the table in Figure 20–12.

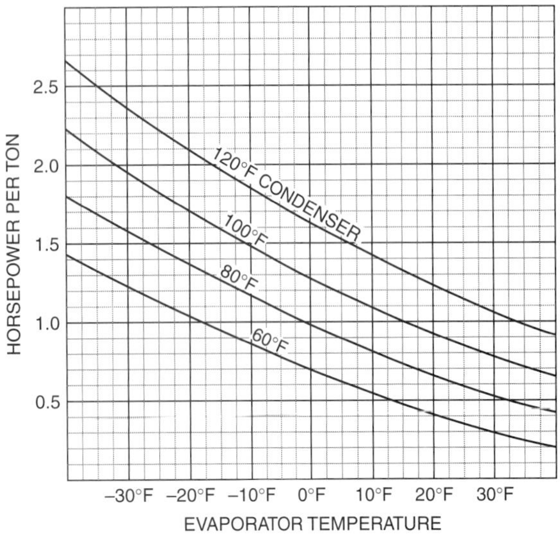

Figure 20–11 Theoretical horsepower per ton, ammonia.

EXAMPLE 6 What is the displacement of a two cylinder compressor with a 1⅝-inch bore and a 1¾-inch stroke that turns at 575 rpm?

From Figure 20–12, a 158-inch-bore compressor pumps 0.1196 cubic feet per inch of stroke per 100 rpm. The displacement is

$$D = 0.1196 \times 1.75 \times 2 \times 575/100$$
$$= 2.41 \text{ cubic feet per minute}$$

Bore	Cubic Feet per Minute	Bore	Cubic Feet per Minute
1	0.0454	2¾	0.344
1⅛	0.0575	3	0.409
1¼	0.0710	3¼	0.481
1⅜	0.0857	3½	0.557
1½	0.1022	3¾	0.639
1⅝	0.1196	4	0.727
1¾	0.1390	4½	0.920
1⅞	0.160	5	1.136
2	0.182	5½	1.373
2¼	0.230	6	1.636
2½	0.2841		

Figure 20–12 Displacement per inch stroke per cylinder per 100 rpm.

Most manufacturers furnish tables that show the actual capacities of their equipment under various operating conditions. Figure 20–13 is a generic table for illustrative purposes. This table is for air-cooled condensing units rated at a condensing temperature assumed to be average. At the bottom of the table is a correction factor to be applied for other condensing temperatures. Evaporator temperatures are listed across the top. All one has to do is to find the required evaporator temperature, then go down that column until the required rating is found.

EXAMPLE 7 What size machine is required for the conditions of Example 6 if a 20°F temperature difference is used for the evaporator?

The total required capacity is 3236 Btu per hour at an ambient temperature of 105°F. The evaporator temperature is

$$40 - 20 = 20°\text{F}$$

From Figure 20–13, in the 20°F column, a ½-horsepower machine turning at 300 rpm delivers 3220 Btu per hour, and turning at 355 rpm it delivers 3780 Btu per hour. Because a correction must be made for a higher condensing temperature, the 3220 Btu figure is too small. The ambient temperature of 105°F is 15°F above the rated condensing temperature. A 6 percent correction is required for each 10°F; in this case we need a 9 percent correction, or the machine will be good for only 91 percent of 3780 Btu per hour.

$$0.91 \times 3780 = 3440 \text{ Btu per hour}$$

Now we have more than the 3236 Btu per hour required, so this unit, ½ horsepower turning at 355 rpm, is satisfactory.

EXAMPLE 8 The A-5 unit chosen in Example 7 is a two-cylinder, 1¾-inch × 2-inch compressor. What is the volumetric efficiency of this unit at the conditions of Example 7? Is this volumetric efficiency reasonable?

From Figure 20–12, the displacement per inch of stroke per cylinder per 100 rpm is 0.139 cubic

Sizing Compressors 249

Model No.	HP	rpm	Suction Temperature, °F / Suction Pressure, psi												
			−20 / 0.6	−15 / 2.5	−10 / 4.5	−5 / 6.7	0 / 9.2	5 / 11.8	10 / 14.6	15 / 17.7	20 / 21.0	25 / 24.6	30 / 28.5	35 / 32.6	40 / 37.0
A-2-L	1/4	300	675	790	890	1080	1310								
A-2-S	1/4	285					1230	1440	1750	2060	2450				
A-2-H	1/4	275									2200	2450	2540	2630	2700
A-3-L	1/3	360	840	1100	1240	1540	1790								
A-3-S	1/3	345					1680	1680	2220	2530	2940				
A-3-H	1/3	335									2800	2960	3100	3240	3380
A-5-L	1/2	410	1450	1600	1880	2240	2520								
A-5-S	1/2	355					2180	2450	2860	3260	3780				
A-5-H	1/2	300									3220	3480	3880	4100	4430
A-7-L	3/4	290	2430	2750	3200	3750	4050								
A-7-S	3/4	255					3570	4100	4700	5300	6070				
A-7-H	3/4	220									5170	5520	5900	6360	6650
A-10-L	1	390	3870	4380	5100	5950	6500								
A-10-S	1	340					5500	6200	7150	8500	8650				
A-10-H	1	290									7350	7850	8400	8960	9600
A-15-L	1.5	295	6280	7100	8200	10200	10700								
A-15-S	1.5	255					8800	9600	10200	11700	13000				
A-15-H	1.5	220									11000	11900	13000	14100	15300
A-20-L	2	390	8200	9200	10600	13200	13750								
A-20-S	2	340					11900	12800	14300	15600	17300				
A-20-H	2	290									14600	16000	17600	19200	21000

For each 10° increase in room temperature, decrease capacity 6%
For each 10° decrease in room temperature, decrease capacity 6%

Figure 20–13 Capacities of air-cooled condensing units (BTU/hour for 90°F room temperature).

Refrigerating capacities and BHP requirements at 1200 rpm.

At reduced speeds, capacities and BHP requirements are decreased in direct proportion to reduction in speed.

Condensing pressure, psig and Corresponding Temperature, °F		Suction Pressures, psig, and Corresponding Temperatures, °F									
		0 / −28°	5 / −17.2°	10 / −8.4°	15 / −1.0°	20 / 5.5°	25 / 11.3°	30 / 16.6°	35 / 21.4°	40 / 25.8°	45 / 30.0°	50 / 33.8°
95 / 61.1°	Tons	5.5	9.0	12.6	16.0	19.6	23.2	26.7	30.0	33.4	36.9	40.4
	BHP	13.1	15.0	16.9	18.6	19.9	20.8	20.7	19.9	18.8	17.2	15.4
115 / 70.4°	Tons	5.1	8.5	11.9	15.3	18.7	22.1	25.5	28.7	32.0	35.3	38.6
	BHP	13.6	15.9	18.2	20.4	21.6	22.9	23.1	22.9	22.4	21.6	20.5
135 / 78.7°	Tons	4.7	8.0	11.3	14.5	17.7	21.0	24.2	27.4	30.6	33.7	36.9
	BHP	14.3	17.1	19.6	21.7	23.1	24.4	25.3	25.6	25.7	25.8	25.9
155 / 86.1°	Tons	4.3	7.4	10.6	13.7	16.9	20.0	23.1	26.3	29.4	32.3	35.4
	BHP	14.7	17.6	20.4	22.4	23.9	25.5	26.1	27.0	27.2	27.2	27.3
165 / 89.6°	Tons		7.2	10.3	13.3	16.5	19.5	22.6	25.7	28.7	31.6	34.6
	BHP		18.0	20.6	22.8	24.6	26.1	27.0	27.6	27.9	27.9	28.0
175 / 93.0°	Tons		6.9	9.9	13.0	16.1	19.1	22.2	25.3	28.2	31.1	34.0
	BHP		18.4	21.3	23.3	25.2	26.6	27.6	28.4	28.8	29.4	29.7
185 / 96.2°	Tons		6.7	9.5	12.7	15.8	18.7	21.8	24.7	27.7	30.6	33.5
	BHP		19.0	21.5	23.7	25.7	27.2	28.2	29.2	29.9	30.6	31.3
205 / 102.3°	Tons			9.0	12.1	15.2	18.1	21.0	23.9	26.9	29.8	32.6
	BHP			22.4	24.8	26.6	28.2	29.4	30.5	31.5	32.4	33.5
225 / 108.0°	Tons							20.3	23.1	26.1	28.9	31.8
	BHP							30.9	32.0	33.1	34.0	35.1

Figure 20–14 Two-cylinder ammonia multi-cylinder compressor (BHP = brake horsepower). *Courtesy of Vilter Manufacturing Corporation.*

feet per minute. For a two-cylinder compressor with a 2-inch stroke turning at 355 rpm,

$$D = 0.139 \times 2 \times 355/100 = 1.97 \text{ cubic feet per minute}$$

For an air-cooled condenser, a condensing temperature 20°F warmer than the condensing air is assumed. For a 90°F room this is a 110°F condensing temperature. From Figure 20–10, for a 20°F evaporator and a 110°F condenser, the required displacement per ton is 4.98 cubic feet per minute. Thus, a theoretical displacement of 4.98 cubic feet per minute produces 1 ton or 12,000 Btu per hour of refrigeration. A displacement of 1.97 cubic feet per minute gives

$$1.97/4.98 \times 12,000 = 4747 \text{ Btu per hour}$$

This manufacturer gives this machine a rating of 3780 Btu per hour. Therefore the volumetric efficiency at this condition is

$$3780/4747 = 0.796 \text{ or } 79.6 \text{ percent}$$

The evaporator temperature of 20°F gives a suction pressure of 33.13 psia. The condensing temperature of 110°F gives a condensing pressure of 161.23 psia. Therefore, the compression ratio is

$$161.23/33.13 = 4.36$$

Figure 20–5 shows an average volumetric efficiency of 71 percent with a possible maximum of 83 percent for a compression ratio of 4.36. Therefore, this compressor has a rating above average worn machines but conservative below top maxima.

Tables for ammonia machines often provide even more detail. Figure 20–14 lists the displacement and brake horsepower per ton at various suction and discharge pressures for a particular two-cylinder compressor. Four-, eight-, and twelve-cylinder compressors in this model line will produce two, four, and six times the refrigeration specified here.

Questions

1. What are the theoretical and actual displacements per ton for an R-22 compressor with a 5°F evaporator and an 85°F condenser?
2. Check the answer to Question 1 on Figure 20–10.
3. What is the required power per ton for the unit in Question 1?
4. Check your answer to Question 4 on Figure 20–11.
5. What is the percent flash gas at the expansion valve for the unit in Question 1?
6. What is the actual displacement of a 2-inch by 2¼-inch, two-cylinder compressor turning at 425 rpm? Calculate using the displacement formula.
7. Check your answer to Question 6 in Figure 20–12.
8. What is the volumetric efficiency of the compressor of Figure 20–14 operating at 25 psig suction pressure, 185 psig condensing pressure?
9. What is the volumetric efficiency of the A-5 unit of Example 8 with a −20°F evaporator?
10. Calculate the theoretical horsepower of the A-5 unit for a −20°F evaporator.

Chapter 21 Compressor Calculations II

Capacity Variation

Chapter 20 introduced various tables and charts to help choose the proper size equipment to do a given refrigeration job. In this chapter, refrigerant tables and pressure–enthalpy diagrams are used to analyze the operation of a refrigeration system under various operating conditions, and to analyze the operation of more complex refrigeration circuits.

One thing that is very apparent in Figure 20–9 and Table 20-2 is the reduction in capacity at lower evaporator temperatures. This is due mainly to the increase in volume of the refrigerant vapor at lower pressures. A worked-out problem can be used to illustrate the difference.

EXAMPLE 1 If Example 1 in Chapter 20 is repeated for a $-20°F$ evaporator, what is the required displacement per ton?

From the ammonia tables, we find the following:

Heat of vapor at $-20°F$:
$$h_g = 605.0 \text{ Btu per pound}$$
Volume of vapor at $-20°F$:
$$V_g = 14.68 \text{ cubic feet per pound}$$
Heat of liquid at $86°F$:
$$h_f = 138.9 \text{ Btu per pound}$$

Using this information,

$$RE = h_g - h_f = 605.0 - 138.9 = 466.1 \text{ Btu per pound}$$
$$w = 200/RE = 200/466.1$$
$$= 0.429 \text{ pounds per minute per ton}$$

$$\text{Displacement} = D_t = wV = 0.429 \times 14.68$$
$$= 6.30 \text{ cubic feet per minute per ton}$$

Compare the volume of vapor at $5°F$ with the volume at $-20°F$. Although the refrigeration effect is a little less at $-20°F$, the volume is the factor that makes the greatest difference. It must be remembered that the actual displacement varies more than the figure shown above, because the volumetric efficiency also declines at lower temperatures. The variations of course affect the performance of a given size compressor. The variation in volume makes the greatest difference, so a rough check on the change of capacity of a machine can be made by taking only this into account. The capacity is inversely proportional to this volume, but the volume is also inversely proportional to the absolute suction pressure. So this can be further simplified by using a direct proportion with the absolute pressures.

In formula form,

$$R_1/R_2 = P_1/P_2$$

and thus

$$R_2 = R_1 \times P_2/P_1$$

where

R_1 = refrigeration produced at original condition
R_2 = refrigeration produced at final condition
P_1 = absolute suction pressure at original condition
P_2 = absolute suction pressure at final condition

EXAMPLE 2 An ammonia machine is chosen to give 10 tons of refrigeration with a 5°F evaporator and an 86°F condenser. What is its capacity when the evaporator is running at −15°F?

The absolute pressures are

34.27 psia at 5°F
20.88 psia at −15°F

Then

$R_2 = R_1 \times P_2/P_1 = 10 \times 20.88/34.27 = 6.09$ tons

Effect of Changing Suction Temperature

Occasionally it is desirable to make up a table or chart showing the effects of variations in operating conditions. This might seem to be quite a task, but if the work is properly systematized, it can be reduced to a minimum.

EXAMPLE 3 A 4-inch × 4½-inch × 3-cylinder R-134a compressor operates at 300 rpm. The condensing pressure is to be held constant at 90°F. How will the capacity and horsepower of this machine vary with evaporator temperatures from −40°F to +40°F?

The actual displacement is found by formula:

$$D_a = 0.7854B^2LN \text{ rpm}/1728$$
$$= 0.7854 \times 4^2 \times 4.5 \times 3 \times 300/1728$$
$$= 29. \text{ cubic feet per minute}$$

Using the tables, the following data are found from the condensing temperature:

Absolute condensing pressure = 119 psia
Heat of liquid at 90°F = 41.6 Btu per pound

For the different evaporator conditions a table such as Figure 21–1 is set up. The various evaporator temperatures to be checked are listed in the first row. Then the information needed from the tables follows. Letters are assigned to each item to simplify formulas. At the end of each row is given the formula or method of finding the data required. By setting up the work in this manner, it is routinized so that the arithmetic involved becomes the only real chore to the problem. It is suggested that you check several of these figures for yourself.

Displacement = 29.4 cu ft/min; Condensing Temperature t_c = 90°F;
Condensing Pressure P_c = 114 psia; Heat of Liquid at 90°F h_c = 28.7 Btu/lb

Evaporator temperature t, °F	−40	−20	0	20	40	
Evaporator pressure P, psia	7.4	12.9	21.2	33.1	49.8	From table
Heat of vapor h_g, Btu	97.2	100.2	103.1	106	108.8	From table
Volume per pound V_g, cu ft	5.8	3.5	2.16	1.40	0.95	From table
Heat of superheated vapor H_s, Btu	123.12	120.85	119.5	117.63	117.05	From table
Refrigerating effect RE, Btu	55.6	58.6	61.5	64.4	67.2	$h_g - h_t$
w, lb/min/ton	3.6	3.41	3.25	3.1	2.98	200/RE
Theoretical displacement per ton D_t, cu ft	20.9	11.9	7	4.34	2.83	wV
Compression ratio CR	16	9.22	5.61	3.6	2.39	P_c/P_e
Volumetric efficiency E_v, %	37	53	66	73	79	From Figure 20–5
Actual displacement D_a, cu ft	10.9	15.6	19.4	21.5	23.2	29.4 × E_v
Refrigeration R, tons	0.52	1.31	2.77	4.95	8.19	D/D_t
Heat of compression, Btu	25.92	20.65	16.45	11.63	8.25	$h_s - h_g$
Theoretical hp/ton	2.2	1.66	1.26	0.85	0.58	$w(h_s - h_g)/42.42$
Actual hp	1.35	2.56	4.1	4.95	5.59	$R \times thp/0.85$

Figure 21–1 Effect of varying evaporator temperature (three-cylinder, 4-inch × 4½-inch, R-134a compressor, running at 300 rpm).

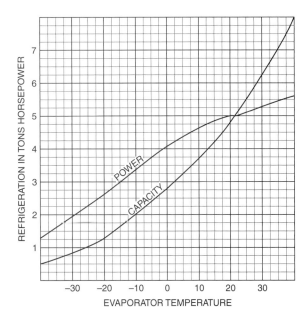

Figure 21–2 Effect of varying evaporator temperature (three-cylinder, 4-inch × 4½-inch, R-134a compressor, running at 300 rpm).

The relationships of the data worked out above is more apparent if a curve is plotted of the results. Figure 21–2 shows such a curve, giving the capacity and horsepower for different suction temperatures. If some point such as 0°F evaporator on such a curve is taken as 100 percent, variations each way can be figured in percents and then applied to any compressor working under the same conditions. Figure 14–1 is just such a curve.

Notice that both the capacity and horsepower decrease at lower temperatures, and the capacity decreases more quickly than the power. This means we need more power for a given amount of refrigeration. If the capacities at these lower temperatures are divided by the power, data for a curve similar to Figure 14–2 can be obtained. These data also help explain why equipment that has been properly chosen for a freezer job may be so badly overloaded as to open a motor overload when it is first started up with a warm evaporator. It also shows why a unit that is properly chosen for a freezer job will be overloaded on a higher-temperature job.

Effect of Changing Condensing Temperature

There are many factors besides suction temperature that, if varied, can change the compressor capacity and power. Although they do not produce as great a change as does suction temperature, their effects should be recognized. In Chapter 14 it was pointed out that, for economy, condensing pressure should be as low as possible. It is of interest to analyze the effect of changing condensing temperatures.

EXAMPLE 4 How will the capacity and horsepower of the compressor of Example 3 vary with condensing temperatures varying from 60°F to 120°F?

To study the effect of a varying condensing temperature, all other conditions including the evaporator temperature must be held constant. Therefore the problem is worked with the evaporator temperature constant at 0°F. Figure 21–3 shows the data for this problem, and Figure 21–4 shows the results on a curve.

Detailed Standard Ton Conditions

Both the previous examples assume no subcooling of the liquid entering the expansion valve and no superheating of the vapor entering the compressor. Actually, this is a highly theoretical condition that is seldom attained in practice. Small variations from this theoretical ideal usually do not make enough difference that they need to be taken into account when deciding on the size of equipment. However, one should know how to determine these differences and understand their effects. Superheat and subcooling are both important in the process because, without superheat, saturated refrigerant will enter the compressor, and without subcooling, a full column of liquid will not enter the metering device.

Displacement = 29.4 cu ft/min; Evaporator Temperature t_e = 0°F; Evaporator Pressure P_e = 21.2 psia; Heat of Vapor at 0°F, h_g = 103.1; Volume per Pound at 0°F = 2.16 cu ft/lb

Condensing temperature t_c, °F	60	80	100	120
Condensing pressure P_c, psia	72.2	101.5	139	186
Heat of liquid h_f, Btu	31.4	38.1	45.1	52.4
Heat of superheated vapor H_s, Btu	114.4	117.5	120.3	122.9
Refrigerating effect RE, Btu	71.7	65	58	50.7
w, lb/min/ton	2.79	3.1	3.45	3.94
Theoretical displacement/ton D_t, cu ft	6.03	6.7	7.45	8.5
Compression ratio CR	3.4	4.79	6.55	8.77
Volumetric efficiency E_v, %	75	68	62.5	55
Actual displacement D, cu ft	22	20	18.4	16.2
Refrigeration R, tons	3.65	3	2.47	1.9
Heat of compression, Btu	11.25	14.36	17.2	19.8
Theoretical hp/ton	.74	1.05	1.4	1.84
Actual hp	3.2	3.7	4	4.1

Figure 21–3 Effect of varying condensing temperature (three-cylinder, 4-inch × 4½-inch, R-134a compressor, running at 300 rpm).

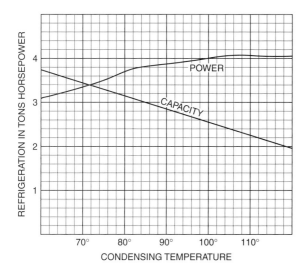

Figure 21–4 Effect of varying condensing temperature (three-cylinder, 4-inch × 4½-inch, R-134a compressor, running at 300 rpm).

In Chapter 20, all standard ton calculations were based on a 5°F evaporator and an 86°F condenser. In Chapter 2, the complete standard ton was described as including 9°F of subcooling in the liquid entering the expansion valve and 9°F of superheating in the vapor entering the compressor. These conditions are plotted in Figure 21–5. The heat of liquid at 77°F can be taken either from a chart or a table, but it must be taken from the temperature rather than the pressure, because it is the temperature that determines the heat. The condition entering the compressor with 9°F of superheat is most easily found on a chart such as Figure 21–5. The heat is 107 Btu and the volume is 1.9 cubic feet per. Following the entropy lines up to the discharge pressure of 112 psia gives the condition leaving the compressor. This has a heat of 120 Btu.

It is assumed that the 9°F of superheat is picked up in the suction line, so we still need the heat of saturated vapor at 5°F, 103.9 Btu. Now we can find:

$$RE = h_g - h_c = 103.9 - 40.2 = 63.7 \text{ Btu per pound}$$
$$w = 200/RE = 200/52.1$$
$$= 3.14 \text{ pounds per minute per ton}$$
$$D_t = 3.14 \times 1.93 = 6.06 \text{ cubic feet per minute per ton}$$
$$\text{Horsepower per ton} = w(H_s - h_g)/42.42$$
$$= 3.14(118.18 - 103.9)/42.42$$
$$= 1.057 \text{ horsepower per ton}$$

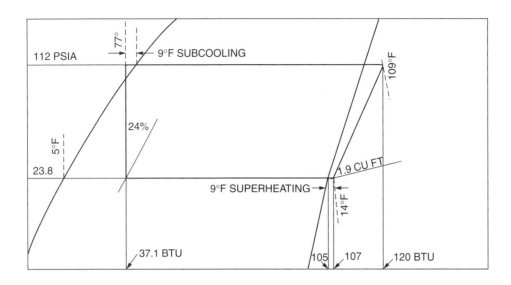

Figure 21-5 Standard ton conditions on pressure–enthalpy diagram, R-134a.

Notice the small difference between these figures and those in Chapter 20.

EXAMPLE 5 What are the capacity and power of ammonia at detailed standard ton conditions?

Figure 21-6 shows standard ton conditions on a pressure–enthalpy diagram with the quantities required for calculations. From this information:

$RE = h_g - h_c = 613.3 - 128.5 = 484.8$ Btu per pound
$w = 200/RE = 200/484.8$
$\quad = 0.413$ pound per minute per ton

$D_t = wV = 0.413 \times 8.3$
$\quad = 3.43$ cubic feet per minute per ton

Horsepower per ton $= w(H_s - h_g)/42.42 = 0.413$
$(721 - 619)/42.42 = 0.995$ horsepower per ton

Effect of Subcooling

Any subcooling of the liquid to the expansion valve decreases the flash gas and increases the refrigeration effect. This increases the refrigeration produced in the evaporator without requiring an

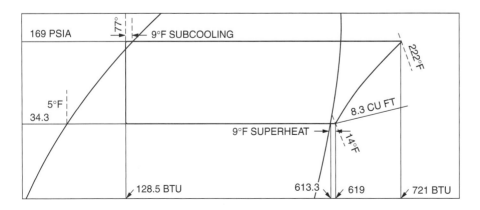

Figure 21-6 Standard ton conditions on pressure–enthalpy diagram, ammonia.

Condensing Temperature t_c = 100°F; Condensing Pressure P_c = 139 psia;
Evaporating Temperature t_e = 0°F; Evaporating Pressure P_e = 21.2 psia;
Compression Ratio CR = 5.55; Displacement = 29.4 cu ft/min;
Volumetric Efficiency E_v = 0.655; Actual Displacement D = 19.25 cu ft/min;
Heat of Vapor at 0°F, hg = 91 Btu/lb
Heat of Superheat H_s = 104.9 Btu/lb

Temperature of liquid to expansion valve, °F	100	80	60	40
Heat of liquid h_f, Btu	45.1	38.1	31.4	24.8
Refrigerating effect RE, Btu	45.9	52.9	59.6	66.2
w, lb/min/ton	4.36	3.78	3.36	3.02
Theoretical displacement/ton D_t, cu ft	9.42	8.16	7.26	6.52
Refrigeration in R, tons	2.04	2.36	2.65	2.95
Theoretical hp/ton	1.43	1.24	1.18	1.06
Actual hp	3.43	3.44	3.68	3.68

Figure 21–7 Effect of varying liquid subcooling with constant condensing pressure (three-cylinder, 4-inch × 4½-inch, R-134a compressor, running at 300 rpm).

increase in power. Figure 21-7 shows this worked out, and Figure 21-8 shows it in graphic form.

Effect of Superheated Suction Vapor

Superheating the suction vapor increases its volume, which reduces the capacity of the compressor. This takes place with little or no change in horsepower. Also notice that the temperature of the discharge vapor can be read directly from the pressure–enthalpy diagram for any condition entering the compressor. This method of determining the discharge temperature is recommended over the method given in Chapter 14. Figure 21-9 and Figure 21-10 show the variations for these varying superheats.

Wet Compression

Another check that can be made on the pressure–enthalpy diagram is the result of wet compression. This is the situation in which some liquid comes back up the suction line and enters the compressor. Figure 21-11 shows this plotted on a skeleton diagram. If a small amount of liquid returns, such as at A–B–C, the heat generated by the first compression stroke evaporates this liquid. However, this evaporation keeps the vapor from superheating until all the liquid is evaporated, as at B. Then, from B to C the vapor superheats. Naturally, the temperature will be lower at C than if dry vapor entered the

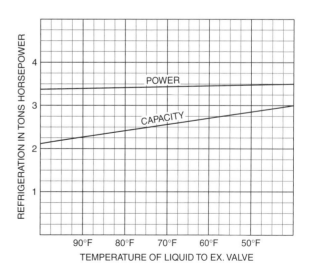

Figure 21–8 Effect of varying liquid subcooling (three-cylinder, 4-inch × 4½-inch, R-134a compressor, running at 300 rpm).

Condensing Temperature t_c = 90°F; Condensing Pressure P_c = 119.14 psia; Evaporating Temperature t_e = 0°F; Evaporating Pressure P_e = 21.2 psia; Compression Ratio CR = 5.62; Displacement = 29.4 cu ft/min; Volumetric Efficiency E_v = 0.66; Actual Displacement D = 19.4 cu ft/min; Heat of Vapor at 0°F, h_g = 103.1 Btu/lb; Heat of Liquid at 90°F, h_f = 41.6 Btu/lb; Refrigerating Effect RE = 61.5 Btu/lb

Superheated suction temperature t_s, °F	0	20	40	60
Volume per pound V, cu ft	2.16	2.3	2.4	2.5
Heat of vapor h_g, Btu	103.1	106	108.8	111.5
Heat of superheated vapor H_s, Btu	119.4	122.7	125.9	128.9
Theoretical displacement per ton D_t, cu ft	7.02	7.48	7.8	8.12
Refrigeration in R, tons	2.76	2.6	2.5	2.34
Heat of compression $H_s - hg$, Btu	16.3	16.7	17.1	17.36
Theoretical hp/ton	1.25	1.28	1.31	1.33
Actual hp	4.06	3.92	3.85	3.66

Figure 21-9 Effect of superheated suction vapor with constant evaporating and condensing pressure (three-cylinder, 4-inch × 4½-inch, R-134a compressor, running at 300 rpm).

compressor. The compression line D–E illustrates the maximum quantity of liquid that can be returned without trouble. Here there is just enough cooling effect in the liquid refrigerant present to match the heat of compression that would cause superheating. Thus a dry but saturated vapor is left in the cylinder at the end of the stroke. However, if more than this quantity of liquid is returned, as at F–G, the quantity G–E of liquid is left at the end of the stroke. Because a compressor is designed to pump vapor, not liquid, this noncompressable, unyielding liquid will be pounded against the valve plate, with damaging results. However, because about 5 percent liquid refrigerant can be returned and still be vaporized, it is easy to see why a small amount of refrigerant is not as damaging to a compressor as a small amount of oil.

Wet compression was at one time used to a considerable extent in ammonia compressors. It was impossible to return exactly the right amount of liquid from the evaporator to get the desired cooling, so a small expansion valve was added to the suction line at the compressor inlet. If a problem is worked out, two results are noticed. The first is a reduction in refrigeration capacity due to the shortening of the refrigerating effect line. The second result is a reduction in horsepower. In theory, the savings in power is greater than the loss in capacity, so there should be a net gain. In practice, however, the results did not seem to bear this out. It is thought that the heat transfer from the compressing vapor to the liquid was slower than the compression and discharge. Therefore part of this theoretical cooling took place in the discharge line on the way to the condenser, instead of inside the cylinder where it would be effective in reducing power requirements. Because wet compression does reduce discharge temperatures, it did serve a useful purpose here. If discharge temperatures got much above 250°F with normal compression, wet compression helped reduce or eliminate oil breakdown. However, wet compression is very seldom used at the present time.

Changing Refrigerant

In Chapter 3, it was mentioned that although a machine always works better if it is used with the refrigerant for which it was designed, there are times when consideration is given to the advisability of changing to another refrigerant. The tables or capacity curves provide a way to determine the capacity changes involved. The job could be completely recalculated for the new refrigerant, but this is not necessary. The biggest changes will be in the refrigeration effect and the volume of the vapor. The change in capacity is proportional to a change in the refrigeration

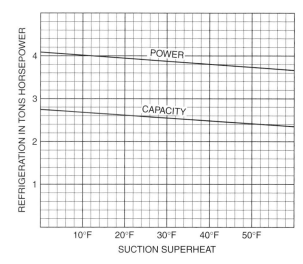

Figure 21–10 Effect of varying superheated suction vapor with constant evaporating and condensing pressure (three-cylinder, 4-inch × 4½-inch, R-134a compressor, running at 300 rpm).

effect, but inversely proportional to a change of volume. The following formula can be used:

$$R_2 = RE_2/RE_1 \times V_1/V_2 \times R_1$$

where

R_1 = refrigerating capacity with the first refrigerant
R_2 = refrigerating capacity with the second refrigerant
RE_1 = refrigeration effect with the first refrigerant
RE_2 = refrigeration effect with the second refrigerant

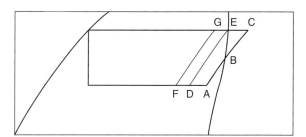

Figure 21–11 Outline of wet compression on a pressure–enthalpy diagram.

V_1 = volume of the first refrigerant vapor
V_2 = volume of the second refrigerant vapor

This gives the capacity at the new condition if no other changes are made. If R_1 is not definitely known, 100 percent may be used instead and the answer obtained in percent of the first capacity. If the compressor speed is divided by this percentage, a driving speed is obtained that will give the same capacity as the compressor working with the original speed and refrigerant.

EXAMPLE 6 An R-134a condensing unit operates with a 15°F evaporator and a 100°F condenser. It rotates at 660 rpm. What is the proper speed if the refrigerant is changed to R-22?

From the tables:

Heat of R-134a liquid at 100°F = 45.1
Heat of R-134a vapor at 15°F = 105.3
Volume of R-134a vapor at 15°F = 1.56
Heat of R-22 liquid at 100°F = 39.27
Heat of R-22 vapor at 15°F = 105.92
Volume of R-22 vapor at 15°F = 1.0272

Then,

$$\begin{aligned}R_2 &= RE_2/RE_1 \times V_1/V_2 \times R_1 \\ &= (105.92 - 39.27)/(105.3 - 45.1) \\ &\quad \times (1.56/1.027) \times R_1 = 1.68 R_1\end{aligned}$$

There will be 1.68 times as much refrigeration with R-22 if nothing else is changed. With a correction for speed, the new speed is

$$660/1.68 = 393 \text{ rpm}$$

This example could also be done using Figure 20–10. The following conditions are found:

Required displacement per ton = 5.2 cubic feet per minute with R-134a
Required displacement per ton = 3.1 cubic feet per minute with R-22

The required speed can be obtained from these data using a direct proportion:

$$\begin{aligned}\text{rpm}_1/\text{rpm}_2 &= D_1/D_2 \\ \text{rpm}_2 &= \text{rpm}_1 \times D_2/D_1 \\ &= 660 \times 3.1/5.2 \\ &= 393 \text{ rpm}\end{aligned}$$

Heat Exchangers

Heat exchangers, as mentioned in Chapter 8, are commonly used on low-temperature compressors with halocarbon refrigerants. They may be used with other refrigerants or for warmer applications. Many tables have been prepared to show savings made with heat exchangers, but these tables often show widely different results. The reason for this is often that widely different conditions are assumed to start with. To understand fully the effects of a heat exchanger, let us first study individual effects separately.

Figure 21–12 illustrates on a pressure–enthalpy diagram the effect of different compression ratios on the expansion valve and evaporator. A constant 80°F condenser but different evaporator temperatures are assumed. Notice that a lower evaporator temperature means more flash gas and a shorter refrigerating effect line. This reduces the capacity of a given size expansion valve because it has to pass more liquid to produce a given refrigeration effect. Also note that the NRE decreases as the suction pressure decreases. The capacity of the compressor is also reduced because of the flash gas the compressor has to pump, in addition the vapor formed in the evaporator. This decrease in NRE and increase in HOC decrease the coefficient of performance of the system.

For a given compression ratio, the cooler the liquid entering the expansion valve, the less the flash gas, and the greater is the refrigerating effect line, as shown in Figure 21–13. Therefore, it is desirable to cool the liquid entering the expansion valve as much as possible.

At the compressor end, anything that superheats the suction vapor increases its size, so reduces the weight of vapor a given compressor will pump, as shown in Figure 21–10. This reduction in capacity is usually with little or no decrease in power.

A heat exchanger will help where it reduces the liquid temperature, but it is a hindrance where it

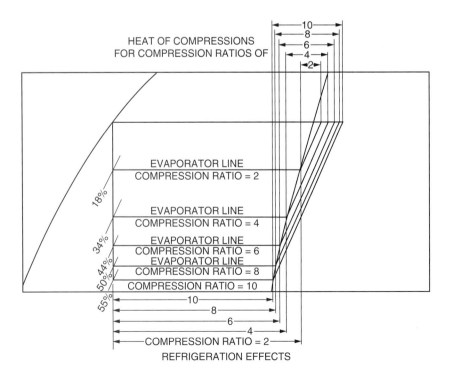

Figure 21–12 Effects of various compression ratios on pressure–enthalpy diagram.

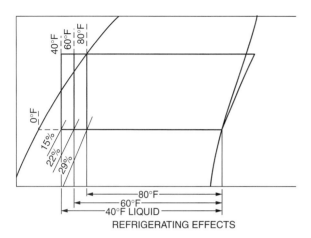

Figure 21-13 Effect of subcooled liquid on pressure–enthalpy diagram

increases the superheat of the suction vapor. In considering a heat exchanger, the gain at one end must be balanced against the loss at the other and the net result found. In addition to the effect on capacities, there are also other factors to be considered. On subzero jobs and on flooded jobs, it is hard to obtain enough superheat to operate a thermostatic expansion valve properly. Uninsulated, cold suction lines may collect frost or dew, which drips on the floor and becomes a nuisance.

For the simplest type of problem, consider a system such as the one shown in Figure 21-14, with and without a heat exchanger. We will calculate the size of machine necessary to give 1 ton of refrigeration with R-134a. A 0°F evaporator and an 80°F condenser are chosen. When the heat exchanger is used, one that will superheat the vapor to 40°F is assumed. The diagrams in Figure 21-14 provide all the necessary data. Following are the figures for the system without the heat exchanger:

Heat of saturated vapor at 0°F = 103.1 Btu per pound
Heat of liquid at 80°F = 38.1 Btu per pound
Refrigerating effect = 65 Btu per
(103.1 − 38.1)

Then,

200/65 = 3.08 pounds per minute per ton
3.08 × 2.16 = 6.65 cubic feet per minute per ton
3.08(118.24 − 103.1)/42.42 = 1.1 horsepower per ton

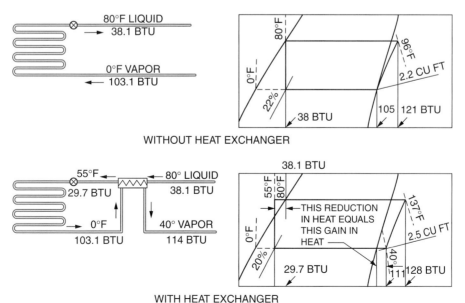

Figure 21-14 Effect of heat exchanger, R-134a.

Thus, for a compressor of 6.65 cubic feet per minute theoretical displacement will be necessary, and it will draw 1.1 horsepower.

For the system with the heat exchanger, it is possible to calculate the reduction in temperature of the liquid and the reduction in flash gas as shown in Figure 21-14. However, it is much easier to use the method of Figure 20-2 and consider the heat exchanger as part of the evaporator. Any gain in total heat beyond that of the liquid entering this part of the system must come from the evaporator. The figures follow:

Heat of vapor superheated to 40°F − 111 Btu per pound
Heat of liquid at 80°F = 38.1 Btu per pound
Refrigerating effect = 72.9 Btu per pound (111 − 38.1)

If we use the same 6.65 cubic feet per minute compressor, it will pump

$$6.65/2.4 = 2.77 \text{ pounds per minute}$$

The new refrigeration capacity is

$$2.77 \times 72.9 = 202 \text{ Btu per minute}$$

This is approximately 1 ton of refrigeration, so we have the same capacity as without the heat exchanger. The power required is

$$2.77(126.54 − 111)/42.42 = 1.01 \text{ horsepower per ton}$$

The superheated suction vapor has a larger volume, so less is pumped. At this particular condition, the extra cooling obtained by superheating the suction vapor is just enough to offset the loss due to the increased volume of vapor. However, because less weight of vapor is pumped, less power is needed to do the job. The compressor discharge temperature is raised to approximately 135°F, but this is not excessively high.

The same job using ammonia as the refrigerant is illustrated in Figure 21-15. The calculations follow.

Heat of saturated vapor at 0°F = 611.8 Btu per pound
Heat of liquid at 80°F = 132.0 Btu per pound
Refrigerating effect = 479.8 Btu per pound (611.8 − 132.0)

Then,

$$200/479.8 = 0.416 \text{ pounds per minute per ton}$$
$$0.416 \times 9.12 = 3.8 \text{ cubic feet per minute per ton}$$
$$0.416(713.0 − 611.8)/42.42 = 0.995 \text{ horsepower per ton}$$

The calculations with the heat exchanger follow:

Heat of vapor superheated to 40°F = 634.5 Btu per pound
Heat of liquid at 80°F = 132.0 Btu per pound

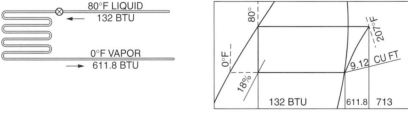

Figure 21-15 Effect of heat exchanger, ammonia.

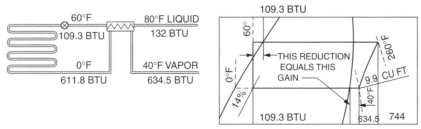

	Without Heat Exchanger		With Heat Exchanger		
Refrigerant	Capacity in Tons	HP	Capacity in Tons	HP	HP/Ton
R-134a	1.00	1.1	1.00	1.01	1.01
Ammonia	1.00	0.995	0.965	0.993	1.03

Figure 21–16 Effects of heat exchangers with two different refrigerants.

Refrigerating effect = 502.5 Btu per pound
(634.5 − 132.0)

Then,

$$3.8/9/9 = 0.384 \text{ pounds per minute}$$
$$0.384 \times 502.5 = 193 \text{ Btu per minute}$$
$$193/200 = 0.965 \text{ tons of refrigeration}$$
$$0.384(744.0 − 634.5)/42.42 = 0.993 \text{ horsepower}$$
$$0.993/0.965 = 1.03 \text{ horsepower per ton}$$

Here we find a condition under which the losses are more than the gains, but with no savings in power. The compressor discharge temperature with ammonia becomes 260°F, which is too high to be desirable.

The results of these calculations are tabulated in Figure 21–16. From this we can draw the following conclusions:

1. With an R-134a system, a heat exchanger increases the operating efficiency.
2. With an ammonia system operating at or near a saturated suction condition, a heat exchanger reduces the operating efficiency. It also raises the compressor discharge temperature to an undesirable level.

Actually, such heat exchangers are never used with ammonia. Carbon in the compressor heads and discharge valves, and burned discharge valves, would result.

Two-Stage Systems and Booster Systems

Low evaporator temperatures result in high compression ratios. High compression ratios require more power, and they increase discharge temperatures to a point at which lubrication problems are aggravated.

If the vapor can be cooled as it is compressed, power usage can be reduced and high-temperature lubrication problems eliminated. Sufficient oil is injected into screw compressors to do this cooling, but there is no way to cool the compressing vapor in a reciprocating compressor continuously. However, some cooling can be obtained between stages in two-stage compression. That is, half the compression is done in one cylinder, the vapor is then run through a cooler, then the compression is finished in another cylinder. These two cylinders may be built into the same machine, or they may be two separate machines.

Figure 21–17 illustrates this system with an outline of the cycle on a pressure-enthalpy diagram. If the vapor can be cooled to the saturation point A between the two stages, then the total distance B–C + A–D, which represents the power required in the two stages, will be less than the distance B–E, which would be the power for single-stage operation. Figure 21–18 shows an indicator diagram of the same set of conditions. The top half of the diagram is corrected for the volume of the second cylinder, which is less than that of the low-pressure cylinder. The cross-hatched section shows the savings in power over a single stage.

There is another place that power can be saved in this type of system. The liquid to the expansion valve is chilled by passing it through the intermediate cooler. This reduces the flash gas by about one-half. The flash gas formed must be compressed from the low pressure to the high pressure. The chilling done in the intermediate cooler evaporates some liquid, which corresponds approximately to the other half of the flash gas. However, this vapor

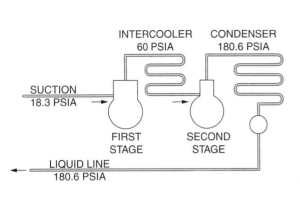
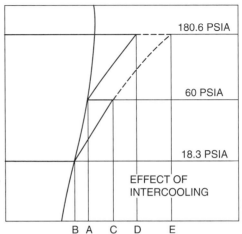

Figure 21–17 Two-stage system, ammonia.

need only be compressed from the intermediate pressure to the high pressure. Thus, there is a savings of about half the power used in compressing about half the flash gas.

Such small savings as these may seem too small to be important. This is true for fractional-horsepower equipment. For large systems, however, such savings begin to pay for themselves when compression ratios are higher than 10 to 1.

The intermediate pressure chosen is such that half the power is used in each stage. This pressure can be found by the following formula:

$$P_I = \sqrt{P_E P_C}$$

where

P_I = intermediate absolute pressure
P_E = evaporator absolute pressure
P_C = condenser absolute pressure

Because of variations from design conditions, this formula is found to work better in actual practice if it is modified as follows:

$$P_I = \sqrt{P_E P_C} + 5$$

Figure 21–19 shows such a system operating on ammonia, with the cycle outlined on a pressure–enthalpy diagram. This cycle is worked out per ton of refrigeration. The intermediate pressure is

$$P_I = \sqrt{P_E P_C} + 5 = \sqrt{10.4 \times 153} + 5 = 44.9 \text{ psia}$$

This is as near 50 psia as can be read on the chart or on pressure gages on the job. From the tables, this gives an intermediate temperature of 21.7°F. It will be assumed that the liquid to the expansion valve is cooled to within 5°F of the intermediate temperature. Then

$$21.7 + 5 = 26.7°F$$

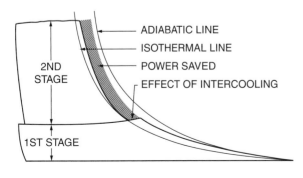

Figure 21–18 Indicator diagram for two-stage system.

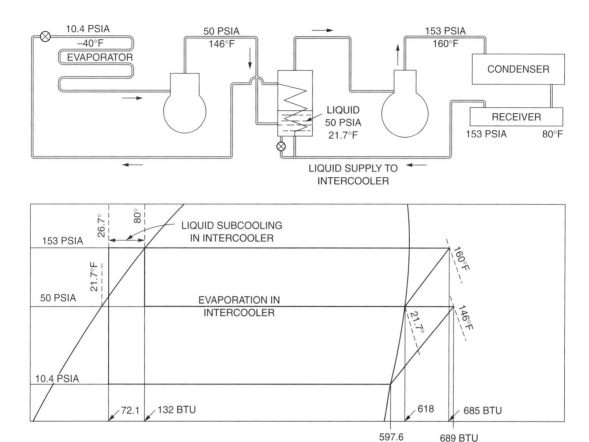

Figure 21–19 Complete two-stage system, ammonia.

Heat of vapor at −40°F = 597.6 Btu per pound
Heat of liquid at 26.7°F = 72.1 Btu per pound
Refrigerating effect = 525.5 Btu per pound
(597.6 − 72.1)

Now,

200/525.5 = 0.381 pounds per minute per ton

So

0.381 × 24.9 = 9.5 cubic feet per minute per ton

is the displacement of the low-pressure cylinder, and

0.381 (689.0 − 597.6)/42.42 = 0.822 horsepower

is on the low-pressure cylinder.

Vapor to the high-pressure cylinder is supplied by three sources: the discharge from the low-pressure cylinder, the liquid evaporated by the superheat in the low-pressure discharge, and the liquid evaporated by the cooling of the liquid to the low-temperature expansion valve.

The first source, the weight discharged from the low-pressure cylinder, is the same as that entering the high-pressure cylinder, 0.381 pounds per minute per ton. This vapor is discharged at 146°F and contains 689 Btu per pound.

To cool it to saturation reduces its heat content to 618 Btu per pound. Thus,

689 − 618 = 71 Btu per pound circulated
to be removed

The liquid supply from the receiver to the intercooler is at 80°F. At the intermediate pressure, it is 21.7°F. Therefore its refrigerating effect is

Heat of vapor at 21.7°F = 618.2 Btu per pound
Heat of liquid at 80°F = 132.0 Btu per pound
Refrigerating effect = 486.2 Btu per pound

To remove 71 Btu requires

71/486.2 = 0.145 pound evaporated per pound of ammonia circulated

Because 0.381 pound per minute per ton is circulated,

0.381 × 0.145 = 0.055 pound per minute per ton

This is the second source of vapor mentioned above.

The liquid to the low-temperature expansion valve is cooled from 80°F to 26.7°F. Thus,

Heat of liquid at 80°F = 132.0 Btu per pound
Heat of liquid at 26.7°F = 72.1 Btu per pound
Heat removed = 59.9 Btu per pound
(132.0 − 72.1)

This means that

59.9/486.2 = 0.123 pound evaporated per pound of ammonia circulated

Because 0.381 pound per minute per ton is circulated,

0.381 × 0.123 = 0.047 pound per minute per ton

This is the third source of vapor.
Now,
Vapor from evaporator = 0.381 pound per minute
Vapor from removing superheat = 0.055 pound per minute
Vapor from subcooling liquid = 0.047 pound per minute
Total high-pressure vapor = 0.483 pound per minute (0.381 + 0.055 + 0.483)

Therefore, 0.483 pound per minute per ton is handled by the high-pressure cylinder.
Next,

0.483 (685 − 618)/42.42 = 0.765 horsepower per ton

on the high-pressure cylinder.

Low-pressure power = 0.822 horsepower per ton
High-pressure power = 0.765 horsepower per ton

Total power = 1.587 horsepower per ton
(0.822 + 0.765)

Therefore, a total of 1.587 horsepower per ton is required.

If such a system is operated as a single stage, the following will be required:

Heat of vapor at −40°F = 597.6 Btu per pound
Heat of liquid at 80°F = 132.0 Btu per pound
Refrigerating effect = 465.6 Btu per pound

Therefore,

200/465.6 = 0.43 pound per minute per ton
0.43(778 − 597.6)/42.42 = 1.83 horsepower per ton
1.83 − 1.587 = 0.243 horsepower per ton saved
0.243/1.83 = 0.133 or 13.3% power savings by using two stages

If the superheated discharge from the first cylinder can be cooled with a water intercooler down to something near the 80°F condensing temperature, this will further reduce the savings made by two staging. Figure 21-20 shows how this water-cooled intercooler could be connected. The only change in the above problem is in the second item supplying vapor to the second stage.

Heat of vapor at 80°F = 630.7 Btu per pound
Heat of vapor at 21.7 F = 618 Btu per pound
Heat to be removed = 12.7 Btu per pound
(630.7 − 618)

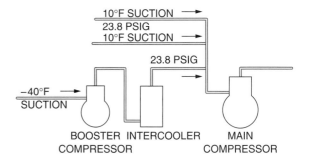

Figure 21-20 Booster compressor system using ammonia.

$12.7/486.2 = 0.026$ pound evaporated per pound of ammonia circulated

$0.381 \times 0.026 = 0.010$ pound per minute per ton

Because 0.047 pound per minute was required for this item in the previous case, this is a savings of

$0.047 - 0.010 = 0.037$ pound per minute

The total high-pressure vapor is thus

$0.483 - 0.037 = 0.446$ pound per minute per ton

The power required is then

$0.446(685 \times 618)/42.42 = 0.706$ horsepower per ton

on the high-pressure pressure cylinder, and the total power is then

$0.822 + 0.706 = 1.528$ horsepower per ton

This problem is for a job for which the $-40°F$ evaporator is the only temperature required. When this is the case, and the second cylinder handles only the vapor from the first cylinder and the intercooler, the system is called a *two-stage* (or *compound*) system. In a cold-storage plant, higher temperatures are also required. Where there is plenty of compressor capacity for this high-pressure suction, a compressor is sometimes added to the system that raises the low-pressure vapor to the higher evaporator pressure (see Figure 21–21). Such a compressor is called a *booster compressor*. The problem is the same as the one we have just done, except that the high-pressure compressor handles suction vapor direct from the warmer evaporators as well as the discharge from the booster.

For the most economical operation, the intercooling done between stages requires consideration. If plenty of cold water is available and intermediate pressures are high, it might be cheaper to use only a water-cooled intercooler with no liquid ammonia intercooling. Sometimes a combination of water cooling and liquid ammonia cooling, as in Figure 21–20, is used. This is usually the most efficient system. In some cases, however, the extra water cooler is not considered worth the potential savings. Each problem must be considered independently, with proper consideration of operating costs, equipment costs, and total operating time per year.

In the above systems, the refrigerant is ammonia. The oil goes to the bottom of the liquid refrigerant in receivers, intercoolers, and so on. It can be drained from these points if appropriate drain valves are supplied.

Oil is more difficult to separate from a R-134a mixture, so it makes an oil-rich mixture on top of the liquid refrigerant. With R-22, the oil and refrigerant separate at low temperatures and the oil floats on top of the liquid refrigerant. In either case, the oil can be trapped in a flooded intercooler and, under many operating conditions, flooded evaporators will trap oil.

To prevent such oil trapping in the intercoolers and at other points, a system such as the one shown in Figure 21–22 is used. Even here, if two compressors are used, the oil levels in each must be watched and corrected if needed. Often certain operating conditions will flush the oil from the low-pressure compressor to the high-pressure compressor. An oil return line must be run from the high-pressure crankcase to the low-pressure crankcase. A valve in this line is opened only when an excess of oil appears in the high-pressure crankcase. The higher second-stage suction pressure then forces oil back to the low-pressure crankcase.

The valve may be hand operated and must be opened and closed by an operator as needed. Automatic operation may include an oil sump

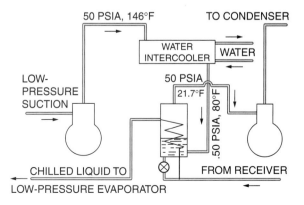

Figure 21–21 Intercooling with water + liquid ammonia.

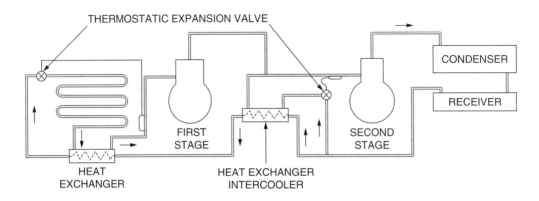

Figure 21-22 Two-stage system for halocarbon refrigerants.

containing a float switch at the proper level installed in the high-pressure crankcase. The float switch operates a solenoid valve in the oil return line. There are also three-stage compound systems for very special, low-temperature jobs.

Cascade Systems

Certain problems inherent in high-compression-ratio systems are eliminated using a cascade system, as shown in Figure 21-23. This configuration uses two separate systems, but the evaporator of the higher-temperature system furnishes the cooling for the condenser of the low-temperature system.

The advantages of such a system over a two-stage system are in getting better separation of the high-pressure, high-temperature end from the low-pressure, low-temperature end. Oil suitable for the low-temperature evaporator does not have to withstand the heat of the high-pressure compression, and vice versa. In cases where temperatures of 50°F to 100°F below zero are required, they are more easily obtained with refrigerants such as R-13. The evaporator of the second stage, using R-22 or R-134a, keeps the condensing pressure of the high-pressure stage within reasonable limits. The disadvantages of such a system are that more equipment is required, which means higher first costs, and the double heat transfers

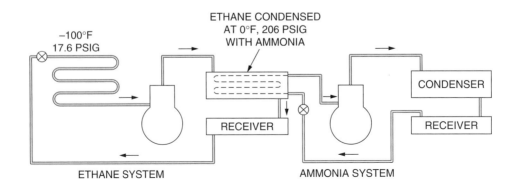

Figure 21-23 Cascade system.

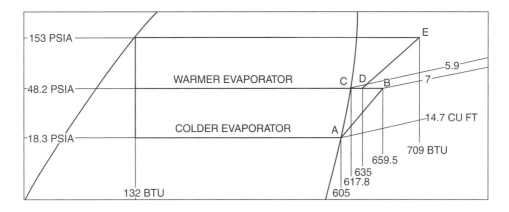

Figure 21-24 Dual-effect compression on pressure–enthalpy diagram.

through the extra condenser–evaporator make it less efficient. There are also three-stage cascade systems for very special, low-temperature jobs.

The system shown in Figure 21–23 uses ethane in the low-temperature system and ammonia in the high temperature system. Other refrigerants suitable for use in the low-temperature system of a cascade system may include:

R-503 (40.1 percent R-23 and 59.9 percent R-13)
R-290 (propane) may be used to help the oil flow at low temperatures
R-13
R-13B1
R-744 (carbon dioxide)
R-508B (46/54) SUVA(R) 95

There are and will continue to be other combinations and possibilities as the refrigerant families are further developed. In any cascade system, the compressors start and stop via the same temperature control device. In other words, you cannot run any one system by itself.

Questions

1. An ammonia machine is chosen to produce 10 tons of refrigeration with a 5°F evaporator and an 86°F condenser. What is its capacity with a 40°F evaporator?

2. Make a table similar to Figure 21–16 showing the capacity and horsepower of the compressor of Question 1 at evaporator temperatures from +40°F to −40°F.

3. An R-22 compressor is operating at −15°F suction and 90°F condensing temperature. Its speed is 825 rpm. If the refrigerant is changed to R-134a, with 0°F suction and the same condensing temperature, what new speed will maintain the same load on the motor?

4. What is the required displacement per ton and the horsepower per ton for a single-stage R-134a system operating with a −40°F evaporator and 100°F condenser and without a heat exchanger?

5. For Question 4, what is the required displacement per ton and the horsepower per ton if a heat exchanger that will heat the suction vapor to +20°F is added?

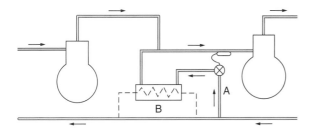

Figure 21-25 Two-stage R-134a system, Question 7.

6. For Question 4, what intermediate pressure would you choose if this system is to be two-staged? Use the formula $P_I = \sqrt{P_E P_C}$ for R-134a.

7. If an intercooler is added to the system of Question 6 as in Figure 21–25, how much refrigerant must be evaporated from line A to bring the vapor discharged from the low-pressure compressor down to saturation, assuming the use of the heat exchanger of Question 5?

8. What are the required displacements and horsepower of the low- and high-pressure compressors of Question 7?

9. What additional refrigerant from line A, Figure 21–25, will be evaporated if line B is passed through the heat exchanger (dotted line), and cooled to within 10°F of the intermediate saturation temperature?

10. How much will this increase the required capacity and horsepower of the high-pressure compressor?

11. How much additional refrigeration will be produced at the low-pressure evaporator?

Chapter 22 Refrigerant Lines and Pressure Drops

We stated previously that the high side of the system has the same pressure from the compressor to the expansion valve and the low side has the same pressure from the expansion valve to the compressor. This is only approximately true. There is a slight pressure drop in all lines that have flow through them. This pressure drop may be only a fraction of a pound per square inch, or it may be several pounds per square inch. The higher the velocity through the lines, the greater is this pressure drop. It is this change of pressure that forces the fluid to flow in the lines. The lines of any refrigeration system should be of sufficient size that these pressure drops do not interfere with efficient operation.

Line sizes are often chosen directly on a velocity basis. Figure 22–1 lists ranges of velocities found in operating systems. However, it should be remembered that the actual pressure drop is of more importance than the velocity. On the other hand, where oil circulates with the refrigerant, vapor velocities must be kept high enough that this oil is carried or blown along with the vapor.

Effect of Pressure Drop

A pressure drop in the suction line reduces the compressor capacity as a result of the conditions discussed in Chapter 21. Any reduction of pressure in the suction line increases the volume of the vapor without the benefit of obtaining a colder evaporator. For a 10°F R-134a evaporator, a pressure of 26.6 psia is required. If there is a 5-psi pressure drop in the suction line, this gives a pressure of 21.6 psia at the compressor. This would reduce the compressor capacity by about 19 percent. Thus, undersized lines that cause such a pressure drop reduce system efficiency and economy. Figure 22–2 lists pressure drops recommended as permissible values. It certainly does no harm to have pressure drops even lower than these if velocities do not get too low. For oil return, a velocity of 750 feet per minute on horizontal runs, and 1500 feet per minute on vertical runs, should be a minimum.

Between the compressor and the condenser, any pressure drop causes a rise in the compressor head pressure. The condensing temperature is fixed by the temperature and the of cooling medium, be it air or water. If the pressure has to fall to reach this point, the compressor has to increase the head pressure by that amount. As shown in Figure 21-4, this decreases the capacity and increases the power. The head temperature will also be found to run much higher. Again, small line sizes that cause much of a pressure drop here are poor economy.

A pressure drop in the liquid line will have very little effect on operating conditions at light loads but can cause considerable trouble at heavy loads. There are two reasons for this. The main reason is the formation of flash gas in the liquid line. If the liquid pressure is reduced below its condensing pressure, its temperature must come down to match the new pressure. Some of the liquid evaporates (flashes to gas) to produce this temperature reduction, just as if the pressure was reduced in an expansion valve. This

272 CHAPTER 22 Refrigerant Lines and Pressure Drops

Refrigerant	Suction	Discharge	Liquid
Ammonia	3000 to 5000	4000 to 6000	100 to 400
Halocarbons	800 to 1800	1800 to 2250	80 to 300

Figure 22-1 Allowable refrigerant velocities in feet per minute.

	Suction Line			Discharge Line	Liquid Line
−20°F	0°F	20°F	40°F		
¼ psi	½ psi	1 psi	2 psi	1 psi	2 to 4 psi

Figure 22-2 Allowable pressure drops.

produces some cooling or refrigeration in the liquid line, where it is wasted. The vapor formed has to pass through the expansion valve, where it takes the place of useful liquid. This reduces the capacity of the expansion valve, especially because vapor refrigerant is less dense than liquid and takes up more space. Also, this warm, high-pressure vapor leaking into the evaporator can start surging, or erratic operation, as gas bubbles push liquid ahead of them as they try to escape from the evaporator. The second effect of a lower liquid line pressure is a reduction in capacity of a given-sized expansion valve. A lower pressure will force less liquid through it, as shown in Figure 4-16.

In addition to losses due to pipe friction, there is a second cause of loss in a liquid line if the evaporator is above the receiver. This is the drop in static head or pressure due to the height of the liquid. Whenever liquid is lifted in a column, as in Figure 22-3, the pressure at the top is less than the pressure at the bottom by the weight of the liquid column. Figure 22-4 lists pressure drops for several common refrigerants.

The permissible pressure drop in the liquid line is greater than that in the suction or the discharge line if the above facts are taken into account and properly sized expansion valves are chosen. However, for good practice, unless liquid lifts make it impossible, this pressure drop should not be more than 4 psi.

The principal effect of a pressure drop in the evaporator is to produce different temperatures in different parts of the evaporator. Actually, this reduces the capacity of the system, similar to a pressure drop in the suction line. This is because

Figure 22-3 Pressure drop in liquid column.

Ammonia	0.26 psi per Foot
R-11	0.63 psi per foot
R-134a	0.51 psi per foot
R-22	0.51 psi per foot
R-502	0.52 psi per foot

Figure 22-4 Pressure drop per foot of lift.

the compressor must operate at the lowest pressure, but the average evaporator temperature is higher than this. Such a pressure drop can be caused by too long a coil of too small a diameter or by an evaporator constructed so that there is a high static head of liquid on the bottom coils.

Sometimes evaporators fed by thermostatic expansion valves cause problems if the velocity is too high. The vapor leaving the coil travels so fast that is returns to the compressor before the expansion valve bulb can respond to its temperature. This condition produces erratic operation, particularly "hunting." The maximum velocities given in Figure 22-1 should not be exceeded.

Calculating Velocity and Pressure Drop

Velocities can be easily calculated once the volume to be circulated is known. The velocity is the volume divided by the area of the pipe used. Instead of calculating it, the velocity can be picked directly from Figure 22-5 once the volume is known. The volume of the suction vapor can be obtained from Figure 20-9 or Figure 20-10, depending on the refrigerant. The volume of the liquid can be obtained from Figure 22-6, and the discharge vapor from Figure 22-7.

EXAMPLE 1 A 10-ton R-134a system is to be operated with a 20°F evaporator. Choose suitable line sizes by the velocity method.

From Figure 20-10, the required displacement per ton (assuming an 80°F condenser) is 4.3 cubic feet per minute. From Figure 22-6, the liquid flow is 0.0395 cubic feet per minute per ton, and the discharge vapor flow is 1.44 cubic feet per minute per ton, as shown in Figure 22-7. This gives the following flow data for 10 tons:

Suction vapor: 43 cubic feet per minute
Discharge vapor: 14.4 cubic feet per minute
Liquid: 0.395 cubic feet per minute

From Figure 22-5, a $2^1/_8$-inch suction line provides a velocity of 2100 feet per minute, which is above the 1800 feet per minute limit of Figure 22-1.

A $2^5/_8$-inch line that drops to 1300 feet per minute would have to be chosen. The discharge line must be $1^5/_8$-inch, and its velocity 1400 feet per minute. A $^5/_8$-inch liquid line will produce a liquid velocity that is less than the 300 feet per minute limit, which is satisfactory.

The pressure drop is affected by many variables. The roughness of the pipe, the velocity of flow, the viscosity or stickiness of the refrigerant, and the density of the refrigerant all have an influence. Formulas have been worked out that take friction factors from special curves, but the number of variables involved makes them tedious to use. For that reason, various tables and charts have been made up to help solve pressure-drop problems. Because there are so many different refrigerants in use, space considerations preclude reprinting all the necessary data here. They may be viewed at the DuPont website, www.DuPont.com /suva/na/usa/index.html. At the bottom of the page, select "technical information" and go from there. Figure 22-8 shows pressure drops in lines for R-134a. Pick the tons of refrigeration in the upper right-hand corner, drop down to the proper line size for the information required, then go horizontally to the left of the pipe size and then down to the pressure drop.

Notice that the pressure drops are given per equivalent 100 feet of line. If the equivalent length of the line of the job being considered is 40 feet, the pressure drop is 40 percent of the figure given in the chart. Often, 100 feet is considered a minimum length to take care of fittings, and so on. To find the equivalent length of line, add the actual length of all pipe in the line to the equivalent length of all fittings in the line. The equivalent length of the fittings can be taken from Figure 22-9.

EXAMPLE 2 Choose the lines for Example 1 by the pressure-drop method instead of by the velocity method.

Line Size	Pressure Drop
Suction $2^1/_8$-inch	0.65 psi per 100 feet
Discharge $1^5/_8$-inch	0.85 psi per 100 feet
Liquid $^3/_4$-inch	1.8 psi per 100 feet

Although the $^3/_4$-inch liquid line seems to have a small pressure drop for a liquid line, $^5/_8$-inch tubing

CHAPTER 22 *Refrigerant Lines and Pressure Drops*

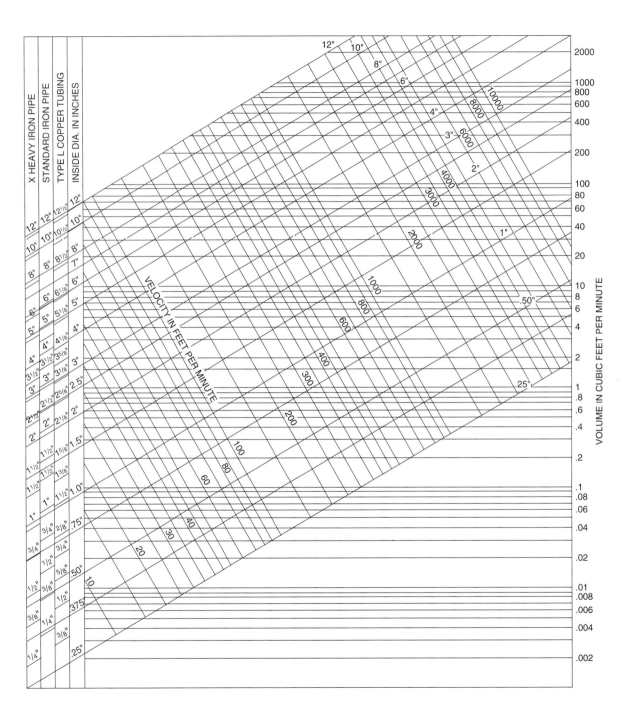

Figure 22–5 Velocity of flow.

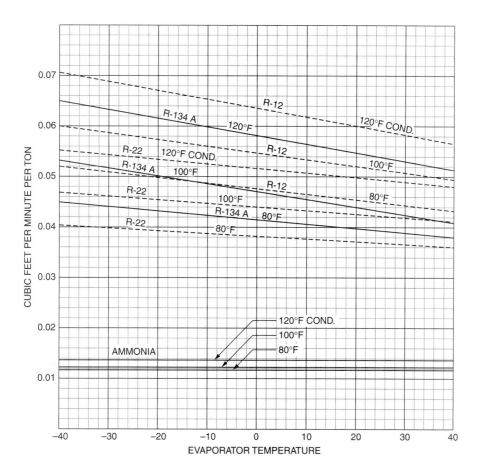

Figure 22-6 Volume of liquid per ton of refrigeration.

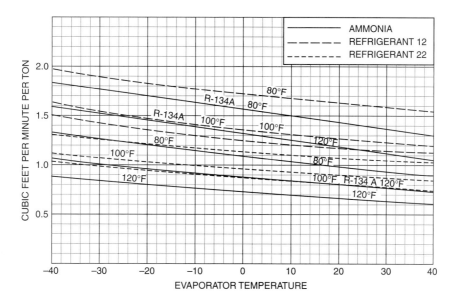

Figure 22-7 Volume of discharge vapor per ton of refrigeration.

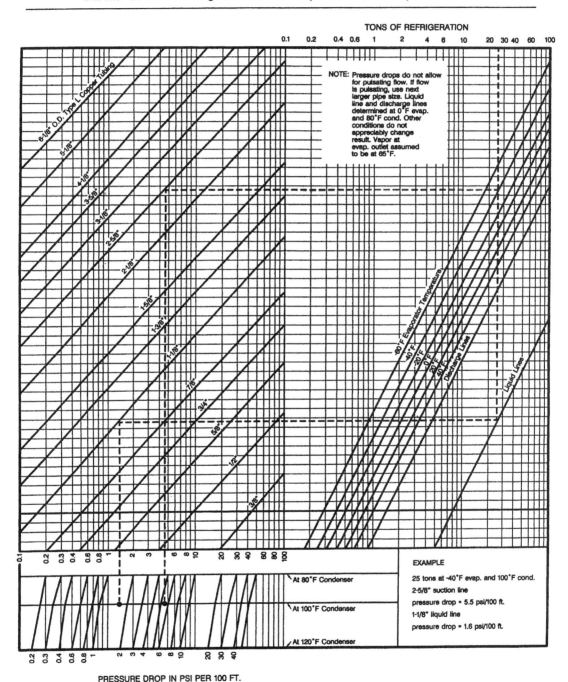

Figure 22-8 Pressure drops in lines for R-134a. *Reprinted with permission of DuPont.*

Line size IPS In inches OD		3/8 1/2	1/2 5/8	3/4 7/8	1 1 1/8	1 1/4 1 3/8	1 1/2 1 5/8	2 2 1/8	2 1/2 2 5/8	3 3 1/8	3 1/2 3 5/8	4 4 1/8	5 5 1/8	6 6 1/8	8 8 1/8	10 10
Globe valve (open)	335	14	16	22	28	36	42	57	69	83	99	118	138	168	225	280
Angle valve	165	7	9	12	15	18	21	28	34	42	49	57	70	83	117	140
Elbow	31	1	2	2	3	4	4	5	7	8	10	12	14	16	20	26
Tee (through side)	65	3	4	5	6	8	9	12	14	17	20	22	28	34	44	56

Figure 22–9 Equivalent feet of pipe for valves and fittings.

gives too much pressure drop. Notice that using this method, we calculate a smaller suction line but should use a larger liquid line than when we use the velocity method. Longer lines or colder temperatures might well calculate to the same size suction line or even a larger one. This method uses pressure drops, which should be the determining factor. So this is the better way of doing the calculation.

EXAMPLE 3 A 3-ton R-134a job works with a 0°F evaporator and a 110°F condenser. It is so placed that 30 feet of liquid and suction line must be run between the condensing unit and the evaporator section. Each line contains four ells and one shutoff valve (globe valve type). What size lines should be chosen, and what is the expected pressure drop in each?

A 1 5/8-inch suction line gives a pressure drop of 0.40 psi per 100 feet. The equivalent length of the 1 5/8-inch line is determined as follows:

4 ells at 4 feet per ell = 16 feet
1 shutoff valve = 42 feet
Length of tubing = 30 feet
Total equivalent length = 16 + 42 + 30
= 88 feet

The pressure drop in the suction line is

$$88/100 \times 0.40 = 0.352 \text{ psi}$$

A 1 1/8-inch discharge line gives a pressure drop of 0.6 psi per 100 feet. If this is a short line from the compressor to the condenser, this will be satisfactory. If the equivalent length is more than 100 feet, make sure the appropriate line size is chosen to avoid excessive pressure drop.

A 1/2-inch liquid line will give a drop of approximately 2.0 psi per 100 feet. The equivalent length is

4 ells at 1 foot per ell = 4 feet
1 shutoff valve = 14 feet
Length of tubing = 30 feet
Total equivalent length = 4 + 14 + 30
= 48 feet

The pressure drop in the liquid line is

$$48/100 \times 2 \text{ approximately} = 1.0 \text{ psi}$$

Notice that a 3/8-inch liquid line will produce excessive velocity.

Effect of Changing Flow Conditions

A few general rules regarding piping and friction should be mentioned before we leave this subject. First, the cross-sectional area of a pipe varies as the square of the diameter. That is, if the diameter of the pipe is doubled, the area is increased four times; if the diameter is tripled, the area is increased nine times. An increase of 50 percent in diameter gives a little more than twice the area.

The velocity is inversely proportional to the area. Thus, if the area is doubled, the velocity will be half the original. If this is combined with the

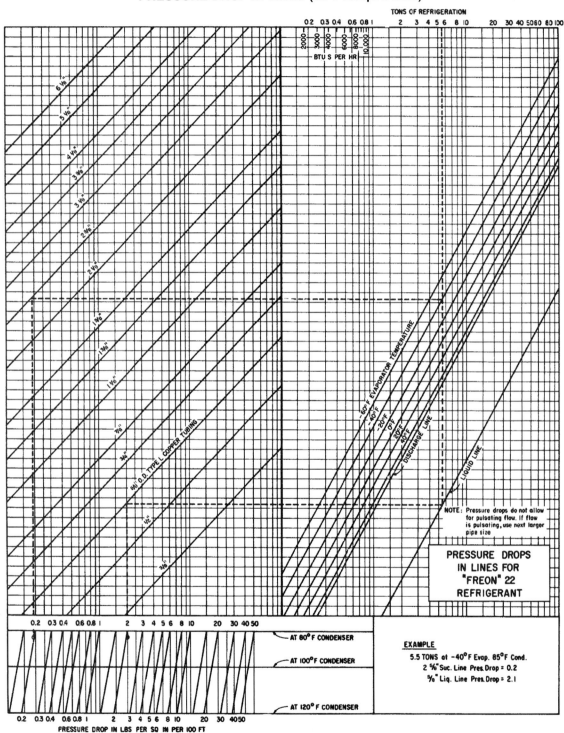

Figure 22-10 Pressure drop in R-22 lines. *Reprinted with permission of DuPont.*

Steel Line Size		Saturated Suction Temperature, °F					
		−60		−40		−20	
IPS	SCH	Δt = 0.25°F Δp = 0.046	Δt = 0.50°F Δp = 0.092	Δt = 0.25°F Δp = 0.077	Δt = 0.50°F Δp = 0.155	Δt = 0.25°F Δp = 0.123	Δt = 0.50°F Δp = 0.245
3/8	80	0.03	0.05	0.06	0.09	0.11	0.16
1/2	80	0.06	0.10	0.12	0.18	0.22	0.32
3/4	80	0.15	0.22	0.28	0.42	0.50	0.73
1	80	0.30	0.45	0.57	0.84	0.99	1.44
1 1/4	40	0.82	1.21	1.53	2.24	2.65	3.84
1 1/2	40	1.25	1.83	2.32	3.38	4.00	5.80
2	40	2.43	3.57	4.54	6.59	7.79	11.26
2 1/2	40	3.94	5.78	7.23	10.56	12.50	18.03
3	40	7.10	10.30	13.00	18.81	22.23	32.09
4	40	14.77	21.21	26.81	38.62	45.66	65.81
5	40	26.66	38.65	48.68	70.07	82.70	119.60
6	40	43.48	62.83	79.18	114.26	134.37	193.44
8	40	90.07	129.79	163.48	235.38	277.80	397.55
10	40	164.26	236.39	297.51	427.71	504.98	721.08
12	ID*	264.07	379.88	477.55	686.10	808.93	1157.59

Steel Line Size		Saturated Suction Temperature, °F					
		0		20		40	
IPS	SCH	Δt = 0.25°F Δp = 0.184	Δt = 0.50°F Δp = 0.368	Δt = 0.25°F Δp = 0.265	Δt = 0.50°F Δp = 0.530	Δt = 0.25°F Δp = 0.366	Δt = 0.50°F Δp = 0.582
3/8	80	0.18	0.26	0.28	0.40	0.41	0.53
1/2	80	0.36	0.52	0.55	0.80	0.82	1.05
3/4	80	0.82	1.18	1.26	1.83	1.87	2.38
1	40	1.62	2.34	2.50	3.60	3.68	4.69
1 1/4	40	4.30	6.21	6.63	9.52	9.76	12.42
1 1/2	40	6.49	9.34	9.98	14.34	14.68	18.64
2	40	12.57	18.12	19.35	27.74	28.45	36.08
2 1/2	40	20.19	28.94	30.98	44.30	45.37	57.51
3	40	35.87	51.35	54.98	78.50	80.40	101.93
4	40	73.56	105.17	112.34	160.57	164.44	208.34
5	40	133.12	190.55	203.53	289.97	296.88	376.18
6	40	216.05	308.62	329.59	469.07	480.96	609.57
8	40	444.56	633.82	676.99	962.47	985.55	1250.34
10	40	806.47	1148.72	1226.96	1744.84	1786.55	2263.99
12	ID*	1290.92	1839.28	1964.56	2790.37	2862.23	3613.23

Note: Capacities are in tons of refrigeration resulting in a line friction loss (Δp in psi per 100 ft equivalent pipe length), with corresponding change (Δt in °F per 100 ft) in saturation temperature.

*The inside diameter of the pipe is the same as the nominal pipe size.

Figure 22–11 Suction line capacities in tons for ammonia with pressure drops of 0.25 and 0.50°f per 100 ft equivalent. Courtesy of 2006 ASHRAE Handbook-Refrigeration.

above data on diameter, twice the pipe size gives one-fourth the velocity.

The friction in a pipe varies as the square of the velocity. Thus, if the velocity is doubled, the pressure drop is increased four times. If twice as much gas is forced through a given pipe, the pressure drop increases four times. So, because so many of these changes affect the results so quickly, small changes in pipe sizes or in amounts of gas handled can have astonishing results. If lines are large enough, the savings obtained by using larger sizes are insignificant. However, if the size is reduced, pressure drops rise very rapidly. So, if there is any doubt in choosing pipe sizes, it is always better to choose the larger size.

One further point to consider is that all the preceding figures were made up on the basis of pure refrigerants, without oil. The presence of oil in the lines or in the refrigerant increases the friction considerably. However, because the amount of oil that is circulating is never known accurately, it can hardly be taken into account in calculations. This is

Steel Line Size		Suction Lines ($\Delta t = 1°F$)					Discharge Lines $\Delta t = 1°F$ $\Delta p = 2.95$	Steel Line Size		Liquid Lines	
		Saturated Suction Temperature, °F								Velocity = 100 fpm	$\Delta p = 2.0$ psi $\Delta t = 0.7°F$
IPS	SCH	−40 $\Delta p = 0.31$	−20 $\Delta p = 0.49$	0 $\Delta p = 0.73$	20 $\Delta p = 1.06$	40 $\Delta p = 1.46$		IPS	SCH		
3/8	80	—	—	—	—	—	—	3/8	80	8.6	12.1
1/2	80	—	—	—	—	—	3.1	1/2	80	14.2	24.0
3/4	80	—	—	—	2.6	3.8	7.1	3/4	80	26.3	54.2
1	80	—	2.1	3.4	5.2	7.6	13.9	1	80	43.8	106.4
1 1/4	40	3.2	5.6	8.9	13.6	19.9	36.5	1 1/4	80	78.1	228.6
1 1/2	40	4.9	8.4	13.4	20.5	29.9	54.8	1 1/2	80	107.5	349.2
2	40	9.5	16.2	26.0	39.6	57.8	105.7	2	40	204.2	811.4
2 1/2	40	15.3	25.9	41.5	63.2	92.1	168.5	2 1/2	40	291.1	1292.6
3	40	27.1	46.1	73.5	111.9	163.0	297.6	3	40	449.6	2287.8
4	40	55.7	94.2	150.1	228.7	333.0	606.2	4	40	774.7	4662.1
5	40	101.1	170.4	271.1	412.4	600.9	1095.2	5	40	—	—
6	40	164.0	276.4	439.2	667.5	971.6	1771.2	6	40	—	—
8	40	337.2	566.8	901.1	1366.6	1989.4	3623.0	8	40	—	—
10	40	611.6	1027.2	1634.3	2474.5	3598.0	—	10	40	—	—
12	ID*	981.6	1644.5	2612.4	3963.5	5764.6	—	12	ID*	—	—

Notes:
1. Table capacities are in tons of refrigeration.
   Δp = pressure drop due to line friction, psi per 100 ft of equivalent line length
   Δt = corresponding change in saturation temperature, °F per 100 ft
2. Line capacity for other saturation temperatures Δt and equivalent lengths L_e

   Line capacity = Table capacity $\left(\dfrac{\text{Table } L_e}{\text{Actual } L_e} \times \dfrac{\text{Actual } \Delta t}{\text{Table } \Delta t}\right)^{0.55}$

3. Saturation temperature Δt for other capacities and equivalent lengths L_e

   $\Delta t = \text{Table } \Delta t \left(\dfrac{\text{Actual } L_e}{\text{Table } L_e}\right) \left(\dfrac{\text{Actual capacity}}{\text{Table capacity}}\right)^{1.8}$

4. Values based on 90°F condensing temperature. Multiply table capacities by the following factors for other condensing temperatures:

Condensing Temperature, °F	Suction Lines	Discharge Lines
70	1.05	0.78
80	1.02	0.89
90	1.00	1.00
100	0.98	1.11

5. Discharge and liquid line capacities based on 20°F suction. Evaporator temperature is 0°F. The capacity is affected less than 3% when applied from −40 to +40°F extremes.
*The inside diameter of the pipe is the same as the nominal pipe size.

Figure 22–12 Suction, discharge, and liquid line capacities in tons for ammonia (single- or high-stage application). Courtesy of 2006 ASHRAE Handbook-Refrigeration.

still another good reason for always erring toward larger rather than smaller pipe.

Figure 22-10 gives pressure drops in refrigeration lines using R-22. Figures 22-11, 22-12 and 22-13 give line capacities in systems using ammonia.

Questions

1. An ammonia plant is operating with a 0°F evaporator and an 8-psi suction line drop. What is the percent loss in capacity of the compressor due to the suction line loss?

2. If the system in Example 3 used R-22 instead of R-134a, what would be the required line sizes? Use the pressure-loss method.

3. If the system in Example 2 used R-22 with a −40°F evaporator, what would be the required line sizes? Use the pressure-loss method.

4. A 50-ton ammonia plant operates with a 0°F evaporator. The discharge line is 25 feet long, and contains three ells and one globe valve. The liquid and suction lines are 65 feet long and contain four ells and a globe valve each. Choose suitable line sizes.

5. An additional 15-ton evaporator is added to the plant in Question 4. How much does this increase the pressure drops if no change is made in line sizes?

Figure 22-13 Liquid Ammonia Line Capacities (Capacity in tons of refrigeration, except as noted)

Nominal Size, in.	Pumped Liquid Overfeed Ratio			High-Pressure Liquid at 3 psi[a]	Hot-Gas Defrost[a]	Equalizer High Side[b]	Thermosiphon Lubricant Colling Lines Gravity Flow,[c] 1000 Btu/h		
	3:1	4:1	5:1				Supply	Return	Vent
1/2	10	7.5	6	30	—	—	—	—	—
3/4	22	16.5	13	69	4	50	—	—	—
1	43	32.5	26	134	8	100	—	—	—
1 1/4	93.5	70	56	286	20	150	—	—	—
1 1/2	146	110	87.5	439	30	225	200	120	203
2	334	250	200	1016	50	300	470	300	362
2 1/2	533	400	320	1616	92	500	850	530	638
3	768	576	461	2886	162	1000	1312	870	1102
4	1365	1024	819		328	2000	2261	1410	2000
5	—	—	—	—	594	—	3550	2214	3624
6	—	—	—	—	970	—	5130	3200	6378
8	—	—	—	—	—	—	8874	5533	11596

Source: Wile (1977).

[a] Hot-gas line sizes are based on 1.5 psi pressure drop per 100 ft of equivalent length at 100 psig discharge pressure and three times evaporator refrigeration capacity.
[b] Line sizes based on experience using total system evaporator tons.
[c] From Frick Co. (1995). Values for line sizes above 4 in. are extrapolated.

Chapter 23 Brine in Refrigeration

Except for centrifugal compressor systems, all the refrigeration systems we have described so far are *direct* systems. An *indirect* system is illustrated in Figure 23-1. Here a brine or other antifreeze solution is chilled, then pumped around to do the required cooling. The biggest disadvantage of such a system is that lower suction pressures are usually needed to maintain a required room temperature. Comparison figures are shown in Figure 23-2 and Figure 23-3. Another disadvantage of a brine system is the corrosion caused by the brine. However, a brine system also has many advantages. All of the refrigerant-containing equipment is in the engine room and thus under the direct supervision of the engineer. A leak in any other part of the building will leak only brine. This will cause considerably less damage than a refrigerant leak, particularly if the refrigerant is ammonia. If there is a leak in the ammonia side of the system, steps can be taken immediately to remedy it because the leak can occur only in the engine rooms. This opens up the possibility of using ammonia in indirect systems for applications such as commercial air conditioning and supermarket refrigeration.

An indirect system has considerable holdover capacity. In case of a temporary shutdown, the cold brine will continue to hold temperatures for a short time. This time may be used to perform some repairs and minor maintenance. If sufficient brine is used, refrigeration equipment can be operated during the day to refrigerate the brine, then turned off and the brine solution used to maintain temperatures overnight. Circulating brine pumps are usually left in operation, but they may be operated with little or no attention. This means it is not necessary to have an operating engineer on the job 24 hours a day. Such systems were very common in small applications before the development of today's automatic control equipment. It is less used at the present time on stationary applications because of the greater simplicity of automatic equipment, but it is still used for special applications.

Another application in which brine works well is in storing refrigeration for heavy intermittent loads. Dairy farms are a classic example. The milk comes out of the cow at approximately 100°F. It must be cooled quickly or quality is lost. Milk cooling loads or something similar may be required only once or twice a day, but an enormous amount of product must be cooled at one time. A direct system would have to be very large in size, which means expensive, yet it would be idle most of the day. A smaller unit can be chosen to run steadily, or on an average of 16 hours out of the 24. When no refrigeration is required, the capacity of the machine is used to chill brine. Then the cold brine plus the machine can be used on the load when required. This combination allows cooling the milk very quickly as compared to a direct system, which may require several hours. The direct system also risks compromising quality or even loss of product.

Brine is also used in some applications because refrigeration from it is considered steadier. Instead of cycling the machine to control temperatures, which results in some temperature variations, the

Brine in Refrigeration   **283**

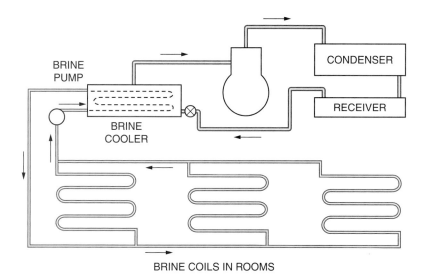

Figure 23–1 Indirect refrigeration system.

Room Temperature (°F)	Direct Expansion	Indirect System	
	Suction Temperature (°F)	Brine Temperature (°F)	Suction Temperature (°F)
40	16	20	8
30	10	12	1
20	3	4	−6
10	−5	−4	−13
0	−12	−12	−20
−10	−20	−20	−27
−20	−30	−28	−34

Figure 23–2 Typical coil temperatures for various room temperatures.

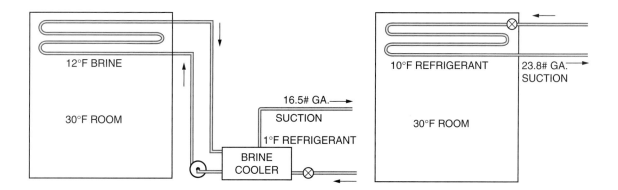

Figure 23–3 Diagram of direct expansion vs. brine.

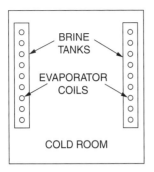

Figure 23-4 Noncirculating brine system.

flow of brine can be controlled to maintain the temperature very nearly constant.

Some smaller, simpler systems are noncirculating. That is, the brine is kept at the point where cooling is required, remaining there to supply holdover capacity, as shown in Figure 23-4. However, brine systems of any size pump the brine from a brine cooler in the engine room to the required point, as in Figure 23-1. With a circulating brine system, a thermostat in the room can be used to control a throttling-type valve in the brine line. This does not close completely, but increases or decreases the flow of brine as required to meet the imposed heat load.

Brines used for holdover where a large refrigeration capacity is needed are sometimes frozen. Frozen brine is used mostly in noncirculating systems such as the one shown in Figure 23-4. Occasionally, where ice-water temperatures are suitable, as in an air conditioning system, water ice may be used instead of frozen brine, as in Figure 23-5. These systems do not work as well with brine, for reasons given later. By using some form of freezing, the latent heat of fusion can be taken advantage of, and a larger refrigeration capacity obtained with less brine and a smaller brine tank.

Brine Chemistry

To understand fully the use of brine, we should know something of its chemical behavior. Pure water freezes at 32°F. As any soluble salt is dissolved in the water, its freezing temperature is lowered. The greater the amount of salt dissolved, the lower will be the freezing point. This is illustrated by the freezing line of Figure 23-6. However, when the solution starts to freeze, the mixture of salt and water does not form ice. Only the water starts to freeze, leaving the salt behind. This salt, forced into the remaining water, increases its concentration, which further lowers its freezing point. Thus, the freezing temperature of a brine becomes continually lower as more of it is frozen. The pure water that freezes out of the brine forms separate ice crystals that float in the brine. If the solution is not cooled to its final freezing point, these ice crystals form a slush in the brine instead of freezing into a solid cake.

The limiting point of this gradual reduction in freezing temperature is the solubility of the particular salt used. Refer to the solubility line of Figure 23-6. The colder the temperature, the less

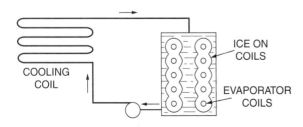

Figure 23-5 Ice storage for holdover capacity.

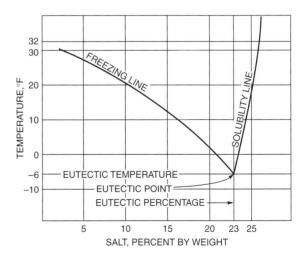

Figure 23-6 Sodium chloride brine chart (illustrating eutectic point).

salt can be dissolved. As temperatures go lower, a point is reached where the solubility line crosses the freezing line. Below this intersection point, any further salt rejected from the freezing fluid is not soluble, so the freezing point is not lowered further. At this point, the entire mass begins to freeze solid. This point is called the *eutectic point,* and a solution with the proper percentage of salt to produce this condition is called a *eutectic solution*. Notice that the eutectic point is the lowest possible freezing point for the brine considered.

Even during freezing at the eutectic point, there is still separation of the salt and water. This creates complications when the frozen brine is thawed. Because the salt is no longer in solution, it drops to the bottom of the tank or container. There it is very slow to dissolve again into the cold solution. Unless agitation is very thorough, the salt may eventually saturate the solution at the bottom. Because this saturated solution is much heavier than a weak solution, it will remain on the bottom. Thus, the entire solution is not reconcentrated to its original percentage, and it will start to freeze at a higher temperature than originally. This is why frozen brine is not more commonly used to store refrigeration capacity.

The two types of brine that are most commonly used for refrigeration are sodium chloride (common table salt), and calcium chloride. The biggest difference between them is that sodium chloride has a eutectic point of −6°F, and calcium chloride has a eutectic point of −60°F. Sodium chloride brine is usually preferred where the temperature range is above 0°F. Below 0°F, calcium chloride brine is usually used. When using calcium chloride, −40°F is generally as cold as we want the process temperature to be. Either brine will cause corrosion if it is not properly controlled.

Glycols, alcohols, and halocarbon refrigerants such as R-11 have also been used instead of brines to carry refrigeration from an engine room to points requiring refrigeration. These fluids do not have the corrosion problems of brine, but they are very expensive compared to brine. Care must be exercised when selecting a secondary refrigerant. Alcohol solutions require special pump seals. Ethylene glycol may not be used around food and/or pharmaceuticals. And, of course, there are now restrictions or outright bans on the use of CFCs and HCFCs.

Corrosion Control

There are three basic factors to be considered in brine corrosion control. These are control of the brine density and alkalinity, and the use of corrosion inhibitors. A brine of high density will dissolve less air, which is usually blamed for a considerable amount of corrosion. A brine that is alkaline is less corrosive to iron and steel than a brine that is neutral or acidic.

In a system such as an ice freezing tank, the alkalinity of the brine should not be too high if the ice cans are galvanized. Brine that is highly alkaline or that contains significant amounts of ammonia will remove the zinc coating from the ice cans.

The value of corrosion-prevention chemicals or inhibitors has been demonstrated over a long period of time. By maintaining a certain amount of chromate in the brine, engineers have found that corrosion and rust are reduced. The chromate can be introduced as chromic acid or sodium dichromate. Calcium chloride brine and sodium chloride brine need a minimum of 1800 and 3600 parts per million of sodium chromate, respectively. Chromic acid is preferred where there is ammonia in the brine or if, for other reasons, the brine is too alkaline. Chromates, however, are a serious environmental hazard. Always check the applicable federal, state, and local regulations before using and disposing of these chemicals. Figure 23–7 gives the recommended dosages of chrome chemicals.

Neutral treatment:
12 lb chromic acid plus 9.6 lb caustic soda is equal to 18 lb sodium dichromate plus 4.8 lb caustic soda.

To neutralize brine contaminated by ammonia:
2.9 lb of chromic acid neutralizes 1 lb ammonia.
8.8 lb of sodium dichromate neutralizes 1 lb ammonia.

Figure 23–7 Chromate dosages for brine treatment, in pounds per 1,080 cubic feet of brine.

The strengths of acids and alkalis are measured by a scale running from 1 to 14. This is called the pH scale, where pH is the $-\log$ of the [H^+] (hydrogen ions) concentration. A pH of 1 means the strongest acid; a pH of 7, which is the center point of the scale, is neutral, and a pH of 14 is the strongest alkali. Various vegetable dyes or indicators are available that show different colors in different pH solutions. For a test, a few drops of the indicator solution are dropped into a test tube of the brine to be checked. For rough checks, the colors are gauged by sight and memory. For the most accurate work, color standards are available with which to make direct comparisons. Phenolphthalein is a common indicator chemical. Phenolphthalein is colorless up to a pH of 8 and red at a pH of 10. From 8 up to 10 it goes from a pale pink to a darker pink and finally red. Litmus is another common indicator. It is red up to a pH of 4.5 and blue at a pH of 8.5. Figure 23–8 lists some common indicators and their color changes. Electronic pH meters with digital readouts are also available. As with any other electronic test equipment, it is important to be sure that the meter is maintained and calibrated properly. The sensing elements of these devices fail quickly if they are not properly maintained and stored when not in use.

For ordinary brine systems, it is recommended that the brine be maintained at a pH of 6.5 to 8.5 with an average of about 8.0. Below pH 7.0 the brine will become more active against iron and steel. Above pH 8.0 it will become active against brass or galvanized (zinc-treated) metals. Ice freezing tanks usually contain galvanized cans as well as uncoated steel. Sometimes, where no galvanizing is present or has all been corroded away, the pH is carried up to 9 or 9.5. This seems to provide better protection for the iron and steel parts, but it may cause pitting if the pH is allowed to get higher.

When the pH must be adjusted, a caustic or alkali must be added to the brine to increase it. Slacked lime or caustic soda may be used. To decrease the pH, muriatic acid or chromic acid may be used. Chromic acid also inhibits corrosion directly. Any acid or alkali should be added very slowly and cautiously, with good circulation maintained. It is important to check that correction is not carried too far. Too much acid, or too rapid addition of acid, at one point will cause more corrosion than if no correction had been attempted. If too much alkali is added to a calcium brine, the calcium chloride may precipitate out, which will allow the brine to freeze. It is also important to check applicable regulations before adding chemicals to the solution.

An operating engineer can learn to make the above checks of the condition of the brine and to maintain it at approximately the proper condition. For many larger systems, however, it is considered worthwhile to engage the services of a water treatment company to make periodic checks of samples furnished. From checks on these samples the representative can prescribe the proper treatment. More accurate control of the chemical condition of the brine can be maintained in this way. In large plants the longer life of all metal parts in contact with the brine well repays the cost of such a service. The operator must still be able to intelligently follow the instructions of the water treatment company to get proper results.

Brine Concentration

The actual choice of the proper brine concentration involves several factors. Figure 23–9 and Figure 23–10 list the properties of sodium chloride and calcium chloride brines, respectively. If ensuring against freezing was the only requirement, the brine could be kept as strong as possible up to the eutectic percentage. However, the stronger the brine, the heavier it is and the lower is its specific heat. When its specific heat is less, more of it must be circulated to perform a given refrigeration job.

Indicator	pH Range	Color Change
Litmus	4.5–8.5	Red–Blue
Phenolphthalein	8.0–10.0	Clear–Red
Bromthymol blue	6.0–7.6	Yellow–Blue
Phenol red	6.8–8.4	Yellow–Red
Cresol red	7.2–8.8	Yellow–Red
Thymol blue	8.0–9.6	Yellow–Blue

Figure 23–8 Common pH indicators.

Brine Concentration

Percent Pure Sodium Chloride by Weight	Specific Gravity	Beume Scale	Specific Heat (BTU/lb)	Freezing Point (°F)	Weight per Gallon (lb)	Weight per Cubic Foot (lb)	Corresponding NH_3 Suction Pressure, Gage
0	1.000	0.00	1.000	32.0	8.34	62.4	47.6
5	1.035	5.1	0.938	27.0	8.65	64.6	41.4
6	1.043	6.1	0.927	25.5	8.71	65.1	39.6
7	1.050	7.0	0.917	24.0	8.76	65.5	37.9
8	1.057	8.0	0.907	23.2	8.82	66.0	37.0
9	1.065	9.0	0.897	21.8	8.89	66.5	35.6
10	1.072	10.1	0.999	20.4	8.95	66.9	34.1
11	1.080	10.8	0.879	18.5	9.02	67.4	32.0
12	1.087	11.8	0.870	17.2	9.08	67.8	30.6
13	1.095	12.7	0.862	15.5	9.14	68.3	28.9
14	1.103	13.6	0.854	13.9	9.22	68.8	27.4
15	1.111	14.5	0.847	12.0	9.28	69.3	25.6
16	1.111	15.4	0.840	10.2	9.33	69.8	24.0
17	1.126	16.3	0.933	8.2	9.40	70.3	20.5
18	1.134	17.2	0.826	6.1	9.47	70.8	20.5
19	1.142	18.1	0.819	4.0	9.54	71.3	18.8
20	1.150	19.0	0.813	1.8	9.60	71.8	17.1
21	1.158	19.9	0.807	−0.8	9.67	72.3	15.1
22	1.166	20.8	0.802	−3.0	9.74	72.8	13.6
23*	1.175	21.7	0.796	−6.0	9.81	73.3	11.6
24	1.183	22.5	0.791	3.8	9.88	73.8	18.6
25	1.191	23.4	0.786	16.1	9.95	74.3	29.5

*Eutectic point. Specific gravity and weights at 59°F. Referred to water at 39°F. Specific heat at 59°F.

Figure 23–9 Properties of sodium chloride brine, pure anhydrous salt NaCl.

When it is heavier, more pumping power must be used to circulate it. Therefore the stronger the brine, the higher will be the pumping costs. For this reason, the brine should be kept only strong enough that it will not freeze out against the refrigerated surfaces in the brine cooler.

A higher-density brine is less corrosive. In ice freezing tanks, it is customary to hold brine density at the maximum that can be carried without floating the cans.

Occasionally, a brine spray system is used to help control humidity. A weak brine (near its freezing point) will tend to keep air in contact with it nearly saturated. If only a small temperature difference is used, such a system will maintain high humidity. If a strong concentration of brine is used, it will absorb moisture from the air and lower its humidity. Refer to Figure 11–7 for one form of brine spray system.

EXAMPLE 1 A single room has a load of 6 tons of refrigeration. It is to be refrigerated by brine coils with not over 6°F rise in brine temperature. What is the difference between using a calcium chloride brine with a specific gravity of 1.14 vs. using one with a specific gravity of 1.23?

From Figure 23–10, the following data are found:

	1.14	1.23
Specific gravity		
Freezing point	10°F	−23.5°F
Percent of salt	16%	25%
Specific heat	0.784	0.692

The formula

$$H = WS(t_2 - t_1)$$

where

H = the quantity of heat
W = weight

Percent Pure Sodium Chloride by Weight	Specific Gravity	Beume Scale	Specific Heat (BTU/lb)	Freezing Point (°F)	Weight per Gallon (lb)	Weight per Cubic Foot (lb)	Corresponding NH$_3$ Suction Pressure, Gage
0	1.000	0.00	1.000	32.0	8.34	62.4	47.6
5	1.044	6.1	0.9246	29.0	8.717	65.15	43.8
6	1.050	7.0	0.9143	28.0	8.760	65.52	42.6
7	1.060	8.2	0.8984	27.0	8.851	66.14	41.4
8	1.069	9.3	0.8842	25.5	8.926	66.70	39.0
9	1.078	10.4	0.8699	24.0	9.001	67.27	37.9
10	1.087	11.6	0.8556	23.0	9.076	67.83	36.8
11	1.096	12.6	0.8429	21.5	9.143	68.33	35.0
12	1.105	13.8	0.8284	19.0	9.227	68.95	32.5
13	1.114	14.8	0.8166	17.0	9.302	69.51	30.4
14	1.124	15.9	0.8043	14.5	9.377	70.08	28.0
15	1.133	16.9	0.7930	12.5	9.452	70.64	26.0
16	1.143	18.0	0.7798	9.5	9.536	71.26	23.4
17	1.152	19.1	0.7672	6.5	9.619	71.89	20.8
18	1.162	20.2	0.7566	3.0	9.703	72.51	18.0
19	1.172	21.3	0.7460	0.0	9.786	73.13	15.7
20	1.182	22.1	0.7375	−3.0	9.853	73.63	13.6
21	1.192	23.0	0.7290	−5.5	9.928	74.19	11.9
22	1.202	24.4	0.7168	−10.5	10.037	75.0	8.8
23	1.212	25.5	0.7076	−15.5	10.120	75.63	5.9
24	1.223	26.4	0.6979	−20.5	10.212	76.32	3.4
25	1.233	27.4	0.6899	−25.0	10.295	76.94	1.3
26	1.244	28.3	0.6820	−30.0	10.379	77.56	1.6*
27	1.254	29.3	0.6735	−36.0	10.471	78.25	6.1*
28	1.265	30.4	0.6657	−43.5	10.563	78.94	10.6*
29	1.276	31.4	0.6584	−53.0	10.655	79.62	15.7*
29.5	1.280	31.7	0.6557	−58.0	10.688	79.87	17.8*

*Inches of mercury (vacuum). Specific gravity and weights at 60°F.
Referred to water at 39°F. Specific heat at 60°F.

Figure 23–10 Properties of calcium chloride brine, pure anhydrous CaCl$_2$.

S = specific heat
$t_2 - t_1$ = temperature difference

can be rearranged to solve for the weight of brine to be circulated:

$$W = H/S(t_2 - t_1)$$

Six tons of refrigeration gives

$$H = 6 \times 200 = 1200 \text{ Btu per minute}$$

For the brine with a specific gravity of 1.14, the circulation

$$W = 1200/0.784 \times 6 = 255 \text{ pounds of brine per minute}$$

For the brine with a specific gravity of 1.234, the circulation is

$$W = 1200/0.6924 \times 6 = 289 \text{ pounds of brine per minute}$$

The latter figure is 13 percent more than the former. The pumping cost is proportional to the weight, so the brine with a specific gravity of 1.23 requires 13 percent additional pumping cost.

EXAMPLE 2 A small brine system is required to maintain a temperature of 0°F. To achieve this, the coils will have to be down to about −5°F. To prevent freezing out on the coils, the freezing point of the brine must be below this temperature.

a. What type of brine should be used?
b. What strength of brine should be used?
c. How many pounds of salt will be necessary to mix this brine?

a. Calcium chloride brine must be used to get below −5°F without freezing.

b. From Figure 23–10, the brine should have a strength of 22 percent or a specific gravity of 1.20 to provide a freezing point that is safely below −5°F.

c. From Figure 23–10, the weight of a gallon of 22 percent brine is 10.0 pounds; 22 percent of 10 pounds is 2.2 pounds per gallon.

EXAMPLE 3 A certain system holds 12,700 gallons of calcium chloride brine. It is supposed to be maintained at 25 percent strength. A hydrometer reading shows a specific gravity of 1.22. How much calcium chloride should be added?

The required brine of 25 percent strength weighs 10.295 pounds per gallon.

$$0.25 \times 10.295 = 2.574 \text{ pounds of salt per gallon is required}$$

Brine with a gravity of 1.22 is 23.8 percent and weighs 10.19 pounds per gallon.

$$0.238 \times 10.19 = 2.425 \text{ pounds per gallon now in the brine}$$

Thus,

$$2.574 - 2.425 = 0.149 \text{ pounds per gallon should be added}$$

$$0.149 \times 12,700 = 1892.3 \text{ pounds of salt are needed}$$

This is almost nineteen 100-pound sacks.

EXAMPLE 4 Five hundred gallons of milk are to be cooled twice a day at a dairy. The milk will be cooled to 70°F using water, and from 70°F to 40°F using refrigeration on a plate cooler similar to the one that will be shown in Figure 24–8. The milk must be cooled immediately. Milking takes about 2 hours, so this gives 2 hours total time in which to cool the 500 gallons.

a. What size condensing unit will be needed if direct expansion is used?
b. What size condensing unit will be needed if a brine storage system is used?
c. What type and density of brine should be used?
d. How much brine will be needed to store the refrigeration required for part **b**?
e. How much salt will be needed to make up the brine?
f. How large a brine storage tank will be needed?

a. From Table A–2 in the Appendix, the specific heat of milk is 0.92, and its weight per gallon is 8.6 pounds.

$$500 \times 8.6 = 4300 \text{ pounds of milk per cooling cycle}$$

The heat to be removed is

$$H = WS(t_2 - t_1) = 4300 \times 0.92 (70 - 40) = 119,000 \text{ Btu}$$

If this heat is to be removed by direct expansion, the compressor can be allowed to run steadily for the 2-hour cooling period.

$$119,000/2 = 59,500 \text{ Btu per hour cooling required}$$

Because there is no safety factor in the running time, at least 15 percent should be added for a factor of safety.

$$1.15 \times 59,500 = 68,400 \text{ Btu per hour equipment capacity}$$

This is

$$68,400/12,000 = 5.7 \text{ tons of refrigeration}$$

b. If a brine storage system is used, the total cooling for the two milkings per day is

$$2 \times 119,000 = 238,000 \text{ Btu per day}$$

On the basis of 16 hours of running time, this is

$$218,000/16 = 14,900 \text{ Btu per hour}$$

Thus,

$$14,900/12,000 = 1.24 \text{ tons of refrigeration}$$

So a little less than one-quarter as large a condensing unit will be needed if a brine system is used.

c. For milk cooling, brine cannot be run too cold or there will be trouble with freezing of the milk in the end passages. To get an adequate temperature difference in the milk cooler, a brine will be chosen that will be cooled to 25°F to begin the milk cooling, and it will be allowed to heat up to 35°F at the end of the cooling period. To cool the brine to 25°F, a refrigerant temperature as low as 15°F might be used. A 14 percent sodium chloride brine will be satisfactory.

d. To give sufficient brine with some factor of safety, the help from the condensing unit during the milk cooling period will be neglected. This will require the total 119,000 Btu to be stored in the brine. The weight required is

$$W = H/S(t_2 - t_1) = 119{,}000/0.854 \times (35 - 25)$$
$$= 13{,}950 \text{ pounds of brine}$$

Brine at 14 percent weighs 9.22 pounds per gallon.

$$13{,}950/9.22 = 1512 \text{ gallons of brine is needed}$$

e. Fourteen percent of 13,950 is

$$0.14 \times 13{,}950 = 1950 \text{ pounds of salt is needed}$$

f. One cubic foot of 14 percent brine weighs 68.8 pounds. Therefore the cubic volume of 13,950 pounds of 14% sodium chloride brine is

$$13{,}910/68.8 = 203 \text{ cubic feet}$$

This is the size of tank needed to hold the brine.

Congealing Tanks and Eutectic Plates

As mentioned earlier, ordinary congealing tanks have not been very successful because the salt separates and settles out upon thawing. However, a eutectic mixture held in a jelled solution will hold the salt whether it separates or not. The salt will not settle out of the jelly, and it has a better chance of redissolving. Figure 23–11 illustrates a refrigeration unit in which the cold evaporator surfaces are eutectic holdover plates.

Different salts are chosen to give different eutectic temperatures. Thus, tanks can have different congealing temperatures. Eutectic plates are commonly used in truck refrigeration units used for daytime deliveries.

Two general systems are used. The first uses a condensing unit mounted on the truck but driven

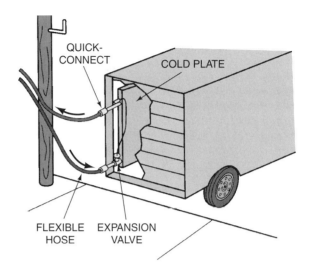

Figure 23–11 Cold plate connected to refrigeration system at the dock.

by 110-volt or 220-volt power. The unit is plugged into the required power source when the truck is garaged for the night. The unit freezes the brine in the plates. When the truck leaves to make daytime deliveries, the unit is disconnected from its power supply, but the frozen plates supply the required refrigeration.

The second system leaves the entire condensing unit in the garage. Flexible liquid and suction lines are connected to the truck by a special connecting block that can be quickly connected or disconnected. This system saves the weight of the condensing unit on the truck, making that much more payload possible. However, refrigerant losses and air leakage into the lines through improperly closed valves are often very troublesome. Most (but not all) refrigerants in use today may not be intentionally released into the atmosphere. This latter system, therefore, requires much more checking and more servicing.

Eutectic plates must be chosen with two considerations in mind: surface area and storage capacity. The heat transfer is usually low because there is no circulation within the tank and only a small temperature differences between the brine and the container. The latent heat of the brine is lower than that of water. Latent heat varies for different substances, but if definite information is not available on the particular substance used, about 110 Btu per pound is a good average to use.

Either natural-convection or forced-draft plates may be used. Plate sizes, heat transfer capacities, and holdover capacities are best taken from manufacturers' catalogs. Example 5 illustrates the principal factors involved in the choice of plates to match a load.

EXAMPLE 5 An ice cream delivery truck is to be maintained at 0°F with congealing plates that have a eutectic temperature of −18°F. The total refrigeration load on the truck body is estimated to be 3000 Btu per hour. Capacity is wanted to last 12 hours to give a factor of safety over a normal 8-hour shift. What area and weight of plates are necessary?

The area is found from the formula

$$A = H/K(t_2 - t_1)$$

where A is the area and K is the K factor.

From Table A-12 in the Appendix, K is 2.25 Btu per square foot. Thus

$$A = 3000/18 \times 2.25 = 74 \text{ square feet}$$

The required weight is found by dividing the total load by the latent heat of 1 pound of the brine:

$$3000 \times 12/110 = 327 \text{ pounds of brine}$$

Therefore tanks must be selected that have a total area of at least 74 square feet and a total weight of at least 327 pounds. In the sizes available, at least three tanks will probably have to be used to get these totals.

Questions

1. What is the difference between a direct and an indirect refrigeration system?
2. Name some of the advantages and disadvantages of a brine system.
3. Do salt solutions necessarily have a constant freezing point?
4. What is a eutectic brine?
5. What is an inhibitor?
6. What effect does the pH of a brine have on its corrosive tendencies?
7. An ice cream popsicle tank is to be maintained at −20°F. What type and strength of brine should be chosen?
8. A room has a heat load of 52,000 Btu per hour. It is cooled with an 18 percent sodium chloride brine. The temperature rise of the brine is not to be higher than 4°F. How much brine must be circulated?
9. Twenty-five hundred gallons of sodium chloride brine is to be maintained at 20 percent strength. A hydrometer reading shows it to have a specific gravity of 1.14. How much salt must be added?
10. Rework Example 4 with the following changes: 300 gallons of milk cooled twice a day. The total cooling period is not to be more than 3 hours.

Chapter 24 Liquid Cooling

Beverage Cooling Requirements

Water or other fluids are often cooled for beverage purposes. For drinking, water should not be cooler than 45°F. If the water is colder than this, enough of it cannot be drunk to quench thirst. For factory work, or in other places where people are hot and perspiring, even 50°F water is too cold. This should be taken into account when putting in bubblers where water is drunk directly. A *bubbler* is the spout of a drinking fountain.

If the water is to be served in glasses, as in a restaurant, it may be cooled to 40°F. The glass is usually warmer than this, and it sits in a warm room after filling but before drinking. This gives the water a chance to warm up to a satisfactory temperature. Beverages that are intended to be sipped slowly can be cooled below 40°F but usually are not. Soft drinks, beer, and other beverages are usually cooled to 40°F. Water cooling requirements, both quantities and recommended temperatures, are listed in Figure 24-1.

Water Coolers

Water coolers are made in a great variety of sizes and shapes. The simplest type of cooler is illustrated in Figure 24-2. A small sealed condensing unit in the base supplies the required refrigeration. The cooler usually is a storage type. It chills and holds in reserve a small tank of water. The refrigeration coil may be fed by a thermostatic expansion valve, an automatic expansion valve, or a capillary tube. A constant-pressure valve is usually placed in the suction line. This is set at a sufficiently high pressure that there is no danger of freezing the water. To freeze up a cooler may burst it. The outlet of the cooler may be a bubbling cup or a glass filler. A thermostatic control with the bulb in a well in the tank is usually used to control temperature.

Figure 24-3 illustrates one type of instantaneous, remote-type cooler. It is remote because refrigerant lines are piped to it from a remote condensing unit. It is an instantaneous type because it holds only a small amount of water at a time, depending on adequate surface and rapid refrigerant evaporation to chill the water as needed. It may be fed with a thermostatic expansion valve. Either a storage or an instantaneous cooler may be used with either the self-contained or remote type.

Where cold water is required at many different places, such as in every room of a hotel, a circulating water system such as that shown in Figure 24-4 is usually used. Refrigeration losses are greater with the cold water flowing through the pipes at all times, but this is the only type of system that has proved satisfactory for such an installation. Such a system is designed so that the water will not pick up more than 5°F during its circuit. Recommended thicknesses of insulation and heat losses from various sized pipes can be found in Table A-16 in the Appendix. To solve such problems, a pipe size must be assumed, then the problem worked out to see if

Usage	Temperature (°F)	Total Water Used and Wasted
Office building—employees	50	1/8 gallon/hour/person
Office building—transients	50	1/2 gallon/hour/250 persons/day
Light manufacturing	50–55	1/5 gallon/hour/person
Heavy manufacturing	50–55	1/4 gallon/hour/person
Restaurant	45–50	1/10 gallon/hour/person
Cafeteria	45–50	1/12 gallon/hour/person
Hotel	50	1/2 gallon/hour/person
Theater	50	1 gallon/hour/75 seats
Store	50	1 gallon/hour/100 customers
School	50–55	1/8 gallon/hour/student
Hospital	45–50	1/12 gallon/day/bed

Figure 24–1 Drinking water table. *Courtesy of Temprite Cooler Corp.*

the assumed size is satisfactory. The speed of water flow through the pipes is limited to 180 feet per minute. Above this speed, churning action turns the water milky as a result of entrained bubbles. Figure 24–5 shows the amount of water that can be circulated through various sizes of pipes at this speed, and the heat load for a 5°F temperature rise.

Most hotels now supply ice on every floor. However, the methods described in this chapter have many applications.

EXAMPLE 1 An office building is to have running ice water to 60 rooms in the building. It is estimated there will be 1800 lineal feet of pipe in the water circuit. The design temperature is 90°F. A check shows city water temperatures to be 70°F in the summer. What size pipe is recommended, and what are the refrigeration requirements based on 1-inch insulation?

Figure 24–2 Self-contained water cooler. *Courtesy of Halsey Taylor.*

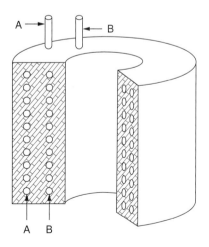

Figure 24–3 Instantaneous remote water cooler. Coil A is the evaporator coil. Coil B is the water coil. Both are embedded in aluminum casting to provide good heat transfer between them.

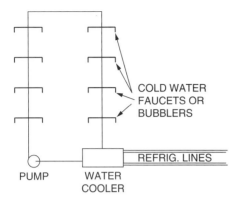

Figure 24-4 Cold water circulating system.

From Figure 24-1, we find that 1/8 gallon per person per hour is required. This table also shows a required temperature of 50°F. A recheck of the job should be made to confirm the average number of persons per office. In this case, assume we find an average of six persons per office, so a total of

$$6 \times 60 \times 1/8 = 45 \text{ gallons per hour}$$

is required to be cooled.

If we assume a 1/2-inch pipe, Figure 24-5 shows that a circulation of 2.79 gallons per minute gives a heat load of 7100 Btu per hour. From Table A-16, the heat picked up by a 1/2-in pipe with 1 inch of insulation is 6 Btu per hour per lineal foot. Therefore the total heat transfer to the water is

$$6 \times 1800 = 10,800 \text{ Btu per hour}$$

This is more than the 7100 Btu that would cause a 5°F rise, so this situation would raise the water temperature more than is permissible. Therefore a 1/2-inch pipe is not satisfactory. A 3/4-inch pipe gives a heat load of 12,500 Btu per hour to heat the water. The heat transfer is still

$$6 \times 1800 = 10,800 \text{ Btu per hour}$$

This is less than the 12,500 Btu above, so we will not heat the water quite 5°F. Therefore the 3/4-inch pipe is satisfactory.

The total refrigeration load will be made up as follows

Cooling 45 gallons of makeup water results in heat removed of

$$H = WS(t_2 - t_1) = (45 \times 8.34) \times 1 \times (70 - 50)$$
$$= 7500 \text{ Btu per hour}$$

Heat leakage through the pipe = 10,800 Btu per hour.

Therefore, the total is

$$7,500 + 10,800 = 18,300 \text{ Btu per hour}$$

To correct for running time,

$$1.5 \times 18,300 = 27,450 \text{ Btu per hour}$$

Therefore, a condensing unit and a water cooler having a capacity of 27,450 Btu per hour should be chosen.

Water or brine coolers for industrial purposes usually require greater quantities and greater loads than beverage coolers. Different types of coolers are available. A refrigerated coil can be placed in an open or closed water tank. For small coolers this coil may be fed by an automatic or thermostatic expansion valve. Large coolers may use a coil like the one shown earlier, in Figure 4-6, fed with a float valve.

Pipe size (in)	Volume		Heat Load (Btu/hr)
	Cubic Feet per Minute	Gallons per Minute	
1/2	0.380	2.79	7,100
3/4	0.668	5.00	12,500
1	1.08	8.10	20,250
1 1/4	1.87	14.0	35,000
1 1/2	2.54	19.0	47,500
2	4.20	31.5	78,500

Figure 24-5 Water cooling table.

Another common type of industrial water cooler is the shell-and-tube type, as shown ealier in Figure 5-6. In this case the shell-and-tube condenser is used as an evaporator. The refrigerant is in the shell space and the liquid to be cooled is in the tubes. The flow of liquid through the tubes and the rapid boiling of the refrigerant with complete separation of the vapor from the liquid give excellent heat transfer. Such coolers are usually preferred for intermediate-size and large jobs. This cooler would be fed by a float valve or float switch and solenoid valve.

Beer Coolers

Figure 24-6 illustrates the common method of storing and cooling draft beer. The kegs are kept in a walk-in cooler. The cooler should be either nearby or in the basement directly under the bar. Best results are obtained by using a refrigerated tunnel to the bar rather than by using insulated lines. Regardless of the amount of insulation, the beer in the line will eventually warm up to the surrounding temperature when there is no flow. With beer, this not only increases the final cooling problem, it complicates the control of foam or head.

From the keg, the beer is piped to a beverage cooler similar to the instantaneous water cooler of Figure 24-3. Here, its temperature is brought down to 40°F as it is delivered to the glass.

Besides the temperature, the control of foam, or head, or bead as it is variously called, presents quite a problem in the dispensing of beer. The foam is caused by bubbles of carbon dioxide. The beer in the keg is under a carbon dioxide pressure that keeps a certain percent dissolved in the beer. It is this carbon dioxide that gives beer its sharp taste. Beer that has lost its carbon dioxide tastes flat. When the pressure is released, some of the carbon dioxide is released. That released under the surface forms bubbles in the beer. The beer is thick or sticky enough that instead of rising to the top and breaking as in soda or pop, the bubbles remain as foam. It is exactly the same kind of action that causes oil foaming when a refrigerant pressure is released on the oil.

Sales appeal requires a certain amount of foam to be served with the beer, but too much foam is wasteful. A glass that is just foam cannot be served a customer. However, when a keg is first opened, several glasses or even pitchers of foam must sometimes be drawn before beer begins to flow. Also, the last beer remaining when the keg is nearly empty may be nearly all foam. And every bit of foam that forms contributes to flattening the beer, in that less of the carbon dioxide "tang" is left in the beer.

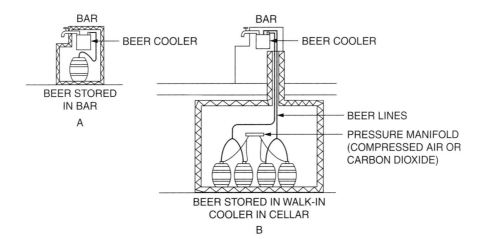

Figure 24-6 Beer cooling system.

Beer foam can be controlled by a combination of two methods, using both temperature and pressure. An increase in pressure reduces the foam, at least until it gets into the glass. A reduction in temperature also decreases the foam, because carbon dioxide remains in solution better at lower temperatures.

Foam control is achieved by adding pressure to the keg as well as keeping the temperature down. The pressure is usually applied by a tank of carbon dioxide through a pressure-reducing valve.

Charged Water, Soda Water, and Seltzer

Charged water, either for soda fountains or for bottling plants, requires refrigeration. The water may be cooled by any of the previously mentioned water coolers. It is then put through a carbonator that agitates the cold water under a carbon dioxide pressure. The cold water will absorb some of the carbon dioxide. Again, as with beer, the lower the temperature and the higher the pressure, the higher the carbon dioxide absorption.

Milk Coolers

Grade A milk must be cooled to 50°F or below on the farm, immediately after being drawn from the cow. In the wholesale dairy it must be pasteurized, then immediately chilled back to 50°F or below. Many different types of refrigerated milk coolers have been used for this purpose over the years.

The earliest farm milk cooler was a chilled-water tank in which milk cans were submerged. The water in the tank was chilled by refrigeration.

The next type used both on the farm and in wholesale dairies was a surface or Baudelot cooler. This was a vertical row of refrigerated pipes, as shown in Figure 24–7. The milk flowed over the outside of the pipes and was chilled by the time it reached the bottom. A water section could be used to start the cooling and reduce the amount of refrigeration required. Although few of these are still in use, some dairy engineers prefer them for

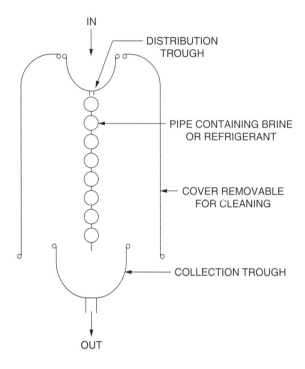

Figure 24–7 Baudelot or surface cooler.

chilling cream. Cold cream gets so thick that it is difficult to force through the passages of more modern closed coolers.

Modern milk coolers are plate-type coolers, Figure 24–8 These are made in small sizes for the farm or large sizes for the industrial dairy. This cooler is made of thin plates of stainless steel formed so that passages are created. The headers are formed and connected so that cooling water or brine flows through the passages formed by every other pair of plates. The milk flows through alternate passages. Thus, the milk has a water-cooled stainless steel wall on each side of it. Excellent, rapid heat transfer is obtained by this method. Such coolers have been used as intercoolers between flash pasteurizers and milk coolers, as in Figure 24–9. The hot milk from the pasteurizers helps heat the milk going to the pasteurizer, and the cool milk entering helps cool the milk leaving. After use, the plate-type cooler can be opened up so that each individual plate is exposed for cleaning and sterilizing.

Figure 24–8 Plate-type milk cooler. *Courtesy of CP Division, St Regis.*

Brewery Refrigeration

The brewing industry is a very large user of industrial refrigeration. The design of the required refrigeration equipment so ties in with the actual design of the rest of the brewing equipment that design data could hardly be included in a book of this scope. However, a brief outline of the brewing process and its use of refrigeration will be of help to anyone in contact with the brewing industry.

Figure 24–10 shows a simplified schematic flow sheet for a brewery. A mash, made chiefly of grains, is cooked in a brew kettle to form a wort. Only one batch a day is cooked up, so it must be the entire daily production. This wort is allowed to cool overnight. This is done in open pans in a carefully controlled or air-conditioned atmosphere. The following morning, the wort is put through a cooler to bring it down to fermenting temperature and then pumped to fermenting tanks. Here the fermenting temperature is maintained accurately to give the required results. Temperatures of from 45°F to 55°F are used, depending on the type of beer or ale to be made. This fermentation generates heat and carbon dioxide. After fermentation, the beer is further cooled, and then pumped to aging tanks. It is aged for several weeks at about 34°F for beer or 45°F for ale. It is then sent to the racking room for barreling or bottling. It is mainly by accurate control of temperature that the brew master gets the results he or she is seeking. Therefore, all temperatures are specified by the brew master and, when need be, altered by the brew master.

Operating costs in a brewery are, probably more than in any other industry, dependent on the design of equipment based on the proper heat balance. A brewery lends itself well to steam power for driving mechanical equipment, with exhaust steam used for cooking and heating. Even where the convenience of pushbutton electric power is desired, a-steam driven generator is sometimes installed to furnish the electric power. In larger breweries, cogeneration systems are sometimes used. Heat exchangers between the brew kettle and the first wort cooler are used to reduce the heating requirements for the batch being prepared and to reduce the cooling requirements for the batch being cooled. Water cooling, usually with water from a cooling tower, is used wherever possible. Refrigeration takes up where the water cooling leaves off. Heavy peak loads such as wort cooling and cooling between fermenting and aging are staggered to reduce total demand loads. Where heavy demands increase equipment or power costs excessively, water tanks, brine tanks, or congealing tanks may be used to store refrigeration for the peaks.

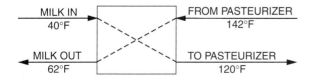

Figure 24–9 Plate cooler as intercooler in flash pasteurizing.

298　CHAPTER 24　*Liquid Cooling*

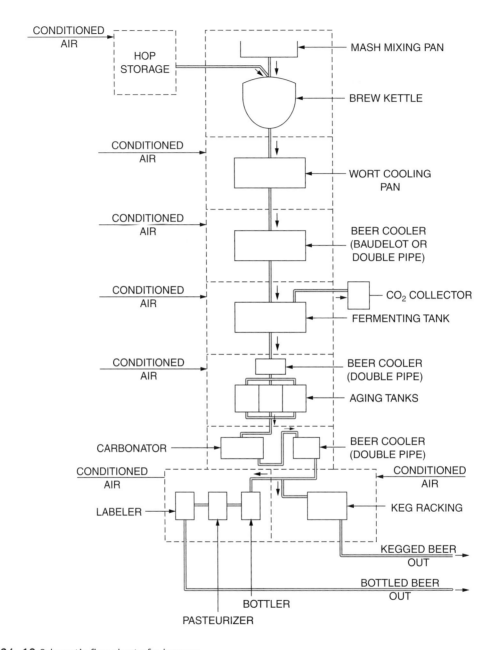

Figure 24–10 Schematic flow sheet of a brewery.

Any cooling that may be necessary in the air conditioning of the wort pans after cooking is usually done with water. The wort is usually cooled in a stainless steel shell-and-tube cooler. The heads on the cooler are so designed that they can be easily opened for cleaning and sterilizing.

The fermenting room presents two problems, cooling and air conditioning. Because fermentation generates heat, this heat, as well as leakage to the room or tanks, must be absorbed. This is sometimes done entirely by coils in the room, but it is better done by *attemperator coils* in each fermenting tank.

Attemperator coils are merely cooling coils inside the tank through which refrigerated water is pumped. The water flow can be accurately controlled to maintain the required temperature. Even with attemperator coils, the fermenting room is also refrigerated. This keeps down the heat transfer load on the tanks.

Fermentation also produces carbon dioxide. If something is not done about this, the carbon dioxide concentration of the fermenting room will increase to a point at which it would be hazardous for work personnel. This carbon dioxide may or may not be reclaimed, but there is always enough that adequate ventilation is necessary to reduce its concentration. Any air delivered to the room must be cooled to room temperature. It is also particularly important to keep this air clean. Fermentation requires the maintenance of conditions that are conducive to the growth of bacteria. Therefore it is important to keep the room, and the air delivered to the room, clean enough that it cannot contaminate the beer.

Some of the carbon dioxide is dissolved directly in the beer, but the beer is often further charged with carbon dioxide when it is bottled or kegged. Therefore, the carbon dioxide escaping is often reclaimed and stored for further use. It may be stored as gas, but this requires gas holders of large volume. Much space can be saved by liquefying the carbon dioxide. This is done by increasing its pressure and reducing its temperature by refrigeration.

After fermentation, the temperature must be further dropped as the beer goes to aging tanks. At this point the beer cannot be aerated, so an enclosed cooler must be used. Either a double-pipe or a shell-and-tube cooler may be used. Fermentation is completed by this time, so attemperator coils are not necessary. The rooms are kept at aging temperature to reduce heat transfer to the beer.

After aging, the beer is pumped to the racking room to be bottled or put in kegs. The racking room is usually kept at about aging temperature to keep the temperature of the beer from rising. If the beer is to be put in kegs, it is usually first put through a small double-pipe or shell- type cooler and its temperature reduced 2°F or 3°F. This offsets any temperature rise due to handling and pumping. If the beer is to be bottled or canned, it is first pasteurized, then cooled to a low temperature. Heat exchangers are commonly used here to save on refrigeration and heat requirements. Whether the beer is put into kegs, bottles, or cans, the filling is usually done in a carbon dioxide atmosphere or under a carbon dioxide pressure. This helps maintain the sharpness of the beer. This is where the carbon dioxide from the fermentation is used.

In addition to the above direct refrigeration requirements, two important ingredients require careful storage at reduced temperatures. Hops, which help flavor the wort, are often purchased soon after harvest time in the fall. They must then be stored under proper conditions to prevent spoilage. About 34°F at a humidity of 65 percent has proved best for this. The other important ingredient is yeast. This must be kept at reduced temperatures, usually about 38°F. It must also be kept in a clean atmosphere to prevent contamination from any airborne organisms.

Refrigeration in Wine Making

In recent years an increasing amount of refrigeration is being used in the making and aging of wine. Here, as in the brewery, fermentation generates heat. This fermentation can be more accurately controlled by controlling the temperature. This is usually at a temperature high enough that water from a cooling tower is sufficient.

The aging process can be greatly speeded up by refrigeration. The purpose of aging is to allow time for sediments to precipitate. The clear wine can then be filtered or drawn off and it will remain clear. Wine that has not been properly aged will precipitate and become cloudy in the bottle. This aging process used to take several years. By using refrigeration, fall wine can be made ready for the Christmas market.

One of two methods may be used. The wine may be cooled to within about a degree of its congealing point, then held for 2 weeks or more. The second method is to congeal the wine and hold it for several days, then allow it to thaw. With either method, the wine may be allowed to stand in tanks so the precipitate will settle out, or it may be pumped through a filter. Some wine makers feel that one such treatment, if accurately controlled, is sufficient. Others cool and filter the wine as many as five times. It is possible, according to some authorities, to force precipitation so far that the wine loses some of its bouquet.

The aging wine is stored in large tanks. It may be kept chilled in any of three different ways. Direct expansion or brine coils may be placed inside the tanks. This method provides the poorest control because of lack of agitation, temperature variations throughout the tank, and sometimes freezing out on the coils.

The second, and probably the most common method, is to use an external cooler, usually a double-pipe or shell-type cooler. Aerating coolers cannot be used because the wine must be protected from the oxygen in the air to prevent secondary fermentation or spoilage. Coolers are usually made of stainless steel. A single central cooler may be arranged with lines and valves to it, so it can be used to operate on one tank at a time. The wine is pumped from the tank, through the cooler, and back to the tank. This circulation is kept up continuously until the tank temperature is brought down to the desired point. Then the valves and pipes or hoses can be used to transfer the cooler to another tank.

The third method is to have all the tanks in a room refrigerated to the desired aging temperature. This alone is too slow to be satisfactory, but when this method is combined with an external cooler, it produces the most accurate and best-controlled results.

The temperature required depends on the type of wine and whether it is to be aged at above or below freezing. Table or dry wines will freeze at temperatures from 20°F to 22°F. Sweet or fortified wines will freeze at temperatures from 7°F to 12°F.

One other practice requiring refrigeration, commonly used with sparkling wines, is to do the final aging in bottles. While aging, the bottles are set upside down. This allows the sediment to settle against the cork. To remove this sediment, the neck of the bottle is frozen by immersing it in refrigerated brine. The cork is then removed with the bottle upright. The sediment is frozen to the cork, so it comes out with the cork. The bottle is then recorked and is ready for market.

Questions

1. a. What is the coldest temperature to which water should be cooled for drinking?

   b. What temperatures are recommended for drinking water?

2. What is the difference between a remote and a self-contained water cooler?

3. What is the difference between a storage and an instantaneous-type cooler?

4. A circulating water system in a hotel has cold water outlets in 80 rooms. The water circuit is to be 1200 lineal feet. The design temperature is 95°F. The summer water temperature is 70°F.

   a. What size circulating pipe should be used?

   b. What is the total refrigeration load?

5. Why is additional pressure sometimes used in beer dispensing?

6. Why is soda water chilled before charging with carbon dioxide, even though it will be allowed to warm up again after bottling?

7. What is a plate cooler?

8. Of what materials are most modern milk coolers made?

9. List the different locations in a brewery that require the cooling of a liquid.

10. List the different locations in a brewery that require room refrigeration.

11. Why is aging wine refrigerated?

12. What three methods are used in cooling wine?

Chapter 25 Complete Systems

Commercial Multiplexing

Previous chapters have all dealt mostly with individual parts of a refrigeration system. In this chapter, these parts are put together into an integrated whole. Most present-day domestic and small commercial equipment is made up entirely of factory-assembled units. Therefore, all that needs to be done to install them is to move them into place and connect them to electric power, and sometimes water lines and drains. Any but the simplest commercial equipment usually leads to *multiplexing*, that is, connecting several cabinets or fixtures to the same compressor.

Where all fixtures are to operate at the same temperature, this is no problem. If different temperatures are wanted in different places, however, a few general rules should be followed as far as possible. Particularly with compressors with no capacity control, the following are necessary to prevent faulty temperature regulation:

1. Place auxiliary valves such as solenoid valves or evaporator pressure-regulating (EPR) valves in the lines of the warmer fixtures, solenoid valves in the liquid lines, and EPR valves in the suction lines.
2. Place a check valve in the suction line of the coldest fixture close to the outlet of the evaporator.
3. Do not connect fixtures having too great a temperature difference to a single compressor.
4. Do not connect fixtures having too great a difference in peak loads to a single compressor.
5. Have at least one-third of the total load open to the compressor with no auxiliary valves. (This must be the suction from the coldest evaporator.)
6. Choose a condensing unit with a capacity equal to the sum of all the fixtures to be connected, and at the suction temperature of the coldest fixture.

The reason for the first and second rules was discussed in Chapter 9. Too wide a temperature variation is not desirable because any condition that maintains the load on the higher-temperature fixtures will maintain such a high suction pressure that refrigeration on the lower-temperature fixtures will be unsatisfactory. If there is a large variation in peak loads, steady compressor operation caused by a large load on one fixture may cause too much refrigeration on more lightly loaded units. At least one-third of the total load should be open to the compressor at all times. If auxiliary valves close off more than two-thirds the compressor capacity, the large capacity on the small remaining load will cause short cycling. The condensing unit must have capacity to handle the total load if it is to refrigerate it all satisfactorily. It must be able to handle this total load at the lowest suction pressure because the auxiliary valves usually throttle the higher pressure down to the lower pressure. Some typical examples illustrate these rules.

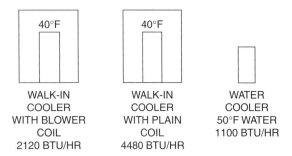

Figure 25–1 Example 1.

EXAMPLE 1 A system such as the one in Figure 25–1 is to be multiplexed. The figures give the calculated heat load for each fixture without a correction for running time. Work out the material list required.

The blower coil is usually operated at about a 10°F temperature difference. It must work on a defrosting cycle to prevent plugging the coil with frost. A solenoid valve operated from a pressure control set for pressures to give temperatures of about 24°F and 36°F will do this. The natural-convection coil will operate at about 20°F temperature difference. This will require a refrigerant temperature of 20°F, which is the coldest of the entire group. So the backpressure control on the condensing unit should be set to give this average temperature. Also, this is the biggest part of the load, so it falls within our rule of having at least one-third of the capacity open to the condensing unit. The water cooler will probably run with about a 5°F temperature difference, so its suction temperature will be about 45°F. This could be obtained with an evaporator pressure-regulating valve.

The required coil sizes will each be 50 percent larger than the data of Figure 25–1 to allow for running time. Therefore the coils and their loads are as follows:

Blower coil: 3180 Btu per hour at 10°F temperature difference
Natural-convection coil: 6720 Btu per hour at 20°F temperature difference
Water cooler: 1650 Btu per hour
Total: 11,550 Btu per hour

Therefore a condensing unit of 11,550 Btu per hour at a 20°F suction temperature should be chosen.

The line sizes are chosen from these loads and Figure 22–7, assuming that R-134a is the refrigerant and 100°F is the condensing temperature. To keep the pressure drops within the proper limits, the following line sizes will be used:

	Suction Line	Liquid Line
Line A	3/8 inch	3/8 inch
Line B	5/8 inch	3/8 inch
Line C	3/8 inch	3/8 inch
Line D	7/8 inch	3/8 inch

Figure 25–2 is a complete schematic diagram of the installation including required settings. For simplicity, liquid lines are omitted. If there is a steady load on the water cooler, the cut-in point on the compressor backpressure control will have to be raised, in some cases to at least 40 psi, to prevent short cycling.

The amount of refrigerant for a complete charge can be estimated using Table A–13 in the Appendix as follows. The length and diameter of the tubing in the coils chosen will be needed. This information can be taken from the catalogs from which they are selected.

Compressor charge: 2.00 pounds
Liquid line A: $20 \times 0.037 = 0.74$ pound
Liquid line B: $35 \times 0.037 = 1.30$ pounds
Liquid line C: $50 \times 0.037 = 1.85$ pounds
Liquid line D: $12 \times 0.037 = 0.44$ pound
Blower coil: $90 \times 0.02 = 2.70$ pounds
Natural-convection coil: $120 \times 0.03 = 3.60$ pounds
Water cooler: $21 \times 0.01 = 0.21$ pound
Total: 10.84 pounds

Additional refrigerant should be available for purging, possible losses, or variations in estimates. Figure on 13 pounds total. Remember the *de minimus* loss rule (see Chapter 3).

All this information can be brought together into a bill of materials for the job.

1 Condensing unit, complete with motor starter, capacity 11,500 Btu per hour at 20°F suction temperature
1 Blower coil, capacity 3180 Btu at 10°F temperature difference

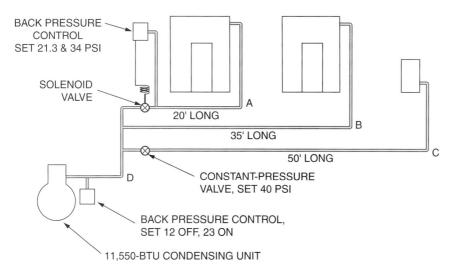

Figure 25-2 Example 1, completed schematic.

1 Natural-convection coil, capacity 6720 Btu at 20°F temperature difference
1 Water cooler, capacity 1100 Btu
1 Thermostatic expansion valve, 3/4-ton capacity
2 Thermostatic expansion valves, 1/2-ton capacity
2 Backpressure controls
1 Solenoid valve
1 Evaporator pressure-regulating valve
1 Check valve
1 Suction manifold with the following hand valves:
  1 3/8-inch valve
  1 5/8-inch valve
  1 7/8-inch valve
1 Liquid manifold with the following hand valves:
  3 3/8-inch valves
165 feet of 3/8-inch copper tubing
65 feet of 3/8-inch copper tubing
25 feet of 5/8-inch copper tubing
60 feet of 7/8-inch copper tubing
Miscellaneous fittings for all above tubing sizes
13 pounds R-134a
1/4 gallon R-134a compressor oil

Electric wiring and plumbing are usually subcontracted, or in a large organization handled by separate departments. Therefore these data are not included.

EXAMPLE 2 A system as outlined in Figure 25-3 is to be multiplexed. Work out the required details for this job.

This system will best be operated using two condensing units because, first, there is too great a difference between temperatures, and second, the colder part of the load does not make up one-third of the total load.

The two walk-in coolers can be hooked together to one condensing unit. This unit should have a refrigeration capacity of

$$1.5(4750 + 6250) = 16{,}500 \text{ Btu per hour}$$

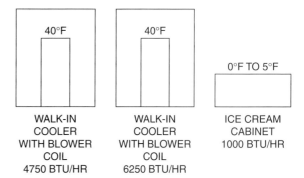

Figure 25-3 Example 2.

A 30°F suction temperature is chosen. No auxiliary valves are needed because both boxes operate at the same temperature. The ice cream cabinet can be connected to a second compressor having a capacity of

$$1.5 \times 1000 = 1500 \text{ Btu per hour}$$

This should be chosen with a suction temperature of −5°F. The line sizes are

	Suction Line	Liquid Line
Line A	3/4 inch	3/8 inch
Line B	7/8 inch	3/8 inch
Line C	1 1/8 inch	1/2 inch
Line D	3/4 inch	3/8 inch

Figure 25-4 illustrates the assembly of the parts to make up the complete system.

EXAMPLE 3 Figure 25-5 shows a job typical of a large restaurant, or an institution such as a boarding school or a hospital. Work out the required details for this job.

Again, the load will best be divided into two groups, with the storage boxes on one condensing unit and the ice and ice cream equipment on the other. To provide even greater flexibility and standby service, these condensing units will be cross-connected as shown in Figure 25-6. These cross connections will normally be left closed, but

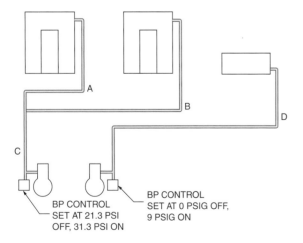

Figure 25-4 Example 2, completed schematic.

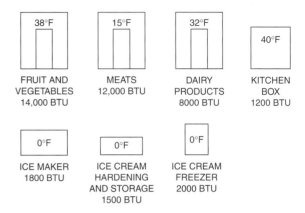

Figure 25-5 Example 3.

in case of failure of one compressor, the other can help hold the other load temporarily. The required capacity of the first condensing unit is

$$1.5(14{,}000 + 12{,}000 + 8{,}000 + 1{,}200) = 52{,}800 \text{ Btu}$$
per hour at 0°F (meat box evaporator)

The ice cream freezer will not be operated continuously. This might be furnished with its own condensing unit. If not, no provision for running time needs to be made, because it will only run for a few hours a day anyway. Therefore, the capacity of the second condensing unit is

$$1.5(1800 + 1500) + 2000 = 6950 \text{ Btu per hour}$$
at −10°F

The temperature of each box or fixture will be regulated by a thermostat that operates a solenoid. When there is a possibility of a suction pressure lower than desired when the solenoid is open, an evaporator pressure-regulating valve is also used. Figure 25-6 illustrates the valving and control of this system. The line sizes are

	Suction Line	Liquid Line
Line A	1-1/8 inch	1/2 inch
Line B	1-3/8 inch	1/2 inch
Line C	1-1/8 inch	3/8 inch
Line D	1/2 inch	3/8 inch
Line E	2-1/8 inch	5/8 inch
Line F	7/8 inch	3/8 inch
Line G	7/8 inch	3/8 inch
Line H	3/4 inch	3/8 inch
Line J	1-1/8 inch	3/8 inch
Line K	1-1/8 inch	3/8 inch

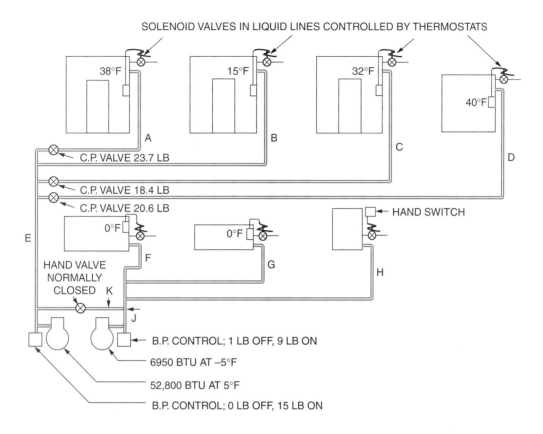

Figure 25-6 Example 3, completed schematic.

The crossover lines K need not be any larger than the lines J, for this is the maximum capacity that will flow through them if one of the compressors is down. Naturally, the small compressor could not carry the load of the big one, but careful conservation of the temperatures already down should make it possible to hold them for a short time.

Cold Storage

Cold-storage warehouses have always been one of the major applications of refrigeration. A cold-storage plant may be a refrigerated collecting station on a large ranch or a small cooperative warehouse that acts as a collecting station for all the growers of a community. Or it may be a large industrial warehouse of many floors covering one or more city blocks. It may be built with only one type of local commodity in mind, or it may precool, freeze, or store all types of perishable commodities required in a large city.

Any refrigeration warehouse, whether large or small, should be designed and built with the following considerations in mind:

1. Who will use it?
2. How will users deliver commodities to it, and how will commodities be shipped out?
3. Is water for condenser cooling and a suitable fuel or power available?
4. What is the cost of land, construction, and other factors?
5. What zoning requirements and what codes might restrict the desired size or application of the structure?
6. What type of commodity or commodities will users want to store?

7. How much commodity will probably be stored at a time?
8. For what length of time will these commodities require storage?
9. What temperatures and humidities will be required for these commodities?
10. Should an ice plant, locker plant, or other load-equalizing factor be included?

The first five of these considerations will dictate the location of the warehouse. It must be convenient to reach for those using it. It must have proper truck loading docks. These should be located so that unloading trucks will not block traffic, either in the street or trying to reach other parts of the plant. Nearly all such warehouses call for a railroad siding. Some operations that thought they would always depend on trucks have since regretted the omission of railroad access.

The other five considerations will determine the type and size of building, and the economics of the entire project. Ice plants or locker plants are sometimes considered in a warehouse to provide a steadier annual load and to increase total income. First costs and estimated operating costs must be justified by estimated income.

EXAMPLE 4 A small cold-storage plant is for the purpose of collecting, precooling, and packing local fruits. Apricots, peaches, and pears are expected. The cold-storage part of the building is shown in Figure 25–7. Four inches of insulation is to be used. It is desired to be able to precool 60,000 pounds of fruit per day. The local design temperature is 90°F. Work out the refrigeration requirements for this job.

A 35°F cooler temperature is common for such fruit. The heat load is calculated as follows:

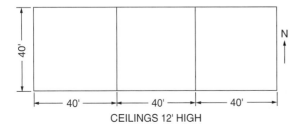

Figure 25–7 Example 4.

Roof area:

$$40 \times 120 = 4800 \text{ square feet}$$

Roof temperature difference including sun effect:

$$(90 - 35) + 20 = 75°F$$

Heat leakage through roof:

$$H = Ak(t_2 - t_1)/x = 4800 \times 0.3 \times 75/4$$
$$= 27{,}000 \text{ Btu per hour}$$

Floor leakage: The floor is up off the ground to provide an air space under it.

$$H = 4800 \times 0.3 \times 55/4 = 19{,}800 \text{ Btu per hour}$$

Heat leakage through south wall:

$$H = 1440 \times 0.3 \times 60/4 = 6480 \text{ Btu per hour}$$

Heat leakage through north wall:

$$H = 1440 \times 0.3 \times 55/4 = 5940 \text{ Btu per hour}$$

Heat leakage through west wall:

$$H = 480 \times 0.3 \times 65/4 = 2340 \text{ Btu per hour}$$

Heat leakage through east wall:

$$H = 480 \times 0.3 \times 55/4 = 1980 \text{ Btu per hour}$$

Total heat leakage = 63,540 Btu per hour

Estimated loss due to door openings, ventilation, and so on, is 127,080 Btu per hour.

Assume an average of six workers in the rooms.

$$6 \times 600 = 3{,}600 \text{ Btu per hour}$$

Lights are estimated at 1/2 watt per square foot of floor area.

$$1/2 \times 4800 \times 3.41 = 8{,}200 \text{ Btu per hour}$$

Total daily steady load is thus 202,420 Btu per hour.

The product load is

$$H = WS(t_2 - t_1)/\text{time} = 60{,}000 \times 0.92 \times 55/24$$
$$= 126{,}500 \text{ Btu per hour}$$

Total leakage and product load = 328,920 Btu per hour

25% for safety factor = 82,230 Btu per hour

Total capacity of equipment required = 411,150 Btu per hour

To choose the required coils for each room it will be assumed that one-third of this total load is in each room.

$$411{,}150/3 = 137{,}050 \text{ Btu per hour}$$

This much capacity is best divided into several blower coils.

The total refrigeration load is

$$411{,}150/12{,}000 = 34.3 \text{ tons}$$

The leakage load will vary somewhat with the outside temperatures. The product load will only be required when products are being cooled. Therefore, the compressor capacity should be divided. At least two compressors should be selected, one to take care of the leakage load and one to take care of the product load. This choice requires compressors of 17.2 tons each. If blower coils are used, suction temperature will be 20°F. A water-cooled condenser will give an 85°F condenser. An air-cooled condenser will give a condensing temperature of 110°F to 120°F.

A halocarbon refrigerant will probably be chosen for a job of this size, perhaps R-22 or R-134a.

Locker Plants

It has been said that a cold-storage warehouse is something like a bank: One puts certain valuable products there for safekeeping. The difference between a bank and a cold-storage warehouse is that the bank customer is satisfied to withdraw any money similar to that deposited. The warehouse the customer, however, wants to withdraw exactly the same product as was deposited. It is fairly simple to keep each customer's product segregated when the quantities are large, but if each customer has at most only a few hundred pounds in storage, and this is apt to be deposited or withdrawn only a few pounds at a time, the segregation and bookkeeping necessary would be so involved as to cost more than the actual storage.

Locker plants have provided an answer to this problem. Each customer rents a separate locker in which the valuable food products are kept. Locker plants act more like the safe deposit vault of a bank. Food is brought to the locker plant by the customer. It may be prepared and packaged at home, but this is more often done right at the plant. The packages are frozen, then transferred to the customer's locker. From there, the customer may withdraw them as required.

Although home freezers have cut into the locker plant business a great deal, many locker plants are still in operation. If they are not operated in conjunction with a cold-storage plant, locker plants should be checked periodically by a good refrigeration service or maintenance technician.

The size of the locker plant and the type of facilities to be included will depend on the probable patronage. All the requirements listed for a good cold-storage warehouse should be considered. A locker plant of less than 400 to 500 lockers is not usually financially self-supporting. If it is built in conjunction with a meat market, cold-storage plant, ice plant, or other similar establishment, it may well pay for the part-time attention it requires. Few locker plants of any size are successful if locker rentals alone are depended on. Processing charges are made for all goods processed and frozen in the plant. Sufficient food should be moved through the plant to make these processing charges pay for the time and labor of the people who must be available to do this work.

Regardless of size, locker plants are most successful in rural communities where the farmer or grower can use it to preserve food for personal or family requirements, or where town dwellers may purchase directly from the farmer and put up their own perishable food requirements. In cities where customers can only buy food through regular retail channels, the savings is not great enough or the food turnover not fast enough to make the plant very successful either from the standpoint of the patron or the operator.

Usually, more meat goes through a locker plant than any other commodity. Therefore a good locker plant becomes a miniature meat-packing plant. Pens and slaughtering chutes may or may not be included. Slaughtering is often done on the farm anyway. However, provision must be made for a chill room, aging room, and cutting room, as well as the freezer and the locker room. In addition, it is desirable to have equipment for making sausage, rendering lard, smoking, pickling, and so on. An experienced butcher, as either a part-time or full-time employee, depending on the size of the plant, is absolutely necessary. In addition to meat-handling equipment, any but the smallest plants usually provide for cleaning, preparing, and packaging fruits and vegetables. Washing tables and trays, with hot and cold water, and provisions for blanching, with either boiling water or steam, are usually all that is required for this.

The amount of food that must be handled in a locker plant will naturally be based on the number of lockers, and to a certain extent on the requirements of the patrons and the local products. Figure 25–8 shows factors that have been worked out to determine the size and refrigeration requirements required once the number of lockers has been determined.

Chill rooms and aging rooms should be maintained at about 35°F. Blower coils are favored. Cutting rooms and fruit- and vegetable-preparing rooms are at normal room temperature. Freezing may be done on plate freezer shelves or in air-blast freezers.

Either shelf or air-blast freezers may be run from −20°F to −40°F. Some freezers have been operated as high as 0°F, but below 0°F gives the best results.

The locker rooms should be held about 0°F or below, as shown in Figure 25–9. Blower coils, pipe coils, and plate coils have all seen considerable use. Defrosting is a problem here that must not be overlooked. Blower coils are usually defrosted daily. Water defrost, electric heat, and hot gas defrost have all been used successfully. If pipe coils or plate coils are used, they are usually grouped over the aisles. This gives better air distribution to all the lockers. It also allows the frost to drop into the aisles instead of onto the tops of the lockers during defrosting. Pipe and plate coils are defrosted as needed, usually at intervals of several months.

Figure 25–10 is a plan for a typical medium-size locker plant suggested by the American Society of Heating, Refrigerating and Air Conditioning Engineers (ASHRAE).

Air Conditioning

Air conditioning is a use of refrigeration technology that has become extremely popular in recent years. However, it should be recognized that refrigeration or cooling is only one part of air conditioning. Air conditioning is heating and ventilating with refinements. Complete air conditioning means controlling

	Room, Size, Floor Area or Volume (sq. ft)	Cubic Feet of Food Products	Pounds of Food Products per Loading	Days Between Loading Periods	Pounds of Food per Day to Be Refrigerated or Handled	Recommended Temperature (°F)
Locker room	1.3 to 1.6	6	250			0
Chill room	0.09 to 1.3	0.163	6.8	2	3.4	35
Aging room	0.38 to 0.55	0.672	28	14	2	35
Freezer	0.08 to 0.11	0.054	2.25	½	4.5	−15 to −40
Meat cutting room and facilities					3.4	Room temp.
Fruit and vegetable preparing facilities					1.1	Room temp.

Figure 25–8 Size of locker plant facilities required per locker.

	Approximate Storage Period (months)		
	At 0°F	At 5°F	At 10°F
Fruit, vegetables, beef, veal	10 to 12	8 to 10	3 to 4
Lamb, mutton, poultry, eggs	8 to 10	6 to 8	3 to 4
Pork, butter	6 to 8	4 to 6	2 to 3
Ground meat (unsalted), cream	4 to 6	3 to 4	1 to 2
Bacon and ham (mild cured, not smoked)	4 to 5	3 to 4	1 to 2
Fish, ground meat with salt in seasoning	3 to 4	2 to 3	1 to 2

Figure 25-9 Approximate storage periods for frozen food in lockers. *Courtesy of American Society of Heating, Refrigerating and Air Conditioning Engineers.*

temperature, heating or cooling as required; controlling humidity, adding or extracting moisture; controlling air distribution; and controlling air purity and cleanliness through filtering. So an air conditioning technician must know more than refrigeration. However, refrigeration is an important part of summer air conditioning, both to cool the air and to chill it below the dew point to reduce the humidity.

To remove humidity, the evaporator coil must be below the dew-point temperature. The control of humidity cannot be treated in detail in a book of this scope. However, if the principles given in Chapter 19 are kept in mind, some understanding of the humidity problem is possible. Briefly, the lower the coil temperature, the drier will be the air; or, the warmer the coil temperature, the less the air will be

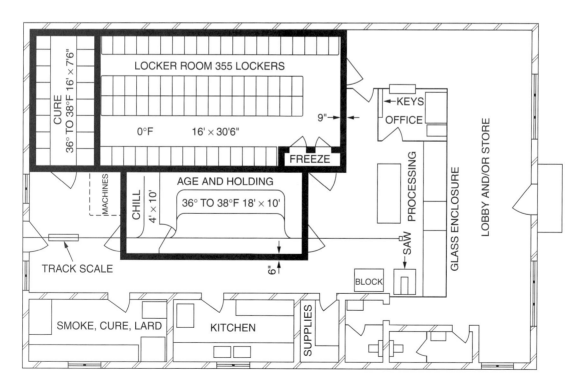

Figure 25-10 Typical frozen food locker plant. *Courtesy of American Society of Heating, Refrigeration and Air Conditioning Engineers.*

dehumidified. It is important to choose a type of coil and operate it at such a temperature that the required dehumidification is accomplished while getting exactly the right amount of air cooling.

Air conditioning loads are different than common refrigeration loads in several respects. Evaporator temperatures are higher. Temperatures of 35°F to 45°F or sometimes even 50°F are used. The evaporator cannot go as low as 32°F because it would plug with frost. With these higher evaporator temperatures, more refrigeration is produced per cubic foot displacement of the compressor and per horsepower. This means more refrigeration and a larger motor for a given sized machine than the average refrigeration technician is used to.

Another difference is that centrifugal blowers can deliver an enormous amount of air with a very large heat load to an air conditioning coil. Air conditioning loads seem surprisingly large to most refrigeration technicians. At times, when there is a great deal of humidity in the air, there can be very heavy refrigeration loads with very little air cooling.

Air conditioning equipment may be self-contained or remote. Figure 25–11 shows a small self-contained window-type room air conditioner. Such equipment is commonly built in $1/2$-horsepower to $2^{1}/_{2}$-horsepower sizes. The units are designed to set in the window so that outside air can be used for condensing purposes.

These units are used not only for residential air conditioning, but also in many older commercial buildings, hotels, and motels, in which individual units can provide air conditioning without the major reconstruction costs that would be necessary to install a central air conditioning system.

Figure 25–12 shows a larger self-contained room cooler. These units are commonly made in 3-horsepower and 5- horsepower sizes, and some manufacturers make them up to 10 horsepower or even 20 horsepower. This type of unit is commonly installed in small stores or office suites. It may include a steam or hot water coil for heating when required. Naturally, such a coil must be supplied by a separate boiler. Figure 25–13 shows several ways in which such a unit can be installed.

Many newer air conditioners chill water in a coil similar to the one shown in Figure 5–6. This chilled water is then pumped to the air cooling coil.

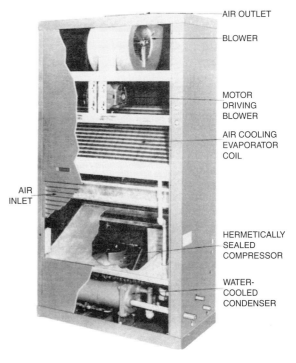

Figure 25–12 Vertical, floor-mounted air conditioner. *Courtesy of Commercial Industrial Air Conditioning Division of Westinghouse Electric Corp.*

Figure 25–11 Self-contained window-type air conditioner. *Photo by Noreen.*

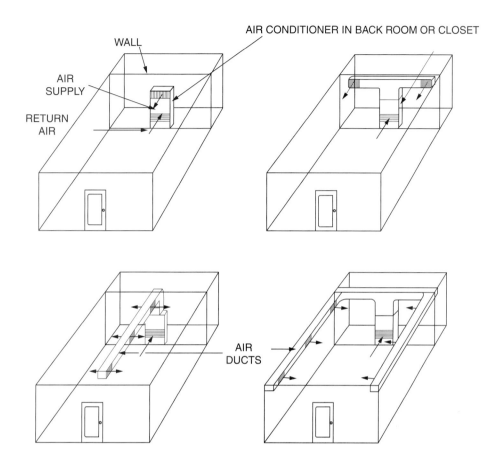

Figure 25–13 Methods of installing vertical air conditioners.

Remote units for either domestic or commercial applications are available. The section containing the evaporator and blower may be designed to fit directly into a duct system, or it may be put into an ornamental cabinet located in the room to be conditioned. The condensing unit may be designed with either an air-cooled or a water-cooled condenser. It may be installed in a basement, a back room, an outside "dog house," or in any other convenient location.

Heat Pumps

An interesting development in heating is the reverse-cycle air conditioning system or heat pump. This is a very special application of refrigeration. We give only a brief introduction here, as discussing it in detail would nearly fill a complete text. There are air-source heat pumps and ground-source (geothermal) heat pumps. Both types of systems employ a special (four-way) valve that reverses the refrigerant flow so the condenser and evaporator functions are reversed. In the summer, the unit cools the room or building. In the winter, the valves are reversed and the room coil becomes the condenser, thus heating the air blown over it.

Air-Source Heat Pumps

Air-source heat pump systems consist of components very similar to a conventional central air conditioning system. As mentioned, in the summer (air conditioning) mode, the coil in the conditioned space is the evaporator and the coil outside the conditioned space is the condenser. Heat is pumped from the conditioned space to the ambient air. In the winter (heating mode), these coils reverse roles. The coil in the conditioned space becomes the condenser and the coil outside the conditioned space is the evaporator. In this case heat is pumped from the ambient air to the conditioned space. The farther north the system is located, the less heat there is available from the ambient, especially in the winter. Therefore heat pumps used in the far north or extreme south regions of the planet require some type of auxiliary heat. Electric resistance heat is readily available in virtually all developed areas. However, cost must be taken into consideration. Where electricity is cost-prohibitive for heating purposes, some other type of auxiliary heat must be considered. Oil, gas, and propane have all been used. The only limit is the design engineer's imagination.

The advantages of a heat pump system can be better understood if reference is made to the coefficient of performance discussed in Chapter 2 and to the total heat in the condenser compared to the motor power of Figure 20–8. Three or four times as much heat can be obtained with a heat pump as with the same amount of power using standard electric heaters. The extra heat comes from the heat in the outside air. In theory, heat is available and recoverable at any temperature above absolute zero. In practice, however, there is a minimum temperature below which heat pumps require auxiliary heat. In some commercial plants, efficiency has been increased by placing the heat-absorbing evaporator in a duct of exhausting warm air.

Heat pump systems should have a high-speed defrost system on the outside coil. If they become plugged with frost, no heating effect will be available. Electric defrost has been used in many cases, but again, operating costs must be considered. These systems also require two-speed or multispeed fans, as the actual number of Btu's pumped in the winter is almost invariably more than that pumped in the summer (depending on geographic location).

Heat pump systems can best be utilized in areas where the heat load in the winter is not more than 25 percent above the cooling load in the summer and where electric power costs are reasonable compared to other fuels. Misapplication of these systems, or any other special system, will leave reputations that will hurt sales in areas where they may fit economically.

Geothermal (Ground-Source/Sink) Heat Pumps

Geothermal or ground-source/sink systems are generally more effective in more northern and southern regions. Eight to ten feet below the ground, the temperature is virtually constant all year. This temperature is the mean of the temperature during the hottest part of the summer and that during the coldest part of the winter. It is generally about 48°F to 55°F, depending on the specific location. Always check with local contractors before starting construction.

In the case of a water-based geothermal system, this constant temperature is tapped by burying some type of tubing at the proper depth and then pumping water through this tubing and the water coil. The coil in the building through which the air is recirculated is called the air coil. As mentioned earlier, the air coil is the evaporator in the summer and the condenser in the winter. The water coil is the condenser in the summer and the evaporator in the winter.

Another type of geothermal system utilizes a pipe buried at the proper depth. It contains part of the system's refrigerant and is an integral part of the system's refrigeration circuit. This pipe is connected directly to the compressor (via the four-way valve) and functions as the water coil described previously. Inside the building (or conditioned space), once again, we find the air coil. The air coil is the evaporator in the summer and the condenser in the winter. The pipe buried in the ground is the condenser in the summer and the evaporator in the winter. This type of system works without the use of a water recirculating pump.

Why use ground-source systems when burying pipe or tubing is costly? Ask yourself whether you would rather condense refrigerant into 95°F air or 50°F dirt or water. The answer is obvious. Over the life of such a system, original installation cost is usually recovered in the form of energy savings. In

Figure 25-14 Sprayed coil-type air conditioner. *Courtesy of Carrier Air Conditioning Co.*

some cases, rebates and incentives are offered to help offset installation cost. The original installation costs may also be incorporated into a mortgage.

Figure 25-14 illustrates a sprayed coil-type conditioner, which is a larger and more elaborate remote type. Sometimes the air washer is omitted. Where the air washer is used, the water temperature may be heated or cooled as required to add to or reduce the moisture content of the air. Or the temperature of the air entering the washer may be so controlled as to give the required condition leaving the washer.

Units such as those shown in Figure 25-11 or 25-12 may be operated by a simple three-point switch with a point for off, a point for fan only, and a point for cooling. Both the fan and the compressor operate on this last point. If this is all the control used, the unit is often small enough that it is impossible to get too much cooling. Or the machine may be turned off manually if it does get too cold.

The simplest automatic control is temperature control. The bulb is usually in the stream of room air drawn into the conditioner. On larger systems, a backpressure control is usually added to eliminate the possibility of too cold a coil.

The control of central station systems is more elaborate. Controls operated by compressed air rather than electricity are common. Air controls can be modulating where required. That is, they can open or close valves or dampers to any required point, without necessarily being wide open or completely closed.

Ice Making

Natural ice or snow has been used in limited amounts since ancient times to aid in the preservation of food or to chill drinks. Natural ice has been harvested in this country since colonial times. That is, ice frozen on ponds or rivers in the winter was sawed into blocks and then packed in sawdust in barns or warehouses to be kept until needed later in the year. During the first half of the nineteenth century, American clipper ships carried natural New England ice to many other parts of the world.

Manufactured ice is an easy, convenient way of obtaining refrigeration or cooling. It may be used for cooling anything from a glass of water (or other beverage) to refrigerator cars and fishing boats. If the ice is purchased, little or no investment in equipment is required. Other users install their own equipment rather than purchase ice.

The ice cube trays of a domestic refrigerator are miniature ice plants. Ice cubes may also be purchased from ice companies that make them by cutting up large blocks of ice.

Many firms have developed automatic icemakers that pump or circulate water through refrigerated ice trays or tubes, then automatically thaw the ice to release it. The finished cubes either drop or are carried into a separate storage bin. When the bin is full, ice cubes in contact with a thermostatic bulb automatically shut down the machine. Some of these icemakers are more or less miniature models of the tube icemaker to be described a little later.

Figure 25-15 (A, B, and C) illustrates the production side of an ice plant. Ice cans are submerged in a refrigerated brine with the required amount of water. These cans are hung in steel frames or baskets so that from 4 to 24 of them can be lifted from the brine by a crane. The brine is most commonly refrigerated by flooded coolers of the type shown in Figure 4-5, Figure 4-6, or Figure 5-8. The brine is circulated by propellers or agitators, and a velocity of about 30 feet per minute must be maintained to secure proper freezing speed. Steel partitions direct the flow of brine through the tank. Air is introduced at the bottom of the cans or through drop tubes down the center to keep the ice clear as it

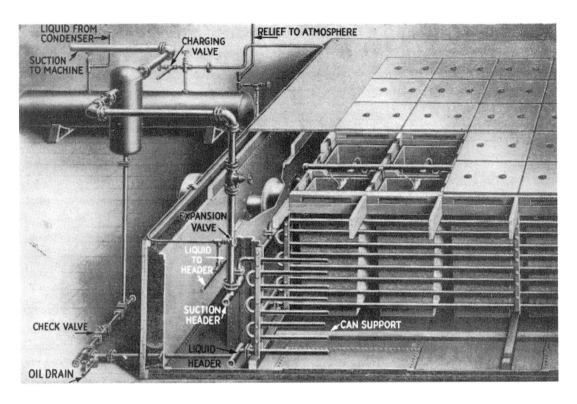

Figure 25–15 (A) Cross section of large ice freezing tank, showing general design and ammonia and air connections. *Courtesy of Frick Company.*

Figure 25–15 (B) 300-pound cans in brine, showing air agitator tubes. *Photo by Noreen, courtesy of Articglacier, Inc.*

Figure 25–15 (C) The harvest: 300-pound clear ice blocks. *Photo by Noreen, courtesy of Articglacier, Inc.*

freezes. The production of clear ice by well-operated air agitation, water conditioning, and freezing speed is a specialty in itself.

Some ice plants have storage rooms only large enough to hold a few days' supply, and make only what can be sold immediately. Other plants have storage for thousands of tons of ice, so peak-season requirements can be built up ahead of time, when sales are light. Ice storage rooms are maintained at about 24°F to 28°F. This prevents melting and keeps the ice dry. Scoring, crushing, or cubing equipment may be installed in the storage room or out on the loading platform. Ice intended for retail sale is usually scored, which marks it for 25-pound or 50-pound cuts. This is done in a jig or guide with circular saws. Where cubes are required, the cakes are run through a cuber, which saws or cuts them up into small-sized ice cubes. More of the ice may be crushed for sale where that type of ice is desired. The crushed ice is usually run through a sizer, which grades it according to size. Thus, coarse, fine, or snow ice can be supplied separately.

After dumping, the cans are turned upright. Directly over them are outlets from measuring cans. A lever is pulled that opens valves and each can in the basket is filled with the required amount of water. These cans are then taken back and replaced in the ice tank. The baskets are pulled in rotation, so the entire tank is covered. Each can remains in place anywhere from 30 to 60 hours to freeze completely.

If plain water were frozen without treatment, the ice would be white—marble ice it is sometimes called. Much of the ice made for car icing and similar applications is made this way. If the ice is intended for retail sale, a crystal clear ice is desired. To make clear ice, the water must first be treated chemically to precipitate dissolved solids or minerals. This is done in a tank large enough to give these solids time to settle out. The water is then run through a sand filter to catch anything that did not settle. The type of treatment used should be specified by an industrial chemist for the types of minerals in the local water. A combination of lime and alum is the most common treatment. The lime precipitates calcium and magnesium carbonates, which are the most common minerals in most waters. The alum coagulates the precipitates so they better settle or filter out.

Even this water will not make clear ice without further treatment because not all solids are removed. Also, any water contains dissolved gases, particularly air. These gases separate out in freezing, making millions of tiny bubbles, which give the ice its white color. Anything that will keep the water agitated will break these bubbles loose from the surface of the ice as they form. The simplest method of agitation is to put an air tube in each can. Air bubbling through the water stirs it up enough to break loose any new bubbles forming.

All this treatment would give an ice that is clear except for the core. It has not been economically feasible until recently to eliminate all minerals from water except by distillation. In the can, pure water freezes first, and the remaining minerals are concentrated in the unfrozen water at the center. If this part is allowed to freeze, it will produce a dirty, cloudy core. This is eliminated by coring the ice. That is, when all but a small core is frozen, the ice is pumped out and the space refilled with fresh treated water. This produces ice with only a thin "feather" down the center.

The air drop tubes are removed before the block freezes solid, or a "needle" or thin tube carrying steam or hot water is inserted inside the drop tube so the latter is melted out. Freezing cans with fixed tubes eliminate these operations, but the agitating air must be dehydrated.

The refrigeration load for an ice plant is calculated just like any other freezing load. To freeze ice, water must first be cooled to 32°F, then the 144 Btu of latent heat removed. This, plus heat losses from the ice tank make necessary roughly 1.5 tons of refrigeration for every ton of ice produced.

The time necessary to freeze ice can be calculated from the formula

$$T = C x^2/32 - t_b$$

where

T = time in hours
C = a constant, 5.5 to 7
x = thickness in inches
32 = freezing point of ice
t_b = brine temperature

The constant C will vary with the condition of the plant. The can submergence and brine velocity have the most effect on it. For older plants, a C of 7 is a good figure to use. The best modern plants have brought this down to 5.5. The thickness x is the total thickness of the ice block.

Notice that the colder the brine, the faster is the freezing. However, if the brine temperature becomes too low, the ice is very brittle and difficult to handle without it cracking. Besides making it unfit for block trade, it is dangerous if someone is handling it or pulling on it with ice tongs. Special chemicals have been added to the water to toughen the ice and make lower brine temperatures practical. Also notice that the time increases as the square of the thickness of the cans. By modifying the constant, this formula can be used to estimate the freezing time of any product. A test under a known condition should establish this constant for that specific condition. Then it is easy to make corrections for changes in thickness or temperature differences.

EXAMPLE 5 How long should a 300-pound cake of ice take to freeze with a 16°F brine temperature?

Assume that $C = 6$. Then

$T = C x^2/32 - t_b = 6 \times 11^2/32 - 16 = 45.3$ hours

Faster Methods for Freezing Ice

The large investment for buildings and equipment, and the length of time required for the can system to produce ice, has encouraged designers to look for other ways to make ice. Several excellent systems have been developed to produce ice in snow, flake, cube, or briquette form, which will start producing ice within a few minutes after refrigeration is applied.

The flake ice machine, shown in Figure 25-16, is one of these systems. The machine shown may be air-cooled or water-cooled and may have a remote condenser. It can produce from 1092 to 1152 pounds of flaked ice in a 24-hour period. The flaker evaporator produces ice in a constant manner by freezing small, thin layers of ice in the evaporator. The ice is then scraped from the wall of the evaporator and moved upward by an internal auger that

Figure 25-16 Flake ice machine. *Reprinted with permission of Ice-O-Matic.*

forces the flakes out of the barrel-shaped evaporator by an extrusion process.

The *Pak-Ice* machine, shown in Figure 25-17, makes ice in snow or crystal form. This machine is very similar in construction to an ice cream freezer. Water is sprayed onto the inside surface and then scraper blades scrape the ice off and out the end. Each ring of this freezer can make 5 tons of ice a day. Up to six rings can be assembled in one unit to provide greater capacity. If the ice is required to be in solid form, a briquette press can be installed on the ice outlet. This compresses the loose snow into hard briquettes that are still small

Figure 25–17 Pack-Ice machine, 30 tons daily. *Courtesy of Vilter Manufacturing Corp.*

enough to be stored in bins or handled by small conveyors or chutes.

The *Tube-Ice* machine is a vertical shell-and-tube cooler operated as a freezer, shown in Figure 5–9. It works in cycles controlled by an automatic timer. During the freezing cycle, water is pumped from the sump to the top of the freezer, from which it circulates down through the tubes. These tubes are submerged in liquid ammonia, so the water freezes to the sides of the tubes. At the end of the freezing cycle, head pressure is admitted to the ammonia side. This forces the liquid ammonia out into the auxiliary drum, and the warm discharge gas thaws the ice loose. The ice drops from the tubes and a revolving cutter cuts it off in short lengths. By changing the time of the freezing cycle, anything from a very thin shell that breaks up into a flake ice to a solid cylindrical ice cube can be produced.

Questions

1. Work out a system and bill of materials for the system shown in Figure 25–18.
2. A 30 feet by 30 feet by 10 feet building with 6 inches of insulation is to be used as an egg-collecting station at a location having a design temperature of 105°F. Cooling capacity to handle 1600 cartons of eggs a day is required.
   a. What is the leakage load?
   b. What is the product load?
   c. What would be the additional load for lights and two workers?
   d. What capacity condensing unit, in Btu, should be chosen?
3. Work out the size of equipment needed for the food locker plant in Figure 25–10. Assume 8 inches of fiberglass in the locker room, 5 inches in the aging and curing room, and 5 inches between the rooms. The refrigerated rooms are protected from sun by other buildings. Assume a design temperature of 90°F.
4. Briefly describe the can ice making system.
5. What other ice making methods are available?
6. a. What is the difference between white ice and clear ice?
   b. How is this difference obtained in the manufacture of ice?

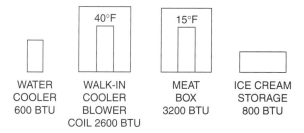

Figure 25–18 Question 1.

Chapter 26 Carbon Dioxide—Dry Ice

Carbon Dioxide Refrigeration

Until the fluorocarbons proved themselves, carbon dioxide was the only available refrigerant that was safe. This led to many installations utilizing carbon dioxide in hospitals, prisons, passenger ships, and other similar applications. At 5°F and 86°F, carbon dioxide pressures are 320 psig and 1024 psig, respectively. Such pressures require double extra-heavy piping throughout. Also compressors, condensers, receivers, and so on, must be designed to withstand such pressures. Carbon dioxide's power per ton is 1.78 horsepower, 78 percent greater than that of other common refrigerants. Its critical temperature is 88°F. Therefore, if an adequate amount of low-temperature cooling water is not available, carbon dioxide cannot be condensed properly.

With these disadvantages, carbon dioxide refrigeration is no longer considered except for the manufacture of dry ice. However, a few old carbon dioxide systems are still in operation.

Manufacture of Dry Ice

The manufacture of dry ice requires carbon dioxide refrigeration equipment. Dry ice is carbon dioxide snow compressed under a tremendous pressure into blocks. The triple point of carbon dioxide is at 60 psig and −70°F. At any pressure less than 60 psig, carbon dioxide cannot exist as a liquid. Any substance that evaporates and melts has such a range, but for most familiar substances, this happens only under high vacuum. The triple point for water, for instance, is at 29.74 inches of vacuum. Actually, the boiling point of water is reduced to 32°F at this pressure, but the freezing point does not change. Therefore water cannot exist as a liquid at a higher vacuum than 29.74 inches: It would be either ice or vapor. In the case of carbon dioxide, the triple point is above atmospheric pressure. Therefore, at atmospheric pressure it cannot exist as a liquid. Only a dry gas is given off, with no liquid to soak into the product being cooled or that has to be drained from the refrigerated space. The gas given off has low toxicity and is not injurious to commonly stored food products in low concentrations. It should be recognized, however, that carbon dioxide is a heavy gas that will settle into low places or closed rooms and there reach a concentration that may be hazardous to life. Thus, care should always be used in entering any place where carbon dioxide may be concentrated, especially because there is no warning odor.

The sources of carbon dioxide to make dry ice are varied. In some places it comes from wells, either in free form or dissolved in water. It is a by-product of many chemical processes such as fermentation and the burning of carbonates. It is present in all flue gases. If a plant is built at the location of wells or a chemical operation, the cost of the carbon dioxide may be little or nothing. However, the temperature of dry ice is such that there is considerable loss if it must be stored for any length of time or shipped for any great distance. And with the power costs necessary to produce it, it is certainly not cheap after it is in frozen form. Therefore, it is usually as economical to manufacture dry ice from more expensive sources of carbon dioxide but near its point of ultimate consumption.

An interesting type of plant burns coal, oil, or gas under a boiler. The carbon dioxide in the flue gas is used to make dry ice. The steam from the boiler drives the compression equipment necessary to freeze the dry ice.

Where the carbon dioxide comes from wells in pure or nearly pure form, it can be run directly to a compressor. If it is dissolved in water, heat is all that is needed to separate it. If it is mixed with flue gases, air, or other contaminating substances, it is dissolved using a caustic solution, then separated from the caustic by heat, as shown in Figure 26–1.

To compress the carbon dioxide from atmospheric pressure to a condensing pressure of about 1000 psig requires a compression ratio of almost 70 to 1. A few cascade systems use ammonia to condense the carbon dioxide and hold its pressure down. More common is a three-stage carbon dioxide system, as shown in Figure 26–2.

The first stage raises the pressure to about 70 psig, the second to about 300 psig, and the third to full condensing temperature. Between the first and second stages is a purifier and drier. It is placed there instead of ahead of the first stage because the gas is easier to dry and purify at higher pressure. Naturally, there is also an intercooler between each stage to remove the heat of compression. Expansion is also done in three stages. The flash gas formed at each expansion is carried back to the proper compressor stage. In this way, one-third of the flash gas need be compressed only from 300 psig to 1000 psig, the second third from 70 psig to 1000 psig, and only the remaining third need be compressed through the entire 70-to-1 compression ratio.

The low-pressure receiver stores the liquid carbon dioxide ready for the snow press. The snowing period is run intermittently. With the press closed, the expansion valve to it is opened. The liquid flows into the press chamber, forming flash gas and snow. The pressure in the press gradually

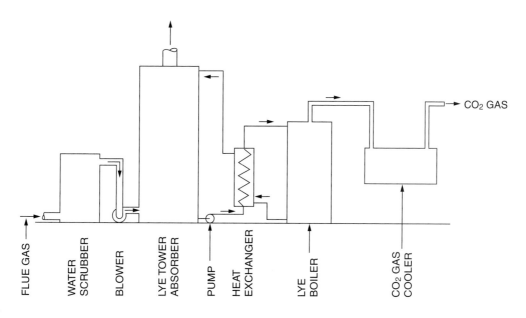

Figure 26–1 Carbon dioxide absorption equipment.

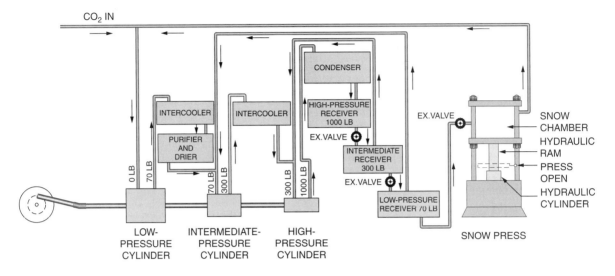

Figure 26–2 Three-stage carbon dioxide compression system.

rises during this period. Sometimes the press or operating cycle is so designed that, toward the end of the snowing period, the pressure gets above the 60-psig critical point. Then liquid carbon dioxide flows into the snow already formed. When this snow is finally pumped down to atmospheric pressure again, it requires less pressing to make a cake of the proper density.

When the press is filled with the proper amount to make the required size cake, the expansion valve is closed, the press chamber is pumped down to atmospheric pressure, and hydraulic pressure is applied to the lower plate of the press. This compresses the snow into a dense mass that can be handled or sawed to size without crumbling. The press is then opened by dropping the lower plate, the ice cake is pushed out, and the press is closed to be ready for the next cake. The blocks of dry ice are wrapped in heavy paper and stored in top-opening, well-insulated boxes.

A cake 10 inches square weighs about 50 pounds. Cakes are sometimes made a little oversize to leave a 50-pound cake after there has been some loss. Some presses make a 10-inch by 20-inch by 20-inch block that is cut into four 10-inch cakes before wrapping.

Advantages and Disadvantages of Dry Ice

Dry ice is a cooling substance, in some respects similar to water ice and in other respects quite different. The greatest difference is in temperature. In a carbon dioxide atmosphere, the temperature of dry ice is $-109.4°F$. In an atmosphere of air, the temperature will drop below $-200°F$. This is a decided advantage in storing things that must be kept frozen or for producing subzero temperatures in small quantities. It can be a disadvantage where the product can be damaged by freezing if the cooling effect of the dry ice is not properly regulated or controlled.

The latent heat of dry ice is 246.3 Btu per pound. If the sublimed gas is allowed to heat up to 32°F before escaping, this will provide an additional 28 Btu of cooling effect or a total of 274.3 Btu. Thus, dry ice has nearly twice the cooling effect per pound of water ice. The density of dry ice runs from 90 to 95 pounds per cubic foot. This is more than half again as heavy as water ice. These factors combine to give dry ice three times the cooling effect of water ice per cubic foot. So, if weight or space is more important than cost, dry ice is the logical choice.

However, the cost of dry ice is 8 or 10 times as much as that of water ice per pound. Increased production has brought dry ice costs down from higher figures, but the greater amount of power necessary for its production will prevent its ever approaching the cost of water ice. Thus, even with twice the cooling per pound, the cost of refrigeration using dry ice is considerably higher. There have been cases where more efficient use of dry ice has brought overall refrigeration costs down to about the same as with water ice.

An ice bunker in a refrigerator car or truck must usually be kept at least half full of water ice to get sufficient heat transfer to maintain temperatures. This means that bunkers half full of ice must be wasted at the end of the trip. Also, cars can often be loaded more heavily and still maintain satisfactory temperatures if dry ice is used. Combining these factors has brought the total refrigeration costs per ton of produce shipped with dry ice down to a figure comparable to that of water ice in some cases. Because of the number of variables, exact comparisons are hard to make for a new set of conditions. Many factors must be considered, and often test runs must be done before a dependable comparison can be made.

Uses of Dry Ice

A great deal of dry ice is used in shipping small packages or shipments that require refrigeration. The 10-inch square blocks are usually sawed into

Based on standard ice cream (average 10 to 12 percent butter fat content) packed at 0°F for summer temperatures (75°F to 100°F) and on 200-lb-test single-corrugated shipping cartons with double-corrugated liners. See variations below for climatic differences.

Outside temperature 0°F to 40°F; chart amounts less 30 percent
Outside temperature 50°F to 74°F; chart amounts less 10 percent
Outside temperature 75°F; to 100°F as shown

Other variations to be considered are consistency and flavor of ice cream; packing temperatures of ice cream above or below zero; care used in sealing cartons; method of handling after packing; and quality of cartons.

Pounds dry ice required to carry package

		4 hr	8 hr	12 hr	16 hr	20 hr	24 hr	36 hr
1 qt	Top	1	1½	1½	2½	3½	4	6½
	Bottom	0	0	½	1	1	2	3
4 qt brick	Top	2	2	2½	3½	4½	6	8
	Bottom	0	½	1	1	1½	2	2½
8 qt brick	Top	2	3	4	6	7	8	12
	Center	0	0	0	0	1	1	2
bottom		1	1	1	2	2½	3½	4½
20 qt brick	Top	3	5	6½	6½	12	15	18
	Center	1	1	2	2	3	4	5
	Bottom	1½	3	3½	4½	5½	7	9
Medium cake	Top	1½	3½	5	6½	7½	9½	14
	Bottom	½	1	1½	2½	3¼	4	6
1 gal bulk	Top	2	2	2½	3½	4½	6	8
	Bottom	0	½	1	1	1½	2	2½
5 gal bulk	Top	3½	5	6½	8	10	12	15
	Bottom	1	2	2½	3	4	5	7

All molds should be iced top and bottom.

Figure 26–3 Ice cream packing chart. *Copyright Liquid Carbonic Corp., 1937. Used by permission.*

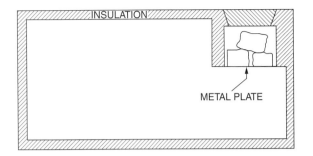

Figure 26–4 Box or truck body with dry ice bunker.

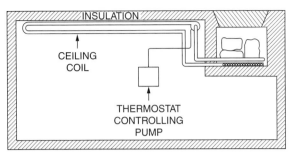

Figure 26–5 Box or truck body with coil and dry ice bunker.

1-inch-thick slabs, then ripped once down the center. This produces a slab that is 1 inch by 5 inches by 10 inches, which weighs approximately 2.5 pounds. These slabs are easily broken into smaller pieces if needed. Flowers or other highly perishable products are shipped in corrugated cardboard cartons with false bottoms and covers to hold dry ice. It is a natural for shipping ice cream. Figure 26–3 shows quantities of dry ice required for ice cream shipments under different conditions.

Figure 26–4 shows a simple system used to refrigerate both boxes and truck bodies. A refrigerated container suitable for LCL (less-than-container-load) freight shipments is made similar to this. The metal plate on the bottom of the dry ice bunker is designed with the proper area to balance the leakage into the box. Notice that there is little difference in the amount of refrigeration produced by this plate surface regardless of whether the bunker is full of dry ice or contains only enough to keep it covered.

The container in Figure 26–5 utilizes a nonfreezing solution such as alcohol or a suitable glycol. This is circulated through a coil in the box or truck body, either by gravity or by a pump. The circulation can be controlled by a thermostat that controls either the flow through a valve or the operation of the pump.

Questions

1. Why was carbon dioxide used as a refrigerant for certain applications in the past?
2. What refrigerants have replaced it?
3. Explain what dry ice is.
4. Under what conditions could dry ice melt to a liquid?
5. Describe the manufacture of dry ice.
6. List the advantages and disadvantages of dry ice over water ice.
7. Why is dry ice used in spite of its being much more expensive than water ice?
8. At 5 cents per pound for dry ice, how much would it cost to pack 2 gallons of brick ice cream for a 24-hour trip?

Chapter 27 — Altitude and Its Effects

Our studies up to this point have made clear the importance of pressure and its measurement in refrigeration. Gage pressure, which is the difference between atmospheric pressure and the pressure in the system, is the pressure considered by the practical technician, yet much engineering information is given in terms of absolute pressure, the pressure above a perfect vacuum. This is because boiling temperatures are always the same for the same absolute pressure, but this is not always true for gage pressures.

For those whose entire work will be at elevations less than 2000 feet above sea level, these variations need not concern them. At elevations above 2000 feet, errors introduced are large enough that they should be taken into account. Certain parts of the United States are well above this elevation, and other parts are near enough to mountain areas that refrigeration service personnel may be called to work on resort, hotel, or other equipment at such elevations.

In Chapter 1 it was stated that absolute pressure can be obtained by adding 14.7 (14.696) to the gage pressure. This is true only at sea level, however. The air pressure is approximately 1 inch or $1/2$ psi less for every 1000-foot rise in elevation. These figures are close enough for practical purposes. Figure 27–1 compares them with the actual variation. Notice that the error is not more than $1/4$ psi or $1/2$ inch until above 10,000 feet. This is an elevation above any normal refrigeration job except in aircraft.

The greatest effect of altitude is the apparent variation in saturation pressure. Ammonia boils at 5°F at 34.3 psia and 19.6 psig at sea level. At 6000 feet elevation, it still boils at 34.3 psia. However, atmospheric pressure at 6000 feet is 3 psi less than at sea level, or $14.7 - 3 = 11.7$ psi. Therefore, for a 5°F evaporator, the low-pressure gage will show $34.3 - 11.7 = 22.6$ psig pressure. This is 3 psi more than the 19.6 psig at sea level.

Thus, *at elevations higher than sea level, a correction must be added to the sea-level gage pressure* to get the corresponding temperature. It is important to realize that 22.6 psig at 6000 feet is not a different pressure: The ammonia pressure inside the system is identical to that at sea level. It is the air pressure surrounding the gage that is different, as shown in Figure 27–2. For the same reason, the condensing pressure at 86°F at 6000 feet is $154.5 + 3 = 157.5$ psig.

EXAMPLE 1 An R-134a system is to have the back-pressure control set for 0°F and 25°F at 5000 feet elevation. What are the gage pressure settings?

First Method: A gage pressure table made up for sea level gives pressures of 6.5 psig and 22.1 psig for these temperatures. The correction per 1000 feet at $1/2$ psi is 2.5 psi. Therefore, the required pressures are

Cut-out point: $6.5 + 2.5 = 9$ psig
Cut-in point: $22.1 + 2.5 = 24.6$ psig

Second Method: An absolute pressure table gives 21.2 psia and 36.8 psia. The atmospheric pressure is

$$14.7 - 2.5 = 12.2 \text{ psi}$$

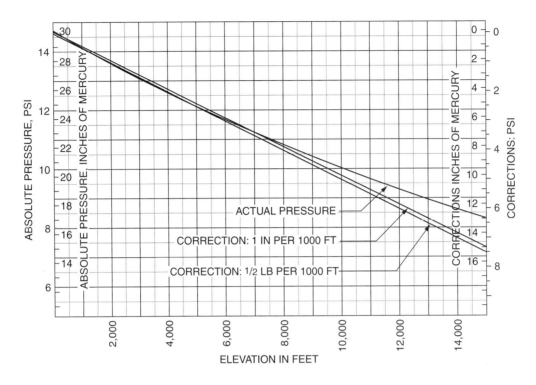

Figure 27–1 Pressure variations with altitude.

Therefore, the required gage pressures are

Cut-out point: 21.2 − 12.2 = 9 psig
Cut-in point: 36.8 − 12.2 = 24.4 psig

EXAMPLE 2 Settings of −35°F and −10°F are asked for on an R-134a job at 6000 feet of elevation. What are the required gage pressure settings?

First Method: The pressure correction is 3 psi. The table gives gage pressures of 8.6 psia and 16.6 psia at sea level.

Cut-out point: 8.6 + 3 = 11.6 psia or 6 inches of mercury
Cut-in point: 16.6 + 3 = 19.6 psia

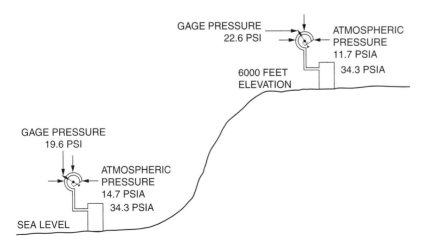

Figure 27–2 Effect of altitude and air pressure on a pressure gage.

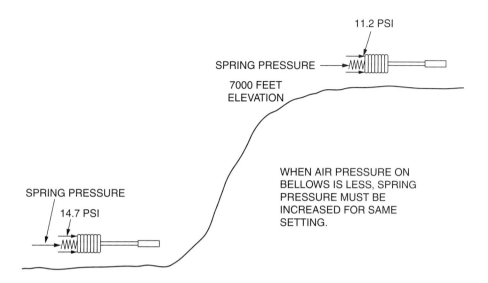

Figure 27-3 Effect of altitude and air pressure on a bellows.

Second Method: The atmospheric pressure at 6,000 feet is 14.7 − 3 = 11.7 psi. The absolute pressures from the table are 8.56 and 16.62.

Cut-out point: 8.56 − 11.7 = −3.14 or 6.3 inches
Cut-in point: 16.62 − 11.7 = 4.92 psig

Technicians who work at elevations high enough to make corrections necessary often make up their own gage pressure tables that are accurate for that altitude. It is well worth the time to do this, because each value then needs to be calculated only once, with a record made for future use.

In addition to changes in gage pressures, altitude can also affect equipment that includes bellows- or diaphragm-operated control devices if the surrounding air is able to act on the control mechanisms. If such equipment is set to gage pressure readings, the corrections discussed above are all that are necessary. Some controls, however, have temperature or pressure scales marked on them. Whether these are temperature or a pressure controls, such scales will not be correct.[1] In Figure 27-3, notice how the air pressure surrounding the bellows helps compress the bellows. If this air pressure is reduced, the spring pressure opposing the bellows must be increased. Controls with a temperature scale will operate colder than the markings show if they are not adjusted. Controls known to be adjusted to a certain temperature in factory tests will operate at a lower temperature at higher altitude.

Automatic expansion valves are also affected by altitude. Again, if they are set to the correct gage setting, they will operate properly. However, this setting will be different than the setting for the same valve set at sea level. Notice in Figure 4-8 how air pressure helps the spring oppose the diaphragm or bellows. Thermostatic expansion valves of the diaphragm type have a refrigerant pressure on each side of the diaphragm. Therefore air pressure will not affect their settings.

Constant-pressure valves are operated by a bellows or diaphragm and so are affected by altitude. In Chapter 24 it was mentioned that constant-pressure

---

[1] One manufacturer makes a control that depends on the expansion and contraction of a liquid with temperature changes. Such a control is not affected by altitude.

valves are sometimes used as temperature-limiting valves in water coolers. These valves are factory adjusted to the required setting, then soldered over so that the adjustment cannot be changed in the field. Therefore, if such equipment is to be used at an elevation above 2000 feet, this must be specified to get the proper factory setting.

Questions

1. What is atmospheric pressure in pounds per square inch and in inches of mercury at the following elevations?
    a. At sea level
    b. At 3600 feet elevation
    c. At 5200 feet elevation
    d. At 8500 feet elevation
2. Give the required pressure settings for all equipment in the system of Example 1 in Chapter 25, with R-134a as the refrigerant, for an elevation of 6600 feet.
3. Will a temperature control operate warmer or colder if it is moved from sea level to an altitude and not readjusted?
4. Why does altitude affect a diaphragm-type automatic expansion valve and not a diaphragm-type thermostatic expansion valve?

Chapter 28 Absorption Systems

In the early years of ammonia refrigeration, absorption systems saw considerable use. Because of certain difficulties, however, and as compression systems were improved, ammonia absorption systems dropped out of the picture for new installations. Some absorption systems are still in operation, however, particularly in oil refineries, where fuel is an economical source of energy.

An absorption system using water as the refrigerant and lithium bromide brine as the absorbent is used in many modern applications. With water as the refrigerant, temperatures approaching 32°F cannot be used. A temperature of 40°F is considered the lower limit. This absorption system was first developed for air conditioning applications, and this is still its biggest use. It is also used in other applications in which chilled water is required.

In its simplest form, an intermittent ammonia absorption system can be made up of two tanks connected with an open pipe, as shown in Figure 28–1. Liquid ammonia is in the left tank and water in the right tank. Ammonia vapor is so highly soluble in water that the vapor over the water is absorbed. This lowers the pressure in the right tank, which causes more vapor to flow from the connecting tube, which in turn is absorbed. This constant absorption of vapor is just as effective as if it was being removed by the suction side of a compressor. A refrigerating cycle is started with the left tank as an evaporator and the right tank, or absorber, taking the place of the compressor. This produces refrigeration in the left tank.

Eventually, the water in the right, absorber tank will become saturated with ammonia and will fail to absorb more vapor. If it is then heated and the left tank is placed in cooling water, the ammonia is distilled from the right tank and condensed in the left tank. This makes a generator out of the right tank and a condenser out of the left tank, completing the refrigerating cycle.

Such systems were once built for domestic use where electric power was not available. The evaporator hung inside an insulated box and produced refrigeration for about 24 hours. Once a day, the entire "dumbbell" was removed from the box. The evaporator-condenser was placed in a bucket of water and the absorber-generator heated with a special burner supplied with the unit. This recharged it so the system was ready to produce refrigeration for another 24 hours. These systems were simple and needed no other source of power, but the quality of refrigeration was poor by today's standards.

Other intermittent ammonia systems did not require removal of the unit from the cabinet. A burner in the unit, containing a measured amount of kerosene, was lighted each day. This charged the unit so it would give approximately 24 hours of cooling.

If water was placed in the evaporator-condenser and lithium bromide in the absorber-generator of Figure 28–1, some absorption of water vapor by the lithium bromide would take place. However, the action was so slow that very little evaporation and cooling would be apparent in the evaporator. The process could be greatly improved by installing pumps

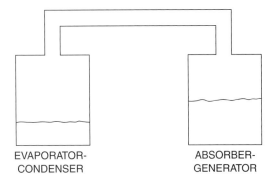

Figure 28-1 Elementary intermittent absorption system.

to force the brine and water through spray nozzles, as shown in Figure 28-2. The lithium bromide spray in the absorber picked up more water vapor and caused a greater reduction in vapor pressure. An absolute pressure of ¼ inch of mercury or a 29.67-inch vacuum is necessary to produce 40°F water; see Figure 3-1. The water spray in the evaporator makes more evaporation and cooling possible in the presence of this reduced vapor pressure. This system produced refrigeration, but only intermittently.

Continuous Absorption System

Figure 28-3 shows, in its simplest form, a continuous-absorption refrigeration system. The condenser and evaporator are similar to those in a compression system. The water vapor from the evaporator is piped to the absorber, where it is absorbed by the

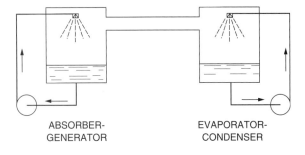

Figure 28-2 Simple intermittent absorption system.

lithium bromide brine. Heat is generated in this absorption process, so a water coil in the absorber keeps the temperature down.

The lower the temperature of the absorber, the more water vapor can be absorbed. The generator is at condensing pressure, usually about 3½ inches of mercury. Therefore a mechanical pump is necessary to deliver the brine to the generator. In the generator, heat is usually applied by a steam coil. The temperature can be controlled by controlling the steam pressure. Water vapor is boiled off and goes to the condenser. The reconcentrated brine is then returned to the absorber. There is restriction enough in this return line to allow for the necessary drop in pressure. A heat exchanger between the concentrated and dilute brines increases efficiency. The water vapor condensed in the condenser is returned to the evaporator. Again, restriction in the line is sufficient that an expansion valve is not needed.

In practice, lithium bromide systems are simplified by consolidating the major components. One design places the condenser and generator in one shell and the evaporator and absorber in a second shell, as in Figure 28-4. The large open spaces in the shells provide ample space for the large volume of water vapor created at such low pressures. The shells, with the necessary heat exchanger, pumps, and controls, are assembled and mounted as a unit. Another design combines the four major parts in a single shell with suitable partitions inside the shell, shown in Figure 28-5.

Direct gas-fired absorption systems have been designed for small air conditioning applications. Larger-sized lithium bromide systems that use steam as a source of power are built in sizes from 100 to 1000 tons capacity. They require from 17 to 20 pounds of steam per hour per ton of refrigeration, depending on conditions. Their coefficient of performance (COP) is much lower than that of a compression system, but this is usually offset by the lower cost of fuel vs. electricity.

The ammonia absorption system used circuits and separate components similar to those shown in Figure 28-3. An expansion valve in the ammonia liquid line and a pressure-reducing valve in the line from the generator to the absorber were necessary to keep the high and low pressures separate. Only

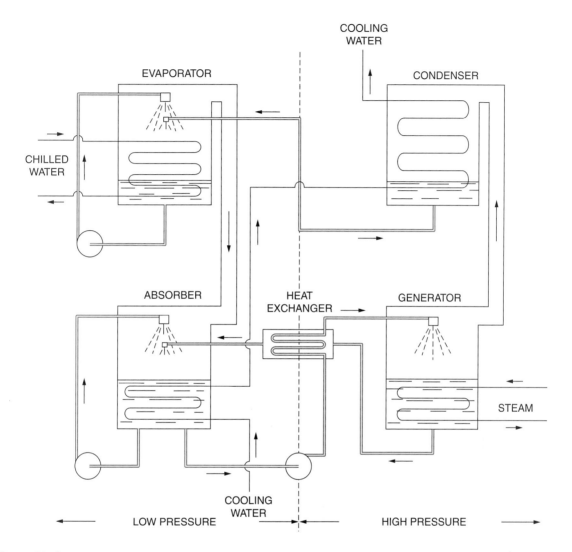

Figure 28-3 Continuous lithium bromide absorption system.

one pump was needed, to raise the pressure of the ammonia–water solution from the absorber to the generator.

Diffusion Absorption System

An interesting modification of the absorption system eliminates the need for a mechanical pump. It is sometimes called a diffusion absorption system. A simplified diagram of this system (the Electrolux system) is shown in Figure 28-6. Instead of an expansion valve, there is a U-tube filled with liquid ammonia. On the condenser side of the U-tube is the ammonia condenser pressure, usually 200 psig or higher with an air-cooled condenser. On the evaporator side, this pressure is balanced by a hydrogen pressure, so there is the same total pressure on each side of the U-tube. Thus, any liquid condensed flows into the U-tube

330 CHAPTER 28 *Absorption Systems*

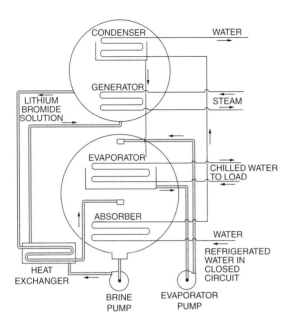

Figure 28-4 Lithium bromide absorption system.

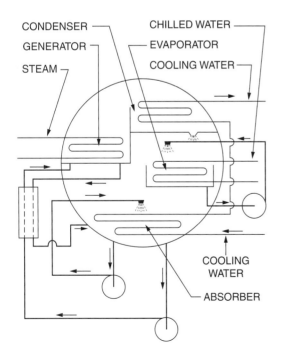

Figure 28-5 Single-chamber lithium bromide absorption system.

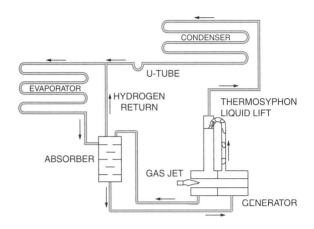

Figure 28-6 Simplified Electrolux system.

and out into the evaporator. The liquid-filled U-tube prevents the hydrogen from backing up into the condenser.

The liquid ammonia in an atmosphere of hydrogen evaporates and creates cooling, just as water evaporates in an atmosphere of air. The ammonia gas diffuses into the hydrogen, thus the name diffusion absorption.

The evaporating temperature of the ammonia depends on the partial pressure of the ammonia vapor that mixes with the hydrogen as it evaporates. The evaporated ammonia mixed with the hydrogen flows to the absorber. The ammonia gas is soluble in water but the hydrogen is not, so the ammonia is dissolved, leaving almost pure hydrogen to return to the evaporator.

The generator acts like a bubble pump. The boiling action lifts the weak ammonia liquor to the highest point in the system containing liquor. From the high point, it can flow down through the absorber and back to the generator.

Small ammonia absorption units are still used in campers. They may be kerosene- or propane-fired, or they may use a 110-volt resistance heater. Many gas-fired units are still in operation. They are popular where electricity is either not available or not dependable.

Figure 28-7 shows a schematic outline of the absorption cycle and diffusion absorption cycle beside a compression cycle. Note there are two circulating

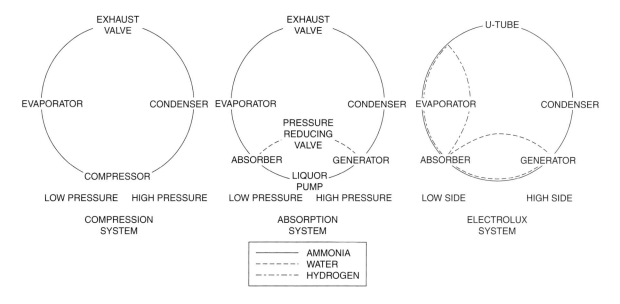

Figure 28–7 Three different refrigeration cycles.

fluids in the absorption system. From the absorber to the generator, they circulate together. From the generator, they divide, and either lithium bromide brine or weak aqueous ammonia returns to the absorber. In the diffusion absorption system, a third fluid, hydrogen, circulates with the ammonia vapor from the evaporator to the absorber.

Questions

1. Why was ammonia used as a refrigerant even in domestic refrigerators in early absorption equipment?
2. In an absorption system, how is refrigerant vapor removed from the evaporator?
3. What is the difference between the intermittent and the continuous absorption systems?
4. What parts of the absorption cycle are similar to the compression cycle?
5. What parts of the absorption cycle are different from the compression cycle?
6. Why is an expansion valve used in an ammonia absorption system but not in a lithium bromide absorption system?
7. What is the suction pressure in a lithium bromide absorption system?
8. What effect would even a slight air leak have on the operation of a lithium bromide absorption system?
9. What are the advantages of a lithium bromide absorption system?
10. What are the disadvantages of a lithium bromide absorption system?

Chapter 29 Refrigeration Codes

The primary purpose of any code is to establish a safety standard to be followed by everyone who installs equipment. Codes have been written to control the installation of buildings, plumbing, electric wiring, and other equipment. Recently, the need for refrigeration codes has received increased attention. The first refrigeration codes, established by different cities, were very different from each other. Each stressed the different concerns of the individuals responsible for the formulation of the code. This made standardization impossible. Eventually a generally recommended code, the Safety Standard for Refrigeration Systems, was created by various national organizations interested in safety and in refrigeration. This code has been revised from time to time as needed and the number of the code usually carries the date of the last revision, such as ANSI/ASHRAE Standard 15-2001.

Various organizations are interested in the safe construction and operation of refrigeration systems and either contribute to writing and updating codes or, at the very least, are interested in the codes. They include but may not be limited to:

- ANSI—American National Standards Institute
- ARI—American Refrigeration Institute
- ASHRAE—American Society of Heating Refrigeration and Air-Conditioning Engineers
- ASME—American Society of Mechanical Engineers
- EPA—Environmental Protection Agency
- HMIS—Hazardous Material Identification System
- IARW—International Association of Refrigerated Warehouses
- IIAR—International Institute of Ammonia Refrigeration
- MFFM—Major Frozen Food Manufacturers
- NBFU—National Board of Fire Underwriters
- NFPA—National Fire Protection Association
- NRSC—National Refrigeration Safety Code
- RETA—Refrigeration Engineer Technician Association
- RSES—Refrigeration Service Engineers Society

The codes written by these various organizations are not national laws enforced throughout the country, but models for jurisdictions that are setting up their own codes. They bear the same relationship to the refrigeration industry as the National Electric Code, which has been adopted by many jurisdictions, has to the electric industry. Some jurisdictions have adopted parts of the national codes but also kept parts of their old codes. Some states have tied their codes to contracting or licensing laws.

Even if you are not working under the jurisdiction of the codes, they should be considered as minimum good practices. For all, they should be recognized as minimum acceptable standards of quality. To boast that a job is "according to code" is to state that it is the least acceptable quality that will get by.

It is highly recommended that everyone in the refrigeration industry push for general adoption of the codes. First, a good code does not hurt the legitimate refrigeration contractor, but it does protect both the contractor and the customer from cut-price jobs in which more or less essential equipment is omitted to meet cost limits.

These codes should be adopted for two reasons. First, they are worked out by committees of experts from all branches of the field. Therefore, they are much better than what could be produced by one person. The intent is to eliminate personal prejudices and beliefs. Such preferences or prejudices are often honestly believed by single individuals who might be instrumental in writing or updating a code. These prejudices are probably more common in refrigeration than in older and better established industries or trades, where traditions of custom or good practice have been built up.

The second reason for adopting these codes is to make them standard throughout the country. Certainly, what is safe in one city will be safe in another. If this is the case, there is no reason for making workers memorize a separate set of rules every time they relocate. This can be particularly troublesome and sometimes costly when one company does business in several different localities. The latest copy of the ANSI/ASHRAE code may be obtained from ASHRAE, 1791 Tullie Circle N.E., Atlanta, Georgia 30329-2305. The codes are far too lengthy to include in this text. They are also periodically updated and rewritten. We encourage you obtain a copy of the latest codes and familiarize yourself with them. The questions at the end of this chapter will assist you in your study of the codes.

Questions

1. Are flames from matches and open cigarette lighters considered to be open flames?
2. What is an Institutional Occupancy? Give examples.
3. What is a Commercial Occupancy? Give examples.
4. What is a Residential Occupancy? Give examples.
5. What is a Public Assembly Occupancy? Give examples.
6. What is the difference between a direct and an indirect system?
7. What is the difference between a high-probability system and a low-probability system?
8. What is the largest size flare connection allowed?
9. What is the difference between soldering and brazing?
10. What are the identification requirements for systems that contain more than 110 pounds of refrigerant?
11. The required sign that is permanently attached to the refrigeration system shall include what information?
12. What valves should be labeled?
13. What is a pressure-relief device?
14. What is a pressure-limiting device?
15. What containers require a pressure-relief device?
16. Where may a fusible plug be substituted for a pressure-relief device?
17. What compressors require a pressure-relief device?
18. At what pressure shall pressure relief devices be set?
19. What relief devices shall be piped to discharge outside the building?
20. What are the pressure-relief requirements for compressors?
21. In the ASHRAE (ASHRAE Standard 34) system of refrigerant grouping (A1, A2, A3, B1, and B2), to what do the letter and number refer?
22. Name some of the refrigerants in the different groups of Question 21.

Appendix

Table A–1 Densities (Weights) and Specific Heats of Various Materials

Material	Density (lb/cu ft)	Specific Heat (Btu/lb)
Air @ 70°F	0.075	0.24
Alcohol	49.6	0.60
Aluminum	166.5	0.214
Bakelite	86.0	0.35
Brass	511–536	0.094
Bronze	531–538	0.090
Copper	552	0.094
Concrete	147	0.19
Cork	15	0.48
Glass (common)	164	0.199
Glycerine		0.576
Ice	57.5	0.504
Iron	480	0.118
Lead	710	0.030
Masonry (brick)	112	0.200
Mercury	847	0.033
Oil (machine)	57.5	0.400
Paper (cardboard)	58	0.324
Rubber	59	0.48
Sand	100	0.195
Steel	492	0.117
Stone	135–200	0.20
Tar	75	0.35
Water	62.4	1.000
Wood		
Fir	37	0.65
Oak	48	0.57
Pine	38	0.47
Zinc	446	0.096

Table A–2 Properties of Food Products

Commodity	Water Content (%)	Specific Heat Above Freezing (Btu/lb)	Specific Heat Frozen (Btu/lb)	Freezing Temperature (°F)	Latent Heat (btu/lb)	Storage Temperature (°F)	Storage Humidity (%)	Storage Life
FRUITS								
Apples	84	0.92	0.39	28.4	120	30–32	85–88	2–7 mo
Apricots	85	0.92	0.46	28.1	122	31–32	80–85	1–2 wk
Avocados	65–82	0.91	0.49	27.2	136	40–55	85–90	4–8 wk
Bananas	75	0.80	0.43	26–30	108	55–60	90	
Blackberries	85	0.92	0.46	28.9	124	31–32	80–85	7–10 days
Cherries	83	0.85	0.45	24–28	118	31–32	80–85	10–14 days
Cranberries	88	0.91	0.47	27.3	128	36–40	85–90	1–3 mo
Currants	85			30.2	123	32–80	85	10–14 days
Dewberries	85	0.92	0.46	29.2	124	31–32	80–85	7–10 days
Dried fruits	15–30		0.29–0.47	21–32	21–43	26–32	70–75	9–12 mo
Figs, dry	24	0.47	0.32		43	40–45	65–75	9–12 mo
Figs, fresh	78	0.88	0.48	27.1	116	28–32	65–75	5–7 days
Gooseberries	88	0.92	0.47	28.9	130	31–32	65–75	3–4 wk
Grapefruit	89	0.92	0.49	28.4	128	32–55	85–90	6–8 wk
Grapes	82	0.92	0.37	27.5	83	31–32	80–85	3–8 wk
Lemons	89	0.91	0.39	28.1	126	55–58	85–90	1–4 wk
Limes	86	0.91	0.49	29.3	126	45–48	85–90	6–8 wk
Melons								
Casaba				29.0		36–40	75–85	4–6 wk
Honey dew				29.0		36–38	75–85	2–4 wk
Honey ball						36–38	75–85	2–4 wk
Muskmelon	81	0.92	0.34	29.0	128	32–34	75–85	7–10 days
Watermelon	92	0.92	0.47	29.2	135	36–40	75–85	2–3 wk
Nuts, dried	3–6	0.21–0.29	0.19–0.24	19–24	4–14	32–50	65–75	8–12 mo
Olives, fresh	75			28.5		45–50	85–90	4–6 wk
Oranges	87	0.89	0.39	28.0	125	34–38	85–90	8–10 wk
Peaches	87	0.92	0.42	29.4	106	31–32	80–85	2–4 wk
Pears								
Bartlett	83	0.90	0.43	28.5	109	29–31	85–90	1–3 mo
Fall and winter	82	0.90	0.43	27–28	108	29–31	85–90	2–7 mo
Persimmons	78	0.80	0.46	28.3	96	31–32	85–90	2–3 wk
Pineapples								
Mature green				29.1		50–60	85–90	3–4 wk
Ripe	85–89	0.90	0.46	29.9	128	40–45	85–90	2–4 wk

(continued)

Table A–2 Properties of Food Products (continued)

Commodity	Water Content (%)	Specific Heat Above Freezing (Btu/lb)	Specific Heat Frozen (Btu/lb)	Freezing Temperature (°F)	Latent Heat (btu/lb)	Storage Temperature (°F)	Storage Humidity (%)	Storage Life
Plums	78	0.83		28.0		31–32	80–85	3–8 wk
Fresh prunes	78	0.83		28.0		31–32	80–85	3–8 wk
Quinces	85			28.1		31–32	80–85	2–3 mo
Raisins						40–45	50–60	9–12 mo
Raspberries	82–86	0.89		30.0		31–32	80–85	7–10 days
Strawberries	90	0.92		29.9		31–32	80–85	7–10 days
VEGETABLES								
Artichokes								
Globe	84	0.90	0.49	29.1	134	31–32	90–95	1–2 wk
Jerusalem	79			27.5		31–32	90–95	2–5 mo
Asparagus	93	0.95	0.48	29.8	134	32	85–90	3–4 wk
Beans, dried	12	0.30	0.24		17	32–36	75	
Beans, green	89	0.91	0.47	29.7	128	32–40	85–90	2–4 wk
Beans, lima	66	0.89	0.41	30.1	100	32–40	85–90	2–4 wk
Beets								
Bunched	88	0.90	0.46	26.9	129	32	85–90	10–14 days
Topped	90	0.93	0.48	29.2	135	32	95–98	1–3 mo
Broccoli	85	0.91	0.49	31.0	136	32–35	90–95	7–10 days
Brussels sprouts						32–35	90–95	3–4 wk
Cabbage	92	0.93	0.47	31.2	131	32	90–95	3–4 mo
Carrots								
Bunched	88	0.87	0.45	29.6	119	32	85–90	10–14 days
Topped	93	0.94	0.48	30.1		32	95–98	4–5 mo
Cauliflower	93	0.95	0.48	29.7	135	32	85–90	2–3 wk
Celery	94	0.95	0.48	28.9	108	31–32	90–95	2–4 mo
Corn, green	74	0.80	0.43	30.5	137	31–32	90–95	1–2 wk
Cucumbers	95	0.93	0.48	30.4	132	45–50	80–85	10–14 days
Eggplant	93	0.91	0.45	25.4		45–50	85–90	7–10 days
Garlic, dry	74			26.4		32	70–75	6–8 mo
Horseradish	73			30.0		32	95–98	10–12 mo
Kohlrabi	90				29.2	32	95–98	2–4 wk
Leek, green	88			31.2	135	32	85–90	1–3 mo
Lettuce	94	0.95	0.48	30.2	126	32	90–95	2–3 wk
Mushrooms	91	0.90	0.47	30.1	126	32–35	80–85	2–3 days
Onions	88	0.91	0.46			32	70–75	6–8 mo

Table A-2 Properties of Food Products (continued)

Commodity	Water Content (%)	Specific Heat Above Freezing (Btu/lb)	Specific Heat Frozen (Btu/lb)	Freezing Temperature (°F)	Latent Heat (btu/lb)	Storage Temperature (°F)	Storage Humidity (%)	Storage Life
Parsnips	79	0.86	0.45	28.9	119	32	90–95	2–4 mo
Peas, green	74	0.85	0.42	30.0	107	32	85–90	1–2 wk
Peppers, green	92	0.90	0.46	30.1		32	85–90	4–6 wk
Peppers, chili						32–50	50–75	6–9 mo
Potatoes	79	0.79	0.42	28.9	106	38–50	85–90	5–12 mo
Pumpkins	90	0.90	0.43	29–30	127	50–55	70–75	2–6 mo
Squash	90	0.90	0.43	29–30	127	50–55	70–75	2–6 mo
Radishes	91	0.90	0.48		132	32	95–98	2–4 mo
Rhubarb	95	0.90	0.46	28.4	134	32	95–98	2–3 wk
Rutabagas	89			29.5		32	90–95	2–4 mo
Spinach	93	0.92	0.51	30.3	129	32	90–95	10–14 days
Sweet Potatoes	78	0.86	0.42	28.5	102	55–60	75–80	4–6 mo
Tomatoes								
Mature green	95	0.92	0.46	30.4	132	55–70	85–90	3–5 wk
Ripe	94	0.92	0.46	30.4	132	50	85–90	7–10 days
Turnips	91	0.90	0.45	30.5	128	32	95–98	4–5 mo
MEATS & FISH								
Beef	62–77	0.77	0.40	28–29	98	32–40	88–92	1–6 wk
Fish, fresh	65–83	0.82	0.41	28	101	33–40	90–95	5–20 days
Fish, dried	45	0.56	0.33		65	23–25	70–75	2–4 mo
Fish, smoked						23–25	70–75	2–4 mo
Fish, brined						26–28	85	2–6 mo
Lamb	60–70	0.66	0.37	28–29	83	32–34	85–90	5–12 days
Poultry	60–65	0.80	0.38	27	86	32	80	1 wk
Pork, fresh	35–46	0.51–0.68	0.38	28–29	86	32–34	85–90	3–7 days
Pork, cured	20	0.50	0.32			32–40	70–80	2–4 mo
Ham, bacon, etc.	20	0.50	0.32			32–40	70–80	2–4 mo
Oysters, shell	80	0.84	0.44	27	116	33–35	90	30–60 days
Oysters, tub	87	0.90	0.46	27	125	32	90	10–15 days
Rabbit						32–34	90–95	1–5 days
Veal	70–80	0.70		28–90		32–34	90–95	5–10 days

(continued)

Table A-2 Properties of Food Products (continued)

Commodity	Water Content (%)	Specific Heat Above Freezing (Btu/lb)	Specific Heat Frozen (Btu/lb)	Freezing Temperature (°F)	Latent Heat (btu/lb)	Storage Temperature (°F)	Storage Humidity (%)	Storage Life
DAIRY PRODUCTS								
Butter	10	0.64	0.34	0–30	15	32–34	65–70	10–12 days
Cheese	35–38	0.64	0.40	4–18	86	32–34	65–70	10 to12 days
Cream	59	0.87	0.37	28	84	30–33		10–12 days
Eggs	68–73	0.76	0.41	28	100	29–31	85–90	8–10 mo
Ice cream	67	0.80	0.46	28	96			
Milk	87	0.92	0.49	31	124	36–38		

Frozen Food Storage

Commodity	Storage Temperature (°F)	Storage Life
Fruits	0 or below	12 mo plus
Vegetables	0 or below	12 mo plus
Meats	0 or below	6–8 mo
Poultry	–10 or below	12 mo
Fish	0 or below	12 mo
Butter	0 or below	8–12 mo
Eggs	0 or below	12 mo plus

Table A–3 Sizes and Weights of Food Containers

Commodity	Container	Contents	Size, Diameter or Width (in)	Height (in)	Approximate Weight of Container (lb unless otherwise noted)	Notes
Milk	Can	10 gal	13	25	26	Milk weighs 8.6 lb/gal
	Can	5 gal	10½	19½	15	
	Can	3 gal	9	19½	5	
	Bottle	1 qt	4	9½	1½	
	Bottle	1 pint	3¼	8½	1	
	Bottle	½ pint	2¾	6¾		
	Carton	1 qt	2¾	8 or 9	3 oz	
	Carton	1 pint	2¼	6½		Square top
			2¾	5½	1 oz	Pinch top
	Carton	½ pint	2¼	3½	¾ oz	

Commodity	Container	Contents	Depth (in)	Width (in)	Length (in)	Weight (lb)	Notes
Milk	Case	12 qt	10¾	14	18½	16	
	Case	20 pints	9	14½	18½	14	
	Case	30½ pints	6¾	14½	18½	11	
Ice cream	Carton	5 gal		9	19	1¾	
		2½ gal		9	10	1	
Eggs	Crate	30 dozen	12	13½	25½	8½	Net wt of eggs, 42 to 44 lb, 400 crates to a carload
	Carton	15 dozen	12	12	13	1½	

	Approximate Weights (lb)			
Commodity	Minimum	Maximum	Average	
Beef	400	1000	500	Handled in halves
Lamb	20	120	40	
Pork	90	450	180	
Veal	70	275	150	

Table A-4 Energy Conversion Equivalents

1 horsepower = 33,000 foot-pounds/minute
              = 550 foot-pounds/second
              = 746 watts
              = 2,545 Btu/hour
              = 42.42 Btu/minute

1 kilowatt = 1,000 watts
           = 1.34 horsepower
1 megawatt = 1,000,000 watts
               = 1,340 horsepower

1 Btu = 778 foot-pounds

1 horsepower-hour = 1 horsepower for 1 hour
                       = 746 watt-hours
                       = 0.746 kilowatt-hours
                       = 2,545 Btu

1 kilowatt-hour = 1 kilowatt expended continuously for 1 hour
                  = 1,000 watt-hours
                  = 1.34 horsepower-hours
                  = 3,415 Btu

1 ton of refrigeration = 288,000 Btu/24 hours
                         = 12,000 Btu/hour
                         = 200 Btu/minute
                         = 83.3 pounds of IME

Numerical Prefixes/Scientific Notation

kilo = × 1,000 = 1×10^{3} to the third power
mega = × 1,000,000 = 1×10^{6} to the sixth power
giga = × 1,000,000,000 = 1×10^{9} to the ninth power
tera = × 1,000,000,000,000 = 1×10^{12} to the twelfth power
peta = × 1,000,000,000,000,000 = 1×10^{15} to the fifteenth power
milli = × 1/1,000 = 1×10^{-3} to the minus third power
micro = × 1/1,000,000 = 1×10^{-6} to the minus sixth power
nano = × 1/1,000,000,000 = 1×10^{-9} to the minus ninth power
pico = × 1/1,000,000,000,000 = 1×10^{-12} to the minus twelfth power
femto = × 1/1,000,000,000,000,000 = 1×10^{-15} to the minus fifteenth power

Table A–5 Pressure Ratios

1 pound per square inch (psi) = 0.068 atmosphere
    = 2.04 inches of mercury
    = 2.31 feet of water
    = 27.7 inches of water

1 inch of mercury = 0.0334 atmosphere
    = 0.491 psi
    = 1.13 feet of water
    = 13.6 inches of water

1 atmosphere = 29.92 inches of mercury
    = 33.94 feet of water
    = 14.696 psi

1 ounce/inch^2 = 0.128 inches of mercury
    = 1.73 inches of water

1 foot of water = 0.0295 atmosphere
    = 0.433 psi
    = 62.4 pounds/foot^2
    = 0.883 inches of mercury

Table A–6 Weights and Measures

Weight	Linear Measure
7,000 grains = 1 pound	12 inches = 1 foot
16 ounces = 1 pound	3 feet = 1 yard
2,000 pounds = 1 ton	2.54 centimeters = 1 inch
2.2 pounds = 1 kilogram	

Square Measure (for measuring areas)

144 square inches = 1 square foot
9 square feet = 1 square yard

Area of a rectangle = length × width ($A = LW$)

Area of a triangle = half the base × height ($A = BH/2$)

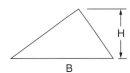

Area of a trapezoid = average length × height ($A = ab/2 \times H$)

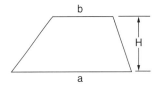

Area of a circle = 3.1416 × radius squared or 0.7854 × diameter squared

($A = \pi R^2 = 0.7854 D^2$)

Surface area of a sphere = 4 × 3.1416 × (radius)^2

($A = 4\pi R^2$)

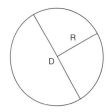

Cubic Measure (for measuring volumes)

1,728 cubic inches = 1 cubic foot
27 cubic feet = 1 cubic yard

(continued)

Table A–6 Weights and Measures (*continued*)

Volume of a rectangular solid = length × width × height

$$(V = L \times W \times H)$$

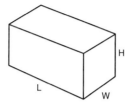

Volume of a cylinder = 3.1416 × radius squared × height
or 0.7854 × diameter squared × height

$$(V = \pi R^2 H = 0.7854 D^2 H)$$

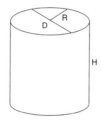

Volume of a sphere = (4/3) × 3.1416 × radius cubed

$$[V = (4/3)\pi R^2]$$

Volume of a cone = (1/3) × 3.1416 × radius squared × height

$$[V = (1/3)\pi R^2 H]$$

Liquid Measure (for measuring the volume of a liquid)

16 ounces = 1 pint

2 pints = 1 quart

32 ounces = 1 quart

4 quarts = 1 gallon

8 pints = 1 gallon

231 cubic inches = 1 gallon

7.48 gallons = 1 cubic foot

8.34 pounds = the weight of 1 gallon of water

Table A-7 Sizes and Weights of Copper Tubing

Outside Diameter (in)	Wall Thickness (in)	Inside Diameter (in)	Weight (lb/ft)	External Surface (length/ft^2)	Internal (in^2)	Surface (ft^2)	External Sectional Area (in^2)	Contents (cu ft/ lineal ft)
Soft-Drawn Copper Tubing								
3/16[1]	0.030	0.1275	0.0526	20.6	0.0127	0.0000883	0.0276	0.0000883
1/4	0.030	0.1900	0.0804	15.3	0.0284	0.000197	0.0491	0.000197
5/16[1]	0.032	0.2485	0.109	12.2	0.0485	0.000337	0.767	0.000337
3/8	0.032	0.311	0.134	10.2	0.0760	0.000527	0.110	0.000527
1/2	0.032	0.431	0.182	7.64	0.1460	0.00101	0.196	0.00101
5/8	0.035	0.555	0.251	6.11	0.242	0.00168	0.307	0.00169
3/4	0.035	0.680	0.305	5.10	0.363	0.00252	0.442	0.00252
Hard-Drawn Copper Tubing Type K Copper Tube (Heavy Wall)								
3/8	0.035	0.305	0.145	10.2	0.0731	0.000508	0.110	0.000508
1/2	0.049	0.402	0.269	7.64	0.127	0.000882	0.196	0.000882
5/8	0.049	0.527	0.344	6.11	0.218	0.001514	0.307	0.001514
3/4	0.049	0.652	0.418	5.10	0.334	0.00232	0.442	0.00232
7/8	0.065	0.745	0.641	4.37	0.436	0.00303	0.601	0.00303
1-1/8	0.065	0.995	0.839	3.40	0.778	0.00540	0.994	0.00540
1-3/8	0.065	1.245	1.04	2.78	1.217	0.00845	1.485	0.00845
1-5/8	0.072	1.481	1.36	2.35	1.722	0.01195	2.07	0.01195
2-1/8	0.083	1.959	2.06	1.80	3.014	0.0209	3.55	0.0209
25/8	0.095	2.435	2.93	1.46	4.656	0.0323	5.41	0.0323
3-1/8	0.109	2.907	4.00	1.22	6.637	0.0471	7.67	0.0471
3-5/8	0.120	3.385	5.12	1.05	9.00	0.0625	10.32	0.0625
4-1/8	0.134	3.857	6.51	0.927	11.68	0.0811	130.36	0.0811
5-1/8	0.160	4.805	9.67	0.746	18.13	0.1259	200.63	0.1259
6-1/8	0.192	5.741	13.87	0.625	25.9	0.1795	290.5	0.1795
8-1/8	0.271	7.583	25.92	0.470	45.2	0.313	510.8	0.313
10-1/8	0.338	9.449	40.28	0.387	70.1	0.487	800.5	0.487
12-1/8	0.405	11.315	57.80	0.316	100.5	0.698	115.5	0.698
Type L Copper Tube (Medium Wall)								
3/8	0.030	0.315	0.126	10.2	0.078	0.00054	0.110	0.00054
1/2	0.035	0.430	0.198	7.64	0.146	0.00101	0.196	0.00101
5/8	0.040	0.545	0.285	6.11	0.233	0.00162	0.307	0.00162
3/4	0.042	0.666	0.362	5.10	0.348	0.00242	0.442	0.00242
7/8	0.045	0.785	0.455	4.37	0.484	0.00336	0.601	0.00336
1-1/4	0.050	1.025	0.655	3.40	0.825	0.00572	0.994	0.00572
1-3/8	0.055	1.265	0.884	2.78	1.256	0.00872	1.485	0.00872
1-5/8	0.060	1.505	1.14	2.35	1.78	0.01235	2.07	0.01235
2-1/8	0.070	1.985	1.75	1.80	3.09	0.0215	3.55	0.0215
2-5/8	0.080	2.465	2.48	1.46	4.77	0.0331	5.41	0.0331
3-1/8	0.090	2.945	3.33	1.22	6.81	0.0473	7.67	0.0473
3-5/8	0.100	3.425	4.29	1.05	9.21	0.0640	10.32	0.0640
4-1/8	0.110	3.905	5.38	0.927	11.97	0.0831	13.36	0.0831
5-1/8	0.125	4.875	7.61	0.746	18.67	0.1295	20.63	0.1295
6-1/8	0.140	5.845	10.20	0.625	26.83	0.1862	29.46	0.1862
8-1/8	0.200	7.725	19.30	0.470	46.87	0.325	51.85	0.325
10-1/8	0.250	9.625	30.06	0.387	72.76	0.505	80.52	0.505
12-1/8	0.280	11.56	40.39	0.316	105.0	0.729	115.5	0.729

(continued)

Table A-7 Sizes and Weights of Copper Tubing (*continued*)

Outside Diameter (in)	Wall Thickness (in)	Inside Diameter (in)	Weight (lb/ft)	External Surface (length/ft^2)	Internal (in^2)	Surface (ft^2)	External Sectional Area (in^2)	Contents (cu ft/ lineal ft)
\multicolumn{9}{c}{Type M Copper Tube (Light Wall)}
3/8	0.025	0.325	0.107	10.2	0.083	0.00576	0.110	0.000576
1/2	0.025	0.450	0.145	7.64	0.159	0.00110	0.196	0.00110
5/8	0.028	0.569	0.204	6.11	0.254	0.00176	0.307	0.00176
3/4	0.030	0.690	0.263	5.10	0.374	0.00260	0.442	0.00260
7/8	0.032	0.811	0.328	4.37	0.516	0.00359	0.601	0.00359
1-1/8	0.035	1.055	0.465	3.40	0.874	0.00606	0.994	0.00606
1-3/8	0.042	1.291	0.682	2.78	1.309	0.00908	1.485	0.00908
1-5/8	0.049	1.527	0.940	2.35	1.831	0.0127	2.07	0.0127
2-1/8	0.058	2.009	1.46	1.80	3.17	0.0220	3.55	0.0220
2-5/8	0.065	2.495	2.03	1.46	4.89	0.0339	5.41	0.0339
3-1/8	0.072	2.981	2.68	1.22	6.98	0.0484	7.67	0.0484
3-5/8	0.083	3.459	3.58	1.05	9.40	0.0652	10.32	0.0652
4-1/8	0.095	3.935	4.66	0.927	12.16	0.0868	13.36	0.0848
5-1/8	0.109	4.907	6.66	0.746	18.91	0.1313	20.63	0.1313
6-1/8	0.122	5.881	8.92	0.625	27.16	0.1885	29.46	0.1885
8-1/8	0.170	7.785	16.47	0.470	47.6	0.330	51.85	0.330
10-1/8	0.212	9.701	25.59	0.387	73.9	0.513	80.52	0.513
12-1/8	0.254	11.617	36.72	0.316	105.6	0.733	115.5	0.733

[1] These sizes are used on some domestic units, but not on commercial equipment.

Table A-8 Dimensions of Standard-Weight Iron Pipe (Schedule 40)

Size (in)	Diameter External (in)	Diameter Internal, Approx. (in)	Nominal Weight/Foot, Plain Ends (lb)	Length/Foot^2 External Surface (ft)	Sectional Area, Internal (in^2)	Sectional Area, Internal (ft^2)	Contents of Pipe/ Lineal Foot (gal)	Contents of Pipe/ Lineal Foot (ft^2)
1/8	0.405	0.269	0.244	9.431	0.056	0.0004	0.003	0.0004
1/4	0.540	0.364	0.424	7.073	0.102	0.0007	0.005	0.0007
3/8	0.675	0.493	0.567	5.658	0.191	0.0013	0.010	0.0013
1/2	0.840	0.622	0.850	4.547	0.300	0.0021	0.016	0.0021
3/4	1.050	0.824	1.130	3.637	0.530	0.0037	0.027	0.0037
1	1.315	1.049	1.678	2.904	0.854	0.0060	0.044	0.0060
1-1/4	1.660	1.380	2.272	2.301	1.482	0.0104	0.077	0.0104
1-1/2	1.900	1.610	2.717	2.010	2.020	0.0141	0.105	0.0141
2	2.375	2.067	3.652	1.608	3.334	0.0233	0.174	0.0233
2-1/2	2.875	2.469	5.793	1.328	4.775	0.0332	0.248	0.0332
3	3.500	3.068	7.575	1.091	7.348	0.0513	0.384	0.0513
3-1/2	4.000	3.548	9.109	0.954	9.833	0.0687	0.513	0.0687
4	4.500	4.026	10.790	0.848	12.667	0.0887	0.661	0.0887
5	5.563	5.047	14.617	0.686	19.921	0.1389	1.039	0.1389
6	6.625	6.065	18.974	0.576	28.779	0.2006	1.500	0.2006
8	8.625	7.981	28.554	0.442	50.027	0.3474	2.598	0.3474
10	10.750	10.020	40.483	0.355	78.618	0.5476	4.096	0.5476
12	12.750	12.000	49.562	0.299	113.087	0.7854	5.875	0.7854

Table A-9 Thermal Conductivities of Various Materials

Material	Weight (lb/ft³)	Conductivity k (Btu/hr/ft²/in/deg)
Aluminum foil		
Spaced	2.36	0.22
Crumpled	0.196	0.28
Asbestos		
Packed	43.8	1.52
Loose	29.3	0.94
Corrugated paper	16.2	0.58
Bagasse	2.4	0.25
Bagasse	8.9	0.30
Coconut husk fiber	3.0	0.33
Corkboard, average	10.5	0.28
Light	6.5	0.25
Cork, granulated	8.1	0.34
Corn stalk pulp	3.86	0.26
Corn stalk pulp	6.44	0.31
Cotton	5.0	0.39
Eel grass mats	9.4	0.31
Eiderdown	4.92	0.33
Felt	16.9	0.25
Glass, cellular	9.0	0.38
Glass, wool	4.0	0.24
Hulls		
Buckwheat	12.24	0.32
Cotton seed	4.43	0.31
Kapok, loose	0.87	0.24
Lith board	12.5	0.38
Magnesium 85%, asbestos 15%	17.4	0.50
Mineral wool	4.0	0.27
Paper, cardboard	43.0	0.49
Polystyrene foam	1.9	0.25
Polyurethane foam	1.5–2.5	0.17
Redwood bark, shredded	5.04	0.25
Rubber		
Hard	74.3	1.10
Soft	68.6	1.21
Sponge	14.0	0.37
Cellular	3.7–7.5	0.22
Sawdust, shavings	0.8–15	0.45
Sisal	6.84	0.24
Steel plates, polished	6.0	0.22
Straw		
Wheat	4.56	0.27
Pressed	8.67	0.31
Wood pulp	4.92	0.26
Wood pulp	9.0	0.28
Wool	4.99	0.26
Wood		
Balsa	7.05	0.32
Cypress	28.7	0.67
Fir	34.1	0.80
Maple	44.3	1.10
Pine, Virginia	34.3	0.96

(continued)

Table A–9 Thermal Conductivities of Various Materials (*continued*)

Material	Weight (lb/ft^3)	Conductivity k (Btu/hr/ft^2/in/deg)
Pine, white	31.2	0.78
Pine, Oregon	37.0	0.80
Redwood	22.0	0.70
Brick		
Average wall	93.0	5.0
Fire	145	9.0
Concrete, Average	150	12.0
Concrete, Cinder	110	4.9
Concrete, Sawdust	0.97	
Glass	164	5.5
Gypsum plaster	5.0	
Mortar, cement	103	2.3
Mortar, lime	90	3.3
Porcelain	250	10.0
Stone		
Granite	179	13.0
Limestone	160	5.8
Marble	170	3.4
Quartz	165	6.0
Sandstone	145	8.5
Slate	175	10.4

Material	Conductivity K-factors (Btu/hr/ft^2/deg (not/inch thick)
Air space (>3/8 in)	1.1
Single glass	1.13
Double glass	0.46
Triple glass	0.29
Four-pane glass	0.21

Table A–10 Other Heat Loads: Sun Effects

	(To be added to TD)		
	Dark (°F)	Average (°F)	Aluminum Paint (°F)
Roof	25	20	10
East or west wall	12	10	5
South wall	6	5	3
Electric load	3.41 Btu/watt-hour		
Mechanical load	2545 Btu/hp-hour		
Occupants	600 Btu/person		

Table A–11 Cooler Usage Factors

	(Do not use for over 1500 cu ft)	
	Usage factor (Btu/hr/cu ft/deg)	
Usage	Reach-in Cabinet	Walk-in Cabinet
Light	0.18	0.06
Medium	0.22	0.08
Heavy	0.30	0.12

Table A–12 Heat Transfer in Refrigeration Equipment

Conductivity K (Btu/ ft²/deg/hr)

Cooling Air

	Still Air	Air Velocity (ft/min)						
		200	300	400	500	600	700	800
Direct expansion coil	2.5	5.0	5.8	6.6	7.7	8.3	8.9	9.3
Brine coil	3.0	8.5	10.0	11.0	12.0	12.8	13.2	13.6
Plate or holdover tank	2.25							

Cooling Liquid: Refrigeration Coil in Liquid Tank

	Still liquid	Agitated Liquid Velocity (ft/min)					
		20	30	40	50	100	150
Direct expansion	12	18	22	25	28	40	49

Brine Coolers

Velocity (ft/min)	Shell-and-Tube	Double-Pipe
50	55	
100	80	75
150	100	85
200	115	97.5
250	125	110
300		120
400		135
Congealing tank, still brine		8
Shell-and-coil liquid ammonia cooler		45

Baudelot Coolers

Water or milk, direct expansion	60
Water or milk, brine	80
Cream, direct expansion	50
Oil, direct expansion	10

Condenser Data

Velocity (ft/min)	Horizontal Shell-and-Tube	Double-Pipe	Shell-and-Coil	Water Flow (gpm/tube)	Vertical Shell-and-Tube	
					2-in tube	2½-in tube
50	75	125	160	1	115	80
100	120	195	225	2	175	120
150	150	235	260	3	235	150
200	175	260	285	4	290	185
250	190	280	305	5		215
300	210	290	320	6		245
350	220		335	7		280
400	225					

Evaporative condenser	125–250
Shell-and-tube desuperheater	25
Double-pipe heat exchanger	80

Table A–13 Refrigerant Required to Charge a System

Pipe Size (in)	Pounds of Refrigerant per Lineal Foot of Type L Copper Tubing					
	Liquid Line		Evaporator Coil			
			Dry Expansion (25% liquid)		Flooded (50% liquid)	
	R-12	R-22	R-12	R-22	R-12	R-22
1/4	0.01	0.01				
3/8	0.04	0.04	0.01	0.01	0.02	0.02
1/2	0.08	0.08	0.02	0.02	0.04	0.04
5/8	0.13	0.12	0.03	0.03	0.06	0.06
7/8	0.27	0.25	0.07	0.06	0.14	0.12
1-1/8	0.47	0.42	0.12	0.10	0.24	0.21
1-3/8	0.71	0.65	0.18	0.16	0.36	0.32
1-5/8	1.01	0.92	0.25	0.23	0.50	0.46
2-1/8	1.75	1.59	0.44	0.40	0.88	0.80
2-5/8	2.70	2.46	0.68	0.62	1.35	1.23
3-1/8	3.86	3.51	0.96	0.88	1.93	1.76
3-5/8	5.23	4.76	1.31	1.19	2.62	2.38
4-1/8	6.72	6.12	1.68	1.53	3.36	3.06
5-1/8	10.60	9.67	2.65	2.42	5.30	4.84
6-1/8	15.10	13.80	3.78	3.45	7.55	6.90

Table A–13 Refrigerant Required to Charge a System (*continued*)

Pipe Size (in)	Pounds of Refrigerant per Lineal Foot of Iron Pipe									
	Liquid Line			Evaporator Coil					
				Dry Expansion (25% liquid)			Flooded (50% liquid)		
	R-12	R-22	NH$_3$	R-12	R-22	NH$_3$	R-12	R-22	NH$_3$
1/4 X	0.04	0.04	0.02						
1/4 Std.	0.06	0.05	0.03						
3/8 X	0.08	0.07	0.04						
3/8 Std.	0.11	0.10	0.05						
1/2 X	0.13	0.12	0.06						
1/2 Std.	0.17	0.16	0.08	0.04	0.04	0.02	0.08	0.08	0.04
3/4 X	0.24	0.22	0.11	0.06	0.06	0.03	0.12	0.11	0.05
3/4 Std.	0.30	0.27	0.14	0.08	0.07	0.04	0.15	0.13	0.07
1 X	0.40	0.37	0.19	0.10	0.09	0.05	0.20	0.18	0.10
1 Std.	0.49	0.44	0.22	0.12	0.11	0.06	0.24	0.22	0.11
1-1/4 X	0.72	0.66	0.33	0.18	0.16	0.08	0.36	0.33	0.16
1-1/4 Std.	0.86	0.77	0.39	0.22	0.19	0.10	0.43	0.38	0.20
1-1/2 X	1.00	0.90	0.46	0.25	0.22	0.12	0.50	0.45	0.23
1-1/2 Std.	1.15	1.04	0.53	0.29	0.26	0.13	0.58	0.52	0.26
2 X	1.67	1.52	0.77	0.42	0.38	0.19	0.84	0.76	0.38
2 Std.	1.90	1.72	0.87	0.48	0.43	0.22	0.95	0.86	0.44
2-1/2 X	2.40	2.17	1.10	0.60	0.54	0.28	1.20	1.08	0.55
2-1/2 Std.	2.70	2.46	1.24	0.68	0.62	0.31	1.35	1.23	0.62
3 X	3.73	3.40	1.71	0.92	0.85	0.43	1.86	1.70	0.86
3 Std.	4.19	3.80	1.92	1.05	0.95	0.48	2.10	1.90	0.96
3-1/2 X	5.00	4.57	2.30				2.50	2.28	1.15
3-1/2 Std.	5.58	5.10	2.57				2.79	2.55	1.28
4 X	6.50	5.91	2.99				3.25	2.96	1.50
4 Std.	7.21	6.58	3.31				3.60	3.29	1.66
4-1/2 Std.	9.05	8.24	4.15				4.52	4.12	2.06
5 X	10.2	9.37	4.71						
5 Std.	11.3	10.3	5.20						
6 X	14.7	13.4	6.78						
6 Std.	16.3	14.9	7.53						

To calculate the weight of any refrigerant in 1 foot of pipe, calculate the internal volume by using the cylinder formula from Table A–6. You will probably use inches, which will give you volume in cubic inches. Convert to cubic feet by dividing cubic inches by 1,728. Next, multiply cubic feet by density (from the thermodynamic tables). Pay attention to liquid, vapor, and the mixes. Information is also available from vendors and manufacturers.

Table A–14 Log-Mean Temperature Difference

	\multicolumn{20}{c	}{D1 or D2}
	1	2	3	4	5	6	7	8	9	10	11	12	13	14	15	16	17	18	19	20
1	1.00	1.44	1.82	2.16	2.48	2.79	3.08	3.37	3.64	3.91	4.17	4.43	4.68	4.93	5.17	5.41	5.65	5.88	6.11	6.34
2	1.44	2.00	2.47	2.89	3.28	3.64	3.99	4.33	4.65	4.97	5.28	5.58	5.88	6.17	6.45	6.73	7.01	7.28	7.55	7.82
3	1.82	2.47	3.00	3.51	3.95	4.33	4.73	5.11	5.40	5.82	6.17	6.49	6.82	7.15	7.46	7.77	8.08	8.37	8.67	8.97
4	2.16	2.89	3.51	4.00	4.48	4.93	5.36	5.77	6.17	6.55	6.92	7.28	7.64	8.00	8.32	8.66	8.98	9.31	9.63	9.94
5	2.48	3.28	3.95	4.48	5.00	5.49	5.94	6.38	6.81	7.21	7.61	8.00	8.37	8.74	9.10	9.46	9.81	10.15	10.49	10.82
6	2.79	3.64	4.33	4.93	5.49	6.00	6.37	7.01	7.40	7.85	8.27	8.70	9.08	9.47	9.98	10.22	10.61	10.96	11.30	11.67
7	3.08	3.99	4.73	5.36	5.94	6.37	7.00	7.63	7.86	8.39	8.87	9.32	9.67	10.10	10.52	10.86	11.26	11.65	12.04	12.37
8	3.37	4.33	5.10	5.77	6.38	7.01	7.63	8.00	8.49	8.96	9.42	9.86	10.30	10.72	11.13	11.54	11.94	12.33	12.72	13.10
9	3.64	4.65	5.40	6.17	6.81	7.40	7.86	8.49	9.00	9.58	10.06	10.52	10.97	11.24	11.70	12.14	12.57	12.99	13.39	13.92
10	3.91	4.97	5.82	6.55	7.21	7.85	8.39	8.96	9.58	10.00	10.49	10.97	11.43	11.89	12.33	12.77	13.19	13.61	14.02	14.43
11	4.17	5.28	6.17	6.92	7.61	8.27	8.87	9.42	10.06	10.49	11.00	11.49	11.96	12.42	12.94	13.33	13.79	14.22	14.65	15.06
12	4.43	5.58	6.49	7.28	8.00	8.70	9.32	9.86	10.52	10.97	11.49	12.00	12.50	12.99	13.45	13.90	14.45	14.80	15.23	15.66
13	4.68	5.88	6.82	7.64	8.37	9.08	9.67	10.30	10.97	11.43	11.96	12.50	13.00	13.48	13.91	14.44	14.90	15.35	15.80	16.26
14	4.93	6.17	7.15	8.00	8.74	9.47	10.10	10.72	11.24	11.89	12.42	12.99	13.48	14.00	14.58	14.93	15.46	15.90	16.38	16.81
15	5.17	6.45	7.46	8.32	9.10	9.98	10.52	11.13	11.70	12.33	12.94	13.45	13.91	14.58	15.00	15.87	16.00	16.46	16.90	17.39
16	5.41	6.73	7.77	8.66	9.46	10.22	10.86	11.54	12.14	12.77	13.33	13.901444	14.90	14.93	15.87	16.00	16.29	16.98	17.31	17.93
17	5.65	7.01	8.08	8.98	9.81	10.61	11.26	11.94	12.57	13.19	13.79	14.45	14.90	15.46	16.00	16.29	17.00	17.51	18.07	18.51
18	5.88	7.28	8.37	9.31	10.15	10.96	11.65	12.33	1.299	13.61	14.22	14.80	15.35	15.90	16.46	16.98	17.51	18.00	18.35	18.99
19	6.11	7.55	8.67	9.63	10.49	11.30	12.04	12.72	13.39	14.02	14.65	15.23	15.80	16.38	16.90	17.31	18.07	18.35	19.00	19.23
20	6.34	7.82	8.95	9.94	10.82	11.67	12.37	13.10	13.92	14.43	15.06	15.66	16.26	16.81	17.39	17.93	18.51	18.99	19.23	20.00
21	6.57	8.08	9.25	10.25	11.15	12.00	12.74	13.47	14.19	14.83	15.47	16.08	16.69	17.26	17.83	18.35	18.96	19.43	20.24	20.49
22	6.79	8.34	9.54	10.56	11.47	12.35	13.11	13.84	14.57	15.22	15.87	16.50	17.11	17.71	18.28	18.84	19.40	19.96	20.45	20.99
23	7.02	8.60	9.82	10.86	11.79	12.68	13.44	14.20	14.89	15.61	16.27	16.92	17.53	18.12	18.72	18.27	19.90	20.38	20.90	21.46

Table A–14 Log-Mean Temperature Difference (*continued*)

										D1 or D2										
	1	2	3	4	5	6	7	8	9	10	11	12	13	14	15	16	17	18	19	20
24	7.24	8.85	10.01	11.16	12.11	13.02	13.79	14.56	15.27	15.99	16.64	17.31	17.95	18.55	19.15	19.73	20.33	20.86	21.48	21.94
25	7.46	9.11	10.38	11.46	12.43	13.34	14.14	14.92	15.65	16.37	17.05	17.74	18.35	18.95	19.58	20.14	20.76	21.30	21.86	22.41
26	7.67	9.36	10.65	11.75	12.74	13.67	14.46	15.26	16.06	16.75	17.43	18.11	18.76	19.38	20.01	20.60	21.20	21.77	22.34	22.87
27	7.89	9.61	10.92	12.05	13.05	13.99	14.81	15.62	16.38	17.11	17.82	18.50	19.20	19.79	20.42	21.01	21.63	22.19	22.76	23.33
28	8.10	9.85	11.19	12.33	13.35	14.31	15.15	15.96	16.75	17.48	18.20	18.89	19.55	20.20	20.83	21.44	22.04	22.62	23.20	23.77
29	8.32	10.01	11.46	12.62	13.65	14.63	15.49	16.31	17.10	17.85	18.57	19.27	19.94	20.60	21.24	21.85	22.49	23.07	23.66	24.22
30	8.53	10.34	11.73	12.90	13.95	14.94	15.79	16.64	17.46	18.20	18.94	19.64	20.33	20.99	21.64	22.27	22.90	23.48	24.08	24.66
31	8.74	10.58	11.98	13.19	14.25	15.25	16.12	16.98	17.81	18.56	19.31	20.02	20.71	21.27	22.09	22.67	23.31	23.92	24.50	25.01
32	8.94	10.82	12.26	13.47	14.55	15.57	16.45	17.31	18.11	18.91	19.66	20.39	21.09	21.77	22.45	23.08	23.72	24.33	24.94	25.53
33	9.15	11.06	12.51	13.74	14.84	15.87	16.75	17.64	18.46	19.26	20.03	20.76	21.47	22.18	22.83	23.47	24.13	24.75	25.35	25.96
34	9.36	11.29	12.76	14.02	15.13	16.17	17.08	17.97	18.80	19.61	20.37	21.12	21.85	22.53	23.22	23.88	24.53	25.15	25.79	26.39
35	9.56	11.53	13.03	14.29	15.47	16.48	17.40	18.29	19.14	19.96	20.72	21.48	22.22	22.92	23.60	24.27	24.94	25.58	26.19	26.80
36	9.77	11.76	13.28	14.56	15.70	16.77	17.71	18.62	19.48	20.30	21.08	21.85	22.58	23.30	23.99	24.66	25.33	25.97	26.62	27.22
37	9.97	12.00	13.53	14.83	15.99	17.07	18.01	18.94	19.81	20.64	21.43	22.20	22.95	23.66	24.37	25.04	25.72	26.36	27.01	27.63
38	10.17	12.23	13.78	15.10	16.27	17.36	18.32	19.25	20.14	20.97	21.78	22.55	23.30	24.05	24.73	25.43	26.11	26.77	27.41	28.04
39	10.37	12.45	14.04	15.37	16.55	17.67	18.63	19.57	20.47	21.31	22.13	22.91	23.67	24.41	25.12	25.81	26.50	27.16	27.80	28.45
40	10.57	12.68	14.29	15.63	16.83	17.95	18.92	19.88	20.80	21.64	22.46	23.26	24.02	24.77	25.49	26.19	26.89	27.56	28.21	28.86

Courtesy McQuay Incorporated.

Table A-15

SATURATED AMMONIA—TEMPERATURE TABLE

Temp °F t	Pressure Absolute lbs./in.² p	Pressure Gage lb./in.² g.p.	Volume vapor ft³/lb. V	Density vapor lbs./ft.³ 1/V	Heat content Liquid Btu./lb. h	Heat content Vapor Btu./lb. H	Latent heat Btu./lb. L	Entropy Liquid Btu./lb.°F s	Entropy Vapor Btu./lb.°F S	Temp °F t
−60	5.55	*18.6	44.73	0.02235	−21.2	589.6	610.8	−0.0517	1.4769	−60
−59	5.74	*18.2	43.37	.02306	−20.1	590.0	610.1	− .0490	.4741	−59
−58	5.93	*17.8	42.05	.02378	−19.1	590.4	609.5	− .0464	.4713	−58
−57	6.13	*17.4	40.79	.02452	−18.0	590.8	608.8	− .0438	.4686	−57
−56	6.33	*17.0	39.56	.02528	−17.0	691.2	608.2	− .0412	.4658	−56
−55	6.54	*16.6	38.38	0.02605	−15.9	591.6	607.5	−0.0386	1.4631	−55
−54	6.75	*16.2	37.24	.02685	−14.8	592.1	606.9	− .0360	.4604	−54
−53	6.97	*15.7	36.15	.02766	−13.8	592.4	606.2	− .0334	.4577	−53
−52	7.20	*15.3	35.09	.02850	−12.7	592.9	605.6	− .0307	.4551	−52
−51	7.43	*14.8	34.06	.02936	−11.7	593.2	604.9	− .0281	.4524	−51
−50	7.67	*14.3	33.08	0.03023	−10.6	593.7	604.3	−0.0256	1.4497	−50
−49	7.91	*13.8	32.12	.03113	−9.6	594.0	603.6	− .0230	.4471	−49
−48	8.16	*13.3	31.20	.03205	−8.5	594.4	602.9	− .0204	.4445	−48
−47	8.42	*12.8	30.31	.03299	−7.6	594.9	602.3	− .0179	.4419	−47
−46	8.68	*12.2	29.45	.03395	−6.4	595.2	601.6	− .0153	.4393	46
−45	8.95	*11.7	28.62	0.03494	−5.3	595.6	600.9	−0.0127	1.4368	−45
−44	9.23	*11.1	27.82	.03595	−4.3	596.0	600.3	− .0102	.4342	−44
−43	9.51	*10.6	27.04	.03698	−3.2	596.4	599.6	− .0076	.4317	−43
−42	9.81	*10.0	26.29	.03804	−2.1	596.8	598.9	− .0051	.4292	−42
−41	10.10	*9.3	25.56	.03912	−1.1	597.2	598.3	− .0025	.4267	−41
−40	10.41	*8.7	24.86	0.04022	0.0	597.6	597.6	0.0000	1.4242	−40
−39	10.72	*8.1	24.18	.04135	1.1	598.0	596.9	.0025	.4217	−39
−38	11.04	*7.4	23.53	.04251	2.1	598.3	596.2	.0051	.4193	−38
−37	11.37	*6.8	22.89	.04369	3.2	598.7	595.5	.0076	.4169	−37
−36	11.71	*6.1	22.27	.04489	4.3	599.1	594.8	.0101	.4144	−36
−35	12.05	*5.4	21.68	0.04613	5.3	599.5	594.2	0.0126	1.4120	−35
−34	12.41	*4.7	21.10	.04739	6.4	599.9	593.5	.0151	.4096	−34
−33	12.77	*3.9	20.54	.04868	7.4	600.2	592.8	.0176	.4072	−33
−32	13.14	*3.2	20.00	.04999	8.5	600.6	592.1	.0201	.4048	−32
−31	13.52	*2.4	19.48	.05134	9.6	601.0	591.4	.0226	.4025	−31
−30	13.90	*1.6	18.97	0.05271	10.7	601.4	590.7	0.0250	1.4001	−30
−29	14.30	*0.8	18.48	.05411	11.7	601.7	590.0	.0275	.3978	−29
−28	14.71	0.0	18.00	.05555	12.8	602.1	589.3	.0300	.3955	−28
−27	15.12	0.4	17.54	.05701	13.9	602.5	588.6	.0325	.3932	−27
−26	15.55	0.8	17.09	.05850	14.9	602.8	587.9	.0350	.3909	−26
−25	15.98	1.3	16.66	0.06003	16.0	603.2	587.2	0.0374	1.3886	−25
−24	16.42	1.7	16.24	.06158	17.1	603.6	586.5	.0399	.3863	−24
−23	16.88	2.2	15.83	.06317	18.1	603.9	585.8	.0423	.3840	−23
−22	17.34	2.6	15.43	.06479	19.2	604.3	585.1	.0448	.3818	−22
−21	17.81	3.1	15.05	.06644	20.3	604.6	584.3	.0472	.3796	−21
−20	18.30	3.6	14.68	0.06813	21.4	605.0	583.6	0.0497	1.3774	−20
−19	18.79	4.1	14.32	.06985	22.4	605.3	582.9	.0521	.3752	−19
−18	19.30	4.6	13.97	.07161	23.5	605.7	582.2	.0545	.3729	−18
−17	19.81	5.1	13.62	.07340	24.6	606.1	581.5	.0570	.3708	−17
−16	20.34	5.6	13.29	.07522	25.6	606.4	580.8	.0594	.3686	−16
−15	20.88	6.2	12.97	0.07709	26.7	606.7	580.0	0.0618	1.3664	−15
−14	21.43	6.7	12.66	.07898	27.8	607.1	579.3	.0642	.3643	−14
−13	21.99	7.3	12.36	.08092	28.9	607.5	578.6	.0666	.3621	−13
−12	22.56	7.9	12.06	.08289	30.0	607.8	577.8	.0690	.3600	−12
−11	23.15	8.5	11.78	.08490	31.0	608.1	577.1	.0714	.3579	−11
−10	23.74	9.0	11.50	0.08695	32.1	608.5	576.4	0.0738	1.3558	−10

* Inches of mercury below one standard atmosphere (29.92 in.).

Table A–15 (continued)

SATURATED AMMONIA–TEMPERATURE TABLE

Temp °F t	Pressure Absolute lbs./in.² p	Pressure Gage lb./in.² g.p.	Volume vapor ft³/lb V	Density vapor lbs./ft.³ 1/V	Heat content Liquid Btu./lb. h	Heat content Vapor Btu./lb. H	Latent heat Btu./lb. L	Entropy Liquid Btu./lb.°F s	Entropy Vapor Btu./lb.°F S	Temp °F t
−10	23.74	9.0	11.50	0.08695	32.1	608.5	576.4	0.0738	1.3558	−10
−9	24.35	9.7	11.23	.08904	33.2	608.8	575.6	.0762	.3537	−9
−8	24.97	10.3	10.97	.09117	34.3	609.2	574.9	.0786	.3516	−8
−7	25.61	10.9	10.71	.09334	35.4	609.5	574.1	.0809	.3495	−7
−6	26.26	11.6	10.47	.09555	36.4	609.8	573.4	.0833	.3474	−6
−5	26.92	12.2	10.23	0.09780	37.5	610.1	572.6	0.0857	1.3454	−5
−4	27.59	12.9	9.991	.1001	38.6	610.5	571.9	.0880	.3433	−4
−3	28.28	13.6	9.763	.1024	39.7	610.8	571.1	.0904	.3413	−3
−2	28.98	14.3	9.541	.1048	40.7	611.1	570.4	.0928	.3393	−2
−1	29.69	15.0	9.326	.1072	41.8	611.4	569.6	.0951	.3372	−1
0	30.42	15.7	9.116	0.1097	42.9	611.8	568.9	0.0975	1.3352	0
1	31.16	16.5	8.912	.1122	44.0	612.1	568.1	.0998	.3332	1
2	31.92	17.2	8.714	.1148	45.1	612.4	567.3	.1022	.3312	2
3	32.69	18.0	8.521	.1174	46.2	612.7	566.5	.1045	.3292	3
4	33.47	18.8	8.333	.1200	47.2	613.0	565.8	.1069	.3273	4
5	34.27	19.6	8.150	0.1227	48.3	613.3	565.0	0.1092	1.3253	5
6	35.09	20.4	7.971	.1254	49.4	613.6	564.2	.1115	.3234	6
7	35.92	21.2	7.798	.1282	50.5	613.9	563.4	.1138	.3214	7
8	36.77	22.1	7.629	.1311	51.6	614.3	562.7	.1162	.3195	8
9	37.63	22.9	7.464	.1340	52.7	614.6	561.9	.1185	.3176	9
10	38.51	23.8	7.304	0.1369	53.8	614.9	561.1	0.1208	1.3157	10
11	39.40	24.7	7.148	.1399	54.9	615.2	560.3	.1231	.3137	11
12	40.31	25.6	6.996	.1429	56.0	615.5	559.5	.1254	.3118	12
13	41.24	26.5	6.847	.1460	57.1	615.8	558.7	.1277	.3099	13
14	42.18	27.5	6.703	.1492	58.2	616.1	557.9	.1300	.3081	14
15	43.14	28.4	6.562	0.1524	59.2	616.3	557.1	0.1323	1.3062	15
16	44.12	29.4	6.425	.1556	60.3	616.6	556.3	.1346	.3043	16
17	45.12	30.4	6.291	.1590	61.4	616.9	555.5	.1369	.3025	17
18	46.13	31.4	6.161	.1623	62.5	617.2	554.7	.1392	.3006	18
19	47.16	32.5	6.034	.1657	63.6	617.5	553.9	.1415	.2988	19
20	48.21	33.5	5.910	0.1692	64.7	617.8	553.1	0.1437	1.2969	20
21	49.28	34.6	5.789	.1728	65.8	618.0	552.2	.1460	.2951	21
22	50.36	35.7	5.671	.1763	66.9	618.3	551.4	.1483	.2933	22
23	51.47	36.8	5.556	.1800	68.0	618.6	550.6	.1505	.2915	23
24	52.59	37.9	5.443	.1837	69.1	618.9	549.8	.1528	.2897	24
25	53.73	39.0	5.334	0.1875	70.2	619.1	548.9	0.1551	1.2879	25
26	54.90	40.2	5.227	.1913	71.3	619.4	548.1	.1573	.2861	26
27	56.08	41.4	5.123	.1952	72.4	619.7	547.3	.1596	.2843	27
28	57.28	42.6	5.021	.1992	73.5	619.9	546.4	.1618	.2825	28
29	58.50	43.8	4.922	.2032	74.6	620.2	545.6	.1641	.2808	29
30	59.74	45.0	4.825	0.2073	75.7	620.5	544.8	0.1663	1.2790	30
31	61.00	46.3	4.730	.2114	76.8	620.7	543.9	.1686	.2773	31
32	62.29	47.6	4.637	.2156	77.9	621.0	543.1	.1708	.2755	32
33	63.59	48.9	4.547	.2199	79.0	621.2	542.2	.1730	.2738	33
34	64.91	50.2	4.459	.2243	80.1	621.5	541.4	.1753	.2721	34
35	66.26	51.6	4.373	0.2287	81.2	621.7	540.5	0.1775	1.2704	35
36	67.63	52.9	4.289	.2332	82.3	622.0	539.7	.1797	.2686	36
37	69.02	54.3	4.207	.2377	83.4	622.2	538.8	.1819	.2669	37
38	70.43	55.7	4.126	.2423	84.6	622.5	537.9	.1841	.2652	38
39	71.87	57.2	4.048	.2470	85.7	622.7	537.0	.1863	.2635	39
40	73.32	58.6	3.971	0.2518	86.8	623.0	536.2	0.1885	1.2618	40

Table A-15 (continued)

PROPERTIES OF SUPERHEATED AMMONIA VAPOR

(V = Volume in ft^3/lb; H = heat content in Btu/lb; S = entropy in Btu/lb °F)

Absolute pressure in lbs./in.^{2} (Saturation temperature in Bold.)

Temp. °F.	125 68.31°			130 70.53°			135 72.69°			Temp. °F.
	V	H	S	V	H	S	V	H	S	
Sat.	2.380	628.8	1.2168	2.291	629.3	1.2132	2.209	629.6	1.2100	Sat.
80	2.461	637.2	1.2322	2.355	636.0	1.2260	2.257	634.9	1.2199	80
90	2.528	644.0	.2448	2.421	643.0	.2388	2.321	642.0	.2329	90
100	2.593	650.7	1.2568	2.484	649.7	1.2509	2.382	648.8	1.2452	100
110	2.657	657.1	.2682	2.546	656.3	.2625	2.442	655.4	.2569	110
120	2.719	663.5	.2792	2.606	662.7	.2736	2.501	661.9	.2681	120
130	2.780	669.7	.2899	2.665	668.9	.2843	2.559	668.2	.2790	130
140	2.840	675.8	.3002	2.724	675.1	.2947	2.615	674.4	.2894	140
150	2.900	681.8	1.3102	2.781	681.2	1.3048	2.671	680.5	1.2996	150
160	2.958	687.8	.3199	2.838	687.2	.3146	2.726	686.6	.3094	160
170	3.016	693.7	.3294	2.894	693.2	.3241	2.780	692.6	.3191	170
180	3.074	699.6	.3387	2.949	699.1	.3335	2.834	698.6	.3284	180
190	3.131	705.5	.3478	3.004	705.0	.3426	2.887	704.5	.3376	190
200	3.187	711.3	1.3567	3.059	710.9	1.3516	2.940	710.4	1.3466	200
210	3.243	717.2	.3654	3.113	716.7	.3604	2.992	716.2	.3554	210
220	3.299	723.0	.3740	3.167	722.5	.3690	3.044	722.1	.3641	220
230	3.354	728.8	.3825	3.220	728.3	.3775	3.096	727.9	.3726	230
240	3.409	734.5	.3908	3.273	734.1	.3858	3.147	733.7	.3810	240
250	3.434	740.3	1.3990	3.326	739.9	1.3941	3.198	739.6	1.3893	250
260	3.519	746.1	.4071	3.379	745.7	.4022	3.249	745.4	.3974	260
270	3.573	751.9	.4151	3.431	751.5	.4102	3.300	751.2	.4054	270
280	3.627	757.7	.4230	3.483	757.3	.4181	3.350	757.0	.4133	280
290	3.681	763.5	.4308	3.535	763.1	.4259	3.400	762.8	.4212	290
300	3.735	769.3	1.4385	3.587	769.0	1.4336	3.450	768.6	1.4289	300
320	3.842	780.9	.4536	3.690	780.6	.4487	3.550	780.3	.4441	320

Temp. °F.	150 78.81°			160 82.64°			170 86.89°			Temp. °F.
	V	H	S	V	H	S	V	H	S	
Sat.	1.994	630.5	1.2009	1.372	631.1	1.1952	1.764	631.6	1.1900	Sat.
90	2.061	638.8	1.2161	1.914	636.6	1.2055	1.784	634.4	1.1952	90
100	2.118	645.9	1.2289	1.969	643.9	1.2186	1.837	641.9	1.2087	100
110	2.174	652.8	.2410	2.023	651.0	.2311	1.889	649.1	.2215	110
120	2.228	659.4	.2526	2.075	657.8	.2429	1.939	656.1	.2336	120
130	2.281	665.9	.2638	2.125	664.4	.2542	1.988	662.8	.2452	130
140	2.334	672.3	.2745	2.175	670.9	.2652	2.035	669.4	.2563	140
150	2.385	678.6	1.2849	2.224	677.2	1.2757	2.081	675.9	1.2669	150
160	2.435	684.8	.2949	2.272	683.5	.2859	2.127	682.3	.2773	160
170	2.485	690.9	.3047	2.319	689.7	.2958	2.172	688.5	.2873	170
180	2.534	696.9	.3142	2.365	695.8	.3054	2.216	694.7	.2971	180
190	2.583	702.9	.3236	2.411	701.9	.3148	2.260	700.8	.3066	190
200	2.631	708.9	1.3327	2.457	707.9	1.3240	2.303	706.9	1.3159	200
210	2.679	714.8	.3416	2.502	713.9	.3331	2.346	713.0	.3249	210
220	2.726	720.7	.3504	2.547	719.9	.3419	2.389	719.0	.3338	220
230	2.773	726.6	.3590	2.591	725.8	.3506	2.431	724.9	.3426	230
240	2.820	732.5	.3675	2.635	731.7	.3591	2.473	730.9	.3512	240
250	2.866	738.4	1.3758	2.679	737.6	1.3675	2.514	736.8	1.3596	250
260	2.912	744.3	.3840	2.723	743.5	.3757	2.555	742.8	.3679	260
270	2.958	750.1	.3921	2.766	749.4	.3838	2.596	748.7	.3761	270
280	3.004	756.0	.4001	2.809	755.3	.3919	2.637	754.6	.3841	280
290	3.049	761.8	.4079	2.852	761.2	.3998	2.678	760.5	.3921	290
300	3.095	767.7	1.4157	2.895	767.1	1.4076	2.718	766.4	1.3999	300
320	3.185	779.4	.4310	2.980	778.9	.4229	2.798	778.3	.4153	320
340	3.274	791.2	.4459	3.064	790.7	.4379	2.878	790.1	.4303	340

Table A-15 (continued)

(V = volume in ft³/lb; H = heat content in Btu/lb; S = entropy in Btu/lb °F)

Temp. °F.	Absolute pressure in lbs./in.² (Saturation temperature in Bold.)									Temp. °F.
	140 74.79°			145 76.85°			150 78.81°			
	V	H	S	V	H	S	V	H	S	
Sat.	2.132	629.9	1.2068	2.061	630.2	1.2038	1.994	630.5	1.2009	Sat.
80	2.166	633.8	1.2140	2.080	632.6	1.2082	2.001	631.4	1.2025	80
90	2.228	640.9	.2272	2.141	639.9	.2216	2.061	638.9	.2161	90
100	2.288	647.8	1.2396	2.200	646.9	1.2342	2.118	645.9	1.2289	100
110	2.347	654.5	.2515	2.257	653.6	.2462	2.174	652.8	.2410	110
120	2.404	661.1	.2628	2.313	660.2	.2577	2.228	659.4	.2526	120
130	2.460	667.4	.2738	2.368	666.7	.2687	2.281	665.9	.2638	130
140	2.515	673.7	.2843	2.421	673.0	.2793	2.334	672.3	.2745	140
150	2.569	679.9	1.2945	2.474	679.2	1.2896	2.385	678.6	1.2849	150
160	2.622	686.0	.3045	2.526	685.4	.2996	2.435	684.8	.2949	160
170	2.675	692.0	.3141	2.577	691.4	.3093	2.485	690.9	.3047	170
180	2.727	698.0	.3236	2.627	697.5	.3188	2.534	696.9	.3142	180
190	2.779	704.0	.3328	2.677	703.4	.3281	2.583	702.9	.3236	190
200	2.830	709.9	1.3418	2.727	709.4	1.3372	2.631	708.9	1.3327	200
210	2.880	715.8	.3507	2.776	715.3	.3461	2.679	714.8	.3416	210
220	2.931	721.6	.3594	2.825	721.2	.3548	2.726	720.7	.3504	220
230	2.981	727.5	.3679	2.873	727.1	.3634	2.773	726.6	.3590	230
240	3.030	733.3	.3763	2.921	732.9	.3718	2.820	732.6	.3675	240
250	3.080	739.2	1.3846	2.969	738.8	1.3801	2.866	738.4	1.3758	250
260	3.129	745.0	.3928	3.017	744.6	.3883	2.912	744.3	.3840	260
270	3.179	750.8	.4008	3.064	750.5	.3964	2.958	750.1	.3921	270
280	3.227	756.7	.4088	3.111	756.3	.4043	3.004	756.0	.4001	280
290	3.275	762.5	.4166	3.158	762.2	.4122	3.049	761.8	.4079	290
300	3.323	768.3	1.4243	3.205	768.0	1.4199	3.095	767.7	1.4157	300
320	3.420	780.0	.4395	3.298	779.7	.4352	3.185	779.4	.4310	320
	180 89.78°			190 93.15°			200 96.54°			
Sat.	1.667	632.0	1.1850	1.581	632.4	1.1802	1.502	632.7	1.1756	Sat.
90	1.668	632.2	1.1853	. . . .	. . . .	. . . .	. . . .	. . . .	. . . .	90
100	1.720	639.9	1.1992	1.615	637.8	1.1899	1.520	635.6	1.1809	100
110	1.770	647.3	.2123	1.663	645.4	.2034	1.567	643.4	.1947	110
120	1.818	654.4	.2247	1.710	652.6	.2160	1.612	650.9	.2077	120
130	1.865	661.3	.2364	1.755	659.7	.2281	1.656	658.1	.2200	130
140	1.910	668.0	.2477	1.799	666.5	.2396	1.698	665.0	.2317	140
150	1.955	674.6	1.2586	1.842	673.2	1.2506	1.740	671.8	1.2429	150
160	1.999	681.0	.2691	1.884	679.7	.2612	1.780	678.4	.2537	160
170	2.042	687.3	.2792	1.925	686.1	.2715	1.820	684.9	.2641	170
180	2.084	693.6	.2891	1.966	692.5	.2815	1.859	691.3	.2742	180
190	2.126	699.8	.2987	2.005	698.7	.2912	1.897	697.7	.2840	190
200	2.167	705.9	1.3081	2.045	704.9	1.3007	1.935	703.9	1.2935	200
210	2.208	712.0	.3172	2.084	711.1	.3099	1.972	710.1	.3029	210
220	2.248	718.1	.3262	2.123	717.2	.3189	2.009	716.3	.3120	220
230	2.288	724.1	.3350	2.161	723.2	.3278	2.046	722.4	.3209	230
240	2.328	730.1	.3436	2.199	729.3	.3365	2.082	728.4	.3296	240
250	2.367	736.1	1.3521	2.236	735.3	1.3450	2.118	734.5	1.3382	250
260	2.407	742.0	.3605	2.274	741.3	.3534	2.154	740.5	.3467	260
270	2.446	748.0	.3687	2.311	747.3	.3617	2.189	746.5	.3550	270
280	2.484	753.9	.3768	2.348	753.2	.3698	2.225	752.5	.3631	280
290	2.523	759.9	.3847	2.384	759.2	.3778	2.260	758.5	.3712	290
300	2.561	765.8	1.3926	2.421	765.2	1.3857	2.295	764.5	1.3791	300
320	2.637	777.7	.4081	2.493	777.1	.4012	2.364	776.5	.3947	320
340	2.713	789.6	.4231	2.565	789.0	.4163	2.432	788.5	.4099	340

360 Appendix

Table A-15 (continued)

(V = volume in ft³/lb; H = heat content in Btu/lb;
S = entropy in Btu/lb °F)

Temp. °F.	Absolute pressure in lbs./in.² (Saturation temperature in Bold.)									Temp. °F.
	200 96.34°			210 99.43°			220 102.42°			
	V	H	S	V	H	S	V	H	S	
Sat.	1.502	632.7	1.1756	1.431	633.0	1.1713	1.367	633.2	1.1671	Sat.
110	1.567	643.4	1.1947	1.480	641.5	1.1863	1.400	639.4	1.1781	110
120	1.612	650.9	.2077	1.524	649.1	.1996	1.443	647.3	.1917	120
130	1.656	658.1	.2200	1.566	656.4	.2121	1.485	654.8	.2045	130
140	1.698	665.0	.2317	1.608	663.5	.2240	1.525	662.0	.2167	140
150	1.740	671.8	1.2429	1.648	670.4	1.2354	1.564	669.0	1.2281	150
160	1.780	678.4	.2537	1.687	677.1	.2464	1.601	675.8	.2394	160
170	1.820	684.9	.2641	1.725	683.7	.2569	1.638	682.5	.2501	170
180	1.859	691.3	.2742	1.762	690.2	.2672	1.675	689.1	.2604	180
190	1.897	697.7	.2840	1.799	696.6	.2771	1.710	695.5	.2704	190
200	1.935	703.9	1.2935	1.836	702.9	1.2867	1.745	701.9	1.2801	200
210	1.972	710.1	.3029	1.872	709.2	.2961	1.780	708.2	.2896	210
220	2.009	716.3	.3120	1.907	715.3	.3053	1.814	714.4	.2989	220
230	2.046	722.4	.3209	1.942	721.5	.3143	1.848	720.6	.3079	230
240	2.082	728.4	.3296	1.977	727.6	.3231	1.881	726.8	.3168	240
250	2.118	734.5	1.3382	2.011	733.7	1.3317	1.914	732.9	1.3255	250
260	2.154	740.5	.3467	2.046	739.8	.3402	1.947	739.0	.3340	260
270	2.189	746.5	.3550	2.080	745.8	.3486	1.980	745.1	.3424	270
280	2.225	752.5	.3631	2.113	751.8	.3568	2.012	751.1	.3507	280
290	2.260	758.5	.3712	2.147	757.9	.3649	2.044	757.2	.3588	290
300	2.295	764.5	1.3791	2.180	763.9	1.3728	2.076	763.2	1.3668	300
320	2.364	776.5	.3947	2.246	775.9	.3884	2.140	775.3	.3825	320
340	2.432	788.5	.4099	2.312	787.9	.4037	2.203	787.4	.3978	340
360	2.500	800.5	.4247	2.377	800.0	.4186	2.265	799.5	.4127	360
380	2.568	812.5	.4392	2.442	812.0	.4331	2.327	811.6	.4273	380
	250 110.80°			260 113.42°			270 115.97°			
Sat.	1.202	633.8	1.1555	1.155	633.9	1.1518	1.1182	633.9	1.1483	Sat.
120	1.240	641.5	1.1690	1.182	639.5	1.1617	1.128	637.5	1.1544	120
130	1.278	649.6	.1827	1.220	647.8	.1757	1.166	645.9	.1689	130
140	1.316	657.2	.1956	1.257	655.6	.1889	1.202	653.9	.1823	140.
150	1.352	664.6	1.2078	1.292	663.1	1.2014	1.236	661.6	1.1950	150
160	1.386	671.8	.2195	1.326	670.4	.2132	1.269	669.0	.2071	160
170	1.420	678.7	.2306	1.359	677.5	.2245	1.302	676.2	.2185	170
180	1.453	685.5	.2414	1.391	684.4	.2354	1.333	683.2	.2296	180
190	1.486	692.2	.2517	1.422	691.1	.2458	1.364	690.0	.2401	190
200	1.518	698.8	1.2617	1.453	697.7	1.2560	1.394	696.7	1.2504	200
210	1.549	705.3	.2715	1.484	704.3	.2658	1.423	703.3	.2603	210
220	1.580	711.7	.2810	1.514	710.7	.2754	1.452	709.8	.2700	220
230	1.610	718.0	.2902	1.543	717.1	.2847	1.481	716.2	.2794	230
240	1.640	724.3	.2993	1.572	723.4	.2938	1.509	722.6	.2885	240
250	1.670	730.5	1.3081	1.601	729.7	1.3027	1.537	728.9	1.2975	250
260	1.699	736.7	.3168	1.630	736.0	.3115	1.565	735.2	.3063	260
270	1.729	742.9	.3253	1.658	742.2	.3200	1.592	741.4	.3149	270
280	1.758	749.1	.3337	1.686	748.4	.3285	1.620	747.7	.3234	280
290	1.786	755.2	.3420	1.714	754.5	.3367	1.646	753.9	.3317	290
300	1.815	761.3	1.3501	1.741	760.7	1.3449	1.673	760.0	1.3399	300
320	1.872	773.5	.3659	1.796	772.9	.3608	1.726	772.3	.3559	320
340	1.928	785.7	.3814	1.850	785.2	.3763	1.778	784.6	.3714	340
360	1.983	797.9	.3964	1.904	797.4	.3914	1.830	796.9	.3866	360
380	2.038	810.1	.4111	1.957	809.6	.4062	1.881	809.1	.4014	380
400	2.093	822.3	1.4255	2.009	821.9	1.4206	1.932	821.4	1.4158	400

Table A-15 *(continued)*

(V = volume in ft³/lb; H = heat content in Btu/lb;
S = entropy in Btu/lb °F)

Temp. °F.	Absolute pressure in lbs./in.² (Saturation temperature in Bold)									Temp. °F.
	230 105.30°			240 108.09°			250 110.80°			
	V	H	S	V	H	S	V	H	S	
Sat.	1.307	633.4	1.1631	1.253	633.6	1.1592	1.202	633.8	1.1555	Sat.
110	1.328	637.4	1.1700	1.261	635.3	1.1621	. . . .	. . . .	. . . .	110
120	1.370	645.4	.1840	1.302	643.5	.1764	1.240	641.5	1.1690	120
130	1.140	653.1	.1971	1.342	651.3	.1898	1.278	649.6	.1827	130
140	1.449	660.4	.2095	1.380	658.8	.2025	1.316	657.2	.1956	140
150	1.487	667.6	1.2213	1.416	666.1	1.2145	1.352	664.6	1.2078	150
160	1.524	674.5	.2325	1.452	673.1	.2259	1.386	671.8	.2195	160
170	1.559	681.3	.2434	1.487	680.0	.2369	1.420	678.7	.2306	170
180	1.594	687.9	.2538	1.521	686.7	.2475	1.453	685.5	.2414	180
190	1.629	694.4	.2640	1.554	693.3	.2577	1.486	692.2	.2517	190
200	1.663	700.9	1.2738	1.587	699.8	1.2677	1.518	698.8	1.2617	200
210	1.696	707.2	.2834	1.619	706.2	.2773	1.549	705.3	.2715	210
220	1.729	713.5	.2927	1.651	712.6	.2867	1.580	711.7	.2810	220
230	1.762	719.8	.3018	1.683	718.9	.2959	1.610	718.0	.2902	230
240	1.794	726.0	.3107	1.714	725.1	.3049	1.640	724.3	.2993	240
250	1.826	732.1	1.3195	1.745	731.1	1.3137	1.670	730.5	1.3081	250
260	1.857	738.3	.3281	1.775	737.5	.3224	1.699	736.7	.3168	260
270	1.889	744.4	.3365	1.805	743.6	.3308	1.729	742.9	.3253	270
280	1.920	750.5	.3448	1.835	749.8	.3392	1.758	749.1	.3337	280
290	1.951	756.5	.3530	1.865	755.9	.3474	1.786	755.2	.3420	290
300	1.982	762.6	1.3610	1.895	762.0	1.3554	1.815	761.3	1.3501	300
320	2.043	774.7	.3767	1.954	774.1	.3712	1.872	773.5	.3659	320
340	2.103	786.8	.3921	2.012	786.3	.3866	1.928	785.7	.3814	340
360	2.163	798.9	.4070	2.069	798.4	.4016	1.983	797.9	.3964	360
380	2.222	811.1	.4217	2.126	810.6	.4163	2.038	810.1	.4111	380
	280 118.45°			290 120.86°			300 123.21°			
Sat.	1.072	634.0	1.1449	1.034	634.0	1.1415	0.999	634.0	1.1383	Sat.
120	1.078	635.4	1.1473	. . . .	. . . .	. . . .	. . . .	. . . .	. . . .	120
130	1.115	644.0	.1621	1.068	642.1	1.1554	1.023	640.1	1.1487	130
140	1.151	652.2	.1759	1.103	650.5	.1695	1.058	648.7	.1632	140
150	1.184	660.1	1.1888	1.136	658.5	1.1827	1.091	656.9	1.1767	150
160	1.217	667.6	.2011	1.168	666.1	.1952	1.123	664.7	.1894	160
170	1.249	674.9	.2127	1.199	673.5	.2070	1.153	672.2	.2014	170
180	1.279	681.9	.2239	1.229	680.7	.2183	1.183	679.5	.2129	180
190	1.309	688.9	.2346	1.259	687.7	.2292	1.211	686.5	.2239	190
200	1.339	695.6	1.2449	1.287	694.6	1.2396	1.239	693.5	1.2344	200
210	1.367	702.3	.2550	1.315	701.3	.2497	1.267	700.3	.2447	210
220	1.396	708.8	.2647	1.343	707.9	.2596	1.294	706.9	.2546	220
230	1.424	715.3	.2742	1.370	714.4	.2691	1.320	713.5	.2642	230
240	1.451	721.8	.2834	1.397	720.9	.2784	1.346	720.0	.2736	240
250	1.478	728.1	1.2924	1.423	727.3	1.2875	1.372	726.5	1.2827	250
260	1.505	734.4	.3013	1.449	733.7	.2964	1.397	732.9	.2917	260
270	1.532	740.7	.3099	1.475	740.0	.3051	1.422	739.2	.3004	270
280	1.558	747.0	.3184	1.501	746.3	.3137	1.447	745.5	.3090	280
290	1.584	753.2	.3268	1.526	752.5	.3221	1.472	751.8	.3175	290
300	1.610	759.4	1.3350	1.551	758.7	1.3303	1.496	758.1	1.3257	300
320	1.661	771.7	.3511	1.601	771.1	.3464	1.544	770.5	.3419	320
340	1.712	784.0	.3667	1.650	783.5	.3621	1.592	782.9	.3576	340
360	1.762	796.3	.3819	1.698	795.8	.3773	1.639	795.3	.3729	360
380	1.811	808.7	1.3967	1.747	808.2	.3922	1.686	807.7	.3878	380
400	1.861	821.0	1.4112	1.794	820.5	1.4067	1.732	820.1	1.4024	400

Table A-15 (continued)

Properties of Liquid Ammonia

Temp °F. t	Saturation Pressure (abs.) lbs./in.² p	Saturation Volume ft.³/lb. v	Saturation Density lbs./ft.³ $\frac{1}{v}$	Specific heat. Btu./lb. °F c	Heat content. Btu./lb. h	Latent heat. Btu./lb. L	Latent heat of pressure variation. Btu./lb. lb./in.² l	Variation of h with p (t constant). Btu./lb. lb./in.² $\left(\frac{\partial h}{\partial p}\right)_t$	Compressibility. per lb./in.² ×10⁵ $-\frac{1}{v}\left(\frac{\partial v}{\partial p}\right)_t$	Temp. °F. t
Triple point.	.....	0.01961*	51.00*	......	......	......	......	......	......	......
	0.88	.02182	45.83	......	......	......	......	......	......	—107.86
—100	1.24	0.02197	45.52	(1.040)	(—63.0)	(633)				—100
—95	1.52	.02207	45.32	(1.042)	(—57.8)	(631)				—95
—90	1.86	.02216	45.12	(1.043)	(—52.6)	(628)				—90
—85	2.27	.02226	44.92	(1.045)	(—47.4)	(625)				—85
—80	2.74	.02236	44.72	(1.046)	(—42.2)	(622)				—80
—75	3.29	0.02246	44.52	(1.048)	(—36.9)	(619)				—75
—70	3.94	.02256	44.32	(1.050)	(—31.7)	(616)				—70
—65	4.69	.02267	44.11	(1.052)	(—26.4)	(613)				—65
—60	5.55	.02278	43.91	1.054	—21.18	610.8	—0.0016	0.0026	4.4	—60
—55	6.54	.02288	43.70	1.056	—15.90	607.5	— .0026	.0016	4.5	—55
—50	7.67	0.02299	43.49	1.058	—10.61	604.3	—0.0017	0.0026	4.6	—50
—45	8.95	.02310	43.28	1.060	—5.31	600.9	— .0017	.0026	4.7	—45
—40	10.41	.02322	43.08	1.062	0.00	597.6	— .0018	.0025	4.8	—40
—35	12.05	.02333	42.86	1.064	5.32	594.2	— .0018	.0025	5.0	—35
—30	13.90	.02345	42.65	1.066	10.66	590.7	— .0019	.0025	5.1	—30
—25	15.98	0.02357	42.44	1.068	16.00	587.2	—0.0019	0.0024	5.2	—25
—20	18.30	.02369	42.22	1.070	21.36	583.6	— .0020	.0024	5.4	—20
—15	20.88	.02381	42.00	1.073	26.73	580.0	— .0020	.0024	5.5	—25
—10	23.74	.02393	41.78	1.075	32.11	576.4	— .0021	.0023	5.7	—10
— 5	26.92	.02406	41.56	1.078	37.51	572.6	— .0022	.0023	5.8	— 5
0	30.42	0.02419	41.34	1.080	42.92	568.9	—0.0022	0.0022	6.0	0
5	34.27	.02432	41.11	1.083	48.35	565.0	— .0023	.0022	6.2	5
10	38.51	.02446	40.89	1.085	53.79	561.1	— .0024	.0021	6.4	10
15	43.14	.02460	40.66	1.088	59.24	557.1	— .0025	.0021	6.6	15
20	48.21	.02474	40.43	1.091	64.71	553.1	— .0025	.0020	6.8	20
25	53.73	0.02488	40.20	1.094	70.20	548.9	—0.0026	0.0020	7.0	25
30	59.74	.02503	39.96	1.097	75.71	544.8	— .0027	.0019	7.3	30
35	66.26	.02518	39.72	1.100	81.23	540.5	— .0028	.0019	7.5	35
40	73.32	.02533	39.49	1.104	86.77	536.2	— .0029	.0018	7.8	40
45	80.96	.02548	39.24	1.108	92.34	531.8	— .0030	.0017	8.1	45
50	89.19	0.02564	39.00	1.112	97.93	527.3	—0.0031	0.0017	8.4	50
55	98.06	.02581	38.75	1.116	103.54	522.8	— .0032	.0016	8.8	55
60	107.6	.02597	38.50	1.120	109.18	518.1	— .0033	.0015	9.1	60
65	117.8	.02614	38.25	1.125	114.85	513.4	— .0034	.0014	9.5	65
70	128.8	.02632	38.00	1.129	120.54	508.6	— .0035	.0013	10.0	70
75	140.5	0.02650	37.74	1.133	126.25	503.7	—0.0037	0.0012	10.4	75

* Properties of solid ammonia at the triple point (—107.86 F.)

Table A-15 (continued)

Properties of Liquid Ammonia

Temp °F.	Saturation						Latent heat of pressure variation.	Variation of h with p (t constant).	Compressibility.	Temp °F.
	Pressure (abs.). lbs./in.²	Volume. ft.³/lb.	Density. lbs./ft.³	Specific heat. Btu./lb. °F	Heat content. Btu./lb.	Latent heat. Btu./lb.	Btu./lb. / lb./in.²	Btu./lb. / lb./in.²	per lb./in.² ×10⁵	
t	p	v	$\frac{1}{v}$	c	h	L	l	$\left(\frac{h}{p}\right)_t$	$\frac{1}{v}\left(\frac{v}{p}\right)_t$	t
75	140.5	0.02650	37.74	1.133	126.25	503.7	— .0037	0.0012	10.4	75
80	153.0	.02668	37.48	1.138	131.99	498.7	— .0038	.0011	10.9	80
85	166.4	.02687	37.21	1.142	137.75	493.6	— .0040	.0010	11.4	85
90	180.6	.02707	36.95	1.147	143.54	488.5	— .0041	.0009	12.0	90
95	195.8	.02727	36.67	1.151	149.36	483.2	— .0043	.0008	12.6	95
100	211.9	0.02747	36.40	1.156	155.21	477.8	—-0.0045	0.0006	13.3	100
105	228.9	.02769	36.12	1.162	161.09	472.3	— .0047	.0005	14.1	105
110	247.0	.02790	35.84	1.168	167.01	466.7	— .0049	.0003	14.9	110
115	266.2	.02813	35.55	1.176	172.97	460.9	— .0051	.0001	15.8	115
120	286.4	.02836	35.26	1.183	178.98	455.0	— .0053	.0000	16.7	120
125	307.8	0.02860	34.96	(1.189)	(185)	(449)	. . . . .	. . . . .	. . . . .	125
130	330.3	.02885	34.66	(1.197)	(191)	(443)	. . . . .	. . . . .	. . . . .	130
135	354.1	.02911	34.35	(1.205)	(197)	(436)	. . . . .	. . . . .	. . . . .	135
140	379.1	.02938	34.04	(1.213)	(203)	(430)	. . . . .	. . . . .	. . . . .	140
145	405.5	.02966	33.72	(1.222)	(210)	(423)	. . . . .	. . . . .	. . . . .	145
150	433.2	0.02995	33.39	(1.23)	(216)	(416)	. . . . .	. . . . .	. . . . .	150
155	462.3	.03025	33.06	(1.24)	(222)	(409)	. . . . .	. . . . .	. . . . .	155
160	492.8	.03056	32.72	(1.25)	(229)	(401)	. . . . .	. . . . .	. . . . .	160
165	524.8	.03089	32.37	(1.26)	(235)	(394)	. . . . .	. . . . .	. . . . .	165
170	558.4	.03124	32.01	(1.27)	(241)	(386)	. . . . .	. . . . .	. . . . .	170
175	593.5	0.03160	31.65	(1.29)	(248)	(377)	. . . . .	. . . . .	. . . . .	175
180	630.3	.03198	31.27	(1.30)	(255)	(369)	. . . . .	. . . . .	. . . . .	180
185	668.7	.03238	30.88	(1.32)	(262)	(360)	. . . . .	. . . . .	. . . . .	185
190	708.9	.03281	30.48	(1.34)	(269)	(351)	. . . . .	. . . . .	. . . . .	190
195	750.9	.03326	30.06	(1.36)	(276)	(342)	. . . . .	. . . . .	. . . . .	195
200	794.7	0.03375	29.63	(1.38)	(283)	(332)	. . . . .	. . . . .	. . . . .	200
210	888.1	.03482	28.72	(1.43)	(297)	(310)	. . . . .	. . . . .	. . . . .	210
220	989.5	.0361	27.7	(1.49)	(313)	(287)	. . . . .	. . . . .	. . . . .	220
230	1099.5	.0376	26.6	(1.57)	(329)	(260)	. . . . .	. . . . .	. . . . .	230
240	1218.5	.0395	25.3	(1.70)	(346)	(229)	. . . . .	. . . . .	. . . . .	240
250	1347	.0422	23.7	(1.90)	(365)	(192)	. . . . .	. . . . .	. . . . .	250
260	1486	.0463	21.6	(2.33)	(387)	(142)	. . . . .	. . . . .	. . . . .	260
270	1635	.0577	17.3	(5.30)	(419)	(52)	. . . . .	. . . . .	. . . . .	270
Critical.	1657	.0686	14.6		(433)	0				271.4

NOTE.—The figures in parentheses were calculated from empirical equations given in Bureau of Standards Scientific Papers Nos. 313 and 315 and represent values obtained by extrapolation beyond the range covered in the experimental work.

Table A-15 (continued)

Properties of Liquid Ammonia

Temp °F.	Saturation. Pressure (abs.). lbs./in.²	Saturation. Volume. ft.³/lb.	Saturation. Density. lbs./ft.³ $\frac{1}{v}$	Saturation. Specific heat. Btu./lb. °F	Saturation. Heat content. Btu./lb.	Saturation. Latent heat. Btu./lb.	Latent heat of pressure variation. Btu./lb. lb./in.²	Variation of h with p (t constant). Btu./lb. lb./in.² $\left(\frac{\partial h}{\partial p}\right)_t$	Compressibility. per lb./in.² ×10⁵ $-\frac{1}{v}\left(\frac{\partial v}{\partial p}\right)_t$	Temp. °F.
t	p	v		c	h	L	l			t
Triple point.	.....	0.01961*	51.00*	.....	.....	.....	.....	.....	.....	.....
	0.88	.02182	45.83	.....	.....	.....	.....	.....	.....	—107.86
—100	1.24	0.02197	45.52	(1.040)	(—63.0)	(633)				—100
—95	1.52	.02207	45.32	(1.042)	(—57.8)	(631)				—95
—90	1.86	.02216	45.12	(1.043)	(—52.6)	(628)				—90
—85	2.27	.02226	44.92	(1.045)	(—47.4)	(625)				—85
—80	2.74	.02236	44.72	(1.046)	(—42.2)	(622)				—80
—75	3.29	0.02246	44.52	(1.048)	(—36.9)	(619)				—75
—70	3.94	.02256	44.32	(1.050)	(—31.7)	(616)				—70
—65	4.69	.02267	44.11	(1.052)	(—26.4)	(613)				—65
—60	5.55	.02278	43.91	1.054	—21.18	610.8	—0.0016	0.0026	4.4	—60
—55	6.54	.02288	43.70	1.056	—15.90	607.5	— .0026	.0016	4.5	—55
—50	7.67	0.02299	43.49	1.058	—10.61	604.3	—0.0017	0.0026	4.6	—50
—45	8.95	.02310	43.28	1.060	—5.31	600.9	— .0017	.0026	4.7	—45
—40	10.41	.02322	43.08	1.062	0.00	597.6	— .0018	.0025	4.8	—40
—35	12.05	.02333	42.86	1.064	5.32	594.2	— .0018	.0025	5.0	—35
—30	13.90	.02345	42.65	1.066	10.66	590.7	— .0019	.0025	5.1	—30
—25	15.98	0.02357	42.44	1.068	16.00	587.2	—0.0019	0.0024	5.2	—25
—20	18.30	.02369	42.22	1.070	21.36	583.6	— .0020	.0024	5.4	—20
—15	20.88	.02381	42.00	1.073	26.73	580.0	— .0020	.0024	5.5	—25
—10	23.74	.02393	41.78	1.075	32.11	576.4	— .0021	.0023	5.7	—10
—5	26.92	.02406	41.56	1.078	37.51	572.6	— .0022	.0023	5.8	—5
0	30.42	0.02419	41.34	1.080	42.92	568.9	—0.0022	0.0022	6.0	0
5	34.27	.02432	41.11	1.083	48.35	565.0	— .0023	.0022	6.2	5
10	38.51	.02446	40.89	1.085	53.79	561.1	— .0024	.0021	6.4	10
15	43.14	.02460	40.66	1.088	59.24	557.1	— .0025	.0021	6.6	15
20	48.21	.02474	40.43	1.091	64.71	553.1	— .0025	.0020	6.8	20
25	53.73	0.02488	40.20	1.094	70.20	548.9	—0.0026	0.0020	7.0	25
30	59.74	.02503	39.96	1.097	75.71	544.8	— .0027	.0019	7.3	30
35	66.26	.02518	39.72	1.100	81.23	540.5	— .0028	.0019	7.5	35
40	73.32	.02533	39.49	1.104	86.77	536.2	— .0029	.0018	7.8	40
45	80.96	.02548	39.24	1.108	92.34	531.8	— .0030	.0017	8.1	45
50	89.19	0.02564	39.00	1.112	97.93	527.3	—0.0031	0.0017	8.4	50
55	98.06	.02581	38.75	1.116	103.54	522.8	— .0032	.0016	8.8	55
60	107.6	.02597	38.50	1.120	109.18	518.1	— .0033	.0015	9.1	60
65	117.8	.02614	38.25	1.125	114.85	513.4	— .0034	.0014	9.5	65
70	128.8	.02632	38.00	1.129	120.54	508.6	— .0035	.0013	10.0	70
75	140.5	0.02650	37.74	1.133	126.25	503.7	—0.0037	0.0012	10.4	75

* Properties of solid ammonia at the triple point (—107.86 F.)

Table A-15 (continued)

Properties of Liquid Ammonia

| Temp °F. | Saturation. | | | | | | Latent heat of pressure variation. | Variation of h with p (t constant). | Compressibility. | Temp. °F. |
| | Pressure (abs.). lbs./in.^{2} | Volume. ft.^{3}/lb. | Density. lbs./ft.^{3} | Specific heat. Btu./lb. °F | Heat content. Btu./lb. | Latent heat. Btu./lb. | Btu./lb. / lb./in.^{2} | Btu./lb. / lb./in.^{2} | per lb./in.^{2} ×10^5 | |
t	p	v	$\dfrac{1}{v}$	c	h	L	l	$\left(\dfrac{h}{p}\right)_t$	$\dfrac{1}{v}\left(\dfrac{v}{p}\right)_t$	t
75	140.5	0.02650	37.74	1.133	126.25	503.7	— .0037	0.0012	10.4	75
80	153.0	.02668	37.48	1.138	131.99	498.7	— .0038	.0011	10.9	80
85	166.4	.02687	37.21	1.142	137.75	493.6	— .0040	.0010	11.4	85
90	180.6	.02707	36.95	1.147	143.54	488.5	— .0041	.0009	12.0	90
95	195.8	.02727	36.67	1.151	149.36	483.2	— .0043	.0008	12.6	95
100	211.9	0.02747	36.40	1.156	155.21	477.8	—.0045	0.0006	13.3	100
105	228.9	.02769	36.12	1.162	161.09	472.3	— .0047	.0005	14.1	105
110	247.0	.02790	35.84	1.168	167.01	466.7	— .0049	.0003	14.9	110
115	266.2	.02813	35.55	1.176	172.97	460.9	— .0051	.0001	15.8	115
120	286.4	.02836	35.26	1.183	178.98	455.0	— .0053	.0000	16.7	120
125	307.8	0.02860	34.96	(1.189)	(185)	(449)				125
130	330.3	.02885	34.66	(1.197)	(191)	(443)				130
135	354.1	.02911	34.35	(1.205)	(197)	(436)				135
140	379.1	.02938	34.04	(1.213)	(203)	(430)				140
145	405.5	.02966	33.72	(1.222)	(210)	(423)				145
150	433.2	0.02995	33.39	(1.23)	(216)	(416)				150
155	462.3	.03025	33.06	(1.24)	(222)	(409)				155
160	492.8	.03056	32.72	(1.25)	(229)	(401)				160
165	524.8	.03089	32.37	(1.26)	(235)	(394)				165
170	558.4	.03124	32.01	(1.27)	(241)	(386)				170
175	593.5	0.03160	31.65	(1.29)	(248)	(377)				175
180	630.3	.03198	31.27	(1.30)	(255)	(369)				180
185	668.7	.03238	30.88	(1.32)	(262)	(360)				185
190	708.9	.03281	30.48	(1.34)	(269)	(351)				190
195	750.9	.03326	30.06	(1.36)	(276)	(342)				195
200	794.7	0.03375	29.63	(1.38)	(283)	(332)				200
210	888.1	.03482	28.72	(1.43)	(297)	(310)				210
220	989.5	.0361	27.7	(1.49)	(313)	(287)				220
230	1099.5	.0376	26.6	(1.57)	(329)	(260)				230
240	1218.5	.0395	25.3	(1.70)	(346)	(229)				240
250	1347	.0422	23.7	(1.90)	(365)	(192)				250
260	1486	.0463	21.6	(2.33)	(387)	(142)				260
270	1635	.0577	17.3	(5.30)	(419)	(52)				270
Critical.	1657	.0686	14.6		(433)	0				271.4

Note.—The figures in parentheses were calculated from empirical equations given in Bureau of Standards Scientific Papers Nos. 313 and 315 and represent values obtained by extrapolation beyond the range covered in the experimental work.

Table A–15

HCFC–123 Saturation Properties—Temperature Table

TEMP. °F	PRESSURE psia	VOLUME ft³/lb		DENSITY lb/ft³		ENTHALPY Btu/lb			ENTROPY Btu/(lb)(°R)		TEMP. °F
		LIQUID v_f	VAPOR v_g	LIQUID $1/v_f$	VAPOR $1/v_g$	LIQUID h_f	LATENT h_{fg}	VAPOR h_g	LIQUID s_f	VAPOR s_g	
−150	0.001	0.0092	——	108.8	0.000	−101.9	169.2	67.3	−0.3012	0.2451	−150
−149	0.001	0.0092	——	108.8	0.000	−98.8	166.2	67.4	−0.2912	0.2438	−149
−148	0.001	0.0092	——	108.7	0.000	−95.8	163.3	67.5	−0.2815	0.2424	−148
−147	0.001	0.0092	——	108.6	0.000	−92.8	160.5	67.6	−0.2722	0.2411	−147
−146	0.001	0.0092	——	108.5	0.000	−90.0	157.8	67.8	−0.2631	0.2399	−146
−145	0.001	0.0092	——	108.5	0.000	−87.3	155.2	67.9	−0.2544	0.2387	−145
−144	0.001	0.0092	10000.0000	108.4	0.000	−84.6	152.6	68.0	−0.2459	0.2375	−144
−143	0.001	0.0092	10000.0000	108.3	0.000	−82.0	150.2	68.1	−0.2378	0.2364	−143
−142	0.001	0.0092	10000.0000	108.2	0.000	−79.5	147.8	68.2	−0.2299	0.2353	−142
−141	0.002	0.0092	10000.0000	108.2	0.000	−77.1	145.5	68.4	−0.2223	0.2342	−141
−140	0.002	0.0093	10000.0000	108.1	0.000	−74.8	143.2	68.5	−0.2149	0.2332	−140
−139	0.002	0.0093	10000.0000	108.0	0.000	−72.5	141.1	68.6	−0.2078	0.2322	−139
−138	0.002	0.0093	10000.0000	107.9	0.000	−70.3	139.0	68.7	−0.2010	0.2312	−138
−137	0.002	0.0093	10000.0000	107.9	0.000	−68.1	137.0	68.8	−0.1943	0.2302	−137
−136	0.003	0.0093	10000.0000	107.8	0.000	−66.1	135.0	69.0	−0.1879	0.2293	−136
−135	0.003	0.0093	10000.0000	107.7	0.000	−64.1	133.2	69.1	−0.1817	0.2284	−135
−134	0.003	0.0093	10000.0000	107.6	0.000	−62.1	131.3	69.2	−0.1758	0.2275	−134
−133	0.004	0.0093	5000.0000	107.6	0.000	−60.2	129.6	69.3	−0.1700	0.2267	−133
−132	0.004	0.0093	5000.0000	107.5	0.000	−58.4	127.9	69.5	−0.1644	0.2258	−132
−131	0.004	0.0093	5000.0000	107.4	0.000	−56.6	126.2	69.6	−0.1590	0.2250	−131
−130	0.005	0.0093	5000.0000	107.3	0.000	−54.9	124.6	69.7	−0.1538	0.2242	−130
−129	0.005	0.0093	5000.0000	107.3	0.000	−53.3	123.1	69.8	−0.1487	0.2235	−129
−128	0.006	0.0093	5000.0000	107.2	0.000	−51.6	121.6	69.9	−0.1438	0.2227	−128
−127	0.006	0.0093	3333.3333	107.1	0.000	−50.1	120.1	70.1	−0.1391	0.2220	−127
−126	0.007	0.0093	3333.3333	107.0	0.000	−48.6	118.7	70.2	−0.1346	0.2213	−126
−125	0.007	0.0093	3333.3333	107.0	0.000	−47.1	117.4	70.3	−0.1302	0.2206	−125
−124	0.008	0.0094	3333.3333	106.9	0.000	−45.7	116.1	70.4	−0.1259	0.2199	−124
−123	0.009	0.0094	2500.0000	106.8	0.000	−44.3	114.8	70.6	−0.1218	0.2193	−123
−122	0.009	0.0094	2500.0000	106.7	0.000	−42.9	113.6	70.7	−0.1178	0.2186	−122
−121	0.010	0.0094	2500.0000	106.7	0.000	−41.6	112.4	70.8	−0.1140	0.2180	−121
−120	0.011	0.0094	2500.0000	106.6	0.000	−40.4	111.3	70.9	−0.1102	0.2174	−120
−119	0.012	0.0094	2000.0000	106.5	0.001	−39.1	110.2	71.0	−0.1066	0.2168	−119
−118	0.012	0.0094	2000.0000	106.5	0.001	−38.0	109.1	71.2	−0.1032	0.2162	−118
−117	0.013	0.0094	1666.6667	106.4	0.001	−36.8	108.1	71.3	−0.0998	0.2157	−117
−116	0.014	0.0094	1666.6667	106.3	0.001	−35.7	107.1	71.4	−0.0965	0.2151	−116
−115	0.015	0.0094	1666.6667	106.2	0.001	−34.6	106.1	71.5	−0.0934	0.2146	−115
−114	0.016	0.0094	1428.5714	106.2	0.001	−33.5	105.2	71.7	−0.0903	0.2140	−114
−113	0.017	0.0094	1428.5714	106.1	0.001	−32.5	104.3	71.8	−0.0874	0.2135	−113
−112	0.019	0.0094	1250.0000	106.0	0.001	−31.5	103.5	71.9	−0.0845	0.2130	−112
−111	0.020	0.0094	1250.0000	105.9	0.001	−30.6	102.6	72.0	−0.0818	0.2125	−111
−110	0.021	0.0094	1111.1111	105.9	0.001	−29.6	101.8	72.2	−0.0791	0.2121	−110
−109	0.023	0.0095	1111.1111	105.8	0.001	−28.7	101.0	72.3	−0.0765	0.2116	−109
−108	0.024	0.0095	1000.0000	105.7	0.001	−27.9	100.3	72.4	−0.0740	0.2111	−108
−107	0.026	0.0095	1000.0000	105.6	0.001	−27.0	99.5	72.5	−0.0716	0.2107	−107
−106	0.027	0.0095	909.0909	105.6	0.001	−26.2	98.8	72.7	−0.0692	0.2102	−106
−105	0.029	0.0095	833.3333	105.5	0.001	−25.4	98.2	72.8	−0.0670	0.2098	−105
−104	0.031	0.0095	833.3333	105.4	0.001	−24.6	97.5	72.9	−0.0648	0.2094	−104
−103	0.033	0.0095	769.2308	105.3	0.001	−23.8	96.9	73.0	−0.0626	0.2090	−103
−102	0.035	0.0095	714.2857	105.3	0.001	−23.1	96.3	73.2	−0.0606	0.2086	−102
−101	0.037	0.0095	666.6667	105.2	0.002	−22.4	95.7	73.3	−0.0586	0.2082	−101
−100	0.039	0.0095	666.6667	105.1	0.002	−21.7	95.1	73.4	−0.0567	0.2078	−100
−99	0.041	0.0095	625.0000	105.1	0.002	−21.0	94.6	73.6	−0.0548	0.2074	−99
−98	0.043	0.0095	588.2353	105.0	0.002	−20.4	94.0	73.7	−0.0530	0.2071	−98
−97	0.046	0.0095	555.5556	104.9	0.002	−19.7	93.5	73.8	−0.0512	0.2067	−97
−96	0.049	0.0095	526.3158	104.8	0.002	−19.1	93.0	73.9	−0.0495	0.2063	−96
−95	0.051	0.0095	500.0000	104.8	0.002	−18.5	92.6	74.1	−0.0478	0.2060	−95
−94	0.054	0.0096	476.1905	104.7	0.002	−17.9	92.1	74.2	−0.0462	0.2056	−94
−93	0.057	0.0096	454.5455	104.6	0.002	−17.3	91.7	74.3	−0.0447	0.2053	−93
−92	0.060	0.0096	434.7826	104.5	0.002	−16.8	91.2	74.5	−0.0431	0.2050	−92
−91	0.063	0.0096	416.6667	104.5	0.002	−16.2	90.8	74.6	−0.0417	0.2047	−91

Table A–15 (continued)
HCFC–123 Saturation Properties—Temperature Table

TEMP. °F	PRESSURE psia	VOLUME ft³/lb		DENSITY lb/ft³		ENTHALPY Btu/lb			ENTROPY Btu/(lb)(°R)		TEMP. °F
		LIQUID v_f	VAPOR v_g	LIQUID $1/v_f$	VAPOR $1/v_g$	LIQUID h_f	LATENT h_{fg}	VAPOR h_g	LIQUID s_f	VAPOR s_g	
−90	0.067	0.0096	384.6154	104.4	0.003	−15.7	90.4	74.7	−0.0403	0.2043	−90
−89	0.070	0.0096	370.3704	104.3	0.003	−15.2	90.0	74.8	−0.0389	0.2040	−89
−88	0.074	0.0096	357.1429	104.2	0.003	−14.7	89.7	75.0	−0.0375	0.2037	−88
−87	0.077	0.0096	333.3333	104.2	0.003	−14.2	89.3	75.1	−0.0362	0.2034	−87
−86	0.081	0.0096	322.5806	104.1	0.003	−13.7	89.0	75.2	−0.0350	0.2031	−86
−85	0.085	0.0096	312.5000	104.0	0.003	−13.3	88.6	75.4	−0.0337	0.2028	−85
−84	0.090	0.0096	294.1176	104.0	0.003	−12.8	88.3	75.5	−0.0325	0.2026	−84
−83	0.094	0.0096	277.7778	103.9	0.004	−12.4	88.0	75.6	−0.0314	0.2023	−83
−82	0.099	0.0096	270.2703	103.8	0.004	−12.0	87.7	75.7	−0.0302	0.2020	−82
−81	0.103	0.0096	256.4103	103.7	0.004	−11.5	87.4	75.9	−0.0291	0.2017	−81
−80	0.108	0.0096	243.9024	103.7	0.004	−11.1	87.1	76.0	−0.0280	0.2015	−80
−79	0.113	0.0097	232.5581	103.6	0.004	−10.7	86.9	76.1	−0.0270	0.2012	−79
−78	0.119	0.0097	227.2727	103.5	0.004	−10.4	86.6	76.3	−0.0260	0.2010	−78
−77	0.124	0.0097	217.3913	103.4	0.005	−10.0	86.4	76.4	−0.0250	0.2007	−77
−76	0.130	0.0097	208.3333	103.4	0.005	−9.6	86.1	76.5	−0.0240	0.2005	−76
−75	0.136	0.0097	200.0000	103.3	0.005	−9.2	85.9	76.7	−0.0231	0.2002	−75
−74	0.142	0.0097	188.6792	103.2	0.005	−8.9	85.7	76.8	−0.0222	0.2000	−74
−73	0.149	0.0097	181.8182	103.1	0.006	−8.5	85.5	76.9	−0.0213	0.1998	−73
−72	0.155	0.0097	175.4386	103.1	0.006	−8.2	85.3	77.1	−0.0204	0.1995	−72
−71	0.162	0.0097	166.6667	103.0	0.006	−7.9	85.1	77.2	−0.0195	0.1993	−71
−70	0.169	0.0097	161.2903	102.9	0.006	−7.6	84.9	77.3	−0.0187	0.1991	−70
−69	0.177	0.0097	153.8462	102.8	0.007	−7.2	84.7	77.4	−0.0179	0.1989	−69
−68	0.185	0.0097	149.2537	102.8	0.007	−6.9	84.5	77.6	−0.0171	0.1986	−68
−67	0.193	0.0097	142.8571	102.7	0.007	−6.6	84.3	77.7	−0.0163	0.1984	−67
−66	0.201	0.0097	136.9863	102.6	0.007	−6.3	84.2	77.8	−0.0156	0.1982	−66
−65	0.209	0.0098	131.5789	102.6	0.008	−6.0	84.0	78.0	−0.0148	0.1980	−65
−64	0.218	0.0098	126.5823	102.5	0.008	−5.7	83.8	78.1	−0.0141	0.1978	−64
−63	0.227	0.0098	121.9512	102.4	0.008	−5.5	83.7	78.2	−0.0134	0.1976	−63
−62	0.237	0.0098	117.6471	102.3	0.009	−5.2	83.6	78.4	−0.0127	0.1974	−62
−61	0.247	0.0098	113.6364	102.3	0.009	−4.9	83.4	78.5	−0.0120	0.1972	−61
−60	0.257	0.0098	108.6957	102.2	0.009	−4.6	83.3	78.6	−0.0113	0.1970	−60
−59	0.267	0.0098	105.2632	102.1	0.010	−4.4	83.2	78.8	−0.0107	0.1969	−59
−58	0.278	0.0098	101.0101	102.0	0.010	−4.1	83.0	78.9	−0.0100	0.1967	−58
−57	0.290	0.0098	97.0874	102.0	0.010	−3.9	82.9	79.0	−0.0094	0.1965	−57
−56	0.301	0.0098	93.4579	101.9	0.011	−3.6	82.8	79.2	−0.0088	0.1963	−56
−55	0.313	0.0098	90.0901	101.8	0.011	−3.4	82.7	79.3	−0.0082	0.1961	−55
−54	0.326	0.0098	86.9565	101.7	0.012	−3.1	82.6	79.4	−0.0076	0.1960	−54
−53	0.339	0.0098	84.0336	101.7	0.012	−2.9	82.5	79.6	−0.0070	0.1958	−53
−52	0.352	0.0098	81.3008	101.6	0.012	−2.6	82.4	79.7	−0.0064	0.1956	−52
−51	0.365	0.0099	78.1250	101.5	0.013	−2.4	82.3	79.8	−0.0058	0.1955	−51
−50	0.380	0.0099	75.1880	101.5	0.013	−2.2	82.2	80.0	−0.0052	0.1953	−50
−49	0.394	0.0099	72.9927	101.4	0.014	−1.9	82.1	80.1	−0.0047	0.1951	−49
−48	0.409	0.0099	70.4225	101.3	0.014	−1.7	82.0	80.3	−0.0041	0.1950	−48
−47	0.425	0.0099	68.0272	101.2	0.015	−1.5	81.9	80.4	−0.0036	0.1948	−47
−46	0.441	0.0099	65.3595	101.2	0.015	−1.3	81.8	80.5	−0.0031	0.1947	−46
−45	0.457	0.0099	63.2911	101.1	0.016	−1.1	81.7	80.7	−0.0025	0.1945	−45
−44	0.474	0.0099	61.3497	101.0	0.016	−0.8	81.6	80.8	−0.0020	0.1944	−44
−43	0.492	0.0099	59.1716	100.9	0.017	−0.6	81.6	80.9	−0.0015	0.1942	−43
−42	0.510	0.0099	57.1429	100.9	0.018	−0.4	81.5	81.1	−0.0010	0.1941	−42
−41	0.529	0.0099	55.2486	100.8	0.018	−0.2	81.4	81.2	−0.0005	0.1940	−41
−40	0.548	0.0099	53.4759	100.7	0.019	0.0	81.3	81.3	0.0000	0.1938	−40
−39	0.568	0.0099	51.8135	100.6	0.019	0.2	81.3	81.5	0.0005	0.1937	−39
−38	0.589	0.0099	50.0000	100.6	0.020	0.4	81.2	81.6	0.0010	0.1935	−38
−37	0.610	0.0100	48.3092	100.5	0.021	0.6	81.1	81.7	0.0015	0.1934	−37
−36	0.631	0.0100	46.7290	100.4	0.021	0.8	81.1	81.9	0.0019	0.1933	−36
−35	0.654	0.0100	45.2489	100.3	0.022	1.0	81.0	82.0	0.0024	0.1931	−35
−34	0.677	0.0100	43.8596	100.3	0.023	1.2	80.9	82.2	0.0029	0.1930	−34
−33	0.701	0.0100	42.5532	100.2	0.024	1.4	80.9	82.3	0.0033	0.1929	−33
−32	0.725	0.0100	41.1523	100.1	0.024	1.6	80.8	82.4	0.0038	0.1928	−32
−31	0.750	0.0100	39.8406	100.0	0.025	1.8	80.8	82.6	0.0043	0.1927	−31

Table A–15 (continued)
HCFC–123 Saturation Properties—Temperature Table

TEMP. °F	PRESSURE psia	VOLUME ft³/lb		DENSITY lb/ft³		ENTHALPY Btu/lb			ENTROPY Btu/(lb)(°R)		TEMP. °F
		LIQUID v_f	VAPOR v_g	LIQUID $1/v_f$	VAPOR $1/v_g$	LIQUID h_f	LATENT h_{fg}	VAPOR h_g	LIQUID s_f	VAPOR s_g	
−30	0.776	0.0100	38.6100	100.0	0.026	2.0	80.7	82.7	0.0047	0.1925	−30
−29	0.803	0.0100	37.4532	99.9	0.027	2.2	80.6	82.8	0.0052	0.1924	−29
−28	0.830	0.0100	36.2319	99.8	0.028	2.4	80.6	83.0	0.0056	0.1923	−28
−27	0.859	0.0100	35.0877	99.7	0.029	2.6	80.5	83.1	0.0061	0.1922	−27
−26	0.888	0.0100	34.0136	99.7	0.029	2.8	80.5	83.3	0.0065	0.1921	−26
−25	0.918	0.0100	33.0033	99.6	0.030	3.0	80.4	83.4	0.0070	0.1920	−25
−24	0.948	0.0100	32.0513	99.5	0.031	3.2	80.4	83.5	0.0074	0.1919	−24
−23	0.980	0.0101	31.0559	99.4	0.032	3.4	80.3	83.7	0.0078	0.1918	−23
−22	1.012	0.0101	30.1205	99.4	0.033	3.5	80.3	83.8	0.0083	0.1917	−22
−21	1.046	0.0101	29.2398	99.3	0.034	3.7	80.2	83.9	0.0087	0.1916	−21
−20	1.080	0.0101	28.3286	99.2	0.035	3.9	80.2	84.1	0.0091	0.1915	−20
−19	1.115	0.0101	27.4725	99.1	0.036	4.1	80.1	84.2	0.0096	0.1914	−19
−18	1.151	0.0101	26.6667	99.1	0.038	4.3	80.1	84.4	0.0100	0.1913	−18
−17	1.188	0.0101	25.9067	99.0	0.039	4.5	80.0	84.5	0.0104	0.1912	−17
−16	1.227	0.0101	25.1889	98.9	0.040	4.7	80.0	84.6	0.0108	0.1911	−16
−15	1.266	0.0101	24.4499	98.8	0.041	4.9	79.9	84.8	0.0113	0.1910	−15
−14	1.306	0.0101	23.7530	98.8	0.042	5.1	79.9	84.9	0.0117	0.1909	−14
−13	1.347	0.0101	23.0415	98.7	0.043	5.2	79.8	85.1	0.0121	0.1908	−13
−12	1.390	0.0101	22.3714	98.6	0.045	5.4	79.8	85.2	0.0125	0.1907	−12
−11	1.433	0.0102	21.7391	98.5	0.046	5.6	79.7	85.3	0.0130	0.1906	−11
−10	1.478	0.0102	21.1416	98.4	0.047	5.8	79.7	85.5	0.0134	0.1905	−10
−9	1.524	0.0102	20.5339	98.4	0.049	6.0	79.6	85.6	0.0138	0.1905	−9
−8	1.571	0.0102	19.9601	98.3	0.050	6.2	79.6	85.8	0.0142	0.1904	−8
−7	1.619	0.0102	19.4175	98.2	0.052	6.4	79.5	85.9	0.0146	0.1903	−7
−6	1.669	0.0102	18.8679	98.1	0.053	6.6	79.5	86.0	0.0151	0.1902	−6
−5	1.720	0.0102	18.3486	98.1	0.055	6.8	79.4	86.2	0.0155	0.1901	−5
−4	1.772	0.0102	17.8571	98.0	0.056	6.9	79.4	86.3	0.0159	0.1901	−4
−3	1.825	0.0102	17.3611	97.9	0.058	7.1	79.3	86.5	0.0163	0.1900	−3
−2	1.880	0.0102	16.8919	97.8	0.059	7.3	79.3	86.6	0.0167	0.1899	−2
−1	1.936	0.0102	16.4474	97.8	0.061	7.5	79.2	86.7	0.0171	0.1898	−1
0	1.993	0.0102	16.0000	97.7	0.063	7.7	79.2	86.9	0.0176	0.1898	0
1	2.052	0.0102	15.5521	97.6	0.064	7.9	79.1	87.0	0.0180	0.1897	1
2	2.113	0.0103	15.1515	97.5	0.066	8.1	79.1	87.2	0.0184	0.1896	2
3	2.174	0.0103	14.7493	97.4	0.068	8.3	79.0	87.3	0.0188	0.1896	3
4	2.238	0.0103	14.3472	97.4	0.070	8.5	79.0	87.4	0.0192	0.1895	4
5	2.303	0.0103	13.9860	97.3	0.072	8.7	78.9	87.6	0.0196	0.1894	5
6	2.369	0.0103	13.6054	97.2	0.074	8.9	78.9	87.7	0.0200	0.1894	6
7	2.437	0.0103	13.2626	97.1	0.075	9.1	78.8	87.9	0.0205	0.1893	7
8	2.507	0.0103	12.9199	97.0	0.077	9.3	78.7	88.0	0.0209	0.1893	8
9	2.578	0.0103	12.5786	97.0	0.080	9.4	78.7	88.1	0.0213	0.1892	9
10	2.651	0.0103	12.2549	96.9	0.082	9.6	78.6	88.3	0.0217	0.1892	10
11	2.726	0.0103	11.9474	96.8	0.084	9.8	78.6	88.4	0.0221	0.1891	11
12	2.802	0.0103	11.6414	96.7	0.086	10.0	78.5	88.6	0.0225	0.1890	12
13	2.880	0.0103	11.3507	96.7	0.088	10.2	78.5	88.7	0.0230	0.1890	13
14	2.960	0.0104	11.0619	96.6	0.090	10.4	78.4	88.9	0.0234	0.1889	14
15	3.042	0.0104	10.7759	96.5	0.093	10.6	78.4	89.0	0.0238	0.1889	15
16	3.126	0.0104	10.5152	96.4	0.095	10.8	78.3	89.1	0.0242	0.1888	16
17	3.212	0.0104	10.2459	96.3	0.098	11.0	78.3	89.3	0.0246	0.1888	17
18	3.299	0.0104	10.0000	96.3	0.100	11.2	78.2	89.4	0.0251	0.1888	18
19	3.389	0.0104	9.7466	96.2	0.103	11.4	78.1	89.6	0.0255	0.1887	19
20	3.480	0.0104	9.5147	96.1	0.105	11.6	78.1	89.7	0.0259	0.1887	20
21	3.574	0.0104	9.2764	96.0	0.108	11.8	78.0	89.8	0.0263	0.1886	21
22	3.669	0.0104	9.0498	95.9	0.111	12.0	78.0	90.0	0.0267	0.1886	22
23	3.767	0.0104	8.8339	95.9	0.113	12.2	77.9	90.1	0.0272	0.1885	23
24	3.867	0.0104	8.6207	95.8	0.116	12.4	77.8	90.3	0.0276	0.1885	24
25	3.969	0.0105	8.4104	95.7	0.119	12.6	77.8	90.4	0.0280	0.1885	25
26	4.073	0.0105	8.2102	95.6	0.122	12.9	77.7	90.6	0.0284	0.1884	26
27	4.180	0.0105	8.0192	95.5	0.125	13.1	77.6	90.7	0.0288	0.1884	27
28	4.288	0.0105	7.8309	95.5	0.128	13.3	77.6	90.8	0.0293	0.1884	28
29	4.400	0.0105	7.6453	95.4	0.131	13.5	77.5	91.0	0.0297	0.1883	29

Table A-15 (continued)

HCFC-123 Saturation Properties—Temperature Table

TEMP. °F	PRESSURE psia	VOLUME ft³/lb LIQUID v_f	VOLUME ft³/lb VAPOR v_g	DENSITY lb/ft³ LIQUID $1/v_f$	DENSITY lb/ft³ VAPOR $1/v_g$	ENTHALPY Btu/lb LIQUID h_f	ENTHALPY Btu/lb LATENT h_{fg}	ENTHALPY Btu/lb VAPOR h_g	ENTROPY Btu/(lb)(°R) LIQUID s_f	ENTROPY Btu/(lb)(°R) VAPOR s_g	TEMP. °F
30	4.513	0.0105	7.4627	95.3	0.134	13.7	77.5	91.1	0.0301	0.1883	30
31	4.629	0.0105	7.2886	95.2	0.137	13.9	77.4	91.3	0.0305	0.1883	31
32	4.747	0.0105	7.1225	95.1	0.140	14.1	77.3	91.4	0.0310	0.1882	32
33	4.868	0.0105	6.9541	95.0	0.144	14.3	77.3	91.6	0.0314	0.1882	33
34	4.991	0.0105	6.7935	95.0	0.147	14.5	77.2	91.7	0.0318	0.1882	34
35	5.117	0.0105	6.6401	94.9	0.151	14.7	77.1	91.9	0.0322	0.1882	35
36	5.245	0.0105	6.4893	94.8	0.154	14.9	77.1	92.0	0.0327	0.1881	36
37	5.376	0.0106	6.3412	94.7	0.158	15.1	77.0	92.1	0.0331	0.1881	37
38	5.509	0.0106	6.1958	94.6	0.161	15.4	76.9	92.3	0.0335	0.1881	38
39	5.646	0.0106	6.0569	94.6	0.165	15.6	76.9	92.4	0.0339	0.1881	39
40	5.785	0.0106	5.9207	94.5	0.169	15.8	76.8	92.6	0.0344	0.1881	40
41	5.927	0.0106	5.7904	94.4	0.173	16.0	76.7	92.7	0.0348	0.1880	41
42	6.071	0.0106	5.6593	94.3	0.177	16.2	76.6	92.9	0.0352	0.1880	42
43	6.219	0.0106	5.5340	94.2	0.181	16.4	76.6	93.0	0.0357	0.1880	43
44	6.369	0.0106	5.4142	94.1	0.185	16.7	76.5	93.2	0.0361	0.1880	44
45	6.522	0.0106	5.2938	94.1	0.189	16.9	76.4	93.3	0.0365	0.1880	45
46	6.679	0.0106	5.1787	94.0	0.193	17.1	76.4	93.4	0.0370	0.1880	46
47	6.838	0.0107	5.0659	93.9	0.197	17.3	76.3	93.6	0.0374	0.1879	47
48	7.000	0.0107	4.9554	93.8	0.202	17.5	76.2	93.7	0.0378	0.1879	48
49	7.166	0.0107	4.8497	93.7	0.206	17.7	76.1	93.9	0.0383	0.1879	49
50	7.334	0.0107	4.7461	93.6	0.211	18.0	76.1	94.0	0.0387	0.1879	50
51	7.506	0.0107	4.6425	93.6	0.215	18.2	76.0	94.2	0.0391	0.1879	51
52	7.681	0.0107	4.5455	93.5	0.220	18.4	75.9	94.3	0.0396	0.1879	52
53	7.860	0.0107	4.4484	93.4	0.225	18.6	75.8	94.5	0.0400	0.1879	53
54	8.041	0.0107	4.3535	93.3	0.230	18.9	75.7	94.6	0.0404	0.1879	54
55	8.226	0.0107	4.2626	93.2	0.235	19.1	75.7	94.7	0.0409	0.1879	55
56	8.415	0.0107	4.1736	93.1	0.240	19.3	75.6	94.9	0.0413	0.1879	56
57	8.607	0.0107	4.0866	93.1	0.245	19.5	75.5	95.0	0.0417	0.1879	57
58	8.802	0.0108	4.0016	93.0	0.250	19.8	75.4	95.2	0.0422	0.1879	58
59	9.001	0.0108	3.9185	92.9	0.255	20.0	75.3	95.3	0.0426	0.1879	59
60	9.203	0.0108	3.8388	92.8	0.261	20.2	75.3	95.5	0.0430	0.1879	60
61	9.410	0.0108	3.7594	92.7	0.266	20.4	75.2	95.6	0.0435	0.1879	61
62	9.619	0.0108	3.6832	92.6	0.272	20.7	75.1	95.8	0.0439	0.1879	62
63	9.833	0.0108	3.6075	92.6	0.277	20.9	75.0	95.9	0.0443	0.1879	63
64	10.050	0.0108	3.5348	92.5	0.283	21.1	74.9	96.1	0.0448	0.1879	64
65	10.272	0.0108	3.4638	92.4	0.289	21.4	74.8	96.2	0.0452	0.1879	65
66	10.497	0.0108	3.3944	92.3	0.295	21.6	74.8	96.3	0.0457	0.1879	66
67	10.726	0.0108	3.3267	92.2	0.301	21.8	74.7	96.5	0.0461	0.1879	67
68	10.958	0.0109	3.2605	92.1	0.307	22.0	74.6	96.6	0.0465	0.1879	68
69	11.195	0.0109	3.1959	92.0	0.313	22.3	74.5	96.8	0.0470	0.1879	69
70	11.436	0.0109	3.1328	92.0	0.319	22.5	74.4	96.9	0.0474	0.1879	70
71	11.682	0.0109	3.0713	91.9	0.326	22.7	74.3	97.1	0.0479	0.1879	71
72	11.931	0.0109	3.0111	91.8	0.332	23.0	74.2	97.2	0.0483	0.1879	72
73	12.184	0.0109	2.9525	91.7	0.339	23.2	74.2	97.4	0.0487	0.1879	73
74	12.442	0.0109	2.8952	91.6	0.345	23.4	74.1	97.5	0.0492	0.1880	74
75	12.704	0.0109	2.8393	91.5	0.352	23.7	74.0	97.7	0.0496	0.1880	75
76	12.970	0.0109	2.7840	91.4	0.359	23.9	73.9	97.8	0.0501	0.1880	76
77	13.241	0.0109	2.7307	91.3	0.366	24.2	73.8	98.0	0.0505	0.1880	77
78	13.517	0.0110	2.6788	91.3	0.373	24.4	73.7	98.1	0.0509	0.1880	78
79	13.796	0.0110	2.6274	91.2	0.381	24.6	73.6	98.2	0.0514	0.1880	79
80	14.081	0.0110	2.5780	91.1	0.388	24.9	73.5	98.4	0.0518	0.1880	80
81	14.369	0.0110	2.5291	91.0	0.395	25.1	73.4	98.5	0.0523	0.1881	81
82	14.663	0.0110	2.4814	90.9	0.403	25.3	73.3	98.7	0.0527	0.1881	82
83	14.961	0.0110	2.4355	90.8	0.411	25.6	73.2	98.8	0.0531	0.1881	83
84	15.264	0.0110	2.3901	90.7	0.418	25.8	73.1	99.0	0.0536	0.1881	84
85	15.572	0.0110	2.3452	90.6	0.426	26.1	73.1	99.1	0.0540	0.1881	85
86	15.885	0.0110	2.3020	90.6	0.434	26.3	73.0	99.3	0.0545	0.1882	86
87	16.203	0.0111	2.2594	90.5	0.443	26.6	72.9	99.4	0.0549	0.1882	87
88	16.525	0.0111	2.2183	90.4	0.451	26.8	72.8	99.6	0.0553	0.1882	88
89	16.853	0.0111	2.1777	90.3	0.459	27.0	72.7	99.7	0.0558	0.1882	89

Table A–15 (continued)

HCFC–123 Saturation Properties—Temperature Table

TEMP. °F	PRESSURE psia	VOLUME ft³/lb		DENSITY lb/ft³		ENTHALPY Btu/lb			ENTROPY Btu/(lb)(°R)		TEMP. °F
		LIQUID v_f	VAPOR v_g	LIQUID $1/v_f$	VAPOR $1/v_g$	LIQUID h_f	LATENT h_{fg}	VAPOR h_g	LIQUID s_f	VAPOR s_g	
90	17.186	0.0111	2.1381	90.2	0.468	27.3	72.6	99.9	0.0562	0.1883	90
91	17.523	0.0111	2.0991	90.1	0.476	27.5	72.5	100.0	0.0567	0.1883	91
92	17.866	0.0111	2.0610	90.0	0.485	27.8	72.4	100.1	0.0571	0.1883	92
93	18.215	0.0111	2.0239	89.9	0.494	28.0	72.3	100.3	0.0576	0.1883	93
94	18.568	0.0111	1.9877	89.8	0.503	28.3	72.2	100.4	0.0580	0.1884	94
95	18.927	0.0111	1.9524	89.8	0.512	28.5	72.1	100.6	0.0584	0.1884	95
96	19.291	0.0112	1.9175	89.7	0.522	28.7	72.0	100.7	0.0589	0.1884	96
97	19.661	0.0112	1.8836	89.6	0.531	29.0	71.9	100.9	0.0593	0.1884	97
98	20.036	0.0112	1.8501	89.5	0.541	29.2	71.8	101.0	0.0598	0.1885	98
99	20.417	0.0112	1.8179	89.4	0.550	29.5	71.7	101.2	0.0602	0.1885	99
100	20.803	0.0112	1.7857	89.3	0.560	29.7	71.6	101.3	0.0606	0.1885	100
101	21.195	0.0112	1.7547	89.2	0.570	30.0	71.5	101.5	0.0611	0.1886	101
102	21.593	0.0112	1.7241	89.1	0.580	30.2	71.4	101.6	0.0615	0.1886	102
103	21.997	0.0112	1.6943	89.0	0.590	30.5	71.3	101.8	0.0620	0.1886	103
104	22.406	0.0112	1.6650	88.9	0.601	30.7	71.2	101.9	0.0624	0.1887	104
105	22.821	0.0113	1.6364	88.8	0.611	31.0	71.1	102.0	0.0628	0.1887	105
106	23.242	0.0113	1.6082	88.8	0.622	31.2	71.0	102.2	0.0633	0.1887	106
107	23.670	0.0113	1.5808	88.7	0.633	31.5	70.9	102.3	0.0637	0.1888	107
108	24.103	0.0113	1.5540	88.6	0.644	31.7	70.8	102.5	0.0642	0.1888	108
109	24.542	0.0113	1.5277	88.5	0.655	32.0	70.6	102.6	0.0646	0.1888	109
110	24.988	0.0113	1.5017	88.4	0.666	32.2	70.5	102.8	0.0650	0.1889	110
111	25.440	0.0113	1.4765	88.3	0.677	32.5	70.4	102.9	0.0655	0.1889	111
112	25.898	0.0113	1.4518	88.2	0.689	32.7	70.3	103.1	0.0659	0.1890	112
113	26.362	0.0113	1.4276	88.1	0.701	33.0	70.2	103.2	0.0664	0.1890	113
114	26.833	0.0114	1.4037	88.0	0.712	33.2	70.1	103.4	0.0668	0.1890	114
115	27.310	0.0114	1.3805	87.9	0.724	33.5	70.0	103.5	0.0672	0.1891	115
116	27.794	0.0114	1.3576	87.8	0.737	33.7	69.9	103.6	0.0677	0.1891	116
117	28.284	0.0114	1.3353	87.7	0.749	34.0	69.8	103.8	0.0681	0.1891	117
118	28.781	0.0114	1.3134	87.6	0.761	34.3	69.7	103.9	0.0686	0.1892	118
119	29.285	0.0114	1.2918	87.6	0.774	34.5	69.6	104.1	0.0690	0.1892	119
120	29.796	0.0114	1.2708	87.5	0.787	34.8	69.5	104.2	0.0694	0.1893	120
121	30.313	0.0114	1.2502	87.4	0.800	35.0	69.4	104.4	0.0699	0.1893	121
122	30.837	0.0115	1.2300	87.3	0.813	35.3	69.2	104.5	0.0703	0.1894	122
123	31.368	0.0115	1.2102	87.2	0.826	35.5	69.1	104.7	0.0707	0.1894	123
124	31.906	0.0115	1.1908	87.1	0.840	35.8	69.0	104.8	0.0712	0.1894	124
125	32.451	0.0115	1.1716	87.0	0.854	36.0	68.9	105.0	0.0716	0.1895	125
126	33.004	0.0115	1.1529	86.9	0.867	36.3	68.8	105.1	0.0721	0.1895	126
127	33.563	0.0115	1.1346	86.8	0.881	36.6	68.7	105.2	0.0725	0.1896	127
128	34.130	0.0115	1.1166	86.7	0.896	36.8	68.6	105.4	0.0729	0.1896	128
129	34.703	0.0115	1.0989	86.6	0.910	37.1	68.5	105.5	0.0734	0.1897	129
130	35.285	0.0116	1.0817	86.5	0.925	37.3	68.3	105.7	0.0738	0.1897	130
131	35.873	0.0116	1.0646	86.4	0.939	37.6	68.2	105.8	0.0742	0.1898	131
132	36.470	0.0116	1.0480	86.3	0.954	37.8	68.1	106.0	0.0747	0.1898	132
133	37.073	0.0116	1.0317	86.2	0.969	38.1	68.0	106.1	0.0751	0.1899	133
134	37.684	0.0116	1.0156	86.1	0.985	38.4	67.9	106.3	0.0755	0.1899	134
135	38.303	0.0116	0.9999	86.0	1.000	38.6	67.8	106.4	0.0760	0.1899	135
136	38.930	0.0116	0.9844	85.9	1.016	38.9	67.7	106.5	0.0764	0.1900	136
137	39.565	0.0117	0.9693	85.8	1.032	39.1	67.5	106.7	0.0768	0.1900	137
138	40.207	0.0117	0.9545	85.7	1.048	39.4	67.4	106.8	0.0773	0.1901	138
139	40.857	0.0117	0.9398	85.6	1.064	39.7	67.3	107.0	0.0777	0.1901	139
140	41.515	0.0117	0.9255	85.5	1.081	39.9	67.2	107.1	0.0781	0.1902	140
141	42.181	0.0117	0.9115	85.4	1.097	40.2	67.1	107.3	0.0786	0.1902	141
142	42.856	0.0117	0.8977	85.3	1.114	40.4	67.0	107.4	0.0790	0.1903	142
143	43.538	0.0117	0.8841	85.2	1.131	40.7	66.8	107.5	0.0794	0.1903	143
144	44.229	0.0117	0.8708	85.2	1.148	41.0	66.7	107.7	0.0799	0.1904	144
145	44.928	0.0118	0.8578	85.1	1.166	41.2	66.6	107.8	0.0803	0.1904	145
146	45.635	0.0118	0.8449	85.0	1.184	41.5	66.5	108.0	0.0807	0.1905	146
147	46.351	0.0118	0.8323	84.9	1.202	41.7	66.4	108.1	0.0812	0.1905	147
148	47.075	0.0118	0.8199	84.8	1.220	42.0	66.2	108.3	0.0816	0.1906	148
149	47.808	0.0118	0.8078	84.7	1.238	42.3	66.1	108.4	0.0820	0.1906	149

Table A-15 (continued)

HCFC-123 Saturation Properties—Temperature Table

TEMP. °F	PRESSURE psia	VOLUME ft³/lb		DENSITY lb/ft³		ENTHALPY Btu/lb			ENTROPY Btu/(lb)(°R)		TEMP. °F
		LIQUID v_f	VAPOR v_g	LIQUID $1/v_f$	VAPOR $1/v_g$	LIQUID h_f	LATENT h_{fg}	VAPOR h_g	LIQUID s_f	VAPOR s_g	
150	48.549	0.0118	0.7959	84.6	1.257	42.5	66.0	108.5	0.0824	0.1907	150
151	49.299	0.0118	0.7841	84.5	1.275	42.8	65.9	108.7	0.0829	0.1907	151
152	50.058	0.0119	0.7726	84.4	1.294	43.1	65.8	108.8	0.0833	0.1908	152
153	50.825	0.0119	0.7613	84.3	1.314	43.3	65.6	109.0	0.0837	0.1909	153
154	51.602	0.0119	0.7502	84.2	1.333	43.6	65.5	109.1	0.0842	0.1909	154
155	52.387	0.0119	0.7393	84.1	1.353	43.9	65.4	109.2	0.0846	0.1910	155
156	53.181	0.0119	0.7285	84.0	1.373	44.1	65.3	109.4	0.0850	0.1910	156
157	53.985	0.0119	0.7180	83.8	1.393	44.4	65.1	109.5	0.0854	0.1911	157
158	54.797	0.0119	0.7077	83.7	1.413	44.6	65.0	109.7	0.0859	0.1911	158
159	55.619	0.0120	0.6974	83.6	1.434	44.9	64.9	109.8	0.0863	0.1912	159
160	56.450	0.0120	0.6875	83.5	1.455	45.2	64.8	109.9	0.0867	0.1912	160
161	57.290	0.0120	0.6776	83.4	1.476	45.4	64.6	110.1	0.0871	0.1913	161
162	58.140	0.0120	0.6680	83.3	1.497	45.7	64.5	110.2	0.0876	0.1913	162
163	58.999	0.0120	0.6585	83.2	1.519	46.0	64.4	110.4	0.0880	0.1914	163
164	59.868	0.0120	0.6491	83.1	1.541	46.2	64.3	110.5	0.0884	0.1914	164
165	60.746	0.0120	0.6399	83.0	1.563	46.5	64.1	110.6	0.0888	0.1915	165
166	61.634	0.0121	0.6309	82.9	1.585	46.8	64.0	110.8	0.0892	0.1916	166
167	62.532	0.0121	0.6220	82.8	1.608	47.0	63.9	110.9	0.0897	0.1916	167
168	63.439	0.0121	0.6133	82.7	1.631	47.3	63.8	111.1	0.0901	0.1917	168
169	64.357	0.0121	0.6047	82.6	1.654	47.6	63.6	111.2	0.0905	0.1917	169
170	65.284	0.0121	0.5963	82.5	1.677	47.8	63.5	111.3	0.0909	0.1918	170
171	66.221	0.0121	0.5880	82.4	1.701	48.1	63.4	111.5	0.0914	0.1918	171
172	67.169	0.0122	0.5798	82.3	1.725	48.4	63.2	111.6	0.0918	0.1919	172
173	68.126	0.0122	0.5718	82.2	1.749	48.6	63.1	111.7	0.0922	0.1919	173
174	69.094	0.0122	0.5639	82.1	1.773	48.9	63.0	111.9	0.0926	0.1920	174
175	70.072	0.0122	0.5561	82.0	1.798	49.2	62.8	112.0	0.0930	0.1921	175
176	71.060	0.0122	0.5485	81.9	1.823	49.4	62.7	112.2	0.0934	0.1921	176
177	72.059	0.0122	0.5410	81.8	1.849	49.7	62.6	112.3	0.0939	0.1922	177
178	73.068	0.0122	0.5336	81.7	1.874	50.0	62.5	112.4	0.0943	0.1922	178
179	74.088	0.0123	0.5263	81.6	1.900	50.2	62.3	112.6	0.0947	0.1923	179
180	75.118	0.0123	0.5192	81.4	1.926	50.5	62.2	112.7	0.0951	0.1923	180
181	76.159	0.0123	0.5121	81.3	1.953	50.8	62.1	112.8	0.0955	0.1924	181
182	77.211	0.0123	0.5052	81.2	1.980	51.0	61.9	113.0	0.0959	0.1924	182
183	78.274	0.0123	0.4984	81.1	2.007	51.3	61.8	113.1	0.0964	0.1925	183
184	79.348	0.0123	0.4916	81.0	2.034	51.6	61.6	113.2	0.0968	0.1926	184
185	80.432	0.0124	0.4850	80.9	2.062	51.9	61.5	113.4	0.0972	0.1926	185
186	81.528	0.0124	0.4785	80.8	2.090	52.1	61.4	113.5	0.0976	0.1927	186
187	82.635	0.0124	0.4721	80.7	2.118	52.4	61.2	113.6	0.0980	0.1927	187
188	83.753	0.0124	0.4659	80.6	2.147	52.7	61.1	113.8	0.0984	0.1928	188
189	84.882	0.0124	0.4596	80.5	2.176	52.9	61.0	113.9	0.0988	0.1928	189
190	86.023	0.0124	0.4536	80.4	2.205	53.2	60.8	114.0	0.0993	0.1929	190
191	87.175	0.0125	0.4475	80.2	2.234	53.5	60.7	114.2	0.0997	0.1929	191
192	88.339	0.0125	0.4416	80.1	2.264	53.7	60.6	114.3	0.1001	0.1930	192
193	89.514	0.0125	0.4358	80.0	2.295	54.0	60.4	114.4	0.1005	0.1931	193
194	90.700	0.0125	0.4301	79.9	2.325	54.3	60.3	114.6	0.1009	0.1931	194
195	91.899	0.0125	0.4244	79.8	2.356	54.6	60.1	114.7	0.1013	0.1932	195
196	93.109	0.0126	0.4189	79.7	2.387	54.8	60.0	114.8	0.1017	0.1932	196
197	94.331	0.0126	0.4134	79.6	2.419	55.1	59.9	115.0	0.1021	0.1933	197
198	95.565	0.0126	0.4080	79.5	2.451	55.4	59.7	115.1	0.1025	0.1933	198
199	96.811	0.0126	0.4027	79.3	2.483	55.6	59.6	115.2	0.1029	0.1934	199
200	98.069	0.0126	0.3975	79.2	2.516	55.9	59.4	115.3	0.1034	0.1934	200
201	99.340	0.0126	0.3923	79.1	2.549	56.2	59.3	115.5	0.1038	0.1935	201
202	100.622	0.0127	0.3872	79.0	2.583	56.5	59.1	115.6	0.1042	0.1935	202
203	101.917	0.0127	0.3822	78.9	2.616	56.7	59.0	115.7	0.1046	0.1936	203
204	103.224	0.0127	0.3773	78.8	2.650	57.0	58.8	115.9	0.1050	0.1937	204
205	104.544	0.0127	0.3725	78.7	2.685	57.3	58.7	116.0	0.1054	0.1937	205
206	105.876	0.0127	0.3677	78.5	2.720	57.6	58.6	116.1	0.1058	0.1938	206
207	107.221	0.0128	0.3630	78.4	2.755	57.8	58.4	116.2	0.1062	0.1938	207
208	108.578	0.0128	0.3583	78.3	2.791	58.1	58.3	116.4	0.1066	0.1939	208
209	109.948	0.0128	0.3538	78.2	2.827	58.4	58.1	116.5	0.1070	0.1939	209

Table A-15 (continued)
HCFC-123 Saturation Properties—Temperature Table

TEMP. °F	PRESSURE psia	VOLUME ft³/lb		DENSITY lb/ft³		ENTHALPY Btu/lb			ENTROPY Btu/(lb)(°R)		TEMP. °F
		LIQUID v_f	VAPOR v_g	LIQUID $1/v_f$	VAPOR $1/v_g$	LIQUID h_f	LATENT h_{fg}	VAPOR h_g	LIQUID s_f	VAPOR s_g	
210	111.331	0.0128	0.3492	78.1	2.863	58.6	58.0	116.6	0.1074	0.1940	210
211	112.727	0.0128	0.3448	78.0	2.900	58.9	57.8	116.7	0.1078	0.1940	211
212	114.136	0.0128	0.3404	77.8	2.937	59.2	57.7	116.9	0.1082	0.1941	212
213	115.558	0.0129	0.3361	77.7	2.975	59.5	57.5	117.0	0.1086	0.1941	213
214	116.993	0.0129	0.3319	77.6	3.013	59.7	57.4	117.1	0.1090	0.1942	214
215	118.442	0.0129	0.3277	77.5	3.052	60.0	57.2	117.2	0.1094	0.1942	215
216	119.903	0.0129	0.3236	77.4	3.091	60.3	57.1	117.4	0.1098	0.1943	216
217	121.378	0.0129	0.3195	77.2	3.130	60.6	56.9	117.5	0.1102	0.1943	217
218	122.867	0.0130	0.3155	77.1	3.170	60.8	56.8	117.6	0.1106	0.1944	218
219	124.369	0.0130	0.3115	77.0	3.210	61.1	56.6	117.7	0.1110	0.1944	219
220	125.885	0.0130	0.3076	76.9	3.251	61.4	56.5	117.8	0.1114	0.1945	220
221	127.414	0.0130	0.3038	76.8	3.292	61.7	56.3	118.0	0.1118	0.1945	221
222	128.957	0.0130	0.3000	76.6	3.333	61.9	56.1	118.1	0.1122	0.1946	222
223	130.514	0.0131	0.2963	76.5	3.376	62.2	56.0	118.2	0.1126	0.1946	223
224	132.085	0.0131	0.2926	76.4	3.418	62.5	55.8	118.3	0.1130	0.1947	224
225	133.670	0.0131	0.2889	76.3	3.461	62.8	55.7	118.4	0.1134	0.1947	225
226	135.269	0.0131	0.2854	76.1	3.504	63.0	55.5	118.6	0.1138	0.1948	226
227	136.882	0.0132	0.2818	76.0	3.548	63.3	55.4	118.7	0.1142	0.1948	227
228	138.510	0.0132	0.2783	75.9	3.593	63.6	55.2	118.8	0.1146	0.1949	228
229	140.152	0.0132	0.2749	75.8	3.638	63.9	55.0	118.9	0.1150	0.1949	229
230	141.808	0.0132	0.2715	75.6	3.683	64.2	54.9	119.0	0.1154	0.1950	230
231	143.479	0.0132	0.2682	75.5	3.729	64.4	54.7	119.1	0.1158	0.1950	231
232	145.164	0.0133	0.2649	75.4	3.775	64.7	54.5	119.3	0.1162	0.1951	232
233	146.864	0.0133	0.2616	75.3	3.822	65.0	54.4	119.4	0.1166	0.1951	233
234	148.579	0.0133	0.2584	75.1	3.870	65.3	54.2	119.5	0.1170	0.1952	234
235	150.308	0.0133	0.2552	75.0	3.918	65.5	54.1	119.6	0.1174	0.1952	235
236	152.053	0.0134	0.2521	74.9	3.966	65.8	53.9	119.7	0.1178	0.1953	236
237	153.812	0.0134	0.2490	74.7	4.015	66.1	53.7	119.8	0.1182	0.1953	237
238	155.587	0.0134	0.2460	74.6	4.065	66.4	53.6	119.9	0.1186	0.1953	238
239	157.376	0.0134	0.2430	74.5	4.115	66.7	53.4	120.1	0.1190	0.1954	239
240	159.180	0.0134	0.2400	74.4	4.166	66.9	53.2	120.2	0.1194	0.1954	240
241	161.000	0.0135	0.2371	74.2	4.217	67.2	53.0	120.3	0.1198	0.1955	241
242	162.836	0.0135	0.2342	74.1	4.269	67.5	52.9	120.4	0.1202	0.1955	242
243	164.688	0.0135	0.2314	74.0	4.322	67.8	52.7	120.5	0.1205	0.1956	243
244	166.555	0.0135	0.2286	73.8	4.375	68.1	52.5	120.6	0.1209	0.1956	244
245	168.438	0.0136	0.2258	73.7	4.428	68.3	52.4	120.7	0.1213	0.1956	245
246	170.336	0.0136	0.2231	73.6	4.483	68.6	52.2	120.8	0.1217	0.1957	246
247	172.250	0.0136	0.2204	73.4	4.538	68.9	52.0	120.9	0.1221	0.1957	247
248	174.181	0.0136	0.2177	73.3	4.593	69.2	51.8	121.0	0.1225	0.1958	248
249	176.127	0.0137	0.2151	73.2	4.649	69.5	51.7	121.1	0.1229	0.1958	249
250	178.090	0.0137	0.2125	73.0	4.706	69.8	51.5	121.2	0.1233	0.1958	250
251	180.068	0.0137	0.2099	72.9	4.764	70.0	51.3	121.3	0.1237	0.1959	251
252	182.063	0.0137	0.2074	72.7	4.822	70.3	51.1	121.4	0.1241	0.1959	252
253	184.075	0.0138	0.2049	72.6	4.881	70.6	50.9	121.5	0.1245	0.1959	253
254	186.102	0.0138	0.2024	72.5	4.940	70.9	50.8	121.6	0.1249	0.1960	254
255	188.147	0.0138	0.2000	72.3	5.000	71.2	50.6	121.8	0.1253	0.1960	255
256	190.208	0.0139	0.1976	72.2	5.061	71.5	50.4	121.9	0.1256	0.1960	256
257	192.286	0.0139	0.1952	72.0	5.123	71.8	50.2	122.0	0.1260	0.1961	257
258	194.381	0.0139	0.1929	71.9	5.185	72.0	50.0	122.0	0.1264	0.1961	258
259	196.492	0.0139	0.1905	71.8	5.248	72.3	49.8	122.1	0.1268	0.1961	259
260	198.621	0.0140	0.1883	71.6	5.312	72.6	49.6	122.2	0.1272	0.1962	260
261	200.767	0.0140	0.1860	71.5	5.377	72.9	49.4	122.3	0.1276	0.1962	261
262	202.930	0.0140	0.1838	71.3	5.442	73.2	49.3	122.4	0.1280	0.1962	262
263	205.110	0.0141	0.1816	71.2	5.508	73.5	49.1	122.5	0.1284	0.1963	263
264	207.308	0.0141	0.1794	71.0	5.575	73.8	48.9	122.6	0.1288	0.1963	264
265	209.523	0.0141	0.1772	70.9	5.643	74.0	48.7	122.7	0.1292	0.1963	265
266	211.755	0.0141	0.1751	70.7	5.711	74.3	48.5	122.8	0.1295	0.1964	266
267	214.006	0.0142	0.1730	70.6	5.781	74.6	48.3	122.9	0.1299	0.1964	267
268	216.274	0.0142	0.1709	70.4	5.851	74.9	48.1	123.0	0.1303	0.1964	268
269	218.560	0.0142	0.1689	70.3	5.922	75.2	47.9	123.1	0.1307	0.1964	269

Table A–15 (continued)

HCFC–123 Saturation Properties—Temperature Table

TEMP. °F	PRESSURE psia	VOLUME ft³/lb		DENSITY lb/ft³		ENTHALPY Btu/lb			ENTROPY Btu/(lb)(°R)		TEMP. °F
		LIQUID v_f	VAPOR v_g	LIQUID $1/v_f$	VAPOR $1/v_g$	LIQUID h_f	LATENT h_{fg}	VAPOR h_g	LIQUID s_f	VAPOR s_g	
270	220.865	0.0143	0.1668	70.1	5.994	75.5	47.7	123.2	0.1311	0.1965	270
271	223.187	0.0143	0.1648	70.0	6.067	75.8	47.5	123.3	0.1315	0.1965	271
272	225.527	0.0143	0.1628	69.8	6.141	76.1	47.3	123.4	0.1319	0.1965	272
273	227.886	0.0144	0.1609	69.7	6.216	76.4	47.1	123.4	0.1323	0.1965	273
274	230.263	0.0144	0.1590	69.5	6.291	76.7	46.9	123.5	0.1327	0.1966	274
275	232.658	0.0144	0.1570	69.4	6.368	77.0	46.7	123.6	0.1331	0.1966	275
276	235.072	0.0145	0.1551	69.2	6.445	77.3	46.4	123.7	0.1335	0.1966	276
277	237.505	0.0145	0.1533	69.0	6.524	77.5	46.2	123.8	0.1338	0.1966	277
278	239.957	0.0145	0.1514	68.9	6.604	77.8	46.0	123.9	0.1342	0.1966	278
279	242.427	0.0146	0.1496	68.7	6.684	78.1	45.8	123.9	0.1346	0.1966	279
280	244.917	0.0146	0.1478	68.6	6.766	78.4	45.6	124.0	0.1350	0.1967	280
281	247.425	0.0146	0.1460	68.4	6.849	78.7	45.4	124.1	0.1354	0.1967	281
282	249.953	0.0147	0.1442	68.2	6.933	79.0	45.2	124.2	0.1358	0.1967	282
283	252.500	0.0147	0.1425	68.1	7.018	79.3	44.9	124.3	0.1362	0.1967	283
284	255.067	0.0147	0.1408	67.9	7.104	79.6	44.7	124.3	0.1366	0.1967	284
285	257.653	0.0148	0.1391	67.7	7.191	79.9	44.5	124.4	0.1370	0.1967	285
286	260.259	0.0148	0.1374	67.6	7.280	80.2	44.3	124.5	0.1374	0.1967	286
287	262.884	0.0148	0.1357	67.4	7.370	80.5	44.0	124.6	0.1378	0.1967	287
288	265.530	0.0149	0.1340	67.2	7.461	80.8	43.8	124.6	0.1382	0.1968	288
289	268.195	0.0149	0.1324	67.1	7.553	81.1	43.6	124.7	0.1386	0.1968	289
290	270.881	0.0150	0.1308	66.9	7.647	81.4	43.3	124.8	0.1389	0.1968	290
291	273.586	0.0150	0.1292	66.7	7.742	81.7	43.1	124.8	0.1393	0.1968	291
292	276.312	0.0150	0.1276	66.5	7.838	82.0	42.9	124.9	0.1397	0.1968	292
293	279.059	0.0151	0.1260	66.4	7.936	82.3	42.6	125.0	0.1401	0.1968	293
294	281.826	0.0151	0.1245	66.2	8.035	82.6	42.4	125.0	0.1405	0.1968	294
295	284.614	0.0152	0.1229	66.0	8.135	83.0	42.1	125.1	0.1409	0.1968	295
296	287.423	0.0152	0.1214	65.8	8.237	83.3	41.9	125.2	0.1413	0.1968	296
297	290.252	0.0152	0.1199	65.6	8.341	83.6	41.6	125.2	0.1417	0.1968	297
298	293.103	0.0153	0.1184	65.4	8.446	83.9	41.4	125.3	0.1421	0.1968	298
299	295.975	0.0153	0.1169	65.3	8.553	84.2	41.1	125.3	0.1425	0.1967	299
300	298.868	0.0154	0.1155	65.1	8.662	84.5	40.9	125.4	0.1429	0.1967	300
301	301.783	0.0154	0.1140	64.9	8.772	84.8	40.6	125.4	0.1433	0.1967	301
302	304.719	0.0155	0.1126	64.7	8.884	85.1	40.4	125.5	0.1437	0.1967	302
303	307.677	0.0155	0.1111	64.5	8.998	85.4	40.1	125.5	0.1441	0.1967	303
304	310.658	0.0156	0.1097	64.3	9.113	85.8	39.8	125.6	0.1445	0.1967	304
305	313.660	0.0156	0.1083	64.1	9.231	86.1	39.6	125.6	0.1449	0.1967	305
306	316.684	0.0157	0.1070	63.9	9.350	86.4	39.3	125.7	0.1453	0.1966	306
307	319.730	0.0157	0.1056	63.7	9.472	86.7	39.0	125.7	0.1457	0.1966	307
308	322.800	0.0158	0.1042	63.5	9.595	87.0	38.7	125.8	0.1461	0.1966	308
309	325.891	0.0158	0.1029	63.3	9.721	87.4	38.5	125.8	0.1465	0.1966	309
310	329.006	0.0159	0.1015	63.1	9.849	87.7	38.2	125.8	0.1469	0.1965	310
311	332.143	0.0159	0.1002	62.9	9.979	88.0	37.9	125.9	0.1474	0.1965	311
312	335.304	0.0160	0.0989	62.6	10.112	88.3	37.6	125.9	0.1478	0.1965	312
313	338.488	0.0160	0.0976	62.4	10.247	88.6	37.3	125.9	0.1482	0.1964	313
314	341.695	0.0161	0.0963	62.2	10.385	89.0	37.0	126.0	0.1486	0.1964	314
315	344.926	0.0161	0.0950	62.0	10.525	89.3	36.7	126.0	0.1490	0.1964	315
316	348.181	0.0162	0.0937	61.8	10.668	89.6	36.4	126.0	0.1494	0.1963	316
317	351.460	0.0163	0.0925	61.5	10.814	90.0	36.1	126.0	0.1498	0.1963	317
318	354.763	0.0163	0.0912	61.3	10.963	90.3	35.8	126.1	0.1502	0.1962	318
319	358.090	0.0164	0.0900	61.1	11.115	90.6	35.4	126.1	0.1507	0.1962	319
320	361.442	0.0164	0.0887	60.8	11.270	91.0	35.1	126.1	0.1511	0.1961	320
321	364.818	0.0165	0.0875	60.6	11.428	91.3	34.8	126.1	0.1515	0.1961	321
322	368.220	0.0166	0.0863	60.3	11.590	91.7	34.4	126.1	0.1519	0.1960	322
323	371.647	0.0166	0.0851	60.1	11.755	92.0	34.1	126.1	0.1523	0.1959	323
324	375.099	0.0167	0.0839	59.8	11.924	92.3	33.8	126.1	0.1528	0.1959	324
325	378.577	0.0168	0.0827	59.6	12.097	92.7	33.4	126.1	0.1532	0.1958	325
326	382.081	0.0169	0.0815	59.3	12.274	93.0	33.1	126.1	0.1536	0.1957	326
327	385.611	0.0169	0.0803	59.1	12.456	93.4	32.7	126.1	0.1541	0.1956	327
328	389.167	0.0170	0.0791	58.8	12.642	93.7	32.3	126.1	0.1545	0.1955	328
329	392.751	0.0171	0.0779	58.5	12.832	94.1	32.0	126.0	0.1549	0.1955	329

Table A-15 (continued)
HCFC-123 Saturation Properties—Temperature Table

TEMP. °F	PRESSURE psia	VOLUME ft³/lb		DENSITY lb/ft³		ENTHALPY Btu/lb			ENTROPY Btu/(lb)(°R)		TEMP. °F
		LIQUID v_f	VAPOR v_g	LIQUID $1/v_f$	VAPOR $1/v_g$	LIQUID h_f	LATENT h_{fg}	VAPOR h_g	LIQUID s_f	VAPOR s_g	
330	396.361	0.0172	0.0768	58.2	13.028	94.5	31.6	126.0	0.1554	0.1954	330
331	399.998	0.0173	0.0756	57.9	13.229	94.8	31.2	126.0	0.1558	0.1953	331
332	403.663	0.0173	0.0744	57.7	13.435	95.2	30.8	126.0	0.1563	0.1952	332
333	407.355	0.0174	0.0733	57.4	13.648	95.5	30.4	125.9	0.1567	0.1950	333
334	411.076	0.0175	0.0721	57.1	13.866	95.9	30.0	125.9	0.1572	0.1949	334
335	414.825	0.0176	0.0710	56.7	14.091	96.3	29.5	125.8	0.1576	0.1948	335
336	418.604	0.0177	0.0698	56.4	14.323	96.7	29.1	125.8	0.1581	0.1947	336
337	422.411	0.0178	0.0687	56.1	14.563	97.0	28.7	125.7	0.1585	0.1945	337
338	426.248	0.0179	0.0675	55.8	14.811	97.4	28.2	125.6	0.1590	0.1944	338
339	430.115	0.0180	0.0664	55.4	15.067	97.8	27.7	125.6	0.1595	0.1942	339
340	434.013	0.0182	0.0652	55.1	15.332	98.2	27.3	125.5	0.1599	0.1940	340
341	437.942	0.0183	0.0641	54.7	15.607	98.6	26.8	125.4	0.1604	0.1939	341
342	441.902	0.0184	0.0629	54.4	15.893	99.0	26.3	125.3	0.1609	0.1937	342
343	445.884	0.0185	0.0618	54.0	16.191	99.4	25.8	125.2	0.1614	0.1935	343
344	449.911	0.0187	0.0606	53.6	16.501	99.8	25.2	125.0	0.1619	0.1933	344
345	453.970	0.0188	0.0594	53.2	16.825	100.2	24.7	124.9	0.1624	0.1930	345
346	458.063	0.0190	0.0583	52.8	17.164	100.7	24.1	124.8	0.1629	0.1928	346
347	462.190	0.0191	0.0571	52.3	17.520	101.1	23.5	124.6	0.1634	0.1926	347
348	466.352	0.0193	0.0559	51.9	17.894	101.5	22.9	124.4	0.1640	0.1923	348
349	470.550	0.0195	0.0547	51.4	18.290	102.0	22.2	124.2	0.1645	0.1920	349
350	474.785	0.0197	0.0535	50.9	18.709	102.4	21.6	124.0	0.1650	0.1917	350
351	479.057	0.0199	0.0522	50.4	19.154	102.9	20.8	123.8	0.1656	0.1913	351
352	483.368	0.0201	0.0509	49.8	19.630	103.4	20.1	123.5	0.1662	0.1909	352
353	487.719	0.0203	0.0496	49.2	20.142	103.9	19.3	123.2	0.1668	0.1905	353
354	492.111	0.0206	0.0483	48.6	20.695	104.4	18.4	122.9	0.1674	0.1901	354
355	496.545	0.0209	0.0470	47.9	21.299	105.0	17.5	122.5	0.1680	0.1896	355
356	501.024	0.0212	0.0455	47.2	21.964	105.5	16.6	122.1	0.1687	0.1890	356
357	505.549	0.0215	0.0440	46.4	22.706	106.1	15.5	121.6	0.1694	0.1884	357
358	510.122	0.0220	0.0425	45.5	23.549	106.8	14.3	121.1	0.1702	0.1877	358
359	514.745	0.0225	0.0408	44.5	24.532	107.5	12.9	120.4	0.1711	0.1868	359
360	519.422	0.0231	0.0389	43.3	25.722	108.3	11.3	119.6	0.1720	0.1858	360
361	524.157	0.0240	0.0367	41.7	27.267	109.3	9.2	118.5	0.1733	0.1844	361
362	528.954	0.0255	0.0337	39.2	29.644	110.8	6.1	116.9	0.1750	0.1824	362

© 2004. E.I. du Pont de Nemours and Company (DuPont). All rights reserved, used under license of DuPont.

Table A–15 (continued)

HFC-134a Saturation Properties—Temperature Table

TEMP. °F	PRESSURE psia	VOLUME ft³/lb LIQUID v_f	VOLUME ft³/lb VAPOR v_g	DENSITY lb/ft³ LIQUID $1/v_f$	DENSITY lb/ft³ VAPOR $1/v_g$	ENTHALPY Btu/lb LIQUID h_f	ENTHALPY Btu/lb LATENT h_{fg}	ENTHALPY Btu/lb VAPOR h_g	ENTROPY Btu/(lb)(°R) LIQUID s_f	ENTROPY Btu/(lb)(°R) VAPOR s_g	TEMP. °F
−150	0.073	0.0101	454.5455	98.86	0.0022	−31.2	112.1	80.9	−0.0859	0.2761	−150
−149	0.077	0.0101	416.6667	98.76	0.0024	−30.9	112.0	81.1	−0.0850	0.2754	−149
−148	0.082	0.0101	400.0000	98.67	0.0025	−30.6	111.8	81.2	−0.0841	0.2747	−148
−147	0.087	0.0101	384.6154	98.57	0.0026	−30.3	111.7	81.4	−0.0832	0.2740	−147
−146	0.092	0.0102	357.1429	98.48	0.0028	−30.1	111.6	81.5	−0.0823	0.2733	−146
−145	0.098	0.0102	333.3333	98.38	0.0030	−29.8	111.4	81.6	−0.0815	0.2726	−145
−144	0.104	0.0102	322.5806	98.29	0.0031	−29.5	111.3	81.8	−0.0806	0.2720	−144
−143	0.110	0.0102	303.0303	98.19	0.0033	−29.2	111.1	81.9	−0.0797	0.2713	−143
−142	0.116	0.0102	285.7143	98.10	0.0035	−28.9	111.0	82.1	−0.0788	0.2706	−142
−141	0.123	0.0102	270.2703	98.00	0.0037	−28.7	110.9	82.2	−0.0780	0.2700	−141
−140	0.130	0.0102	256.4103	97.91	0.0039	−28.4	110.7	82.3	−0.0771	0.2693	−140
−139	0.137	0.0102	243.9024	97.81	0.0041	−28.1	110.6	82.5	−0.0763	0.2687	−139
−138	0.145	0.0102	232.5581	97.72	0.0043	−27.8	110.5	82.6	−0.0754	0.2681	−138
−137	0.153	0.0102	222.2222	97.62	0.0045	−27.6	110.3	82.8	−0.0745	0.2674	−137
−136	0.162	0.0103	208.3333	97.53	0.0048	−27.3	110.2	82.9	−0.0737	0.2668	−136
−135	0.171	0.0103	200.0000	97.43	0.0050	−27.0	110.1	83.1	−0.0728	0.2662	−135
−134	0.180	0.0103	188.6792	97.34	0.0053	−26.7	109.9	83.2	−0.0720	0.2656	−134
−133	0.190	0.0103	181.8182	97.24	0.0055	−26.5	109.8	83.3	−0.0712	0.2650	−133
−132	0.200	0.0103	172.4138	97.15	0.0058	−26.2	109.7	83.5	−0.0703	0.2644	−132
−131	0.211	0.0103	163.9344	97.05	0.0061	−25.9	109.6	83.6	−0.0695	0.2639	−131
−130	0.222	0.0103	156.2500	96.96	0.0064	−25.6	109.4	83.8	−0.0686	0.2633	−130
−129	0.234	0.0103	149.2537	96.86	0.0067	−25.4	109.3	83.9	−0.0678	0.2627	−129
−128	0.246	0.0103	140.8451	96.77	0.0071	−25.1	109.2	84.1	−0.0670	0.2621	−128
−127	0.259	0.0103	135.1351	96.67	0.0074	−24.8	109.0	84.2	−0.0661	0.2616	−127
−126	0.273	0.0104	128.2051	96.58	0.0078	−24.5	108.9	84.4	−0.0653	0.2610	−126
−125	0.287	0.0104	121.9512	96.48	0.0082	−24.3	108.8	84.5	−0.0645	0.2605	−125
−124	0.301	0.0104	116.2791	96.39	0.0086	−24.0	108.6	84.6	−0.0637	0.2600	−124
−123	0.317	0.0104	111.1111	96.29	0.0090	−23.7	108.5	84.8	−0.0628	0.2594	−123
−122	0.333	0.0104	106.3830	96.20	0.0094	−23.4	108.4	84.9	−0.0620	0.2589	−122
−121	0.349	0.0104	102.0408	96.10	0.0098	−23.2	108.2	85.1	−0.0612	0.2584	−121
−120	0.366	0.0104	97.0874	96.01	0.0103	−22.9	108.1	85.2	−0.0604	0.2579	−120
−119	0.385	0.0104	92.5926	95.91	0.0108	−22.6	108.0	85.4	−0.0596	0.2574	−119
−118	0.403	0.0104	88.4956	95.82	0.0113	−22.3	107.8	85.5	−0.0588	0.2569	−118
−117	0.423	0.0104	84.7458	95.72	0.0118	−22.1	107.7	85.7	−0.0580	0.2564	−117
−116	0.443	0.0105	81.3008	95.63	0.0123	−21.8	107.6	85.8	−0.0571	0.2559	−116
−115	0.465	0.0105	77.5194	95.53	0.0129	−21.5	107.4	86.0	−0.0563	0.2554	−115
−114	0.487	0.0105	74.6269	95.44	0.0134	−21.2	107.3	86.1	−0.0555	0.2549	−114
−113	0.510	0.0105	71.4286	95.34	0.0140	−20.9	107.2	86.2	−0.0547	0.2545	−113
−112	0.534	0.0105	68.0272	95.25	0.0147	−20.7	107.1	86.4	−0.0539	0.2540	−112
−111	0.558	0.0105	65.3595	95.15	0.0153	−20.4	106.9	86.5	−0.0531	0.2535	−111
−110	0.584	0.0105	62.5000	95.06	0.0160	−20.1	106.8	86.7	−0.0523	0.2531	−110
−109	0.611	0.0105	60.2410	94.96	0.0166	−19.8	106.7	86.8	−0.0515	0.2526	−109
−108	0.639	0.0105	57.4713	94.87	0.0174	−19.6	106.5	87.0	−0.0507	0.2522	−108
−107	0.668	0.0106	55.2486	94.77	0.0181	−19.3	106.4	87.1	−0.0500	0.2517	−107
−106	0.698	0.0106	52.9101	94.68	0.0189	−19.0	106.3	87.3	−0.0492	0.2513	−106
−105	0.729	0.0106	51.0204	94.58	0.0196	−18.7	106.1	87.4	−0.0484	0.2509	−105
−104	0.761	0.0106	48.7805	94.49	0.0205	−18.4	106.0	87.6	−0.0476	0.2505	−104
−103	0.795	0.0106	46.9484	94.39	0.0213	−18.2	105.9	87.7	−0.0468	0.2500	−103
−102	0.830	0.0106	45.0450	94.30	0.0222	−17.9	105.7	87.9	−0.0460	0.2496	−102
−101	0.866	0.0106	43.2900	94.20	0.0231	−17.6	105.6	88.0	−0.0452	0.2492	−101
−100	0.903	0.0106	41.6667	94.11	0.0240	−17.3	105.5	88.2	−0.0444	0.2488	−100
−99	0.942	0.0106	40.0000	94.01	0.0250	−17.0	105.3	88.3	−0.0437	0.2484	−99
−98	0.982	0.0106	38.4615	93.92	0.0260	−16.7	105.2	88.5	−0.0429	0.2480	−98
−97	1.024	0.0107	37.0370	93.82	0.0270	−16.5	105.1	88.6	−0.0421	0.2476	−97
−96	1.067	0.0107	35.5872	93.73	0.0281	−16.2	104.9	88.8	−0.0413	0.2472	−96
−95	1.111	0.0107	34.2466	93.63	0.0292	−15.9	104.8	88.9	−0.0406	0.2468	−95
−94	1.157	0.0107	33.0033	93.54	0.0303	−15.6	104.7	89.1	−0.0398	0.2465	−94
−93	1.205	0.0107	31.7460	93.44	0.0315	−15.3	104.5	89.2	−0.0390	0.2461	−93
−92	1.254	0.0107	30.5810	93.34	0.0327	−15.1	104.4	89.4	−0.0383	0.2457	−92
−91	1.305	0.0107	29.4985	93.25	0.0339	−14.8	104.3	89.5	−0.0375	0.2454	−91

Table A-15 (continued)
HFC-134a Saturation Properties—Temperature Table

TEMP. °F	PRESSURE psia	VOLUME ft³/lb		DENSITY lb/ft³		ENTHALPY Btu/lb			ENTROPY Btu/(lb)(°R)		TEMP. °F
		LIQUID v_f	VAPOR v_g	LIQUID $1/v_f$	VAPOR $1/v_g$	LIQUID h_f	LATENT h_{fg}	VAPOR h_g	LIQUID s_f	VAPOR s_g	
−90	1.358	0.0107	28.4091	93.15	0.0352	−14.5	104.1	89.6	−0.0367	0.2450	−90
−89	1.413	0.0107	27.3973	93.06	0.0365	−14.2	104.0	89.8	−0.0360	0.2446	−89
−88	1.469	0.0108	26.3852	92.96	0.0379	−13.9	103.9	89.9	−0.0352	0.2443	−88
−87	1.527	0.0108	25.4453	92.87	0.0393	−13.6	103.7	90.1	−0.0344	0.2439	−87
−86	1.587	0.0108	24.5098	92.77	0.0408	−13.4	103.6	90.2	−0.0337	0.2436	−86
−85	1.649	0.0108	23.6407	92.68	0.0423	−13.1	103.5	90.4	−0.0329	0.2433	−85
−84	1.713	0.0108	22.8311	92.58	0.0438	−12.8	103.3	90.5	−0.0322	0.2429	−84
−83	1.779	0.0108	22.0264	92.49	0.0454	−12.5	103.2	90.7	−0.0314	0.2426	−83
−82	1.848	0.0108	21.2766	92.39	0.0470	−12.2	103.1	90.8	−0.0306	0.2423	−82
−81	1.918	0.0108	20.5339	92.29	0.0487	−11.9	102.9	91.0	−0.0299	0.2419	−81
−80	1.991	0.0108	19.8413	92.20	0.0504	−11.6	102.8	91.1	−0.0291	0.2416	−80
−79	2.066	0.0109	19.1571	92.10	0.0522	−11.4	102.7	91.3	−0.0284	0.2413	−79
−78	2.143	0.0109	18.5185	92.01	0.0540	−11.1	102.5	91.4	−0.0276	0.2410	−78
−77	2.222	0.0109	17.8891	91.91	0.0559	−10.8	102.4	91.6	−0.0269	0.2407	−77
−76	2.304	0.0109	17.3010	91.82	0.0578	−10.5	102.3	91.7	−0.0261	0.2404	−76
−75	2.389	0.0109	16.7224	91.72	0.0598	−10.2	102.1	91.9	−0.0254	0.2401	−75
−74	2.476	0.0109	16.1812	91.62	0.0618	−9.9	102.0	92.0	−0.0246	0.2398	−74
−73	2.565	0.0109	15.6495	91.53	0.0639	−9.6	101.8	92.2	−0.0239	0.2395	−73
−72	2.658	0.0109	15.1286	91.43	0.0661	−9.4	101.7	92.4	−0.0232	0.2392	−72
−71	2.753	0.0109	14.6413	91.33	0.0683	−9.1	101.6	92.5	−0.0224	0.2389	−71
−70	2.850	0.0110	14.1844	91.24	0.0705	−8.8	101.4	92.7	−0.0217	0.2386	−70
−69	2.951	0.0110	13.7174	91.14	0.0729	−8.5	101.3	92.8	−0.0209	0.2383	−69
−68	3.055	0.0110	13.2802	91.05	0.0753	−8.2	101.2	93.0	−0.0202	0.2381	−68
−67	3.161	0.0110	12.8700	90.95	0.0777	−7.9	101.0	93.1	−0.0195	0.2378	−67
−66	3.271	0.0110	12.4688	90.85	0.0802	−7.6	100.9	93.3	−0.0187	0.2375	−66
−65	3.384	0.0110	12.0773	90.76	0.0828	−7.3	100.7	93.4	−0.0180	0.2373	−65
−64	3.499	0.0110	11.6959	90.66	0.0855	−7.0	100.6	93.6	−0.0173	0.2370	−64
−63	3.619	0.0110	11.3379	90.56	0.0882	−6.8	100.5	93.7	−0.0165	0.2367	−63
−62	3.741	0.0111	10.9890	90.47	0.0910	−6.5	100.3	93.9	−0.0158	0.2365	−62
−61	3.867	0.0111	10.6610	90.37	0.0938	−6.2	100.2	94.0	−0.0151	0.2362	−61
−60	3.996	0.0111	10.3306	90.27	0.0968	−5.9	100.0	94.2	−0.0143	0.2360	−60
−59	4.129	0.0111	10.0200	90.17	0.0998	−5.6	99.9	94.3	−0.0136	0.2357	−59
−58	4.265	0.0111	9.7182	90.08	0.1029	−5.3	99.8	94.5	−0.0129	0.2355	−58
−57	4.405	0.0111	9.4340	89.98	0.1060	−5.0	99.6	94.6	−0.0122	0.2352	−57
−56	4.549	0.0111	9.1491	89.88	0.1093	−4.7	99.5	94.8	−0.0114	0.2350	−56
−55	4.696	0.0111	8.8810	89.78	0.1126	−4.4	99.3	94.9	−0.0107	0.2348	−55
−54	4.848	0.0111	8.6207	89.69	0.1160	−4.1	99.2	95.1	−0.0100	0.2345	−54
−53	5.003	0.0112	8.3752	89.59	0.1194	−3.8	99.1	95.2	−0.0093	0.2343	−53
−52	5.162	0.0112	8.1301	89.49	0.1230	−3.5	98.9	95.4	−0.0085	0.2341	−52
−51	5.326	0.0112	7.8989	89.39	0.1266	−3.2	98.8	95.5	−0.0078	0.2339	−51
−50	5.493	0.0112	7.6687	89.30	0.1304	−3.0	98.6	95.7	−0.0071	0.2336	−50
−49	5.665	0.0112	7.4516	89.20	0.1342	−2.7	98.5	95.8	−0.0064	0.2334	−49
−48	5.841	0.0112	7.2411	89.10	0.1381	−2.4	98.3	96.0	−0.0057	0.2332	−48
−47	6.022	0.0112	7.0373	89.00	0.1421	−2.1	98.2	96.1	−0.0050	0.2330	−47
−46	6.207	0.0112	6.8399	88.90	0.1462	−1.8	98.0	96.3	−0.0043	0.2328	−46
−45	6.397	0.0113	6.6489	88.81	0.1504	−1.5	97.9	96.4	−0.0035	0.2326	−45
−44	6.591	0.0113	6.4683	88.71	0.1546	−1.2	97.8	96.6	−0.0028	0.2324	−44
−43	6.790	0.0113	6.2893	88.61	0.1590	−0.9	97.6	96.7	−0.0021	0.2322	−43
−42	6.994	0.0113	6.1162	88.51	0.1635	−0.6	97.5	96.9	−0.0014	0.2320	−42
−41	7.203	0.0113	5.9488	88.41	0.1681	−0.3	97.3	97.0	−0.0007	0.2318	−41
−40	7.417	0.0113	5.7904	88.31	0.1727	0.0	97.2	97.2	0.0000	0.2316	−40
−39	7.636	0.0113	5.6338	88.21	0.1775	0.3	97.0	97.3	0.0007	0.2314	−39
−38	7.860	0.0113	5.4825	88.11	0.1824	0.6	96.9	97.5	0.0014	0.2312	−38
−37	8.090	0.0114	5.3362	88.01	0.1874	0.9	96.7	97.6	0.0021	0.2310	−37
−36	8.325	0.0114	5.1948	87.91	0.1925	1.2	96.6	97.8	0.0028	0.2308	−36
−35	8.565	0.0114	5.0582	87.81	0.1977	1.5	96.4	97.9	0.0035	0.2306	−35
−34	8.811	0.0114	4.9261	87.71	0.2030	1.8	96.3	98.1	0.0042	0.2304	−34
−33	9.062	0.0114	4.7985	87.61	0.2084	2.1	96.1	98.2	0.0049	0.2303	−33
−32	9.319	0.0114	4.6729	87.51	0.2140	2.4	96.0	98.4	0.0056	0.2301	−32
−31	9.582	0.0114	4.5537	87.41	0.2196	2.7	95.8	98.5	0.0063	0.2299	−31

Table A–15 (continued)

HFC–134a Saturation Properties—Temperature Table

TEMP. °F	PRESSURE psia	VOLUME ft³/lb		DENSITY lb/ft³		ENTHALPY Btu/lb			ENTROPY Btu/(lb)(°R)		TEMP. °F
		LIQUID v_f	VAPOR v_g	LIQUID $1/v_f$	VAPOR $1/v_g$	LIQUID h_f	LATENT h_{fg}	VAPOR h_g	LIQUID s_f	VAPOR s_g	
−30	9.851	0.0115	4.4366	87.31	0.2254	3.0	95.7	98.7	0.0070	0.2297	−30
−29	10.126	0.0115	4.3234	87.21	0.2313	3.3	95.5	98.8	0.0077	0.2296	−29
−28	10.407	0.0115	4.2141	87.11	0.2373	3.6	95.4	99.0	0.0084	0.2294	−28
−27	10.694	0.0115	4.1068	87.01	0.2435	3.9	95.2	99.1	0.0091	0.2292	−27
−26	10.987	0.0115	4.0032	86.91	0.2498	4.2	95.1	99.3	0.0098	0.2291	−26
−25	11.287	0.0115	3.9032	86.81	0.2562	4.5	94.9	99.4	0.0105	0.2289	−25
−24	11.594	0.0115	3.8066	86.71	0.2627	4.8	94.8	99.6	0.0112	0.2288	−24
−23	11.906	0.0115	3.7120	86.61	0.2694	5.1	94.6	99.7	0.0119	0.2286	−23
−22	12.226	0.0116	3.6206	86.51	0.2762	5.4	94.5	99.9	0.0126	0.2284	−22
−21	12.553	0.0116	3.5323	86.41	0.2831	5.7	94.3	100.0	0.0132	0.2283	−21
−20	12.885	0.0116	3.4471	86.30	0.2901	6.0	94.2	100.2	0.0139	0.2281	−20
−19	13.225	0.0116	3.3625	86.20	0.2974	6.3	94.0	100.3	0.0146	0.2280	−19
−18	13.572	0.0116	3.2819	86.10	0.3047	6.6	93.9	100.5	0.0153	0.2278	−18
−17	13.927	0.0116	3.2031	86.00	0.3122	6.9	93.7	100.6	0.0160	0.2277	−17
−16	14.289	0.0116	3.1270	85.89	0.3198	7.2	93.6	100.8	0.0167	0.2276	−16
−15	14.659	0.0117	3.0525	85.79	0.3276	7.5	93.4	100.9	0.0174	0.2274	−15
−14	15.035	0.0117	2.9806	85.69	0.3355	7.8	93.3	101.1	0.0180	0.2273	−14
−13	15.420	0.0117	2.9104	85.59	0.3436	8.1	93.1	101.2	0.0187	0.2271	−13
−12	15.812	0.0117	2.8417	85.48	0.3519	8.4	92.9	101.4	0.0194	0.2270	−12
−11	16.212	0.0117	2.7762	85.38	0.3602	8.7	92.8	101.5	0.0201	0.2269	−11
−10	16.620	0.0117	2.7115	85.28	0.3688	9.0	92.6	101.7	0.0208	0.2267	−10
−9	17.037	0.0117	2.6490	85.17	0.3775	9.3	92.5	101.8	0.0214	0.2266	−9
−8	17.461	0.0118	2.5880	85.07	0.3864	9.7	92.3	102.0	0.0221	0.2265	−8
−7	17.893	0.0118	2.5291	84.97	0.3954	10.0	92.1	102.1	0.0228	0.2264	−7
−6	18.334	0.0118	2.4716	84.86	0.4046	10.3	92.0	102.3	0.0235	0.2262	−6
−5	18.784	0.0118	2.4155	84.76	0.4140	10.6	91.8	102.4	0.0241	0.2261	−5
−4	19.242	0.0118	2.3613	84.65	0.4235	10.9	91.7	102.5	0.0248	0.2260	−4
−3	19.709	0.0118	2.3084	84.55	0.4332	11.2	91.5	102.7	0.0255	0.2259	−3
−2	20.184	0.0118	2.2568	84.44	0.4431	11.5	91.3	102.8	0.0262	0.2258	−2
−1	20.669	0.0119	2.2065	84.34	0.4532	11.8	91.2	103.0	0.0268	0.2256	−1
0	21.163	0.0119	2.1580	84.23	0.4634	12.1	91.0	103.1	0.0275	0.2255	0
1	21.666	0.0119	2.1106	84.13	0.4738	12.4	90.9	103.3	0.0282	0.2254	1
2	22.178	0.0119	2.0644	84.02	0.4844	12.7	90.7	103.4	0.0288	0.2253	2
3	22.700	0.0119	2.0194	83.91	0.4952	13.0	90.5	103.6	0.0295	0.2252	3
4	23.231	0.0119	1.9755	83.81	0.5062	13.4	90.4	103.7	0.0302	0.2251	4
5	23.772	0.0119	1.9327	83.70	0.5174	13.7	90.2	103.9	0.0308	0.2250	5
6	24.322	0.0120	1.8911	83.60	0.5288	14.0	90.0	104.0	0.0315	0.2249	6
7	24.883	0.0120	1.8508	83.49	0.5403	14.3	89.9	104.2	0.0322	0.2248	7
8	25.454	0.0120	1.8113	83.38	0.5521	14.6	89.7	104.3	0.0328	0.2247	8
9	26.034	0.0120	1.7727	83.27	0.5641	14.9	89.5	104.4	0.0335	0.2245	9
10	26.625	0.0120	1.7355	83.17	0.5762	15.2	89.4	104.6	0.0342	0.2244	10
11	27.227	0.0120	1.6989	83.06	0.5886	15.5	89.2	104.7	0.0348	0.2244	11
12	27.839	0.0121	1.6633	82.95	0.6012	15.9	89.0	104.9	0.0355	0.2243	12
13	28.462	0.0121	1.6287	82.84	0.6140	16.2	88.9	105.0	0.0362	0.2242	13
14	29.095	0.0121	1.5949	82.73	0.6270	16.5	88.7	105.2	0.0368	0.2241	14
15	29.739	0.0121	1.5620	82.63	0.6402	16.8	88.5	105.3	0.0375	0.2240	15
16	30.395	0.0121	1.5298	82.52	0.6537	17.1	88.3	105.5	0.0381	0.2239	16
17	31.061	0.0121	1.4984	82.41	0.6674	17.4	88.2	105.6	0.0388	0.2238	17
18	31.739	0.0122	1.4678	82.30	0.6813	17.7	88.0	105.7	0.0395	0.2237	18
19	32.428	0.0122	1.4380	82.19	0.6954	18.1	87.8	105.9	0.0401	0.2236	19
20	33.129	0.0122	1.4090	82.08	0.7097	18.4	87.7	106.0	0.0408	0.2235	20
21	33.841	0.0122	1.3806	81.97	0.7243	18.7	87.5	106.2	0.0414	0.2234	21
22	34.566	0.0122	1.3528	81.86	0.7392	19.0	87.3	106.3	0.0421	0.2233	22
23	35.302	0.0122	1.3259	81.75	0.7542	19.3	87.1	106.5	0.0427	0.2233	23
24	36.050	0.0122	1.2995	81.64	0.7695	19.6	87.0	106.6	0.0434	0.2232	24
25	36.810	0.0123	1.2737	81.52	0.7851	20.0	86.8	106.7	0.0440	0.2231	25
26	37.583	0.0123	1.2486	81.41	0.8009	20.3	86.6	106.9	0.0447	0.2230	26
27	38.368	0.0123	1.2240	81.30	0.8170	20.6	86.4	107.0	0.0453	0.2229	27
28	39.166	0.0123	1.2000	81.19	0.8333	20.9	86.2	107.2	0.0460	0.2229	28
29	39.977	0.0123	1.1766	81.08	0.8499	21.2	86.1	107.3	0.0467	0.2228	29

Table A-15 (continued)
HFC-134a Saturation Properties—Temperature Table

TEMP. °F	PRESSURE psia	VOLUME ft³/lb		DENSITY lb/ft³		ENTHALPY Btu/lb			ENTROPY Btu/(lb)(°R)		TEMP. °F
		LIQUID v_f	VAPOR v_g	LIQUID $1/v_f$	VAPOR $1/v_g$	LIQUID h_f	LATENT h_{fg}	VAPOR h_g	LIQUID s_f	VAPOR s_g	
30	40.800	0.0124	1.1538	80.96	0.8667	21.6	85.7	107.4	0.0473	0.2227	30
31	41.636	0.0124	1.1315	80.85	0.8838	21.9	85.7	107.6	0.0480	0.2226	31
32	42.486	0.0124	1.1098	80.74	0.9011	22.2	85.5	107.7	0.0486	0.2226	32
33	43.349	0.0124	1.0884	80.62	0.9188	22.5	85.3	107.9	0.0492	0.2225	33
34	44.225	0.0124	1.0676	80.51	0.9367	22.8	85.2	108.0	0.0499	0.2224	34
35	45.115	0.0124	1.0472	80.40	0.9549	23.2	85.0	108.1	0.0505	0.2223	35
36	46.018	0.0125	1.0274	80.28	0.9733	23.5	84.8	108.3	0.0512	0.2223	36
37	46.935	0.0125	1.0080	80.17	0.9921	23.8	84.6	108.4	0.0518	0.2222	37
38	47.866	0.0125	0.9890	80.05	1.0111	24.1	84.4	108.6	0.0525	0.2221	38
39	48.812	0.0125	0.9705	79.94	1.0304	24.5	84.2	108.7	0.0531	0.2221	39
40	49.771	0.0125	0.9523	79.82	1.0501	24.8	84.1	108.8	0.0538	0.2220	40
41	50.745	0.0125	0.9346	79.70	1.0700	25.1	83.9	109.0	0.0544	0.2219	41
42	51.733	0.0126	0.9173	79.59	1.0902	25.4	83.7	109.1	0.0551	0.2219	42
43	52.736	0.0126	0.9003	79.47	1.1107	25.8	83.5	109.2	0.0557	0.2218	43
44	53.754	0.0126	0.8837	79.35	1.1316	26.1	83.3	109.4	0.0564	0.2217	44
45	54.787	0.0126	0.8675	79.24	1.1527	26.4	83.1	109.5	0.0570	0.2217	45
46	55.835	0.0126	0.8516	79.12	1.1742	26.7	82.9	109.7	0.0576	0.2216	46
47	56.898	0.0127	0.8361	79.00	1.1960	27.1	82.7	109.8	0.0583	0.2216	47
48	57.976	0.0127	0.8210	78.88	1.2181	27.4	82.5	109.9	0.0589	0.2215	48
49	59.070	0.0127	0.8061	78.76	1.2405	27.7	82.3	110.1	0.0596	0.2214	49
50	60.180	0.0127	0.7916	78.64	1.2633	28.0	82.1	110.2	0.0602	0.2214	50
51	61.305	0.0127	0.7774	78.53	1.2864	28.4	81.9	110.3	0.0608	0.2213	51
52	62.447	0.0128	0.7634	78.41	1.3099	28.7	81.8	110.5	0.0615	0.2213	52
53	63.604	0.0128	0.7498	78.29	1.3337	29.0	81.6	110.6	0.0621	0.2212	53
54	64.778	0.0128	0.7365	78.16	1.3578	29.4	81.4	110.7	0.0628	0.2211	54
55	65.963	0.0128	0.7234	78.04	1.3823	29.7	81.2	110.9	0.0634	0.2211	55
56	67.170	0.0128	0.7106	77.92	1.4072	30.0	81.0	111.0	0.0640	0.2210	56
57	68.394	0.0129	0.6981	77.80	1.4324	30.4	80.8	111.1	0.0647	0.2210	57
58	69.635	0.0129	0.6859	77.68	1.4579	30.7	80.6	111.3	0.0653	0.2209	58
59	70.892	0.0129	0.6739	77.56	1.4839	31.0	80.4	111.4	0.0659	0.2209	59
60	72.167	0.0129	0.6622	77.43	1.5102	31.4	80.2	111.5	0.0666	0.2208	60
61	73.459	0.0129	0.6507	77.31	1.5369	31.7	80.0	111.6	0.0672	0.2208	61
62	74.769	0.0130	0.6394	77.19	1.5640	32.0	79.7	111.8	0.0678	0.2207	62
63	76.096	0.0130	0.6283	77.06	1.5915	32.4	79.5	111.9	0.0685	0.2207	63
64	77.440	0.0130	0.6175	76.94	1.6194	32.7	79.3	112.0	0.0691	0.2206	64
65	78.803	0.0130	0.6069	76.81	1.6477	33.0	79.1	112.2	0.0698	0.2206	65
66	80.184	0.0130	0.5965	76.69	1.6764	33.4	78.9	112.3	0.0704	0.2205	66
67	81.582	0.0131	0.5863	76.56	1.7055	33.7	78.7	112.4	0.0710	0.2205	67
68	83.000	0.0131	0.5764	76.44	1.7350	34.0	78.5	112.5	0.0717	0.2204	68
69	84.435	0.0131	0.5666	76.31	1.7649	34.4	78.3	112.7	0.0723	0.2204	69
70	85.890	0.0131	0.5570	76.18	1.7952	34.7	78.1	112.8	0.0729	0.2203	70
71	87.363	0.0131	0.5476	76.05	1.8260	35.1	77.9	112.9	0.0735	0.2203	71
72	88.855	0.0132	0.5384	75.93	1.8573	35.4	77.6	113.0	0.0742	0.2202	72
73	90.366	0.0132	0.5294	75.80	1.8889	35.7	77.4	113.2	0.0748	0.2202	73
74	91.897	0.0132	0.5206	75.67	1.9210	36.1	77.2	113.3	0.0754	0.2201	74
75	93.447	0.0132	0.5119	75.54	1.9536	36.4	77.0	113.4	0.0761	0.2201	75
76	95.016	0.0133	0.5034	75.41	1.9866	36.8	76.8	113.5	0.0767	0.2200	76
77	96.606	0.0133	0.4950	75.28	2.0201	37.1	76.6	113.7	0.0773	0.2200	77
78	98.215	0.0133	0.4868	75.15	2.0541	37.4	76.3	113.8	0.0780	0.2200	78
79	99.844	0.0133	0.4788	75.02	2.0885	37.8	76.1	113.9	0.0786	0.2199	79
80	101.494	0.0134	0.4709	74.89	2.1234	38.1	75.9	114.0	0.0792	0.2199	80
81	103.164	0.0134	0.4632	74.75	2.1589	38.5	75.7	114.1	0.0799	0.2198	81
82	104.855	0.0134	0.4556	74.62	2.1948	38.8	75.4	114.3	0.0805	0.2198	82
83	106.566	0.0134	0.4482	74.49	2.2312	39.2	75.2	114.4	0.0811	0.2197	83
84	108.290	0.0134	0.4409	74.35	2.2681	39.5	75.0	114.5	0.0817	0.2197	84
85	110.050	0.0135	0.4337	74.22	2.3056	39.9	74.8	114.6	0.0824	0.2196	85
86	111.828	0.0135	0.4267	74.08	2.3436	40.2	74.5	114.7	0.0830	0.2196	86
87	113.626	0.0135	0.4198	73.95	2.3821	40.5	74.3	114.9	0.0836	0.2196	87
88	115.444	0.0135	0.4130	73.81	2.4211	40.9	74.1	115.0	0.0843	0.2195	88
89	117.281	0.0136	0.4064	73.67	2.4607	41.2	73.8	115.1	0.0849	0.2195	89

Table A-15 (continued)
HFC-134a Saturation Properties—Temperature Table

TEMP. °F	PRESSURE psia	VOLUME ft³/lb LIQUID v_f	VOLUME ft³/lb VAPOR v_g	DENSITY lb/ft³ LIQUID $1/v_f$	DENSITY lb/ft³ VAPOR $1/v_g$	ENTHALPY Btu/lb LIQUID h_f	ENTHALPY Btu/lb LATENT h_{fg}	ENTHALPY Btu/lb VAPOR h_g	ENTROPY Btu/(lb)(°R) LIQUID s_f	ENTROPY Btu/(lb)(°R) VAPOR s_g	TEMP. °F
90	119.138	0.0136	0.3999	73.54	2.5009	41.6	73.6	115.2	0.0855	0.2194	90
91	121.024	0.0136	0.3935	73.40	2.5416	41.9	73.4	115.3	0.0861	0.2194	91
92	122.930	0.0137	0.3872	73.26	2.5829	42.3	73.1	115.4	0.0868	0.2193	92
93	124.858	0.0137	0.3810	73.12	2.6247	42.6	72.9	115.5	0.0874	0.2193	93
94	126.809	0.0137	0.3749	72.98	2.6672	43.0	72.7	115.7	0.0880	0.2193	94
95	128.782	0.0137	0.3690	72.84	2.7102	43.4	72.4	115.8	0.0886	0.2192	95
96	130.778	0.0138	0.3631	72.70	2.7539	43.7	72.2	115.9	0.0893	0.2192	96
97	132.798	0.0138	0.3574	72.56	2.7981	44.1	71.9	116.0	0.0899	0.2191	97
98	134.840	0.0138	0.3517	72.42	2.8430	44.4	71.7	116.1	0.0905	0.2191	98
99	136.906	0.0138	0.3462	72.27	2.8885	44.8	71.4	116.2	0.0912	0.2190	99
100	138.996	0.0139	0.3408	72.13	2.9347	45.1	71.2	116.3	0.0918	0.2190	100
101	141.109	0.0139	0.3354	71.99	2.9815	45.5	70.9	116.4	0.0924	0.2190	101
102	143.247	0.0139	0.3302	71.84	3.0289	45.8	70.7	116.5	0.0930	0.2189	102
103	145.408	0.0139	0.3250	71.70	3.0771	46.2	70.4	116.6	0.0937	0.2189	103
104	147.594	0.0140	0.3199	71.55	3.1259	46.6	70.2	116.7	0.0943	0.2188	104
105	149.804	0.0140	0.3149	71.40	3.1754	46.9	69.9	116.9	0.0949	0.2188	105
106	152.039	0.0140	0.3100	71.25	3.2256	47.3	69.7	117.0	0.0955	0.2187	106
107	154.298	0.0141	0.3052	71.11	3.2765	47.6	69.4	117.1	0.0962	0.2187	107
108	156.583	0.0141	0.3005	70.96	3.3282	48.0	69.2	117.2	0.0968	0.2186	108
109	158.893	0.0141	0.2958	70.81	3.3806	48.4	68.9	117.3	0.0974	0.2186	109
110	161.227	0.0142	0.2912	70.66	3.4337	48.7	68.6	117.4	0.0981	0.2185	110
111	163.588	0.0142	0.2867	70.51	3.4876	49.1	68.4	117.5	0.0987	0.2185	111
112	165.974	0.0142	0.2823	70.35	3.5423	49.5	68.1	117.6	0.0993	0.2185	112
113	168.393	0.0142	0.2780	70.20	3.5977	49.8	67.8	117.7	0.0999	0.2184	113
114	170.833	0.0143	0.2737	70.05	3.6539	50.2	67.6	117.8	0.1006	0.2184	114
115	173.298	0.0143	0.2695	69.89	3.7110	50.5	67.3	117.9	0.1012	0.2183	115
116	175.790	0.0143	0.2653	69.74	3.7689	50.9	67.0	117.9	0.1018	0.2183	116
117	178.297	0.0144	0.2613	69.58	3.8276	51.3	66.8	118.0	0.1024	0.2182	117
118	180.846	0.0144	0.2573	69.42	3.8872	51.7	66.5	118.1	0.1031	0.2182	118
119	183.421	0.0144	0.2533	69.26	3.9476	52.0	66.2	118.2	0.1037	0.2181	119
120	186.023	0.0145	0.2494	69.10	4.0089	52.4	65.9	118.3	0.1043	0.2181	120
121	188.652	0.0145	0.2456	68.94	4.0712	52.8	65.6	118.4	0.1050	0.2180	121
122	191.308	0.0145	0.2419	68.78	4.1343	53.1	65.4	118.5	0.1056	0.2180	122
123	193.992	0.0146	0.2382	68.62	4.1984	53.5	65.1	118.6	0.1062	0.2179	123
124	196.703	0.0146	0.2346	68.46	4.2634	53.9	64.8	118.7	0.1068	0.2178	124
125	199.443	0.0146	0.2310	68.29	4.3294	54.3	64.5	118.8	0.1075	0.2178	125
126	202.211	0.0147	0.2275	68.13	4.3964	54.6	64.2	118.8	0.1081	0.2177	126
127	205.008	0.0147	0.2240	67.96	4.4644	55.0	63.9	118.9	0.1087	0.2177	127
128	207.834	0.0147	0.2206	67.80	4.5334	55.4	63.6	119.0	0.1094	0.2176	128
129	210.688	0.0148	0.2172	67.63	4.6034	55.8	63.3	119.1	0.1100	0.2176	129
130	213.572	0.0148	0.2139	67.46	4.6745	56.2	63.0	119.2	0.1106	0.2175	130
131	216.485	0.0149	0.2107	67.29	4.7467	56.5	62.7	119.2	0.1113	0.2174	131
132	219.429	0.0149	0.2075	67.12	4.8200	56.9	62.4	119.3	0.1119	0.2174	132
133	222.402	0.0149	0.2043	66.95	4.8945	57.3	62.1	119.4	0.1125	0.2173	133
134	225.405	0.0150	0.2012	66.77	4.9700	57.7	61.8	119.5	0.1132	0.2173	134
135	228.438	0.0150	0.1981	66.60	5.0468	58.1	61.5	119.6	0.1138	0.2172	135
136	231.502	0.0151	0.1951	66.42	5.1248	58.5	61.2	119.6	0.1144	0.2171	136
137	234.597	0.0151	0.1922	66.24	5.2040	58.8	60.8	119.7	0.1151	0.2171	137
138	237.723	0.0151	0.1892	66.06	5.2844	59.2	60.5	119.8	0.1157	0.2170	138
139	240.880	0.0152	0.1864	65.88	5.3661	59.6	60.2	119.8	0.1163	0.2169	139
140	244.068	0.0152	0.1835	65.70	5.4491	60.0	59.9	119.9	0.1170	0.2168	140
141	247.288	0.0153	0.1807	65.52	5.5335	60.4	59.6	120.0	0.1176	0.2168	141
142	250.540	0.0153	0.1780	65.34	5.6192	60.8	59.2	120.0	0.1183	0.2167	142
143	253.824	0.0153	0.1752	65.15	5.7064	61.2	58.9	120.1	0.1189	0.2166	143
144	257.140	0.0154	0.1726	64.96	5.7949	61.6	58.6	120.1	0.1195	0.2165	144
145	260.489	0.0154	0.1699	64.78	5.8849	62.0	58.2	120.2	0.1202	0.2165	145
146	263.871	0.0155	0.1673	64.59	5.9765	62.4	57.9	120.3	0.1208	0.2164	146
147	267.270	0.0155	0.1648	64.39	6.0695	62.8	57.5	120.3	0.1215	0.2163	147
148	270.721	0.0156	0.1622	64.20	6.1642	63.2	57.2	120.4	0.1221	0.2162	148
149	274.204	0.0156	0.1597	64.01	6.2604	63.6	56.8	120.4	0.1228	0.2161	149

Table A-15 (continued)
HFC-134a Saturation Properties—Temperature Table

TEMP. °F	PRESSURE psia	VOLUME ft³/lb LIQUID v_f	VOLUME ft³/lb VAPOR v_g	DENSITY lb/ft³ LIQUID $1/v_f$	DENSITY lb/ft³ VAPOR $1/v_g$	ENTHALPY Btu/lb LIQUID h_f	ENTHALPY Btu/lb LATENT h_{fg}	ENTHALPY Btu/lb VAPOR h_g	ENTROPY Btu/(lb)(°R) LIQUID s_f	ENTROPY Btu/(lb)(°R) VAPOR s_g	TEMP. °F
150	277.721	0.0157	0.1573	63.81	6.3584	64.0	56.5	120.5	0.1234	0.2160	150
151	281.272	0.0157	0.1548	63.61	6.4580	64.4	56.1	120.5	0.1240	0.2159	151
152	284.857	0.0158	0.1525	63.41	6.5593	64.8	55.7	120.6	0.1247	0.2158	152
153	288.477	0.0158	0.1501	63.21	6.6625	65.2	55.4	120.6	0.1253	0.2157	153
154	292.131	0.0159	0.1478	63.01	6.7675	65.6	55.0	120.6	0.1260	0.2156	154
155	295.820	0.0159	0.1455	62.80	6.8743	66.0	54.6	120.7	0.1266	0.2155	155
156	299.544	0.0160	0.1432	62.59	6.9831	66.4	54.3	120.7	0.1273	0.2154	156
157	303.304	0.0160	0.1410	62.38	7.0940	66.9	53.9	120.7	0.1279	0.2153	157
158	307.100	0.0161	0.1388	62.17	7.2068	67.3	53.5	120.8	0.1286	0.2152	158
159	310.931	0.0161	0.1366	61.96	7.3218	67.7	53.1	120.8	0.1293	0.2151	159
160	314.800	0.0162	0.1344	61.74	7.4390	68.1	52.7	120.8	0.1299	0.2150	160
161	318.704	0.0163	0.1323	61.52	7.5584	68.5	52.3	120.9	0.1306	0.2149	161
162	322.646	0.0163	0.1302	61.30	7.6801	69.0	51.9	120.9	0.1312	0.2148	162
163	326.625	0.0164	0.1281	61.08	7.8042	69.4	51.5	120.9	0.1319	0.2146	163
164	330.641	0.0164	0.1261	60.86	7.9308	69.8	51.1	120.9	0.1326	0.2145	164
165	334.696	0.0165	0.1241	60.63	8.0600	70.2	50.7	120.9	0.1332	0.2144	165
166	338.788	0.0166	0.1221	60.40	8.1917	70.7	50.3	120.9	0.1339	0.2142	166
167	342.919	0.0166	0.1201	60.16	8.3262	71.1	49.8	120.9	0.1346	0.2141	167
168	347.089	0.0167	0.1182	59.93	8.4635	71.5	49.4	120.9	0.1352	0.2140	168
169	351.298	0.0168	0.1162	59.69	8.6037	72.0	49.0	120.9	0.1359	0.2138	169
170	355.547	0.0168	0.1143	59.45	8.7470	72.4	48.5	120.9	0.1366	0.2137	170
171	359.835	0.0169	0.1124	59.20	8.8934	72.8	48.1	120.9	0.1373	0.2135	171
172	364.164	0.0170	0.1106	58.95	9.0431	73.3	47.6	120.9	0.1380	0.2133	172
173	368.533	0.0170	0.1087	58.70	9.1961	73.7	47.2	120.9	0.1386	0.2132	173
174	372.942	0.0171	0.1069	58.45	9.3527	74.2	46.7	120.9	0.1393	0.2130	174
175	377.393	0.0172	0.1051	58.19	9.5129	74.6	46.2	120.8	0.1400	0.2128	175
176	381.886	0.0173	0.1033	57.92	9.6770	75.1	45.7	120.8	0.1407	0.2126	176
177	386.421	0.0173	0.1016	57.66	9.8451	75.5	45.2	120.8	0.1414	0.2125	177
178	390.998	0.0174	0.0998	57.39	10.0173	76.0	44.7	120.7	0.1421	0.2123	178
179	395.617	0.0175	0.0981	57.11	10.1939	76.5	44.2	120.7	0.1428	0.2121	179
180	400.280	0.0176	0.0964	56.83	10.3750	76.9	43.7	120.7	0.1435	0.2119	180
181	404.987	0.0177	0.0947	56.55	10.5609	77.4	43.2	120.6	0.1442	0.2116	181
182	409.738	0.0178	0.0930	56.26	10.7518	77.9	42.7	120.5	0.1449	0.2114	182
183	414.533	0.0179	0.0913	55.96	10.9481	78.4	42.1	120.5	0.1456	0.2112	183
184	419.373	0.0180	0.0897	55.66	11.1498	78.8	41.6	120.4	0.1464	0.2109	184
185	424.258	0.0181	0.0880	55.35	11.3575	79.3	41.0	120.3	0.1471	0.2107	185
186	429.189	0.0182	0.0864	55.04	11.5713	79.8	40.4	120.2	0.1478	0.2104	186
187	434.167	0.0183	0.0848	54.72	11.7916	80.3	39.8	120.1	0.1486	0.2102	187
188	439.192	0.0184	0.0832	54.40	12.0189	80.8	39.2	120.0	0.1493	0.2099	188
189	444.264	0.0185	0.0816	54.07	12.2536	81.3	38.6	119.9	0.1501	0.2096	189
190	449.384	0.0186	0.0800	53.73	12.4962	81.8	38.0	119.8	0.1508	0.2093	190
191	454.552	0.0187	0.0784	53.38	12.7472	82.3	37.4	119.7	0.1516	0.2090	191
192	459.757	0.0189	0.0769	53.02	13.0072	82.8	36.7	119.5	0.1523	0.2087	192
193	465.026	0.0190	0.0753	52.65	13.2769	83.4	36.0	119.4	0.1531	0.2083	193
194	470.346	0.0191	0.0738	52.27	13.5570	83.9	35.3	119.2	0.1539	0.2080	194
195	475.717	0.0193	0.0722	51.88	13.8484	84.4	34.6	119.1	0.1547	0.2076	195
196	481.139	0.0194	0.0707	51.48	14.1522	85.0	33.9	118.9	0.1555	0.2072	196
197	486.614	0.0196	0.0691	51.07	14.4693	85.5	33.1	118.7	0.1563	0.2068	197
198	492.142	0.0197	0.0676	50.64	14.8012	86.1	32.3	118.4	0.1572	0.2063	198
199	497.724	0.0199	0.0660	50.20	15.1493	86.7	31.5	118.2	0.1580	0.2059	199
200	503.361	0.0201	0.0645	49.73	15.5155	87.3	30.7	118.0	0.1589	0.2054	200
201	509.054	0.0203	0.0629	49.25	15.9020	87.9	29.8	117.7	0.1597	0.2049	201
202	514.805	0.0205	0.0613	48.75	16.3113	88.5	28.9	117.4	0.1606	0.2043	202
203	520.613	0.0207	0.0597	48.22	16.7466	89.1	27.9	117.0	0.1616	0.2037	203
204	526.481	0.0210	0.0581	47.66	17.2121	89.8	26.9	116.7	0.1625	0.2031	204
205	532.410	0.0212	0.0565	47.06	17.7129	90.5	25.8	116.3	0.1635	0.2024	205
206	538.402	0.0215	0.0548	46.42	18.2558	91.2	24.7	115.8	0.1645	0.2016	206
207	544.458	0.0219	0.0531	45.73	18.8499	91.9	23.4	115.3	0.1656	0.2008	207
208	550.581	0.0222	0.0513	44.98	19.5084	92.7	22.1	114.8	0.1667	0.1998	208
209	556.773	0.0227	0.0494	44.14	20.2504	93.5	20.6	114.1	0.1680	0.1988	209

Table A–15 (continued)

HFC–134a Saturation Properties—Temperature Table

TEMP. °F	PRESSURE psia	VOLUME ft³/lb		DENSITY lb/ft³		ENTHALPY Btu/lb			ENTROPY Btu/(lb)(°R)		TEMP. °F
		LIQUID v_f	VAPOR v_g	LIQUID $1/v_f$	VAPOR $1/v_g$	LIQUID h_f	LATENT h_{fg}	VAPOR h_g	LIQUID s_f	VAPOR s_g	
210	563.037	0.0232	0.0474	43.19	21.1071	94.4	18.9	113.4	0.1693	0.1976	210
211	569.378	0.0238	0.0452	42.07	22.1329	95.5	17.0	112.4	0.1708	0.1961	211
212	575.801	0.0246	0.0427	40.67	23.4420	96.7	14.5	111.2	0.1726	0.1942	212
213	582.316	0.0259	0.0394	38.65	25.3583	98.4	11.1	109.5	0.1750	0.1915	213

Table A-15 (continued)
HFC-134a Superheated Vapor—Constant Pressure Tables

V = Volume in ft³/lb H = Enthalpy in Btu/lb S = Entropy in Btu/(lb) (°R) v_s = Velocity of Sound in ft/sec
Cp = Heat Capacity at Constant Pressure in Btu/(lb) (°F) Cp/Cv = Heat Capacity Ratio (Dimensionless)

TEMP °F	PRESSURE = 60.00 PSIA							PRESSURE = 65.00 PSIA						TEMP °F
	V	H	S	Cp	Cp/Cv	v_s		V	H	S	Cp	Cp/Cv	v_s	
49.8	0.01271	28.0	0.0601	0.3282	1.5343	1901.3	SAT LIQ	0.01280	29.4	0.0629	0.3302	1.5380	1864.8	54.2
49.8	0.79390	110.2	0.2214	0.2232	1.1934	480.5	SAT VAP	0.73400	110.7	0.2211	0.2261	1.1979	479.7	54.2
50	0.79428	110.2	0.2215	0.2232	1.1932	480.6		—	—	—	—	—	—	50
60	0.81793	112.4	0.2258	0.2222	1.1820	488.5		0.74699	112.1	0.2237	0.2253	1.1908	484.5	60
70	0.84104	114.6	0.2300	0.2217	1.1725	496.0		0.76888	114.3	0.2280	0.2243	1.1800	492.3	70
80	0.86356	116.9	0.2342	0.2215	1.1643	503.2		0.79020	116.5	0.2321	0.2239	1.1708	499.8	80
90	0.88566	119.1	0.2382	0.2217	1.1572	510.1		0.81103	118.8	0.2363	0.2238	1.1628	506.9	90
100	0.90728	121.3	0.2422	0.2222	1.1509	516.7		0.83139	121.0	0.2403	0.2240	1.1558	513.8	100
110	0.92868	123.5	0.2462	0.2229	1.1453	523.2		0.85143	123.3	0.2443	0.2245	1.1497	520.4	110
120	0.94967	125.8	0.2500	0.2237	1.1403	529.4		0.87116	125.5	0.2482	0.2252	1.1443	526.8	120
130	0.97040	128.0	0.2539	0.2247	1.1359	535.5		0.89071	127.8	0.2520	0.2260	1.1394	533.1	130
140	0.99098	130.3	0.2577	0.2258	1.1318	541.4		0.90975	130.0	0.2558	0.2270	1.1350	539.1	140
150	1.01133	132.5	0.2614	0.2271	1.1282	547.1		0.92894	132.3	0.2596	0.2282	1.1311	545.0	150
160	1.03135	134.8	0.2651	0.2284	1.1248	552.8		0.94751	134.6	0.2633	0.2294	1.1275	550.7	160
170	1.05130	137.1	0.2688	0.2298	1.1218	558.3		0.96628	136.9	0.2670	0.2307	1.1242	556.4	170
180	1.07101	139.4	0.2724	0.2312	1.1189	563.7		0.98464	139.2	0.2707	0.2321	1.1212	561.9	180
190	1.09075	141.7	0.2760	0.2327	1.1163	569.0		1.00291	141.5	0.2743	0.2335	1.1184	567.3	190
200	1.11012	144.0	0.2796	0.2343	1.1139	574.2		1.02114	143.9	0.2778	0.2350	1.1158	572.5	200
210	1.12943	146.4	0.2831	0.2359	1.1117	579.3		1.03918	146.2	0.2814	0.2365	1.1134	577.7	210
220	1.14903	148.8	0.2866	0.2375	1.1096	584.3		1.05719	148.6	0.2849	0.2381	1.1112	582.8	220
230	1.16795	151.1	0.2901	0.2391	1.1076	589.2		1.07492	151.0	0.2884	0.2397	1.1091	587.8	230
240	1.18723	153.5	0.2936	0.2408	1.1057	594.1		1.09266	153.4	0.2919	0.2413	1.1072	592.8	240
250	1.20627	156.0	0.2970	0.2425	1.1040	598.9		1.11037	155.8	0.2953	0.2430	1.1053	597.7	250
260	1.22519	158.4	0.3004	0.2442	1.1024	603.7		1.12816	158.3	0.2987	0.2446	1.1036	602.5	260
270	1.24409	160.8	0.3038	0.2459	1.1008	608.4		1.14561	160.7	0.3021	0.2463	1.1020	607.2	270
280	1.26295	163.3	0.3071	0.2476	1.0993	613.0		1.16306	163.2	0.3055	0.2480	1.1004	611.9	280
290	1.28172	165.8	0.3105	0.2493	1.0979	617.5		1.18036	165.7	0.3088	0.2497	1.0990	616.5	290
300	1.30039	168.3	0.3138	0.2510	1.0966	622.0		1.19804	168.2	0.3121	0.2514	1.0976	621.0	300
310	1.31891	170.8	0.3171	0.2527	1.0953	626.5		1.21521	170.7	0.3154	0.2531	1.0963	625.5	310
320	1.33761	173.4	0.3204	0.2544	1.0941	630.9		1.23259	173.2	0.3187	0.2548	1.0950	630.0	320
330	1.35630	175.9	0.3236	0.2561	1.0930	635.3		1.24969	175.8	0.3220	0.2564	1.0938	634.4	330
340	1.37457	178.5	0.3268	0.2578	1.0919	639.6		1.26678	178.4	0.3252	0.2581	1.0927	638.8	340
350	1.39334	181.1	0.3301	0.2595	1.0908	643.9		1.28403	181.0	0.3284	0.2598	1.0916	643.1	350
360	—	—	—	—	—	—		1.30124	183.6	0.3316	0.2615	1.0905	647.4	360

TEMP °F	PRESSURE = 70.00 PSIA							PRESSURE = 75.00 PSIA						TEMP °F
	V	H	S	Cp	Cp/Cv	v_s		V	H	S	Cp	Cp/Cv	v_s	
58.3	0.01288	30.8	0.0655	0.3321	1.5417	1830.2	SAT LIQ	0.01296	32.1	0.0680	0.3341	1.5454	1797.5	62.2
58.3	0.68236	111.3	0.2209	0.2288	1.2026	478.9	SAT VAP	0.63739	111.8	0.2207	0.2316	1.2072	478.0	62.2
60	0.68601	111.7	0.2217	0.2285	1.2002	480.3		—	—	—	—	—	—	60
70	0.70691	114.0	0.2260	0.2272	1.1880	488.5		0.65304	113.6	0.2241	0.2302	1.1966	484.6	70
80	0.72717	116.2	0.2302	0.2264	1.1777	496.3		0.67245	115.9	0.2284	0.2289	1.1850	492.7	80
90	0.74694	118.5	0.2344	0.2260	1.1688	503.7		0.69132	118.2	0.2326	0.2282	1.1751	500.4	90
100	0.76628	120.7	0.2385	0.2259	1.1611	510.8		0.70972	120.5	0.2367	0.2279	1.1666	507.7	100
110	0.78524	123.0	0.2425	0.2262	1.1543	517.6		0.72775	122.7	0.2408	0.2280	1.1592	514.8	110
120	0.80386	125.3	0.2464	0.2267	1.1484	524.2		0.74543	125.0	0.2448	0.2283	1.1526	521.6	120
130	0.82217	127.5	0.2503	0.2274	1.1431	530.6		0.76283	127.3	0.2487	0.2288	1.1469	528.1	130
140	0.84034	129.8	0.2541	0.2283	1.1383	536.8		0.77997	129.6	0.2525	0.2295	1.1417	534.5	140
150	0.85807	132.1	0.2579	0.2293	1.1341	542.8		0.79675	131.9	0.2563	0.2304	1.1371	540.6	150
160	0.87573	134.4	0.2617	0.2304	1.1302	548.7		0.81347	134.2	0.2601	0.2314	1.1330	546.6	160
170	0.89326	136.7	0.2654	0.2316	1.1266	554.4		0.83008	136.5	0.2638	0.2326	1.1292	552.5	170
180	0.91058	139.0	0.2690	0.2329	1.1234	560.0		0.84631	138.9	0.2675	0.2338	1.1257	558.2	180
190	0.92773	141.4	0.2726	0.2343	1.1205	565.5		0.86259	141.2	0.2711	0.2351	1.1226	563.8	190
200	0.94473	143.7	0.2762	0.2357	1.1177	570.9		0.87850	143.6	0.2747	0.2364	1.1197	569.2	200
210	0.96172	146.1	0.2798	0.2372	1.1152	576.2		0.89453	145.9	0.2783	0.2379	1.1170	574.6	210
220	0.97847	148.5	0.2833	0.2387	1.1128	581.4		0.91033	148.3	0.2818	0.2393	1.1145	579.9	220
230	0.99512	150.9	0.2868	0.2403	1.1106	586.4		0.92618	150.7	0.2853	0.2408	1.1122	585.0	230
240	1.01184	153.3	0.2903	0.2419	1.1086	591.5		0.94171	153.1	0.2888	0.2424	1.1100	590.1	240
250	1.02828	155.7	0.2937	0.2435	1.1067	596.4		0.95730	155.6	0.2922	0.2440	1.1080	595.1	250
260	1.04482	158.1	0.2971	0.2451	1.1049	601.2		0.97267	158.0	0.2957	0.2456	1.1061	600.0	260
270	1.06135	160.6	0.3005	0.2467	1.1032	606.0		0.98814	160.5	0.2991	0.2472	1.1043	604.9	270
280	1.07747	163.1	0.3039	0.2484	1.1015	610.8		1.00341	163.0	0.3024	0.2488	1.1027	609.6	280
290	1.09385	165.6	0.3072	0.2501	1.1000	615.4		1.01864	165.5	0.3058	0.2504	1.1011	614.4	290
300	1.11012	168.1	0.3106	0.2517	1.0986	620.0		1.03402	168.0	0.3091	0.2521	1.0996	619.0	300
310	1.12625	170.6	0.3139	0.2534	1.0972	624.6		1.04910	170.5	0.3124	0.2537	1.0982	623.6	310
320	1.14233	173.1	0.3172	0.2551	1.0959	629.1		1.06417	173.0	0.3157	0.2554	1.0968	628.2	320
330	1.15835	175.7	0.3204	0.2567	1.0947	633.5		1.07921	175.6	0.3190	0.2571	1.0955	632.6	330
340	1.17426	178.3	0.3237	0.2584	1.0935	637.9		1.09433	178.2	0.3222	0.2587	1.0943	637.1	340
350	1.19033	180.9	0.3269	0.2601	1.0923	642.3		1.10926	180.8	0.3254	0.2604	1.0931	641.5	350
360	1.20642	183.5	0.3301	0.2618	1.0913	646.6		1.12410	183.4	0.3287	0.2620	1.0920	645.8	360
370	—	—	—	—	—	—		1.13908	186.0	0.3318	0.2637	1.0909	650.1	370

Appendix **383**

Table A-15 (continued)

HFC-134a Superheated Vapor—Constant Pressure Tables

V = Volume in ft³/lb H = Enthalpy in Btu/lb S = Entropy in Btu/(lb) (°R) v_s = Velocity of Sound in ft/sec
Cp = Heat Capacity at Constant Pressure in Btu/(lb) (°F)   Cp/Cv = Heat Capacity Ratio (Dimensionless)

TEMP °F	PRESSURE = 80.00 PSIA							PRESSURE = 85.00 PSIA						TEMP °F
	V	H	S	Cp	Cp/Cv	v_s		V	H	S	Cp	Cp/Cv	v_s	
65.9	0.01304	33.3	0.0703	0.3159	1.5492	1766.4	SAT LIQ	0.01311	34.5	0.0725	0.3378	1.5530	1736.6	69.4
65.9	0.59787	112.3	0.2205	0.2342	1.2120	477.0	SAT VAP	0.56284	112.7	0.2204	0.2369	1.2168	476.0	69.4
70	0.60580	113.2	0.2224	0.2333	1.2058	480.7		0.56395	112.9	0.2206	0.2367	1.2158	476.6	70
80	0.62445	115.6	0.2267	0.2317	1.1928	489.1		0.58200	115.2	0.2250	0.2345	1.2011	485.4	80
90	0.64255	117.9	0.2309	0.2306	1.1818	497.1		0.59945	117.6	0.2293	0.2331	1.1888	493.6	90
100	0.66020	120.2	0.2351	0.2300	1.1723	504.7		0.61641	119.9	0.2335	0.2322	1.1784	501.5	100
110	0.67741	122.5	0.2392	0.2298	1.1642	511.9		0.63295	122.2	0.2376	0.2317	1.1695	509.0	110
120	0.69430	124.8	0.2432	0.2299	1.1571	518.9		0.64910	124.5	0.2417	0.2316	1.1617	516.2	120
130	0.71083	127.1	0.2471	0.2303	1.1508	525.6		0.66494	126.8	0.2456	0.2318	1.1549	523.1	130
140	0.72711	129.4	0.2510	0.2308	1.1453	532.1		0.68055	129.1	0.2495	0.2322	1.1489	529.8	140
150	0.74322	131.7	0.2548	0.2316	1.1403	538.4		0.69585	131.5	0.2534	0.2328	1.1436	536.2	150
160	0.75896	134.0	0.2586	0.2325	1.1358	544.6		0.71093	133.8	0.2572	0.2336	1.1388	542.5	160
170	0.77465	136.3	0.2623	0.2335	1.1318	550.5		0.72574	136.1	0.2609	0.2345	1.1345	548.6	170
180	0.79020	138.7	0.2660	0.2347	1.1281	556.4		0.74052	138.5	0.2646	0.2356	1.1306	554.5	180
190	0.80548	141.0	0.2696	0.2359	1.1248	562.0		0.75506	140.9	0.2683	0.2367	1.1270	560.3	190
200	0.82068	143.4	0.2733	0.2372	1.1217	567.6		0.76953	143.2	0.2719	0.2380	1.1237	565.9	200
210	0.83577	145.8	0.2768	0.2386	1.1188	573.0		0.78388	145.6	0.2755	0.2393	1.1207	571.5	210
220	0.85063	148.2	0.2804	0.2400	1.1162	578.4		0.79802	148.0	0.2790	0.2406	1.1180	576.9	220
230	0.86558	150.6	0.2839	0.2414	1.1138	583.6		0.81208	150.4	0.2826	0.2420	1.1154	582.2	230
240	0.88020	153.0	0.2874	0.2429	1.1115	588.8		0.82617	152.9	0.2861	0.2435	1.1130	587.4	240
250	0.89501	155.4	0.2908	0.2445	1.1094	593.8		0.84005	155.3	0.2895	0.2450	1.1108	592.5	250
260	0.90950	157.9	0.2943	0.2460	1.1074	598.8		0.85383	157.8	0.2930	0.2465	1.1087	597.6	260
270	0.92404	160.4	0.2977	0.2476	1.1056	603.7		0.86760	160.2	0.2964	0.2481	1.1068	602.5	270
280	0.93853	162.8	0.3011	0.2492	1.1038	608.5		0.88137	162.7	0.2998	0.2496	1.1050	607.4	280
290	0.95302	165.3	0.3044	0.2508	1.1021	613.3		0.89493	165.2	0.3031	0.2512	1.1032	612.2	290
300	0.96721	167.9	0.3078	0.2525	1.1006	618.0		0.90851	167.7	0.3065	0.2528	1.1016	617.0	300
310	0.98155	170.4	0.3111	0.2541	1.0991	622.6		0.92200	170.3	0.3098	0.2544	1.1001	621.7	310
320	0.99582	172.9	0.3144	0.2557	1.0977	627.2		0.93545	172.8	0.3131	0.2561	1.0986	626.3	320
330	1.01010	175.5	0.3176	0.2574	1.0964	631.8		0.94886	175.4	0.3164	0.2577	1.0972	630.9	330
340	1.02417	178.1	0.3209	0.2590	1.0951	636.2		0.96219	178.0	0.3196	0.2593	1.0959	635.4	340
350	1.03821	180.7	0.3241	0.2606	1.0939	640.7		0.97570	180.6	0.3228	0.2609	1.0947	639.9	350
360	1.05230	183.3	0.3273	0.2623	1.0927	645.0		0.98892	183.2	0.3260	0.2625	1.0935	644.3	360
370	1.06644	185.9	0.3305	0.2639	1.0916	649.4		1.00220	185.8	0.3292	0.2642	1.0923	648.6	370

TEMP °F	PRESSURE = 90.00 PSIA							PRESSURE = 95.00 PSIA						TEMP °F
	V	H	S	Cp	Cp/Cv	v_s		V	H	S	Cp	Cp/Cv	v_s	
72.8	0.01319	35.7	0.0747	0.3396	1.5568	1708.1	SAT LIQ	0.01326	36.8	0.0767	0.3415	1.5607	1680.6	76
72.8	0.53155	113.1	0.2202	0.2395	1.2217	475.0	SAT VAP	0.50345	113.5	0.2200	0.2420	1.2266	473.9	76
80	0.54416	114.9	0.2234	0.2376	1.2100	481.6		0.51020	114.5	0.2218	0.2408	1.2196	477.6	80
90	0.56107	117.2	0.2278	0.2357	1.1964	490.2		0.52662	116.9	0.2262	0.2384	1.2044	486.6	90
100	0.57743	119.6	0.2320	0.2344	1.1849	498.3		0.54248	119.3	0.2305	0.2368	1.1917	495.0	100
110	0.59333	121.9	0.2361	0.2337	1.1751	506.0		0.55785	121.6	0.2347	0.2357	1.1809	503.0	110
120	0.60887	124.3	0.2402	0.2333	1.1666	513.4		0.57284	124.0	0.2388	0.2351	1.1717	510.6	120
130	0.62406	126.6	0.2442	0.2333	1.1592	520.5		0.58751	126.3	0.2428	0.2349	1.1637	517.9	130
140	0.63906	128.9	0.2481	0.2336	1.1527	527.4		0.60183	128.7	0.2468	0.2350	1.1567	524.9	140
150	0.65368	131.3	0.2520	0.2341	1.1470	534.0		0.61588	131.0	0.2507	0.2353	1.1505	531.7	150
160	0.66814	133.6	0.2558	0.2347	1.1418	540.4		0.62980	133.4	0.2545	0.2359	1.1450	538.2	160
170	0.68236	136.0	0.2596	0.2355	1.1372	546.6		0.64346	135.8	0.2583	0.2366	1.1401	544.6	170
180	0.69643	138.3	0.2633	0.2365	1.1331	552.6		0.65690	138.1	0.2620	0.2374	1.1356	550.7	180
190	0.71023	140.7	0.2670	0.2376	1.1293	558.5		0.67024	140.5	0.2657	0.2384	1.1316	556.7	190
200	0.72401	143.1	0.2706	0.2387	1.1258	564.2		0.68339	142.9	0.2694	0.2395	1.1280	562.6	200
210	0.73768	145.5	0.2742	0.2400	1.1227	569.9		0.69643	145.3	0.2730	0.2407	1.1246	568.3	210
220	0.75126	147.9	0.2778	0.2413	1.1197	575.4		0.70932	147.7	0.2765	0.2419	1.1215	573.9	220
230	0.76470	150.3	0.2813	0.2426	1.1170	580.8		0.72218	150.1	0.2801	0.2433	1.1187	579.3	230
240	0.77791	152.7	0.2848	0.2441	1.1145	586.0		0.73486	152.6	0.2836	0.2446	1.1161	584.7	240
250	0.79108	155.2	0.2883	0.2455	1.1122	591.2		0.74744	155.0	0.2871	0.2460	1.1137	589.9	250
260	0.80431	157.6	0.2917	0.2470	1.1101	596.3		0.76005	157.5	0.2905	0.2475	1.1114	595.1	260
270	0.81746	160.1	0.2951	0.2485	1.1080	601.4		0.77250	160.0	0.2940	0.2490	1.1093	600.2	270
280	0.83036	162.6	0.2985	0.2501	1.1061	606.3		0.78493	162.5	0.2974	0.2505	1.1073	605.2	280
290	0.84331	165.1	0.3019	0.2516	1.1043	611.2		0.79726	165.0	0.3007	0.2520	1.1054	610.1	290
300	0.85616	167.6	0.3052	0.2532	1.1026	616.0		0.80952	167.5	0.3041	0.2536	1.1037	614.9	300
310	0.86904	170.2	0.3086	0.2548	1.1010	620.7		0.82169	170.1	0.3074	0.2551	1.1020	619.7	310
320	0.88191	172.7	0.3119	0.2564	1.0995	625.4		0.83389	172.6	0.3107	0.2567	1.1005	624.4	320
330	0.89461	175.3	0.3151	0.2580	1.0981	630.0		0.84595	175.2	0.3140	0.2583	1.0990	629.1	330
340	0.90728	177.9	0.3184	0.2596	1.0967	634.5		0.85815	177.8	0.3173	0.2599	1.0976	633.7	340
350	0.91988	180.5	0.3216	0.2612	1.0955	639.0		0.87017	180.4	0.3205	0.2615	1.0962	638.2	350
360	0.93257	183.1	0.3249	0.2628	1.0942	643.5		0.88215	183.0	0.3237	0.2631	1.0950	642.7	360
370	0.94518	185.7	0.3280	0.2644	1.0930	647.9		0.89405	185.7	0.3269	0.2647	1.0938	647.2	370
380	0.95767	188.4	0.3312	0.2660	1.0919	652.3		0.90613	188.3	0.3301	0.2663	1.0926	651.5	380

Table A-15 (continued)

HFC-134a Superheated Vapor—Constant Pressure Tables

V = Volume in ft³/lb   H = Enthalpy in Btu/lb   S = Entropy in Btu/(lb) (°R) v_s = Velocity of Sound in ft/sec
Cp = Heat Capacity at Constant Pressure in Btu/(lb) (°F)   Cp/Cv = Heat Capacity Ratio (Dimensionless)

TEMP °F	PRESSURE = 100.00 PSIA							PRESSURE = 110.00 PSIA						TEMP °F
	V	H	S	Cp	Cp/Cv	v_s		V	H	S	Cp	Cp/Cv	v_s	
79.1	0.01333	37.8	0.0787	0.3433	1.5646	1654.2	SAT LIQ	0.01347	39.8	0.0824	0.3469	1.5726	1604.1	85
79.1	0.47803	113.9	0.2199	0.2446	1.2317	472.8	SAT VAP	0.43391	114.6	0.2196	0.2496	1.2420	470.4	85
80	0.47952	114.1	0.2203	0.2442	1.2300	473.6		—	—	—	—	—	—	80
90	0.49552	116.6	0.2248	0.2413	1.2129	482.9		0.44156	115.9	0.2219	0.2477	1.2319	475.4	90
100	0.51093	119.0	0.2291	0.2393	1.1989	491.7		0.45627	118.3	0.2264	0.2446	1.2146	484.8	100
110	0.52587	121.4	0.2333	0.2379	1.1871	499.9		0.47043	120.8	0.2307	0.2425	1.2004	493.6	110
120	0.54037	123.7	0.2375	0.2370	1.1770	507.8		0.48414	123.2	0.2349	0.2410	1.1884	502.0	120
130	0.55451	126.1	0.2415	0.2366	1.1683	515.3		0.49746	125.6	0.2390	0.2401	1.1782	509.9	130
140	0.56838	128.5	0.2455	0.2365	1.1608	522.5		0.51044	128.0	0.2430	0.2395	1.1694	517.5	140
150	0.58194	130.8	0.2494	0.2366	1.1541	529.4		0.52309	130.4	0.2470	0.2394	1.1618	524.7	150
160	0.59527	133.2	0.2533	0.2370	1.1482	536.1		0.53562	132.8	0.2509	0.2395	1.1550	531.7	160
170	0.60835	135.6	0.2570	0.2376	1.1430	542.5		0.54786	135.2	0.2547	0.2399	1.1491	538.5	170
180	0.62135	137.9	0.2608	0.2384	1.1383	548.8		0.55979	137.6	0.2585	0.2404	1.1438	545.0	180
190	0.63416	140.3	0.2645	0.2393	1.1340	554.9		0.57166	140.0	0.2622	0.2411	1.1390	551.3	190
200	0.64675	142.7	0.2682	0.2403	1.1301	560.9		0.58340	142.4	0.2659	0.2420	1.1347	557.4	200
210	0.65924	145.1	0.2718	0.2414	1.1266	566.7		0.59503	144.8	0.2696	0.2429	1.1308	563.4	210
220	0.67155	147.6	0.2754	0.2426	1.1234	572.3		0.60643	147.3	0.2732	0.2440	1.1272	569.3	220
230	0.68385	150.0	0.2789	0.2439	1.1204	577.9		0.61774	149.7	0.2768	0.2452	1.1239	575.0	230
240	0.69604	152.4	0.2825	0.2452	1.1177	583.3		0.62893	152.2	0.2803	0.2464	1.1209	580.5	240
250	0.70806	154.9	0.2859	0.2466	1.1151	588.6		0.64012	154.6	0.2838	0.2477	1.1181	586.0	250
260	0.72015	157.4	0.2894	0.2480	1.1128	593.9		0.65121	157.1	0.2873	0.2490	1.1156	591.4	260
270	0.73196	159.9	0.2928	0.2495	1.1106	599.0		0.66212	159.6	0.2907	0.2504	1.1132	596.6	270
280	0.74388	162.4	0.2962	0.2509	1.1085	604.1		0.67308	162.1	0.2941	0.2518	1.1109	601.8	280
290	0.75569	164.9	0.2996	0.2524	1.1066	609.0		0.68385	164.6	0.2975	0.2533	1.1088	606.9	290
300	0.76740	167.4	0.3030	0.2540	1.1047	613.9		0.69464	167.2	0.3009	0.2547	1.1069	611.9	300
310	0.77906	170.0	0.3063	0.2555	1.1030	618.7		0.70542	169.7	0.3042	0.2562	1.1050	616.8	310
320	0.79064	172.5	0.3096	0.2571	1.1014	623.5		0.71613	172.3	0.3076	0.2577	1.1033	621.6	320
330	0.80225	175.1	0.3129	0.2586	1.0999	628.2		0.72680	174.9	0.3109	0.2593	1.1017	626.4	330
340	0.81387	177.7	0.3162	0.2602	1.0984	632.8		0.73725	177.5	0.3141	0.2608	1.1001	631.1	340
350	0.82535	180.3	0.3194	0.2618	1.0970	637.4		0.74783	180.1	0.3174	0.2624	1.0987	635.8	350
360	0.83675	182.9	0.3226	0.2634	1.0957	641.9		0.75832	182.7	0.3206	0.2639	1.0973	640.4	360
370	0.84818	185.6	0.3258	0.2649	1.0945	646.4		0.76870	185.4	0.3238	0.2655	1.0959	644.9	370
380	0.85955	188.2	0.3290	0.2665	1.0933	650.8		0.77924	188.0	0.3270	0.2670	1.0947	649.4	380
390	—	—	—	—	—	—		0.78958	190.7	0.3302	0.2686	1.0935	653.8	390

TEMP °F	PRESSURE = 120.00 PSIA							PRESSURE = 130.00 PSIA						TEMP °F
	V	H	S	Cp	Cp/Cv	v_s		V	H	S	Cp	Cp/Cv	v_s	
90.5	0.01361	41.8	0.0858	0.3504	1.5808	1557.1	SAT LIQ	0.01374	43.6	0.0890	0.3540	1.5893	1512.8	95.6
90.5	0.39689	115.3	0.2194	0.2546	1.2528	467.9	SAT VAP	0.36538	115.8	0.2192	0.2596	1.2640	465.4	95.6
100	0.41044	117.7	0.2238	0.2506	1.2326	477.6		0.37136	117.0	0.2212	0.2573	1.2531	470.1	100
110	0.42402	120.2	0.2282	0.2475	1.2153	487.1		0.38453	119.5	0.2257	0.2531	1.2321	480.3	110
120	0.43710	122.6	0.2324	0.2453	1.2010	496.0		0.39712	122.0	0.2301	0.2501	1.2150	489.8	120
130	0.44976	125.1	0.2366	0.2438	1.1890	504.4		0.40925	124.5	0.2344	0.2479	1.2009	498.7	130
140	0.46206	127.5	0.2407	0.2428	1.1788	512.4		0.42100	127.0	0.2385	0.2464	1.1890	507.1	140
150	0.47405	129.9	0.2447	0.2423	1.1700	520.0		0.43241	129.5	0.2426	0.2454	1.1788	515.1	150
160	0.48577	132.3	0.2487	0.2421	1.1623	527.3		0.44350	131.9	0.2466	0.2448	1.1701	522.7	160
170	0.49724	134.8	0.2525	0.2422	1.1555	534.3		0.45442	134.4	0.2505	0.2446	1.1624	530.0	170
180	0.50860	137.2	0.2564	0.2425	1.1495	541.1		0.46507	136.8	0.2544	0.2447	1.1557	537.1	180
190	0.51967	139.6	0.2601	0.2430	1.1442	547.6		0.47556	139.2	0.2582	0.2450	1.1497	543.9	190
200	0.53064	142.0	0.2639	0.2437	1.1394	554.0		0.48584	141.7	0.2619	0.2455	1.1444	550.5	200
210	0.54139	144.5	0.2675	0.2445	1.1351	560.1		0.49601	144.2	0.2656	0.2461	1.1396	556.8	210
220	0.55206	146.9	0.2712	0.2454	1.1311	566.2		0.50602	146.6	0.2692	0.2469	1.1352	563.0	220
230	0.56268	149.4	0.2747	0.2465	1.1275	572.0		0.51597	149.1	0.2729	0.2478	1.1313	569.1	230
240	0.57307	151.9	0.2783	0.2476	1.1243	577.8		0.52576	151.6	0.2764	0.2489	1.1277	575.0	240
250	0.58350	154.4	0.2818	0.2488	1.1212	583.4		0.53550	154.1	0.2800	0.2500	1.1244	580.7	250
260	0.59379	156.8	0.2853	0.2501	1.1184	588.9		0.54511	156.6	0.2835	0.2511	1.1214	586.3	260
270	0.60394	159.4	0.2888	0.2514	1.1158	594.2		0.55460	159.1	0.2870	0.2524	1.1186	591.8	270
280	0.61406	161.9	0.2922	0.2527	1.1134	599.5		0.56408	161.6	0.2904	0.2536	1.1160	597.2	280
290	0.62414	164.4	0.2956	0.2541	1.1112	604.7		0.57353	164.2	0.2938	0.2550	1.1136	602.5	290
300	0.63408	167.0	0.2990	0.2555	1.1091	609.8		0.58282	166.7	0.2972	0.2563	1.1113	607.7	300
310	0.64404	169.5	0.3023	0.2570	1.1071	614.8		0.59207	169.3	0.3006	0.2577	1.1092	612.8	310
320	0.65389	172.1	0.3057	0.2584	1.1052	619.8		0.60129	171.9	0.3039	0.2592	1.1072	617.9	320
330	0.66375	174.7	0.3090	0.2599	1.1035	624.6		0.61039	174.5	0.3072	0.2606	1.1054	622.8	330
340	0.67354	177.3	0.3122	0.2614	1.1019	629.4		0.61962	177.1	0.3105	0.2621	1.1036	627.7	340
350	0.68329	179.9	0.3155	0.2629	1.1003	634.1		0.62865	179.7	0.3138	0.2635	1.1020	632.5	350
360	0.69300	182.6	0.3187	0.2645	1.0988	638.8		0.63771	182.4	0.3170	0.2650	1.1004	637.2	360
370	0.70264	185.2	0.3220	0.2660	1.0974	643.4		0.64666	185.0	0.3202	0.2665	1.0989	641.9	370
380	0.71225	187.9	0.3252	0.2675	1.0961	648.0		0.65569	187.7	0.3234	0.2680	1.0975	646.5	380
390	0.72192	190.6	0.3283	0.2690	1.0948	652.5		0.66458	190.4	0.3266	0.2695	1.0962	651.1	390
400	0.73142	193.3	0.3315	0.2706	1.0936	656.9		0.67345	193.1	0.3298	0.2710	1.0949	655.6	400

Table A-15 (continued)

HFC-134a Superheated Vapor—Constant Pressure Tables

V = Volume in ft³/lb   H = Enthalpy in Btu/lb   S = Entropy in Btu/(lb) (°R) v_s = Velocity of Sound in ft/sec
Cp = Heat Capacity at Constant Pressure in Btu/(lb) (°F)   Cp/Cv = Heat Capacity Ratio (Dimensionless)

TEMP °F	PRESSURE = 140.00 PSIA							PRESSURE = 150.00 PSIA						TEMP °F
	V	H	S	Cp	Cp/Cv	v_s		V	H	S	Cp	Cp/Cv	v_s	
100.5	0.01388	45.3	0.0921	0.3576	1.5980	1470.8	SAT LIQ	0.01401	46.9	0.0950	0.3612	1.6070	1430.7	105.1
100.5	0.33818	116.4	0.2190	0.2646	1.2757	462.7	SAT VAP	0.31448	116.9	0.2188	0.2697	1.2878	460.0	105.1
110	0.35042	118.9	0.2234	0.2593	1.2512	473.2		0.32062	118.2	0.2211	0.2664	1.2732	465.7	110
120	0.36266	121.4	0.2279	0.2553	1.2307	483.3		0.33259	120.8	0.2257	0.2610	1.2484	476.6	120
130	0.37438	124.0	0.2322	0.2523	1.2140	492.8		0.34400	123.4	0.2301	0.2571	1.2285	486.7	130
140	0.38568	126.5	0.2364	0.2502	1.2001	501.7		0.35495	126.0	0.2344	0.2543	1.2123	496.1	140
150	0.39662	129.0	0.2406	0.2487	1.1884	510.1		0.36550	128.5	0.2386	0.2523	1.1988	505.0	150
160	0.40727	131.5	0.2446	0.2477	1.1784	518.1		0.37573	131.0	0.2427	0.2508	1.1874	513.3	160
170	0.41764	133.9	0.2486	0.2472	1.1697	525.7		0.38565	133.5	0.2467	0.2499	1.1776	521.3	170
180	0.42779	136.4	0.2524	0.2470	1.1622	533.0		0.39538	136.0	0.2506	0.2494	1.1691	528.9	180
190	0.43777	138.9	0.2563	0.2470	1.1555	540.1		0.40491	138.5	0.2545	0.2492	1.1616	536.2	190
200	0.44755	141.3	0.2601	0.2473	1.1496	546.9		0.41418	141.0	0.2583	0.2493	1.1550	543.3	200
210	0.45714	143.8	0.2638	0.2478	1.1443	553.5		0.42337	143.5	0.2621	0.2496	1.1492	550.1	210
220	0.46659	146.3	0.2675	0.2485	1.1395	559.9		0.43232	146.0	0.2658	0.2501	1.1440	556.7	220
230	0.47596	148.8	0.2711	0.2492	1.1352	566.1		0.44125	148.5	0.2694	0.2507	1.1393	563.1	230
240	0.48520	151.3	0.2747	0.2501	1.1313	572.1		0.44996	151.0	0.2730	0.2515	1.1350	569.3	240
250	0.49432	153.8	0.2782	0.2511	1.1277	578.0		0.45861	153.5	0.2766	0.2524	1.1311	575.3	250
260	0.50342	156.3	0.2818	0.2522	1.1244	583.8		0.46718	156.0	0.2801	0.2533	1.1276	581.2	260
270	0.51235	158.8	0.2853	0.2534	1.1214	589.4		0.47574	158.6	0.2836	0.2544	1.1243	587.0	270
280	0.52121	161.4	0.2887	0.2546	1.1186	594.9		0.48412	161.1	0.2871	0.2555	1.1213	592.6	280
290	0.53011	163.9	0.2921	0.2558	1.1160	600.3		0.49242	163.7	0.2906	0.2567	1.1185	598.1	290
300	0.53888	166.5	0.2955	0.2572	1.1136	605.6		0.50073	166.3	0.2940	0.2580	1.1159	603.6	300
310	0.54750	169.1	0.2989	0.2585	1.1113	610.9		0.50893	168.9	0.2973	0.2593	1.1135	608.9	310
320	0.55611	171.7	0.3022	0.2599	1.1092	616.0		0.51706	171.5	0.3007	0.2606	1.1113	614.1	320
330	0.56478	174.3	0.3056	0.2613	1.1073	621.0		0.52521	174.1	0.3040	0.2620	1.1092	619.2	330
340	0.57330	176.9	0.3089	0.2627	1.1054	626.0		0.53319	176.7	0.3073	0.2633	1.1072	624.2	340
350	0.58184	179.5	0.3121	0.2641	1.1036	630.9		0.54121	179.3	0.3106	0.2647	1.1054	629.2	350
360	0.59028	182.2	0.3154	0.2656	1.1020	635.7		0.54918	182.0	0.3139	0.2662	1.1036	634.1	360
370	0.59866	184.8	0.3186	0.2670	1.1004	640.4		0.55710	184.7	0.3171	0.2676	1.1020	638.9	370
380	0.60716	187.5	0.3218	0.2685	1.0989	645.1		0.56500	187.3	0.3203	0.2690	1.1004	643.7	380
390	0.61542	190.2	0.3250	0.2700	1.0975	649.7		0.57290	190.0	0.3235	0.2705	1.0989	648.4	390
400	0.62375	192.9	0.3282	0.2715	1.0962	654.3		0.58069	192.7	0.3267	0.2719	1.0975	653.0	400
410	0.63203	195.6	0.3313	0.2730	1.0949	658.8		0.58851	195.5	0.3298	0.2734	1.0962	657.6	410

TEMP °F	PRESSURE = 160.00 PSIA							PRESSURE = 170.00 PSIA						TEMP °F
	V	H	S	Cp	Cp/Cv	v_s		V	H	S	Cp	Cp/Cv	v_s	
109.5	0.01414	48.5	0.0977	0.3648	1.6163	1392.4	SAT LIQ	0.01427	50.1	0.1004	0.3685	1.6259	1355.6	113.7
109.5	0.29362	117.3	0.2186	0.2748	1.3006	457.2	SAT VAP	0.27512	117.7	0.2184	0.2800	1.3139	454.4	113.7
110	0.29426	117.5	0.2188	0.2744	1.2988	457.9		—	—	—	—	—	—	110
120	0.30608	120.2	0.2235	0.2675	1.2685	469.6		0.28247	119.5	0.2214	0.2748	1.2915	462.3	120
130	0.31726	122.8	0.2281	0.2624	1.2448	480.4		0.29350	122.2	0.2261	0.2683	1.2631	473.8	130
140	0.32792	125.4	0.2324	0.2588	1.2258	490.4		0.30395	124.9	0.2305	0.2636	1.2406	484.4	140
150	0.33817	128.0	0.2367	0.2561	1.2101	499.7		0.31395	127.5	0.2349	0.2602	1.2225	494.3	150
160	0.34806	130.5	0.2409	0.2541	1.1970	508.5		0.32356	130.1	0.2391	0.2577	1.2075	503.5	160
170	0.35767	133.1	0.2449	0.2528	1.1859	516.8		0.33282	132.6	0.2432	0.2559	1.1949	512.2	170
180	0.36700	135.6	0.2489	0.2519	1.1764	524.7		0.34189	135.2	0.2472	0.2546	1.1842	520.5	180
190	0.37611	138.1	0.2528	0.2515	1.1681	532.3		0.35066	137.7	0.2512	0.2538	1.1749	528.4	190
200	0.38504	140.6	0.2566	0.2513	1.1608	539.6		0.35923	140.3	0.2550	0.2534	1.1669	535.9	200
210	0.39378	143.1	0.2604	0.2514	1.1544	546.7		0.36762	142.8	0.2588	0.2533	1.1598	543.2	210
220	0.40238	145.7	0.2641	0.2517	1.1486	553.4		0.37591	145.3	0.2626	0.2534	1.1535	550.2	220
230	0.41085	148.2	0.2678	0.2522	1.1435	560.0		0.38400	147.9	0.2663	0.2538	1.1479	556.9	230
240	0.41916	150.7	0.2715	0.2528	1.1388	566.4		0.39200	150.4	0.2700	0.2543	1.1428	563.5	240
250	0.42746	153.2	0.2751	0.2536	1.1346	572.6		0.39987	152.9	0.2736	0.2549	1.1383	569.9	250
260	0.43554	155.8	0.2786	0.2545	1.1308	578.6		0.40758	155.5	0.2771	0.2557	1.1341	576.0	260
270	0.44364	158.3	0.2821	0.2555	1.1273	584.5		0.41533	158.1	0.2807	0.2566	1.1304	582.1	270
280	0.45165	160.9	0.2856	0.2565	1.1241	590.3		0.42287	160.6	0.2842	0.2575	1.1269	588.0	280
290	0.45954	163.5	0.2891	0.2577	1.1211	595.9		0.43042	163.2	0.2876	0.2586	1.1237	593.7	290
300	0.46729	166.0	0.2925	0.2588	1.1183	601.4		0.43787	165.8	0.2911	0.2597	1.1208	599.3	300
310	0.47508	168.6	0.2959	0.2601	1.1158	606.9		0.44528	168.4	0.2945	0.2609	1.1181	604.8	310
320	0.48281	171.2	0.2992	0.2613	1.1134	612.2		0.45265	171.0	0.2979	0.2621	1.1155	610.3	320
330	0.49053	173.9	0.3026	0.2627	1.1112	617.4		0.45996	173.6	0.3012	0.2634	1.1132	615.6	330
340	0.49816	176.5	0.3059	0.2640	1.1091	622.5		0.46718	176.3	0.3045	0.2647	1.1110	620.8	340
350	0.50566	179.1	0.3092	0.2654	1.1071	627.6		0.47432	178.9	0.3078	0.2660	1.1089	625.9	350
360	0.51324	181.8	0.3124	0.2667	1.1053	632.5		0.48144	181.6	0.3111	0.2673	1.1069	630.9	360
370	0.52067	184.5	0.3157	0.2681	1.1035	637.4		0.48854	184.3	0.3143	0.2687	1.1051	635.9	370
380	0.52818	187.2	0.3189	0.2696	1.1019	642.2		0.49564	187.0	0.3176	0.2701	1.1034	640.8	380
390	0.53565	189.9	0.3221	0.2710	1.1003	647.0		0.50269	189.7	0.3208	0.2715	1.1018	645.6	390
400	0.54301	192.6	0.3253	0.2724	1.0988	651.7		0.50968	192.4	0.3240	0.2729	1.1002	650.4	400
410	0.55030	195.3	0.3284	0.2738	1.0974	656.3		0.51674	195.1	0.3271	0.2743	1.0987	655.0	410
420	—	—	—	—	—	—		0.52362	197.9	0.3303	0.2757	1.0973	659.7	420

© 2004. E.I. du Pont de Nemours and Company (DuPont). All rights reserved, used under license of DuPont.

Table A-15 (continued)

"FREON" 22 SATURATION PROPERTIES—TEMPERATURE TABLE

TEMP. °F	PRESSURE PSIA	PRESSURE PSIG	VOLUME cu ft/lb LIQUID v_f	VOLUME cu ft/lb VAPOR v_g	DENSITY lb/cu ft LIQUID $1/v_f$	DENSITY lb/cu ft VAPOR $1/v_g$	ENTHALPY Btu/lb LIQUID h_f	ENTHALPY Btu/lb LATENT h_{fg}	ENTHALPY Btu/lb VAPOR h_g	ENTROPY Btu/(lb)(°R) LIQUID s_f	ENTROPY Btu/(lb)(°R) VAPOR s_g	TEMP. °F
−100	2.3983	25.0383*	0.010664	18.433	93.770	0.054252	−14.564	107.935	93.371	−0.03734	0.26274	−100
−99	2.4877	24.8563*	0.010675	17.815	93.678	0.056133	−14.331	107.819	93.488	−0.03670	0.26223	−99
−98	2.5798	24.6688*	0.010685	17.222	93.585	0.058066	−14.097	107.702	93.606	−0.03605	0.26173	−98
−97	2.6747	24.4755*	0.010696	16.652	93.493	0.060053	−13.863	107.586	93.723	−0.03540	0.26123	−97
−96	2.7724	24.2765*	0.010707	16.104	93.401	0.062095	−13.628	107.469	93.840	−0.03476	0.26074	−96
−95	2.8731	24.0715*	0.010717	15.578	93.308	0.064192	−13.393	107.351	93.958	−0.03411	0.26025	−95
−94	2.9768	23.8604*	0.010728	15.072	93.215	0.066347	−13.158	107.233	94.075	−0.03347	0.25977	−94
−93	3.0836	23.6430*	0.010739	14.586	93.123	0.068559	−12.923	107.115	94.192	−0.03283	0.25929	−93
−92	3.1934	23.4193*	0.010749	14.118	93.030	0.070831	−12.688	106.997	94.309	−0.03219	0.25881	−92
−91	3.3065	23.1891*	0.010760	13.668	92.937	0.073163	−12.452	106.878	94.426	−0.03155	0.25834	−91
−90	3.4229	22.9522*	0.010771	13.235	92.843	0.075557	−12.216	106.759	94.544	−0.03091	0.25787	−90
−89	3.5426	22.7085*	0.010782	12.818	92.750	0.078014	−11.979	106.640	94.661	−0.03027	0.25741	−89
−88	3.6657	22.4579*	0.010793	12.417	92.657	0.080534	−11.743	106.520	94.777	−0.02963	0.25695	−88
−87	3.7922	22.2002*	0.010803	12.031	92.563	0.083120	−11.506	106.400	94.894	−0.02900	0.25649	−87
−86	3.9224	21.9352*	0.010814	11.659	92.469	0.085772	−11.268	106.279	95.011	−0.02836	0.25604	−86
−85	4.0562	21.6628*	0.010825	11.301	92.376	0.088491	−11.031	106.159	95.128	−0.02773	0.25560	−85
−84	4.1936	21.3829*	0.010836	10.955	92.282	0.091280	−10.793	106.037	95.244	−0.02709	0.25515	−84
−83	4.3349	21.0953*	0.010847	10.623	92.188	0.094138	−10.555	105.916	95.361	−0.02646	0.25471	−83
−82	4.4800	20.7998*	0.010859	10.302	92.093	0.097068	−10.316	105.794	95.478	−0.02583	0.25428	−82
−81	4.6291	20.4963*	0.010870	9.9929	91.999	0.10007	−10.077	105.671	95.594	−0.02520	0.25384	−81
−80	4.7822	20.1846*	0.010881	9.6949	91.905	0.10315	−9.838	105.548	95.710	−0.02457	0.25342	−80
−79	4.9394	19.8646*	0.010892	9.4074	91.810	0.10630	−9.599	105.425	95.827	−0.02394	0.25299	−79
−78	5.1007	19.5361*	0.010903	9.1301	91.715	0.10953	−9.359	105.302	95.943	−0.02331	0.25257	−78
−77	5.2664	19.1989*	0.010915	8.8625	91.620	0.11283	−9.119	105.178	96.059	−0.02269	0.25215	−77
−76	5.4363	18.8528*	0.010926	8.6043	91.525	0.11622	−8.878	105.053	96.175	−0.02206	0.25174	−76
−75	5.6107	18.4977*	0.010937	8.3551	91.430	0.11969	−8.638	104.928	96.290	−0.02143	0.25133	−75
−74	5.7896	18.1334*	0.010949	8.1145	91.335	0.12324	−8.397	104.803	96.406	−0.02081	0.25092	−74
−73	5.9732	17.7598*	0.010960	7.8822	91.240	0.12687	−8.155	104.677	96.522	−0.02019	0.25052	−73
−72	6.1614	17.3766*	0.010972	7.6579	91.144	0.13058	−7.914	104.551	96.637	−0.01956	0.25012	−72
−71	6.3543	16.9837*	0.010983	7.4412	91.048	0.13439	−7.672	104.424	96.753	−0.01894	0.24972	−71
−70	6.5522	16.5809*	0.010995	7.2318	90.952	0.13828	−7.429	104.297	96.868	−0.01832	0.24932	−70
−69	6.7550	16.1680*	0.011006	7.0295	90.856	0.14226	−7.187	104.170	96.983	−0.01770	0.24893	−69
−68	6.9628	15.7449*	0.011018	6.8339	90.760	0.14633	−6.944	104.042	97.098	−0.01708	0.24855	−68
−67	7.1757	15.3113*	0.011030	6.6449	90.664	0.15049	−6.700	103.913	97.213	−0.01646	0.24816	−67
−66	7.3939	14.8671*	0.011041	6.4621	90.568	0.15475	−6.457	103.785	97.328	−0.01584	0.24778	−66
−65	7.6174	14.4120*	0.011053	6.2854	90.471	0.15910	−6.213	103.655	97.443	−0.01522	0.24740	−65
−64	7.8463	13.9460*	0.011065	6.1144	90.374	0.16355	−5.968	103.525	97.557	−0.01460	0.24703	−64
−63	8.0808	13.4687*	0.011077	5.9491	90.277	0.16809	−5.724	103.395	97.672	−0.01399	0.24666	−63
−62	8.3208	12.9800*	0.011089	5.7891	90.180	0.17274	−5.479	103.264	97.786	−0.01337	0.24629	−62
−61	8.5665	12.4798*	0.011101	5.6343	90.083	0.17749	−5.233	103.133	97.900	−0.01276	0.24592	−61
−60	8.8180	11.9677*	0.011113	5.4844	89.986	0.18233	−4.987	103.001	98.014	−0.01214	0.24556	−60
−59	9.0754	11.4436*	0.011125	5.3394	89.888	0.18729	−4.741	102.869	98.128	−0.01153	0.24520	−59
−58	9.3388	10.9074*	0.011137	5.1989	89.791	0.19235	−4.495	102.736	98.241	−0.01092	0.24484	−58
−57	9.6082	10.3587*	0.011149	5.0629	89.693	0.19752	−4.248	102.603	98.355	−0.01030	0.24449	−57
−56	9.8839	9.7975*	0.011161	4.9312	89.595	0.20279	−4.001	102.469	98.468	−0.00969	0.24414	−56
−55	10.166	9.223*	0.011174	4.8036	89.497	0.20818	−3.754	102.335	98.581	−0.00908	0.24379	−55
−54	10.454	8.636*	0.011186	4.6799	89.399	0.21368	−3.506	102.200	98.694	−0.00847	0.24345	−54
−53	10.749	8.036*	0.011198	4.5601	89.300	0.21929	−3.258	102.065	98.807	−0.00786	0.24310	−53
−52	11.051	7.422*	0.011211	4.4440	89.202	0.22502	−3.009	101.929	98.920	−0.00725	0.24276	−52
−51	11.359	6.795*	0.011223	4.3315	89.103	0.23087	−2.761	101.793	99.032	−0.00665	0.24243	−51
−50	11.674	6.154*	0.011235	4.2224	89.004	0.23683	−2.511	101.656	99.144	−0.00604	0.24209	−50
−49	11.996	5.498*	0.011248	4.1166	88.905	0.24292	−2.262	101.519	99.257	−0.00543	0.24176	−49
−48	12.324	4.829*	0.011261	4.0140	88.806	0.24913	−2.012	101.381	99.369	−0.00483	0.24143	−48
−47	12.660	4.144*	0.011273	3.9145	88.707	0.25546	−1.762	101.242	99.480	−0.00422	0.24110	−47
−46	13.004	3.445*	0.011286	3.8179	88.607	0.26192	−1.511	101.103	99.592	−0.00361	0.24078	−46

* Inches of mercury below one atmosphere

Table A-15 (continued)
"FREON" 22 SATURATION PROPERTIES—TEMPERATURE TABLE

TEMP. °F	PRESSURE PSIA	PRESSURE PSIG	VOLUME cu ft/lb LIQUID v_f	VOLUME cu ft/lb VAPOR v_g	DENSITY lb/cu ft LIQUID $1/v_f$	DENSITY lb/cu ft VAPOR $1/v_g$	ENTHALPY Btu/lb LIQUID h_f	ENTHALPY Btu/lb LATENT h_{fg}	ENTHALPY Btu/lb VAPOR h_g	ENTROPY Btu/(lb)(°R) LIQUID s_f	ENTROPY Btu/(lb)(°R) VAPOR s_g	TEMP. °F
−45	13.354	2.732*	0.011298	3.7243	88.507	0.26851	−1.260	100.963	99.703	−0.00301	0.24046	−45
−44	13.712	2.002*	0.011311	3.6334	88.407	0.27523	−1.009	100.823	99.814	−0.00241	0.24014	−44
−43	14.078	1.258*	0.011324	3.5452	88.307	0.28207	−0.757	100.683	99.925	−0.00181	0.23982	−43
−42	14.451	0.498*	0.011337	3.4596	88.207	0.28905	−0.505	100.541	100.036	−0.00120	0.23951	−42
−41	14.833	0.137	0.011350	3.3764	88.107	0.29617	−0.253	100.399	100.147	−0.00060	0.23919	−41
−40	15.222	0.526	0.011363	3.2957	88.006	0.30342	0.000	100.257	100.257	0.00000	0.23888	−40
−39	15.619	0.923	0.011376	3.2173	87.905	0.31082	0.253	100.114	100.367	0.00060	0.23858	−39
−38	16.024	1.328	0.011389	3.1412	87.805	0.31835	0.506	99.971	100.477	0.00120	0.23827	−38
−37	16.437	1.741	0.011402	3.0673	87.703	0.32602	0.760	99.826	100.587	0.00180	0.23797	−37
−36	16.859	2.163	0.011415	2.9954	87.602	0.33384	1.014	99.682	100.696	0.00240	0.23767	−36
−35	17.290	2.594	0.011428	2.9256	87.501	0.34181	1.269	99.536	100.805	0.00300	0.23737	−35
−34	17.728	3.032	0.011442	2.8578	87.399	0.34992	1.524	99.391	100.914	0.00359	0.23707	−34
−33	18.176	3.480	0.011455	2.7919	87.297	0.35818	1.779	99.244	101.023	0.00419	0.23678	−33
−32	18.633	3.937	0.011469	2.7278	87.195	0.36660	2.035	99.097	101.132	0.00479	0.23649	−32
−31	19.098	4.402	0.011482	2.6655	87.093	0.37517	2.291	98.949	101.240	0.00538	0.23620	−31
−30	19.573	4.877	0.011495	2.6049	86.991	0.38389	2.547	98.801	101.348	0.00598	0.23591	−30
−29	20.056	5.360	0.011509	2.5460	86.888	0.39278	2.804	98.652	101.456	0.00657	0.23563	−29
−28	20.549	5.853	0.011523	2.4887	86.785	0.40182	3.061	98.503	101.564	0.00716	0.23534	−28
−27	21.052	6.536	0.011536	2.4329	86.682	0.41103	3.318	98.353	101.671	0.00776	0.23506	−27
−26	21.564	6.868	0.011550	2.3787	86.579	0.42040	3.576	98.202	101.778	0.00835	0.23478	−26
−25	22.086	7.390	0.011564	2.3260	86.476	0.42993	3.834	98.051	101.885	0.00894	0.23451	−25
−24	22.617	7.921	0.011578	2.2746	86.372	0.43964	4.093	97.899	101.992	0.00953	0.23423	−24
−23	23.159	8.463	0.011592	2.2246	86.269	0.44951	4.352	97.746	102.098	0.01013	0.23396	−23
−22	23.711	9.015	0.011606	2.1760	86.165	0.45956	4.611	97.593	102.204	0.01072	0.23369	−22
−21	24.272	9.576	0.011620	2.1287	86.061	0.46978	4.871	97.439	102.310	0.01131	0.23342	−21
−20	24.845	10.149	0.011634	2.0826	85.956	0.48018	5.131	97.285	102.415	0.01189	0.23315	−20
−19	25.427	10.731	0.011648	2.0377	85.852	0.49075	5.391	97.129	102.521	0.01248	0.23289	−19
−18	26.020	11.324	0.011662	1.9940	85.747	0.50151	5.652	96.974	102.626	0.01307	0.23262	−18
−17	26.624	11.928	0.011677	1.9514	85.642	0.51245	5.913	96.817	102.730	0.01366	0.23236	−17
−16	27.239	12.543	0.011691	1.9099	85.537	0.52358	6.175	96.660	102.835	0.01425	0.23210	−16
−15	27.865	13.169	0.011705	1.8695	85.431	0.53489	6.436	96.502	102.939	0.01483	0.23184	−15
−14	28.501	13.805	0.011720	1.8302	85.326	0.54640	6.699	96.344	103.043	0.01542	0.23159	−14
−13	29.149	14.453	0.011734	1.7918	85.220	0.55810	6.961	96.185	103.146	0.01600	0.23133	−13
−12	29.809	15.113	0.011749	1.7544	85.114	0.56999	7.224	96.025	103.250	0.01659	0.23108	−12
−11	30.480	15.784	0.011764	1.7180	85.008	0.58207	7.488	95.865	103.353	0.01717	0.23083	−11
−10	31.162	16.466	0.011778	1.6825	84.901	0.59436	7.751	95.704	103.455	0.01776	0.23058	−10
−9	31.856	17.160	0.011793	1.6479	84.795	0.60685	8.015	95.542	103.558	0.01834	0.23033	−9
−8	32.563	17.867	0.011808	1.6141	84.688	0.61954	8.280	95.380	103.660	0.01892	0.23008	−8
−7	33.281	18.585	0.011823	1.5812	84.581	0.63244	8.545	95.217	103.762	0.01950	0.22984	−7
−6	34.011	19.315	0.011838	1.5491	84.473	0.64555	8.810	95.053	103.863	0.02009	0.22960	−6
−5	34.754	20.058	0.011853	1.5177	84.366	0.65887	9.075	94.889	103.964	0.02067	0.22936	−5
−4	35.509	20.813	0.011868	1.4872	84.258	0.67240	9.341	94.724	104.065	0.02125	0.22912	−4
−3	36.277	21.581	0.011884	1.4574	84.150	0.68615	9.608	94.558	104.166	0.02183	0.22888	−3
−2	37.057	22.361	0.011899	1.4283	84.042	0.70012	9.874	94.391	104.266	0.02241	0.22864	−2
−1	37.850	23.154	0.011914	1.4000	83.933	0.71431	10.142	94.224	104.366	0.02299	0.22841	−1
0	38.657	23.961	0.011930	1.3723	83.825	0.72872	10.409	94.056	104.465	0.02357	0.22817	0
1	39.476	24.780	0.011945	1.3453	83.716	0.74336	10.677	93.888	104.565	0.02414	0.22794	1
2	40.309	25.613	0.011961	1.3189	83.606	0.75822	10.945	93.718	104.663	0.02472	0.22771	2
3	41.155	26.459	0.011976	1.2931	83.497	0.77332	11.214	93.548	104.762	0.02530	0.22748	3
4	42.014	27.318	0.011992	1.2680	83.387	0.78865	11.483	93.378	104.860	0.02587	0.22725	4
5	42.888	28.192	0.012008	1.2434	83.277	0.80422	11.752	93.206	104.958	0.02645	0.22703	5
6	43.775	29.079	0.012024	1.2195	83.167	0.82003	12.022	93.034	105.056	0.02703	0.22680	6
7	44.676	29.980	0.012040	1.1961	83.057	0.83608	12.292	92.861	105.153	0.02760	0.22658	7
8	45.591	30.895	0.012056	1.1732	82.946	0.85237	12.562	92.688	105.250	0.02818	0.22636	8
9	46.521	31.825	0.012072	1.1509	82.835	0.86892	12.833	92.513	105.346	0.02875	0.22614	9

*Inches of mercury below one atmosphere

Table A-15 (continued)
"FREON" 22 SATURATION PROPERTIES—TEMPERATURE TABLE

TEMP. °F	PRESSURE		VOLUME cu ft/lb		DENSITY lb/cu ft		ENTHALPY Btu/lb			ENTROPY Btu/(lb)(°R)		TEMP. °F
	PSIA	PSIG	LIQUID v_f	VAPOR v_g	LIQUID $1/v_f$	VAPOR $1/v_g$	LIQUID h_f	LATENT h_{fg}	VAPOR h_g	LIQUID s_f	VAPOR s_g	
10	47.464	32.768	0.012088	1.1290	82.724	0.88571	13.104	92.338	105.442	0.02932	0.22592	10
11	48.423	33.727	0.012105	1.1077	82.612	0.90275	13.376	92.162	105.538	0.02990	0.22570	11
12	49.396	34.700	0.012121	1.0869	82.501	0.92005	13.648	91.986	105.633	0.03047	0.22548	12
13	50.384	35.688	0.012138	1.0665	82.389	0.93761	13.920	91.808	105.728	0.03104	0.22527	13
14	51.387	36.691	0.012154	1.0466	82.276	0.95544	14.193	91.630	105.823	0.03161	0.22505	14
15	52.405	37.709	0.012171	1.0272	82.164	0.97352	14.466	91.451	105.917	0.03218	0.22484	15
16	53.438	38.742	0.012188	1.0082	82.051	0.99188	14.739	91.272	106.011	0.03275	0.22463	16
17	54.487	39.791	0.012204	0.98961	81.938	1.0105	15.013	91.091	106.105	0.03332	0.22442	17
18	55.551	40.855	0.012221	0.97144	81.825	1.0294	15.288	90.910	106.198	0.03389	0.22421	18
19	56.631	41.935	0.012238	0.95368	81.711	1.0486	15.562	90.728	106.290	0.03446	0.22400	19
20	57.727	43.031	0.012255	0.93631	81.597	1.0680	15.837	90.545	106.383	0.03503	0.22379	20
21	58.839	44.143	0.012273	0.91932	81.483	1.0878	16.113	90.362	106.475	0.03560	0.22358	21
22	59.967	45.271	0.012290	0.90270	81.368	1.1078	16.389	90.178	106.566	0.03617	0.22338	22
23	61.111	46.415	0.012307	0.88645	81.253	1.1281	16.665	89.993	106.657	0.03674	0.22318	23
24	62.272	47.576	0.012325	0.87055	81.138	1.1487	16.942	89.807	106.748	0.03730	0.22297	24
25	63.450	48.754	0.012342	0.85500	81.023	1.1696	17.219	89.620	106.839	0.03787	0.22277	25
26	64.644	49.948	0.012360	0.83978	80.907	1.1908	17.496	89.433	106.928	0.03844	0.22257	26
27	65.855	51.159	0.012378	0.82488	80.791	1.2123	17.774	89.244	107.018	0.03900	0.22237	27
28	67.083	52.387	0.012395	0.81031	80.675	1.2341	18.052	89.055	107.107	0.03958	0.22217	28
29	68.328	53.632	0.012413	0.79604	80.558	1.2562	18.330	88.865	107.196	0.04013	0.22198	29
30	69.591	54.895	0.012431	0.78208	80.441	1.2786	18.609	88.674	107.284	0.04070	0.22178	30
31	70.871	56.175	0.012450	0.76842	80.324	1.3014	18.889	88.483	107.372	0.04126	0.22158	31
32	72.169	57.473	0.012468	0.75503	80.207	1.3244	19.169	88.290	107.459	0.04182	0.22139	32
33	73.485	58.789	0.012486	0.74194	80.089	1.3478	19.449	88.097	107.546	0.04239	0.22119	33
34	74.818	60.122	0.012505	0.72911	79.971	1.3715	19.729	87.903	107.632	0.04295	0.22100	34
35	76.170	61.474	0.012523	0.71655	79.852	1.3956	20.010	87.708	107.719	0.04351	0.22081	35
36	77.540	62.844	0.012542	0.70425	79.733	1.4199	20.292	87.512	107.804	0.04407	0.22062	36
37	78.929	64.233	0.012561	0.69221	79.614	1.4447	20.574	87.316	107.889	0.04464	0.22043	37
38	80.336	65.640	0.012579	0.68041	79.495	1.4697	20.856	87.118	107.974	0.04520	0.22024	38
39	81.761	67.065	0.012598	0.66885	79.375	1.4951	21.138	86.920	108.058	0.04576	0.22005	39
40	83.206	68.510	0.012618	0.65753	79.255	1.5208	21.422	86.720	108.142	0.04632	0.21986	40
41	84.670	69.974	0.012637	0.64643	79.134	1.5469	21.705	86.520	108.225	0.04688	0.21968	41
42	86.153	71.457	0.012656	0.63557	79.013	1.5734	21.989	86.319	108.308	0.04744	0.21949	42
43	87.655	72.959	0.012676	0.62492	78.892	1.6002	22.273	86.117	108.390	0.04800	0.21931	43
44	89.177	74.481	0.012695	0.61448	78.770	1.6274	22.558	85.914	108.472	0.04855	0.21912	44
45	90.719	76.023	0.012715	0.60425	78.648	1.6549	22.843	85.710	108.553	0.04911	0.21894	45
46	92.280	77.584	0.012735	0.59422	78.526	1.6829	23.129	85.506	108.634	0.04967	0.21876	46
47	93.861	79.165	0.012755	0.58440	78.403	1.7112	23.415	85.300	108.715	0.05023	0.21858	47
48	95.463	80.767	0.012775	0.57476	78.280	1.7398	23.701	85.094	108.795	0.05079	0.21839	48
49	97.085	82.389	0.012795	0.56532	78.157	1.7689	23.988	84.886	108.874	0.05134	0.21821	49
50	98.727	84.031	0.012815	0.55606	78.033	1.7984	24.275	84.678	108.953	0.05190	0.21803	50
51	100.39	85.69	0.012836	0.54698	77.909	1.8282	24.563	84.468	109.031	0.05245	0.21785	51
52	102.07	87.38	0.012856	0.53808	77.784	1.8585	24.851	84.258	109.109	0.05301	0.21768	52
53	103.78	89.08	0.012877	0.52934	77.659	1.8891	25.139	84.047	109.186	0.05357	0.21750	53
54	105.50	90.81	0.012898	0.52078	77.534	1.9202	25.429	83.834	109.263	0.05412	0.21732	54
55	107.25	92.56	0.012919	0.51238	77.408	1.9517	25.718	83.621	109.339	0.05468	0.21714	55
56	109.02	94.32	0.012940	0.50414	77.282	1.9836	26.008	83.407	109.415	0.05523	0.21697	56
57	110.81	96.11	0.012961	0.49606	77.155	2.0159	26.298	83.191	109.490	0.05579	0.21679	57
58	112.62	97.93	0.012982	0.48813	77.028	2.0486	26.589	82.975	109.564	0.05634	0.21662	58
59	114.46	99.76	0.013004	0.48035	76.900	2.0818	26.880	82.758	109.638	0.05689	0.21644	59
60	116.31	101.62	0.013025	0.47272	76.773	2.1154	27.172	82.540	109.712	0.05745	0.21627	60
61	118.19	103.49	0.013047	0.46523	76.644	2.1495	27.464	82.320	109.785	0.05800	0.21610	61
62	120.09	105.39	0.013069	0.45788	76.515	2.1840	27.757	82.100	109.857	0.05855	0.21592	62
63	122.01	107.32	0.013091	0.45066	76.386	2.2190	28.050	81.878	109.929	0.05910	0.21575	63
64	123.96	109.26	0.013114	0.44358	76.257	2.2544	28.344	81.656	110.000	0.05966	0.21558	64

Table A-15 (continued)

"FREON" 22 SATURATION PROPERTIES—TEMPERATURE TABLE

TEMP. °F	PRESSURE		VOLUME cu ft/lb		DENSITY lb/cu ft		ENTHALPY Btu/lb			ENTROPY Btu/(lb)(°R)		TEMP. °F
	PSIA	PSIG	LIQUID v_f	VAPOR v_g	LIQUID $1/v_f$	VAPOR $1/v_g$	LIQUID h_f	LATENT h_{fg}	VAPOR h_g	LIQUID s_f	VAPOR s_g	
65	125.93	111.23	0.013136	0.43663	76.126	2.2903	28.638	81.432	110.070	0.06021	0.21541	65
66	127.92	113.22	0.013159	0.42981	75.996	2.3266	28.932	81.208	110.140	0.06076	0.21524	66
67	129.94	115.24	0.013181	0.42311	75.865	2.3635	29.228	80.982	110.209	0.06131	0.21507	67
68	131.97	117.28	0.013204	0.41653	75.733	2.4008	29.523	80.755	110.278	0.06186	0.21490	68
69	134.04	119.34	0.013227	0.41007	75.601	2.4386	29.819	80.527	110.346	0.06241	0.21473	69
70	136.12	121.43	0.013251	0.40373	75.469	2.4769	30.116	80.298	110.414	0.06296	0.21456	70
71	138.23	123.54	0.013274	0.39751	75.336	2.5157	30.413	80.068	110.480	0.06351	0.21439	71
72	140.37	125.67	0.013297	0.39139	75.202	2.5550	30.710	79.836	110.547	0.06406	0.21422	72
73	142.52	127.83	0.013321	0.38539	75.068	2.5948	31.008	79.604	110.612	0.06461	0.21405	73
74	144.71	130.01	0.013345	0.37949	74.934	2.6351	31.307	79.370	110.677	0.06516	0.21388	74
75	146.91	132.22	0.013369	0.37369	74.799	2.6760	31.606	79.135	110.741	0.06571	0.21372	75
76	149.15	134.45	0.013393	0.36800	74.664	2.7174	31.906	78.899	110.805	0.06626	0.21355	76
77	151.40	136.71	0.013418	0.36241	74.528	2.7593	32.206	78.662	110.868	0.06681	0.21338	77
78	153.69	138.99	0.013442	0.35691	74.391	2.8018	32.506	78.423	110.930	0.06736	0.21321	78
79	155.99	141.30	0.013467	0.35151	74.254	2.8449	32.808	78.184	110.991	0.06791	0.21305	79
80	158.33	143.63	0.013492	0.34621	74.116	2.8885	33.109	77.943	111.052	0.06846	0.21288	80
81	160.68	145.99	0.013518	0.34099	73.978	2.9326	33.412	77.701	111.112	0.06901	0.21271	81
82	163.07	148.37	0.013543	0.33587	73.839	2.9774	33.714	77.457	111.171	0.06956	0.21255	82
83	165.48	150.78	0.013569	0.33083	73.700	3.0227	34.018	77.212	111.230	0.07011	0.21238	83
84	167.92	153.22	0.013594	0.32588	73.560	3.0686	34.322	76.966	111.288	0.07065	0.21222	84
85	170.38	155.68	0.013620	0.32101	73.420	3.1151	34.626	76.719	111.345	0.07120	0.21205	85
86	172.87	158.17	0.013647	0.31623	73.278	3.1622	34.931	76.470	111.401	0.07175	0.21188	86
87	175.38	160.69	0.013673	0.31153	73.137	3.2100	35.237	76.220	111.457	0.07230	0.21172	87
88	177.93	163.23	0.013700	0.30690	72.994	3.2583	35.543	75.968	111.512	0.07285	0.21155	88
89	180.50	165.80	0.013727	0.30236	72.851	3.3073	35.850	75.716	111.566	0.07339	0.21139	89
90	183.09	168.40	0.013754	0.29789	72.708	3.3570	36.158	75.461	111.619	0.07394	0.21122	90
91	185.72	171.02	0.013781	0.29349	72.564	3.4073	36.466	75.206	111.671	0.07449	0.21106	91
92	188.37	173.67	0.013809	0.28917	72.419	3.4582	36.774	74.949	111.723	0.07504	0.21089	92
93	191.05	176.35	0.013836	0.28491	72.273	3.5098	37.084	74.690	111.774	0.07559	0.21072	93
94	193.76	179.06	0.013864	0.28073	72.127	3.5621	37.394	74.430	111.824	0.07613	0.21056	94
95	196.50	181.80	0.013893	0.27662	71.980	3.6151	37.704	74.168	111.873	0.07668	0.21039	95
96	199.26	184.56	0.013921	0.27257	71.833	3.6688	38.016	73.905	111.921	0.07723	0.21023	96
97	202.05	187.36	0.013950	0.26859	71.685	3.7232	38.328	73.641	111.968	0.07778	0.21006	97
98	204.87	190.18	0.013979	0.26467	71.536	3.7783	38.640	73.375	112.015	0.07832	0.20989	98
99	207.72	193.03	0.014008	0.26081	71.386	3.8341	38.953	73.107	112.060	0.07887	0.20973	99
100	210.60	195.91	0.014038	0.25702	71.236	3.8907	39.267	72.838	112.105	0.07942	0.20956	100
101	213.51	198.82	0.014068	0.25329	71.084	3.9481	39.582	72.567	112.149	0.07997	0.20939	101
102	216.45	201.76	0.014098	0.24962	70.933	4.0062	39.897	72.294	112.192	0.08052	0.20923	102
103	219.42	204.72	0.014128	0.24600	70.780	4.0651	40.213	72.020	112.233	0.08107	0.20906	103
104	222.42	207.72	0.014159	0.24244	70.626	4.1247	40.530	71.744	112.274	0.08161	0.20889	104
105	225.45	210.75	0.014190	0.23894	70.472	4.1852	40.847	71.467	112.314	0.08216	0.20872	105
106	228.50	213.81	0.014221	0.23549	70.317	4.2465	41.166	71.187	112.353	0.08271	0.20855	106
107	231.59	216.90	0.014253	0.23209	70.161	4.3086	41.485	70.906	112.391	0.08326	0.20838	107
108	234.71	220.02	0.014285	0.22875	70.005	4.3715	41.804	70.623	112.427	0.08381	0.20821	108
109	237.86	223.17	0.014317	0.22546	69.847	4.4354	42.125	70.338	112.463	0.08436	0.20804	109
110	241.04	226.35	0.014350	0.22222	69.689	4.5000	42.446	70.052	112.498	0.08491	0.20787	110
111	244.25	229.56	0.014382	0.21903	69.529	4.5656	42.768	69.763	112.531	0.08546	0.20770	111
112	247.50	232.80	0.014416	0.21589	69.369	4.6321	43.091	69.473	112.564	0.08601	0.20753	112
113	250.77	236.08	0.014449	0.21279	69.208	4.6994	43.415	69.180	112.595	0.08656	0.20736	113
114	254.08	239.38	0.014483	0.20974	69.046	4.7677	43.739	68.886	112.626	0.08711	0.20718	114
115	257.42	242.72	0.014517	0.20674	68.883	4.8370	44.065	68.590	112.655	0.08766	0.20701	115
116	260.79	246.10	0.014552	0.20378	68.719	4.9072	44.391	68.291	112.682	0.08821	0.20684	116
117	264.20	249.50	0.014587	0.20087	68.554	4.9784	44.718	67.991	112.709	0.08876	0.20666	117
118	267.63	252.94	0.014622	0.19800	68.388	5.0506	45.046	67.688	112.735	0.08932	0.20649	118
119	271.10	256.41	0.014658	0.19517	68.221	5.1238	45.375	67.384	112.759	0.08987	0.20631	119

Table A-15 (continued)

"FREON" 22 SATURATION PROPERTIES—TEMPERATURE TABLE

TEMP. °F	PRESSURE PSIA	PRESSURE PSIG	VOLUME cu ft/lb LIQUID v_f	VOLUME cu ft/lb VAPOR v_g	DENSITY lb/cu ft LIQUID $1/v_f$	DENSITY lb/cu ft VAPOR $1/v_g$	ENTHALPY Btu/lb LIQUID h_f	ENTHALPY Btu/lb LATENT h_{fg}	ENTHALPY Btu/lb VAPOR h_g	ENTROPY Btu/(lb)(°R) LIQUID s_f	ENTROPY Btu/(lb)(°R) VAPOR s_g	TEMP. °F
120	274.60	259.91	0.014694	0.19238	68.054	5.1981	45.705	67.077	112.782	0.09042	0.20613	120
121	278.14	263.44	0.014731	0.18963	67.885	5.2734	46.036	66.767	112.803	0.09098	0.20595	121
122	281.71	267.01	0.014768	0.18692	67.714	5.3498	46.368	66.456	112.824	0.09153	0.20578	122
123	285.31	270.62	0.014805	0.18426	67.543	5.4272	46.701	66.142	112.843	0.09208	0.20560	123
124	288.95	274.25	0.014843	0.18163	67.371	5.5058	47.034	65.826	112.860	0.09264	0.20542	124
125	292.62	277.92	0.014882	0.17903	67.197	5.5856	47.369	65.507	112.877	0.09320	0.20523	125
126	296.33	281.63	0.014920	0.17648	67.023	5.6665	47.705	65.186	112.891	0.09375	0.20505	126
127	300.07	285.37	0.014960	0.17396	66.847	5.7486	48.042	64.863	112.905	0.09431	0.20487	127
128	303.84	289.14	0.014999	0.17147	66.670	5.8319	48.380	64.537	112.917	0.09487	0.20468	128
129	307.65	292.95	0.015039	0.16902	66.492	5.9164	48.719	64.208	112.927	0.09543	0.20449	129
130	311.50	296.80	0.015080	0.16661	66.312	6.0022	49.059	63.877	112.936	0.09598	0.20431	130
131	315.38	300.68	0.015121	0.16422	66.131	6.0893	49.400	63.543	112.943	0.09654	0.20412	131
132	319.29	304.60	0.015163	0.16187	65.949	6.1777	49.743	63.206	112.949	0.09711	0.20393	132
133	323.25	308.55	0.015206	0.15956	65.766	6.2674	50.087	62.866	112.953	0.09767	0.20374	133
134	327.23	312.54	0.015248	0.15727	65.581	6.3585	50.432	62.523	112.955	0.09823	0.20354	134
135	331.26	316.56	0.015292	0.15501	65.394	6.4510	50.778	62.178	112.956	0.09879	0.20335	135
136	335.32	320.63	0.015336	0.15279	65.207	6.5450	51.125	61.829	112.954	0.09936	0.20315	136
137	339.42	324.73	0.015381	0.15059	65.017	6.6405	51.474	61.477	112.951	0.09992	0.20295	137
138	343.56	328.86	0.015426	0.14843	64.826	6.7374	51.824	61.123	112.947	0.10049	0.20275	138
139	347.73	333.04	0.015472	0.14629	64.634	6.8359	52.175	60.764	112.940	0.10106	0.20255	139
140	351.94	337.25	0.015518	0.14418	64.440	6.9360	52.528	60.403	112.931	0.10163	0.20235	140
141	356.19	341.50	0.015566	0.14209	64.244	7.0377	52.883	60.038	112.921	0.10220	0.20214	141
142	360.48	345.79	0.015613	0.14004	64.047	7.1410	53.238	59.670	112.908	0.10277	0.20194	142
143	364.81	350.11	0.015662	0.13801	63.848	7.2461	53.596	59.298	112.893	0.10334	0.20173	143
144	369.17	354.48	0.015712	0.13600	63.647	7.3529	53.955	58.922	112.877	0.10391	0.20152	144
145	373.58	358.88	0.015762	0.13402	63.445	7.4615	54.315	58.543	112.858	0.10449	0.20130	145
146	378.02	363.32	0.015813	0.13207	63.240	7.5719	54.677	58.159	112.836	0.10507	0.20109	146
147	382.50	367.81	0.015865	0.13014	63.034	7.6842	55.041	57.772	112.813	0.10564	0.20087	147
148	387.03	372.33	0.015917	0.12823	62.825	7.7985	55.406	57.380	112.787	0.10622	0.20065	148
149	391.59	376.89	0.015971	0.12635	62.615	7.9148	55.774	56.985	112.758	0.10681	0.20042	149
150	396.19	381.50	0.016025	0.12448	62.402	8.0331	56.143	56.585	112.728	0.10739	0.20020	150
151	400.84	386.14	0.016080	0.12265	62.187	8.1536	56.514	56.180	112.694	0.10797	0.19997	151
152	405.52	390.83	0.016137	0.12083	61.970	8.2763	56.887	55.771	112.658	0.10856	0.19974	152
153	410.25	395.56	0.016194	0.11903	61.751	8.4011	57.261	55.358	112.619	0.10915	0.19950	153
154	415.02	400.32	0.016252	0.11726	61.529	8.5284	57.638	54.939	112.577	0.10974	0.19926	154
155	419.83	405.13	0.016312	0.11550	61.305	8.6580	58.017	54.515	112.533	0.11034	0.19902	155
156	424.68	409.99	0.016372	0.11376	61.079	8.7901	58.399	54.087	112.485	0.11093	0.19878	156
157	429.58	414.88	0.016434	0.11205	60.849	8.9247	58.782	53.652	112.435	0.11153	0.19853	157
158	434.52	419.82	0.016497	0.11035	60.617	9.0620	59.168	53.213	112.381	0.11213	0.19828	158
159	439.50	424.80	0.016561	0.10867	60.383	9.2020	59.557	52.767	112.324	0.11273	0.19802	159
160	444.53	429.83	0.016627	0.10701	60.145	9.3449	59.948	52.316	112.263	0.11334	0.19776	160
161	449.59	434.90	0.016693	0.10537	59.904	9.4907	60.341	51.858	112.199	0.11395	0.19750	161
162	454.71	440.01	0.016762	0.10374	59.660	9.6395	60.737	51.394	112.131	0.11456	0.19723	162
163	459.87	445.17	0.016831	0.10213	59.413	9.7915	61.136	50.923	112.060	0.11518	0.19696	163
164	465.07	450.37	0.016902	0.10054	59.163	9.9467	61.538	50.446	111.984	0.11580	0.19668	164
165	470.32	455.62	0.016975	0.098956	58.909	10.106	61.943	49.961	111.904	0.11642	0.19640	165
166	475.61	460.92	0.017050	0.097393	58.651	10.268	62.351	49.469	111.820	0.11705	0.19611	166
167	480.95	466.26	0.017126	0.095844	58.390	10.434	62.763	48.969	111.732	0.11768	0.19581	167
168	486.34	471.65	0.017204	0.094309	58.125	10.603	63.178	48.461	111.639	0.11831	0.19552	168
169	491.78	477.08	0.017285	0.092787	57.855	10.777	63.596	47.945	111.541	0.11895	0.19521	169
170	497.26	482.56	0.017367	0.091279	57.581	10.955	64.019	47.419	111.438	0.11959	0.19490	170
171	502.79	488.09	0.017451	0.089783	57.303	11.138	64.445	46.885	111.330	0.12024	0.19458	171
172	508.37	493.67	0.017538	0.088299	57.019	11.325	64.875	46.340	111.216	0.12089	0.19425	172
173	513.99	499.30	0.017627	0.086827	56.731	11.517	65.310	45.786	111.096	0.12155	0.19392	173
174	519.67	504.97	0.017719	0.085365	56.438	11.714	65.750	45.221	110.970	0.12222	0.19358	174

© 1964. E.I. du Pont de Nemours and Company (DuPont). All rights reserved, used under license of DuPont.

Table A-15 (continued)

R-507 Saturation Properties—Temperature Table

Temp (°F)	Pressure (psia) Liquid	Pressure (psia) Vapor	Density (lb/ft³) Liquid	Density (lb/ft³) Vapor	Volume (ft³/lb) Liquid	Volume (ft³/lb) Vapor	Enthalpy (Btu/lb) Liquid	Enthalpy (Btu/lb) Latent	Enthalpy (Btu/lb) Vapor	Entropy (Btu/lb-R) Liquid	Entropy (Btu/lb-R) Vapor
−140	0.667	0.666	91.1	0.019	0.0110	51.90	−28.20	96.10	67.9	−0.077	0.224
−139	0.699	0.698	91.0	0.020	0.0110	49.60	−27.90	95.90	68.0	−0.076	0.224
−138	0.732	0.731	90.9	0.021	0.0110	47.50	−27.70	95.90	68.2	−0.075	0.223
−137	0.767	0.766	90.8	0.022	0.0110	45.50	−27.40	95.70	68.3	−0.074	0.223
−136	0.803	0.802	90.7	0.023	0.0110	43.60	−27.10	95.60	68.5	−0.073	0.222
−135	0.840	0.839	90.6	0.024	0.0110	41.80	−26.90	95.50	68.6	−0.072	0.222
−134	0.879	0.878	90.5	0.025	0.0110	40.00	−26.60	95.40	68.8	−0.072	0.221
−133	0.919	0.918	90.4	0.026	0.0111	38.40	−26.30	95.20	68.9	−0.071	0.221
−132	0.961	0.960	90.3	0.027	0.0111	36.80	−26.10	95.20	69.1	−0.070	0.220
−131	1.000	1.000	90.2	0.028	0.0111	35.30	−25.80	95.00	69.2	−0.069	0.220
−130	1.050	1.050	90.1	0.030	0.0111	33.90	−25.50	94.80	69.3	−0.068	0.220
−129	1.100	1.100	90.0	0.031	0.0111	32.50	−25.20	94.70	69.5	−0.067	0.219
−128	1.140	1.140	89.9	0.032	0.0111	31.20	−25.00	94.60	69.6	−0.067	0.219
−127	1.190	1.190	89.8	0.033	0.0111	30.00	−24.70	94.50	69.8	−0.066	0.218
−126	1.250	1.250	89.8	0.035	0.0111	28.80	−24.40	94.30	69.9	−0.065	0.218
−125	1.300	1.300	89.7	0.036	0.0112	27.70	−24.20	94.30	70.1	−0.064	0.217
−124	1.360	1.360	89.6	0.038	0.0112	26.60	−23.90	94.10	70.2	−0.063	0.217
−123	1.410	1.410	89.5	0.039	0.0112	25.60	−23.60	94.00	70.4	−0.063	0.217
−122	1.470	1.470	89.4	0.041	0.0112	24.60	−23.30	93.80	70.5	−0.062	0.216
−121	1.540	1.540	89.3	0.042	0.0112	23.70	−23.10	93.70	70.6	−0.061	0.216
−120	1.600	1.600	89.2	0.044	0.0112	22.80	−22.80	93.60	70.8	−0.060	0.215
−119	1.670	1.670	89.1	0.046	0.0112	22.00	−22.50	93.40	70.9	−0.059	0.215
−118	1.740	1.730	89.0	0.047	0.0112	21.20	−22.20	93.30	71.1	−0.058	0.215
−117	1.810	1.810	88.9	0.049	0.0113	20.40	−22.00	93.20	71.2	−0.058	0.214
−116	1.880	1.880	88.8	0.051	0.0113	19.60	−21.70	93.10	71.4	−0.057	0.214
−115	1.960	1.950	88.7	0.053	0.0113	18.90	−21.40	92.90	71.5	−0.056	0.214
−114	2.030	2.030	88.6	0.055	0.0113	18.20	−21.10	92.80	71.7	−0.055	0.213
−113	2.110	2.110	88.5	0.057	0.0113	17.60	−20.90	92.70	71.8	−0.054	0.213
−112	2.200	2.200	88.4	0.059	0.0113	17.00	−20.60	92.60	72.0	−0.054	0.213
−111	2.280	2.280	88.3	0.061	0.0113	16.40	−20.30	92.40	72.1	−0.053	0.212
−110	2.370	2.370	88.2	0.063	0.0113	15.80	−20.00	92.20	72.2	−0.052	0.212
−109	2.460	2.460	88.1	0.066	0.0114	15.20	−19.80	92.20	72.4	−0.051	0.212
−108	2.560	2.560	88.0	0.068	0.0114	14.70	−19.50	92.00	72.5	−0.051	0.211
−107	2.660	2.660	87.9	0.070	0.0114	14.20	−19.20	91.90	72.7	−0.050	0.211
−106	2.760	2.760	87.8	0.073	0.0114	13.70	−18.90	91.70	72.8	−0.049	0.211
−105	2.860	2.860	87.7	0.075	0.0114	13.30	−18.60	91.60	73.0	−0.048	0.210
−104	2.970	2.970	87.6	0.078	0.0114	12.80	−18.40	91.50	73.1	−0.047	0.210
−103	3.080	3.080	87.5	0.081	0.0114	12.40	−18.10	91.40	73.3	−0.047	0.210
−102	3.190	3.190	87.4	0.084	0.0114	12.00	−17.80	91.20	73.4	−0.046	0.209
−101	3.310	3.310	87.3	0.086	0.0115	11.60	−17.50	91.10	73.6	−0.045	0.209
−100	3.430	3.430	87.2	0.089	0.0115	11.20	−17.30	91.00	73.7	−0.044	0.209
−99	3.550	3.550	87.1	0.092	0.0115	10.80	−17.00	90.90	73.9	−0.044	0.208
−98	3.680	3.680	87.0	0.095	0.0115	10.50	−16.70	90.70	74.0	−0.043	0.208
−97	3.810	3.810	86.9	0.099	0.0115	10.10	−16.40	90.50	74.1	−0.042	0.208
−96	3.950	3.950	86.8	0.102	0.0115	9.82	−16.10	90.40	74.3	−0.041	0.208
−95	4.090	4.080	86.7	0.105	0.0115	9.51	−15.90	90.30	74.4	−0.040	0.207
−94	4.230	4.230	86.6	0.109	0.0115	9.21	−15.60	90.20	74.6	−0.040	0.207
−93	4.370	4.370	86.5	0.112	0.0116	8.92	−15.30	90.00	74.7	−0.039	0.207
−92	4.530	4.530	86.4	0.116	0.0116	8.64	−15.00	89.90	74.9	−0.038	0.206
−91	4.680	4.680	86.3	0.119	0.0116	8.37	−14.70	89.70	75.0	−0.037	0.206

Table A–15 (continued)
R-507 Saturation Properties—Temperature Table

Temp (°F)	Pressure (psia) Liquid	Pressure (psia) Vapor	Density (lb/ft³) Liquid	Density (lb/ft³) Vapor	Volume (ft³/lb) Liquid	Volume (ft³/lb) Vapor	Enthalpy (Btu/lb) Liquid	Enthalpy (Btu/lb) Latent	Enthalpy (Btu/lb) Vapor	Entropy (Btu/lb-R) Liquid	Entropy (Btu/lb-R) Vapor
−90	4.840	4.840	86.2	0.123	0.0116	8.11	−14.40	89.60	75.2	−0.037	0.206
−89	5.000	5.000	86.1	0.127	0.0116	7.87	−14.20	89.50	75.3	−0.036	0.206
−88	5.170	5.170	86.0	0.131	0.0116	7.63	−13.90	89.40	75.5	−0.035	0.205
−87	5.340	5.340	85.9	0.135	0.0116	7.40	−13.60	89.20	75.6	−0.034	0.205
−86	5.520	5.520	85.8	0.139	0.0117	7.17	−13.30	89.10	75.8	−0.034	0.205
−85	5.700	5.700	85.7	0.144	0.0117	6.96	−13.00	88.90	75.9	−0.033	0.205
−84	5.890	5.890	85.6	0.148	0.0117	6.75	−12.80	88.80	76.0	−0.032	0.204
−83	6.080	6.080	85.5	0.153	0.0117	6.55	−12.50	88.70	76.2	−0.031	0.204
−82	6.280	6.280	85.4	0.157	0.0117	6.36	−12.20	88.50	76.3	−0.031	0.204
−81	6.480	6.480	85.3	0.162	0.0117	6.18	−11.90	88.40	76.5	−0.030	0.204
−80	6.690	6.690	85.2	0.167	0.0117	6.00	−11.60	88.20	76.6	−0.029	0.203
−79	6.900	6.900	85.1	0.172	0.0118	5.82	−11.30	88.10	76.8	−0.028	0.203
−78	7.120	7.120	85.0	0.177	0.0118	5.66	−11.00	87.90	76.9	−0.028	0.203
−77	7.340	7.340	84.9	0.182	0.0118	5.50	−10.80	87.90	77.1	−0.027	0.203
−76	7.570	7.570	84.8	0.187	0.0118	5.34	−10.50	87.70	77.2	−0.026	0.203
−75	7.810	7.810	84.7	0.193	0.0118	5.19	−10.20	87.60	77.4	−0.025	0.202
−74	8.050	8.050	84.6	0.198	0.0118	5.05	−9.91	87.41	77.5	−0.025	0.202
−73	8.290	8.290	84.5	0.204	0.0118	4.90	−9.62	87.32	77.7	−0.024	0.202
−72	8.540	8.540	84.4	0.210	0.0119	4.77	−9.33	87.13	77.8	−0.023	0.202
−71	8.800	8.800	84.3	0.216	0.0119	4.64	−9.05	86.95	77.9	−0.022	0.201
−70	9.070	9.070	84.2	0.222	0.0119	4.51	−8.76	86.86	78.1	−0.022	0.201
−69	9.340	9.340	84.0	0.228	0.0119	4.39	−8.47	86.67	78.2	−0.021	0.201
−68	9.610	9.610	83.9	0.234	0.0119	4.27	−8.18	86.58	78.4	−0.020	0.201
−67	9.900	9.900	83.8	0.241	0.0119	4.16	−7.90	86.40	78.5	−0.019	0.201
−66	10.200	10.200	83.7	0.247	0.0119	4.04	−7.61	86.31	78.7	−0.019	0.201
−65	10.500	10.500	83.6	0.254	0.0120	3.94	−7.32	86.12	78.8	−0.018	0.200
−64	10.800	10.800	83.5	0.261	0.0120	3.83	−7.03	86.03	79.0	−0.017	0.200
−63	11.100	11.100	83.4	0.268	0.0120	3.73	−6.74	85.84	79.1	−0.017	0.200
−62	11.400	11.400	83.3	0.275	0.0120	3.63	−6.45	85.65	79.2	−0.016	0.200
−61	11.700	11.700	83.2	0.282	0.0120	3.54	−6.16	85.56	79.4	−0.015	0.200
−60	12.100	12.100	83.1	0.290	0.0120	3.45	−5.87	85.37	79.5	−0.014	0.199
−59	12.400	12.400	83.0	0.298	0.0120	3.36	−5.58	85.28	79.7	−0.014	0.199
−58	12.800	12.800	82.9	0.305	0.0121	3.27	−5.29	85.09	79.8	−0.013	0.199
−57	13.100	13.100	82.8	0.313	0.0121	3.19	−5.00	85.00	80.0	−0.012	0.199
−56	13.500	13.500	82.7	0.321	0.0121	3.11	−4.71	84.81	80.1	−0.011	0.199
−55	13.800	13.800	82.6	0.330	0.0121	3.03	−4.42	84.72	80.3	−0.011	0.199
−54	14.200	14.200	82.5	0.338	0.0121	2.96	−4.12	84.52	80.4	−0.010	0.198
−53	14.600	14.600	82.4	0.347	0.0121	2.88	−3.83	84.33	80.5	−0.009	0.198
−52	15.000	15.000	82.2	0.356	0.0122	2.81	−3.54	84.24	80.7	−0.009	0.198
−51	15.400	15.400	82.1	0.365	0.0122	2.74	−3.25	84.05	80.8	−0.008	0.198
−50	15.800	15.800	82.0	0.374	0.0122	2.68	−2.95	83.95	81.0	−0.007	0.198
−49	16.200	16.200	81.9	0.383	0.0122	2.61	−2.66	83.76	81.1	−0.006	0.198
−48	16.700	16.700	81.8	0.393	0.0122	2.55	−2.36	83.66	81.3	−0.006	0.197
−47	17.100	17.100	81.7	0.402	0.0122	2.49	−2.07	83.47	81.4	−0.005	0.197
−46	17.600	17.600	81.6	0.412	0.0123	2.43	−1.78	83.28	81.5	−0.004	0.197
−45	18.000	18.000	81.5	0.422	0.0123	2.37	−1.48	83.18	81.7	−0.004	0.197
−44	18.500	18.500	81.4	0.433	0.0123	2.31	−1.18	82.98	81.8	−0.003	0.197
−43	18.900	18.900	81.3	0.443	0.0123	2.26	−0.89	82.89	82.0	−0.002	0.197
−42	19.400	19.400	81.2	0.454	0.0123	2.20	−0.59	82.69	82.1	−0.001	0.197
−41	19.900	19.900	81.1	0.465	0.0123	2.15	−0.30	82.60	82.3	−0.001	0.196

Table A-15 (continued)

R-507 Saturation Properties—Temperature Table

Temp (°F)	Pressure (psia)		Density (lb/ft³)		Volume (ft³/lb)		Enthalpy (Btu/lb)			Entropy (Btu/lb-R)	
	Liquid	Vapor	Liquid	Vapor	Liquid	Vapor	Liquid	Latent	Vapor	Liquid	Vapor
−40	20.400	20.400	80.9	0.476	0.0124	2.10	0.00	82.40	82.4	0.000	0.196
−39	20.900	20.900	80.8	0.487	0.0124	2.05	0.30	82.20	82.5	0.001	0.196
−38	21.500	21.500	80.7	0.498	0.0124	2.01	0.60	82.11	82.7	0.001	0.196
−37	22.000	22.000	80.6	0.510	0.0124	1.96	0.89	81.91	82.8	0.002	0.196
−36	22.500	22.500	80.5	0.522	0.0124	1.92	1.19	81.81	83.0	0.003	0.196
−35	23.100	23.100	80.4	0.534	0.0124	1.87	1.49	81.61	83.1	0.004	0.196
−34	23.700	23.600	80.3	0.546	0.0125	1.83	1.79	81.41	83.2	0.004	0.196
−33	24.200	24.200	80.2	0.559	0.0125	1.79	2.09	81.31	83.4	0.005	0.195
−32	24.800	24.800	80.1	0.572	0.0125	1.75	2.39	81.11	83.5	0.006	0.195
−31	25.400	25.400	79.9	0.585	0.0125	1.71	2.69	81.01	83.7	0.006	0.195
−30	26.000	26.000	79.8	0.598	0.0125	1.67	2.99	80.81	83.8	0.007	0.195
−29	26.600	26.600	79.7	0.611	0.0125	1.64	3.29	80.71	84.0	0.008	0.195
−28	27.300	27.300	79.6	0.625	0.0126	1.60	3.59	80.51	84.1	0.008	0.195
−27	27.900	27.900	79.5	0.639	0.0126	1.56	3.89	80.31	84.2	0.009	0.195
−26	28.600	28.500	79.4	0.653	0.0126	1.53	4.19	80.21	84.4	0.010	0.195
−25	29.200	29.200	79.3	0.668	0.0126	1.50	4.49	80.01	84.5	0.011	0.195
−24	29.900	29.900	79.2	0.683	0.0126	1.47	4.80	79.80	84.6	0.011	0.194
−23	30.600	30.600	79.0	0.697	0.0127	1.43	5.10	79.70	84.8	0.012	0.194
−22	31.300	31.300	78.9	0.713	0.0127	1.40	5.40	79.50	84.9	0.013	0.194
−21	32.000	32.000	78.8	0.728	0.0127	1.37	5.71	79.39	85.1	0.013	0.194
−20	32.700	32.700	78.7	0.744	0.0127	1.34	6.01	79.19	85.2	0.014	0.194
−19	33.500	33.400	78.6	0.760	0.0127	1.32	6.31	78.99	85.3	0.015	0.194
−18	34.200	34.200	78.5	0.776	0.0127	1.29	6.62	78.88	85.5	0.015	0.194
−17	35.000	35.000	78.3	0.793	0.0128	1.26	6.92	78.68	85.6	0.016	0.194
−16	35.700	35.700	78.2	0.810	0.0128	1.24	7.23	78.47	85.7	0.017	0.194
−15	36.500	36.500	78.1	0.827	0.0128	1.21	7.54	78.36	85.9	0.017	0.194
−14	37.300	37.300	78.0	0.844	0.0128	1.18	7.84	78.16	86.0	0.018	0.193
−13	38.200	38.100	77.9	0.862	0.0128	1.16	8.15	78.05	86.2	0.019	0.193
−12	39.000	39.000	77.8	0.880	0.0129	1.14	8.46	77.84	86.3	0.019	0.193
−11	39.800	39.800	77.6	0.898	0.0129	1.11	8.76	77.64	86.4	0.020	0.193
−10	40.700	40.700	77.5	0.917	0.0129	1.09	9.07	77.53	86.6	0.021	0.193
−9	41.600	41.500	77.4	0.936	0.0129	1.07	9.38	77.32	86.7	0.021	0.193
−8	42.400	42.400	77.3	0.955	0.0129	1.05	9.69	77.11	86.8	0.022	0.193
−7	43.300	43.300	77.2	0.974	0.0130	1.03	10.00	77.00	87.0	0.023	0.193
−6	44.300	44.200	77.1	0.994	0.0130	1.01	10.30	76.80	87.1	0.024	0.193
−5	45.200	45.200	76.9	1.010	0.0130	0.99	10.60	76.60	87.2	0.024	0.193
−4	46.100	46.100	76.8	1.030	0.0130	0.97	10.90	76.50	87.4	0.025	0.193
−3	47.100	47.100	76.7	1.060	0.0130	0.95	11.20	76.30	87.5	0.026	0.193
−2	48.100	48.000	76.6	1.080	0.0131	0.93	11.60	76.00	87.6	0.026	0.192
−1	49.000	49.000	76.5	1.100	0.0131	0.91	11.90	75.90	87.8	0.027	0.192
0	50.000	50.000	76.3	1.120	0.0131	0.89	12.20	75.70	87.9	0.028	0.192
1	51.100	51.000	76.2	1.140	0.0131	0.88	12.50	75.50	88.0	0.028	0.192
2	52.100	52.100	76.1	1.160	0.0131	0.86	12.80	75.40	88.2	0.029	0.192
3	53.200	53.100	76.0	1.190	0.0132	0.84	13.10	75.20	88.3	0.030	0.192
4	54.200	54.200	75.8	1.210	0.0132	0.83	13.40	75.00	88.4	0.030	0.192
5	55.300	55.300	75.7	1.230	0.0132	0.81	13.70	74.80	88.5	0.031	0.192
6	56.400	56.400	75.6	1.260	0.0132	0.80	14.10	74.60	88.7	0.032	0.192
7	57.500	57.500	75.5	1.280	0.0132	0.78	14.40	74.40	88.8	0.032	0.192
8	58.700	58.600	75.4	1.310	0.0133	0.77	14.70	74.20	88.9	0.033	0.192
9	59.800	59.800	75.2	1.330	0.0133	0.75	15.00	74.10	89.1	0.034	0.192

Table A-15 (continued)

R-507 Saturation Properties—Temperature Table

Temp (°F)	Pressure (psia) Liquid	Pressure (psia) Vapor	Density (lb/ft³) Liquid	Density (lb/ft³) Vapor	Volume (ft³/lb) Liquid	Volume (ft³/lb) Vapor	Enthalpy (Btu/lb) Liquid	Enthalpy (Btu/lb) Latent	Enthalpy (Btu/lb) Vapor	Entropy (Btu/lb-R) Liquid	Entropy (Btu/lb-R) Vapor
10	61.000	60.900	75.1	1.360	0.0133	0.74	15.30	73.90	89.2	0.034	0.192
11	62.200	62.100	75.0	1.380	0.0133	0.72	15.60	73.70	89.3	0.035	0.191
12	63.400	63.300	74.9	1.410	0.0134	0.71	16.00	73.40	89.4	0.036	0.191
13	64.600	64.500	74.7	1.440	0.0134	0.70	16.30	73.30	89.6	0.036	0.191
14	65.800	65.800	74.6	1.460	0.0134	0.68	16.60	73.10	89.7	0.037	0.191
15	67.100	67.000	74.5	1.490	0.0134	0.67	16.90	72.90	89.8	0.038	0.191
16	68.300	68.300	74.3	1.520	0.0135	0.66	17.20	72.80	90.0	0.038	0.191
17	69.600	69.600	74.2	1.550	0.0135	0.65	17.60	72.50	90.1	0.039	0.191
18	70.900	70.900	74.1	1.580	0.0135	0.64	17.90	72.30	90.2	0.040	0.191
19	72.300	72.200	74.0	1.600	0.0135	0.62	18.20	72.10	90.3	0.040	0.191
20	73.600	73.500	73.8	1.630	0.0135	0.61	18.50	72.00	90.5	0.041	0.191
21	75.000	74.900	73.7	1.660	0.0136	0.60	18.80	71.80	90.6	0.042	0.191
22	76.400	76.300	73.6	1.690	0.0136	0.59	19.20	71.50	90.7	0.042	0.191
23	77.800	77.700	73.4	1.730	0.0136	0.58	19.50	71.30	90.8	0.043	0.191
24	79.200	79.100	73.3	1.760	0.0136	0.57	19.80	71.10	90.9	0.044	0.191
25	80.600	80.600	73.2	1.790	0.0137	0.56	20.10	71.00	91.1	0.044	0.191
26	82.100	82.000	73.1	1.820	0.0137	0.55	20.50	70.70	91.2	0.045	0.191
27	83.600	83.500	72.9	1.850	0.0137	0.54	20.80	70.50	91.3	0.046	0.190
28	85.100	85.000	72.8	1.890	0.0137	0.53	21.10	70.30	91.4	0.046	0.190
29	86.600	86.500	72.7	1.920	0.0138	0.52	21.40	70.20	91.6	0.047	0.190
30	88.100	88.000	72.5	1.960	0.0138	0.51	21.80	69.90	91.7	0.048	0.190
31	89.700	89.600	72.4	1.990	0.0138	0.50	22.10	69.70	91.8	0.048	0.190
32	91.300	91.200	72.3	2.030	0.0138	0.49	22.40	69.50	91.9	0.049	0.190
33	92.900	92.800	72.1	2.060	0.0139	0.49	22.80	69.20	92.0	0.050	0.190
34	94.500	94.400	72.0	2.100	0.0139	0.48	23.10	69.00	92.1	0.050	0.190
35	96.100	96.100	71.8	2.140	0.0139	0.47	23.40	68.90	92.3	0.051	0.190
36	97.800	97.700	71.7	2.170	0.0139	0.46	23.80	68.60	92.4	0.052	0.190
37	99.500	99.400	71.6	2.210	0.0140	0.45	24.10	68.40	92.5	0.052	0.190
38	101.000	101.000	71.4	2.250	0.0140	0.44	24.40	68.20	92.6	0.053	0.190
39	103.000	103.000	71.3	2.290	0.0140	0.44	24.70	68.00	92.7	0.054	0.190
40	105.000	105.000	71.2	2.330	0.0141	0.43	25.10	67.70	92.8	0.054	0.190
41	106.000	106.000	71.0	2.370	0.0141	0.42	25.40	67.50	92.9	0.055	0.190
42	108.000	108.000	70.9	2.410	0.0141	0.42	25.80	67.30	93.1	0.056	0.190
43	110.000	110.000	70.7	2.450	0.0141	0.41	26.10	67.10	93.2	0.056	0.190
44	112.000	112.000	70.6	2.490	0.0142	0.40	26.40	66.90	93.3	0.057	0.190
45	114.000	114.000	70.4	2.540	0.0142	0.39	26.80	66.60	93.4	0.058	0.190
46	116.000	116.000	70.3	2.580	0.0142	0.39	27.10	66.40	93.5	0.058	0.189
47	118.000	117.000	70.2	2.620	0.0143	0.38	27.40	66.20	93.6	0.059	0.189
48	120.000	119.000	70.0	2.670	0.0143	0.38	27.80	65.90	93.7	0.059	0.189
49	121.000	121.000	69.9	2.710	0.0143	0.37	28.10	65.70	93.8	0.060	0.189
50	123.000	123.000	69.7	2.760	0.0143	0.36	28.50	65.40	93.9	0.061	0.189
51	125.000	125.000	69.6	2.810	0.0144	0.36	28.80	65.20	94.0	0.061	0.189
52	128.000	127.000	69.4	2.860	0.0144	0.35	29.10	65.10	94.2	0.062	0.189
53	130.000	129.000	69.3	2.900	0.0144	0.34	29.50	64.80	94.3	0.063	0.189
54	132.000	132.000	69.1	2.950	0.0145	0.34	29.80	64.60	94.4	0.063	0.189
55	134.000	134.000	69.0	3.000	0.0145	0.33	30.20	64.30	94.5	0.064	0.189
56	136.000	136.000	68.8	3.050	0.0145	0.33	30.50	64.10	94.6	0.065	0.189
57	138.000	138.000	68.7	3.100	0.0146	0.32	30.90	63.80	94.7	0.065	0.189
58	140.000	140.000	68.5	3.160	0.0146	0.32	31.20	63.60	94.8	0.066	0.189
59	142.000	142.000	68.4	3.210	0.0146	0.31	31.50	63.40	94.9	0.067	0.189

Table A–15 (continued)
R-507 Saturation Properties—Temperature Table

Temp (°F)	Pressure (psia)		Density (lb/ft³)		Volume (ft³/lb)		Enthalpy (Btu/lb)			Entropy (Btu/lb-R)	
	Liquid	Vapor	Liquid	Vapor	Liquid	Vapor	Liquid	Latent	Vapor	Liquid	Vapor
60	145.000	145.000	68.2	3.260	0.0147	0.31	31.90	63.10	95.0	0.067	0.189
61	147.000	147.000	68.1	3.320	0.0147	0.30	32.20	62.90	95.1	0.068	0.189
62	149.000	149.000	67.9	3.370	0.0147	0.30	32.60	62.60	95.2	0.069	0.189
63	152.000	151.000	67.8	3.430	0.0148	0.29	32.90	62.40	95.3	0.069	0.189
64	154.000	154.000	67.6	3.480	0.0148	0.29	33.30	62.10	95.4	0.070	0.188
65	156.000	156.000	67.4	3.540	0.0148	0.28	33.60	61.90	95.5	0.071	0.188
66	159.000	159.000	67.3	3.600	0.0149	0.28	34.00	61.60	95.6	0.071	0.188
67	161.000	161.000	67.1	3.660	0.0149	0.27	34.40	61.20	95.6	0.072	0.188
68	164.000	163.000	67.0	3.720	0.0149	0.27	34.70	61.00	95.7	0.073	0.188
69	166.000	166.000	66.8	3.780	0.0150	0.26	35.10	60.70	95.8	0.073	0.188
70	169.000	168.000	66.6	3.840	0.0150	0.26	35.40	60.50	95.9	0.074	0.188
71	171.000	171.000	66.5	3.910	0.0150	0.26	35.80	60.20	96.0	0.075	0.188
72	174.000	173.000	66.3	3.970	0.0151	0.25	36.10	60.00	96.1	0.075	0.188
73	176.000	176.000	66.2	4.040	0.0151	0.25	36.50	59.70	96.2	0.076	0.188
74	179.000	179.000	66.0	4.100	0.0152	0.24	36.90	59.40	96.3	0.077	0.188
75	181.000	181.000	65.8	4.170	0.0152	0.24	37.20	59.20	96.4	0.077	0.188
76	184.000	184.000	65.7	4.240	0.0152	0.24	37.60	58.80	96.4	0.078	0.188
77	187.000	187.000	65.5	4.310	0.0153	0.23	37.90	58.60	96.5	0.079	0.188
78	190.000	189.000	65.3	4.380	0.0153	0.23	38.30	58.30	96.6	0.079	0.188
79	192.000	192.000	65.1	4.450	0.0154	0.23	38.70	58.00	96.7	0.080	0.188
80	195.000	195.000	65.0	4.520	0.0154	0.22	39.00	57.80	96.8	0.081	0.188
81	198.000	198.000	64.8	4.590	0.0154	0.22	39.40	57.50	96.9	0.081	0.187
82	201.000	201.000	64.6	4.670	0.0155	0.21	39.80	57.10	96.9	0.082	0.187
83	204.000	204.000	64.4	4.750	0.0155	0.21	40.10	56.90	97.0	0.083	0.187
84	207.000	206.000	64.3	4.820	0.0156	0.21	40.50	56.60	97.1	0.083	0.187
85	210.000	209.000	64.1	4.900	0.0156	0.20	40.90	56.30	97.2	0.084	0.187
86	213.000	212.000	63.9	4.980	0.0156	0.20	41.20	56.00	97.2	0.085	0.187
87	216.000	215.000	63.7	5.060	0.0157	0.20	41.60	55.70	97.3	0.085	0.187
88	219.000	218.000	63.5	5.150	0.0157	0.19	42.00	55.40	97.4	0.086	0.187
89	222.000	221.000	63.4	5.230	0.0158	0.19	42.40	55.00	97.4	0.087	0.187
90	225.000	225.000	63.2	5.310	0.0158	0.19	42.70	54.80	97.5	0.087	0.187
91	228.000	228.000	63.0	5.400	0.0159	0.19	43.10	54.50	97.6	0.088	0.187
92	231.000	231.000	62.8	5.490	0.0159	0.18	43.50	54.10	97.6	0.089	0.187
93	234.000	234.000	62.6	5.580	0.0160	0.18	43.90	53.80	97.7	0.089	0.187
94	238.000	237.000	62.4	5.670	0.0160	0.18	44.30	53.40	97.7	0.090	0.186
95	241.000	241.000	62.2	5.760	0.0161	0.17	44.70	53.10	97.8	0.091	0.186
96	244.000	244.000	62.0	5.860	0.0161	0.17	45.00	52.90	97.9	0.091	0.186
97	247.000	247.000	61.8	5.960	0.0162	0.17	45.40	52.50	97.9	0.092	0.186
98	251.000	251.000	61.6	6.050	0.0162	0.17	45.80	52.20	98.0	0.093	0.186
99	254.000	254.000	61.4	6.150	0.0163	0.16	46.20	51.80	98.0	0.093	0.186
100	258.000	257.000	61.2	6.250	0.0163	0.16	46.60	51.50	98.1	0.094	0.186
101	261.000	261.000	61.0	6.360	0.0164	0.16	47.00	51.10	98.1	0.095	0.186
102	265.000	264.000	60.8	6.460	0.0164	0.16	47.40	50.80	98.2	0.095	0.186
103	268.000	268.000	60.6	6.570	0.0165	0.15	47.80	50.40	98.2	0.096	0.186
104	272.000	272.000	60.4	6.680	0.0166	0.15	48.20	50.00	98.2	0.097	0.186
105	275.000	275.000	60.2	6.790	0.0166	0.15	48.60	49.70	98.3	0.097	0.185
106	279.000	279.000	60.0	6.910	0.0167	0.15	49.00	49.30	98.3	0.098	0.185
107	283.000	283.000	59.7	7.020	0.0167	0.14	49.40	48.90	98.3	0.099	0.185
108	287.000	286.000	59.5	7.140	0.0168	0.14	49.80	48.60	98.4	0.099	0.185
109	290.000	290.000	59.3	7.260	0.0169	0.14	50.20	48.20	98.4	0.100	0.185

Table A-15 (continued)

R-507 Saturation Properties—Temperature Table

Temp (°F)	Pressure (psia) Liquid	Pressure (psia) Vapor	Density (lb/ft³) Liquid	Density (lb/ft³) Vapor	Volume (ft³/lb) Liquid	Volume (ft³/lb) Vapor	Enthalpy (Btu/lb) Liquid	Enthalpy (Btu/lb) Latent	Enthalpy (Btu/lb) Vapor	Entropy (Btu/lb-R) Liquid	Entropy (Btu/lb-R) Vapor
110	294.000	294.000	59.1	7.380	0.0169	0.14	50.60	47.80	98.4	0.101	0.185
111	298.000	298.000	58.8	7.510	0.0170	0.13	51.00	47.50	98.5	0.102	0.185
112	302.000	302.000	58.6	7.640	0.0171	0.13	51.40	47.10	98.5	0.102	0.185
113	306.000	306.000	58.4	7.770	0.0171	0.13	51.80	46.70	98.5	0.103	0.184
114	310.000	310.000	58.1	7.900	0.0172	0.13	52.20	46.30	98.5	0.104	0.184
115	314.000	314.000	57.9	8.040	0.0173	0.12	52.70	45.80	98.5	0.104	0.184
116	318.000	318.000	57.7	8.170	0.0173	0.12	53.10	45.40	98.5	0.105	0.184
117	322.000	322.000	57.4	8.320	0.0174	0.12	53.50	45.00	98.5	0.106	0.184
118	326.000	326.000	57.2	8.460	0.0175	0.12	53.90	44.60	98.5	0.106	0.184
119	330.000	330.000	56.9	8.610	0.0176	0.12	54.40	44.10	98.5	0.107	0.184
120	334.000	334.000	56.7	8.760	0.0176	0.11	54.80	43.70	98.5	0.108	0.183
121	339.000	338.000	56.4	8.920	0.0177	0.11	55.20	43.30	98.5	0.109	0.183
122	343.000	343.000	56.1	9.080	0.0178	0.11	55.70	42.80	98.5	0.109	0.183
123	347.000	347.000	55.9	9.240	0.0179	0.11	56.10	42.40	98.5	0.110	0.183
124	352.000	351.000	55.6	9.410	0.0180	0.11	56.50	42.00	98.5	0.111	0.183
125	356.000	356.000	55.3	9.580	0.0181	0.10	57.00	41.40	98.4	0.112	0.182
126	361.000	360.000	55.1	9.760	0.0182	0.10	57.40	41.00	98.4	0.112	0.182
127	365.000	365.000	54.8	9.940	0.0183	0.10	57.90	40.50	98.4	0.113	0.182
128	370.000	369.000	54.5	10.100	0.0184	0.10	58.30	40.00	98.3	0.114	0.182
129	374.000	374.000	54.2	10.300	0.0185	0.10	58.80	39.50	98.3	0.115	0.182
130	379.000	379.000	53.9	10.500	0.0186	0.10	59.30	38.90	98.2	0.115	0.181
131	384.000	383.000	53.6	10.700	0.0187	0.09	59.70	38.50	98.2	0.116	0.181
132	388.000	388.000	53.3	10.900	0.0188	0.09	60.20	37.90	98.1	0.117	0.181
133	393.000	393.000	53.0	11.100	0.0189	0.09	60.70	37.30	98.0	0.118	0.181
134	398.000	398.000	52.6	11.400	0.0190	0.09	61.20	36.80	98.0	0.118	0.180
135	403.000	402.000	52.3	11.600	0.0191	0.09	61.60	36.30	97.9	0.119	0.180
136	408.000	407.000	52.0	11.800	0.0192	0.08	62.10	35.70	97.8	0.120	0.180
137	413.000	412.000	51.6	12.100	0.0194	0.08	62.60	35.10	97.7	0.121	0.180
138	418.000	417.000	51.2	12.300	0.0195	0.08	63.10	34.50	97.6	0.122	0.179
139	423.000	422.000	50.9	12.600	0.0197	0.08	63.60	33.80	97.4	0.122	0.179
140	428.000	428.000	50.5	12.900	0.0198	0.08	64.20	33.10	97.3	0.123	0.179
141	433.000	433.000	50.1	13.100	0.0200	0.08	64.70	32.50	97.2	0.124	0.178
142	438.000	438.000	49.7	13.400	0.0201	0.07	65.20	31.80	97.0	0.125	0.178
143	444.000	443.000	49.3	13.800	0.0203	0.07	65.80	31.00	96.8	0.126	0.177
144	449.000	449.000	48.9	14.100	0.0205	0.07	66.30	30.30	96.6	0.127	0.177
145	454.000	454.000	48.4	14.400	0.0207	0.07	66.90	29.50	96.4	0.128	0.176
146	460.000	459.000	47.9	14.800	0.0209	0.07	67.50	28.70	96.2	0.129	0.176
147	465.000	465.000	47.4	15.200	0.0211	0.07	68.10	27.90	96.0	0.130	0.175
148	471.000	471.000	46.9	15.600	0.0213	0.06	68.70	27.00	95.7	0.130	0.175
149	476.000	476.000	46.4	16.000	0.0216	0.06	69.30	26.10	95.4	0.131	0.174
150	482.000	482.000	45.8	16.500	0.0218	0.06	70.00	25.00	95.0	0.133	0.174
151	488.000	488.000	45.2	17.000	0.0221	0.06	70.70	24.00	94.7	0.134	0.173

© 2004. E.I. du Pont de Nemours and Company (DuPont). All rights reserved, used under license of DuPont.

Table A–16 Recommended Thicknesses and Heat Gain—Piping

(90°F ambient, 0 mph wind velocity)

Surface Temperature (°F)

IPS	45 to 60 RT	45 to 60 HG	30 to 44 RT	30 to 44 HG	15 to 29 RT	15 to 29 HG	0 to 14 RT	0 to 14 HG	−25 to −1 RT	−25 to −1 HG	−50 to −26 RT	−50 to −26 HG	−75 to −51 RT	−75 to −51 HG	−100 to −76 RT	−100 to −76 HG
80% Relative Humidity																
1/2	1	6	1	7	1	9	1-1/2	9	1-1/2	11	1-1/2	12	2	14	2	14
3/4	1	6	1	8	1	10	1-1/2	10	1-1/2	12	2	12	2	14	2	15
1	1	7	1	10	1-1/2	9	1-1/2	11	1-1/2	13	2	13	2	16	2	17
1-1/4	1	9	1	11	1-1/2	11	1-1/2	13	1-1/2	16	2	15	2-1/2	18	2-1/2	17
1-1/2	1	10	1	13	1-1/2	12	1-1/2	14	2	14	2	17	2-1/2	19	2-1/2	19
2	1	11	1	15	1-1/2	14	1-1/2	16	2	16	2	19	2	19	2-1/2	21
2-1/2	1	13	1	17	1-1/2	15	1-1/2	18	2	18	2	21	2-1/2	21	2-1/2	24
3	1	15	1	20	1-1/2	18	1-1/2	21	2	21	2	23	2-1/2	24	2-1/2	27
3-1/2	1	17	1-1/2	16	1-1/2	20	1-1/2	23	2	23	2-1/2	25	2-1/2	27	3	26
4	1	19	1-1/2	18	1-1/2	22	1-1/2	25	2	25	2-1/2	25	2-1/2	29	3	28
6	1	26	1-1/2	24	1-1/2	29	2	27	2	34	2-1/2	33	3	34	3	37
8	1	33	1-1/2	30	1-1/2	37	2	34	2-1/2	35	2-1/2	41	3	41	3	45
10	1	40	1-1/2	36	1-1/2	45	2	41	2-1/2	42	2-1/2	49	3	49	3-1/2	48
85% Relative Humidity																
1/2	1	6	1	7	1-1/2	7	1-1/2	9	2	9	2	11	2-1/2	11	2-1/2	12
3/4	1	6	1-1/2	7	1-1/2	8	1-1/2	10	2	10	2	12	2-1/2	12	2-1/2	14
1	1	7	1-1/2	8	1-1/2	9	2	9	2	11	2-1/2	12	2-1/2	14	3	14
1-1/4	1	9	1-1/2	9	1-1/2	11	2	10	2	10	2-1/2	14	2-1/2	16	3	16
1-1/2	1	10	1-1/2	10	1-1/2	12	2	11	2	13	2-1/2	15	3	15	3	17
2	1	11	1-1/2	11	2	11	2	13	2-1/2	14	2-1/2	17	3	17	3	19

Table A-16 Recommended Thicknesses and Heat Gain—Piping (*continued*)

(90°F ambient, 0 mph wind velocity)

Surface Temperature (°F)

IPS	45 to 60 RT	45 to 60 HG	30 to 44 RT	30 to 44 HG	15 to 29 RT	15 to 29 HG	0 to 14 RT	0 to 14 HG	−25 to −1 RT	−25 to −1 HG	−50 to −26 RT	−50 to −26 HG	−75 to −51 RT	−75 to −51 HG	−100 to −76 RT	−100 to −76 HG
2-1/2	1	13	1-1/2	13	2	13	2	15	2-1/2	16	2-1/2	19	3	19	3-1/2	19
3	1-1/2	11	1-1/2	15	2	14	2	17	2-1/2	18	3	19	3	22	3-1/2	22
3-1/2	1-1/2	12	1-1/2	16	2	16	2	19	2-1/2	20	3	20	3-1/2	21	2-1/2	24
4	1-1/2	13	1-1/2	18	2	17	2	20	2-1/2	21	3	22	3-1/2	23	3-1/2	25
6	1-1/2	18	1-1/2	24	2	23	2-1/2	23	3	25	3	29	3-1/2	30	4	30
8	1-1/2	23	2	24	2	29	2-1/2	28	3	30	3-1/2	32	3-1/2	36	4	36
10	1-1/2	28	2	28	2	35	2-1/2	34	3	36	3-1/2	37	4	39	4	43

90% Relative Humidity

IPS	45 to 60 RT	45 to 60 HG	30 to 44 RT	30 to 44 HG	15 to 29 RT	15 to 29 HG	0 to 14 RT	0 to 14 HG	−25 to −1 RT	−25 to −1 HG	−50 to −26 RT	−50 to −26 HG	−75 to −51 RT	−75 to −51 HG	−100 to −76 RT	−100 to −76 HG
1/2	1-1/2	4	1-1/2	6	2	6	2	7	2-1/2	8	3	9	3-1/2	10	3-1/2	11
3/4	1-1/2	5	2	6	2	7	2-1/2	7	2-1/2	9	3	10	3-1/2	11	3-1/2	12
1	1-1/2	6	2	6	2	8	2-1/2	8	3	9	3-1/2	11	3-1/2	12	4	12
1-1/4	1-1/2	7	2	7	2-1/2	8	2-1/2	9	3	10	3-1/2	11	4	13	4	14
1-1/2	1-1/2	7	2	8	2-1/2	9	2-1/2	10	3	11	3-1/2	12	4	13	4	15
2	1-1/2	8	2	9	2-1/2	10	3	10	3	13	3-1/2	14	4	15	4-1/2	15
2-1/2	2	8	2	10	3	11	3	12	3-1/2	13	4	14	4	16	4-1/2	17
3	2	9	2	12	3	12	3	13	3-1/2	16	4	16	4-1/2	17	5	18
3-1/2	2	10	2-1/2	11	2-1/2	14	3	14	3-1/2	16	4	17	4-1/2	18	5	19
4	2	11	2-1/2	12	3	13	3	15	3-1/2	17	4	18	4-1/2	20	5	20
6	2	15	2-1/2	16	3	17	3-1/2	18	4	20	4-1/2	22	5	24	5-1/2	24
8	2	18	2-1/2	20	3	21	3-1/2	22	4	24	4-1/2	26	5-1/2	26	5-1/2	29
10	2	22	2-1/2	23	3	25	3-1/2	26	4-1/2	26	5	28	5-1/2	31	6	32

IPS = iron pipe size
RT = recommended thickness
HG = heat gain = Btu/hour/linear foot

Courtesy Dow Chemical Company.

CALCULATING OVERALL HEAT LEAKAGE FACTORS

On page 255, a simplified method of calculating heat transfer factors through a composite wall was given. The mathematical method more commonly used is to add the conductances of the various materials in the wall. Conductances are the reciprocals of the heat transfer or k factors. The overall figure obtained is called the U factor. It is calculated by the following formula:

$$U = \frac{1}{\frac{x_1}{k_1} + \frac{x_2}{k_2} + \frac{x_3}{k_3}}$$

where U = overall heat transfer factor for the complete wall thickness in btu/sqft/degree/hour.

x_1 = thickness of first material in inches.
x_2 = thickness of second material in inches.
x_3 = thickness of third material in inches.
k_1 = k factor of first material
k_2 = k factor of second material
k_3 = k factor of third material

As many more x/k factors as are needed can be added if the wall contains more different materials.

Example A – 1. – Work example 18-3 by solving for the U factor.

$$U = \frac{1}{\frac{x_1}{k_1} + \frac{x_2}{k_2}}$$

$$= \frac{1}{\frac{1/2 + 3/4}{.80} + \frac{3}{.24}} = \frac{1}{1.56 + 12.50} = 1/14.06$$

$$= 0.0711 \text{ btu/sqft/hour/degree}$$

This may be checked against the answer of example 18-3 by solving for the leakage of 3.38 inches of glass wool.
$$.24/3.38 = .071$$
The total heat loss for the entire wall then could be solved by the following formula:

$$H = A U (t_2 - t_1)$$

Notice this is the same formula given in chapter 18 except U is substituted for k/x.
The U factor formula is sometimes given more completely as follows:

$$U = \frac{1}{\frac{1}{f_i} + \frac{x_1}{k_1} + \frac{x_2}{k_2} + \frac{x_3}{k_3} + \frac{n}{1.1} + \frac{1}{f_o}}$$

Where f_i = inside air film = 1.65

This is the estimated insulation effect of the film of air that clings to the surface. The inside air film is for still air. For a box inside a building, this figure is also used for fo; because the latter is only for a surface exposed to the wind.

n = the number of separate air spaces between surfaces. The thickness of the air space, if more than 1/4 of an inch, makes no difference.

fo = outside air film = 6.5

This figure is used for the outside of the building which is exposed to outside winds. It is not used for the outside surface of a cooler or cabinet enclosed in another building.

Example A-2 — A wall is made up of an eight-inch brick facing outside, an air space and four inches of cork. What is its U factor?

$$U = \frac{1}{\frac{1}{f_1} + \frac{x_1}{k_1} + \frac{x_2}{k_2} + \frac{x_3}{k_3} + \frac{n}{1.1} + \frac{1}{f_o}}$$

$$= \frac{1}{\frac{1}{1.65} + \frac{8}{5.0} + \frac{4}{.28} + \frac{1}{1.1} + \frac{1}{6.5}}$$

$$= \frac{1}{0.61 + 1.60 + 14.28 + 0.91 + 0.15}$$

$$= 0.0570$$

In common insulation problems, the effect of surface air films is neglected because they are so small. Their effect is of value in the following type of problem.

Example A-3 — Figure the U factor of a double glass door with an air space between the two glasses. Assume each pane of glass is 1/8 inch thick:

$$= \frac{1}{\frac{1}{1.65} + \frac{1}{5.0} + \frac{1}{.28} + \frac{2.5}{1.1}}$$

$$= 0.457$$

Compare this answer with the value given for double glass in Table A-9.

Index

Absorption systems, 327
    continuous, 328–329
    continuous lithium bromide, 329f
    diffusion, 329–331
    direct gas-fired, 328
    elementary intermittent, 328f
    heat exchanger in, 328
    intermittent ammonia, 327
    lithium bromide, 330f
    mechanical pump in, 328
    schematic outline of, 330
    simple intermittent, 328f
    single-chamber lithium bromide, 330f
Accumulator, 66
    flooded evaporator and, 64f
    suction, 65
Adiabatic process, 243
Air, 184
    moisture solubility in, 230f
    relative humidity of, 230
    saturated, 229
    washer, 313
    weight of, 231
Air-acetylene torch, 118
Air agitation tubes, 314f
Air conditioning, 308–311
    complete, 308
    methods of installing vertical, 311f
    remote, 310
    reverse-cycle, 311
    self-contained, 310
    self-contained window-type, 310f
    sprayed coil-type, 313f
    steam jets and, 102
    vertical, floor mounted, 310f
Altitude
    automatic expansion valves and, 325
    bellows and, 325f
    pressure variations with, 324f
    saturation pressure and, 323
American Society of Heating, Refrigerating and Air Conditioning Engineers (ASHRAE), 31, 308
    Handbook of Fundamentals, 34

American Society of Refrigeration Engineers, 31
Ammeters, 209
Ammonia, 12, 15, 20f, 32–33
    boiling temperature of liquid, 12, 13, 14f, 32
    breakdown of, 32
    circulating system for liquid, 65f
    compression temperature of, 33, 90
    compressor manifold, 92f
    compressors, 258
    critical pressure for, 28
    critical temperature of, 33
    discharge temperature of, 174
    displacement per ton of, 246f
    evaporating temperature of, 330
    fed to coil, 14f
    freezing point of, 33
    hazards of, 30
    horsepower per ton of, theoretical, 246–247, 248f
    latent heat of, 32
    pressure-temperature characteristics of, 11, 11f
    saturation temperature of, 14
    stuffing box for, 88f
    suction, discharge, and liquid line capacities for, 280f
    suction line capabilities for, 279f
    toxic properties of, 33
    two-cylinder compressor for, 250f
    vapor, 327
Ammonium hydroxide, 33
ASHRAE. *See* American Society of Heating, Refrigerating and Air Conditioning Engineers
Autotransformer starter, 158

Backpressure, 14
    compressor and, 239
    control, 128–130, 131
    maintaining, 167
    valves regulating, 135
    varying, 80f
Bacteria, 164
Baffles, 67f, 70
Barometer, 10
Baudelot cooler, 296

Beer
    controlling foam on, 296
    coolers, 295, 295f, 296
    cooling problems of, 295
    dispensing, 295
    fermenting temperature, 297
    mash, 297
    operation costs of brewing, 297
    schematic flow sheet for brewing, 298f
    wort, 297
Bellows
    altitude on, effects of, 324f
    flexible, 89, 128
Beverage cooling requirements, 292
Blast freezer, 70
Boiler
    double, 10, 11f
    horizontal-return tube, 62
Boilovers, 174
Bourdon tube, 205, 206
Breakers, 157
Breather line, 88
Brine
    advantage of, 282
    alkalinity of, 285
    calcium chloride, 285, 288f
    chemistry, 284–285
    circulating, 282, 284
    circulation, 313
    coils, 300
    concentration, 286–290
    controlling flow of, 284
    corrosion control, 285–286
    density, 285
    direct expansion v., 283f
    disadvantage of, 282
    freezing temperature of, 284
    high-density, 287
    industrial brine coolers, 294
    lithium bromide, 327, 328
    maintaining chromate in, 285
    noncirculating, 284f
    reconcentrator, 153f
    single-pass system, 64f
    sodium chloride, 284, 287f
    sodium chloride, chart of, 284f
    spray, 151
    thawing, 284
British thermal units (Btu), 2
Btu. *See* British thermal units

Bubbler, 292
"Bulls-eyes," 125
Cabinet
    construction, 197
    humidity, 189
    overloaded, 189
Capillary tube, 45–46, 69, 132
    early applications of, 46
    flow factors through, 45
    high side float valve and, differences between, 46
    troubles with, 46
Carbon dioxide, 29, 299
    absorption equipment, 319f
    compressing, 319
    critical pressure for, 28
    disadvantages of, 318
    pressures, 318
    as refrigerant, 318
    safety of, 318
    sources of, 318
    three-stage compression system for, 320
Cascade systems, 268f, 269
    advantages of, 268
    disadvantages of, 268
    refrigerants suitable for, 268
Celsius, 1
Centigrade, 1
Centrifugal blowers, 310
CFCs. *See* Chlorofluorocarbons
Change of state, 5
Chiller barrel, 60
Chlorofluorocarbons (CFCs), 31
    accelerated ban on, 31
    production reduction in, 31
Chromic acid, 285, 286
Circulation, 61
    brine, 314
    coils and, 67f
    in "dead air space," 192f
    heat transfer and, 61
    insufficient, 189
    reducing, 192f
Clearance volume, 74
Clocks, self-starting electric, 209
Coefficient of performance (COP), 19, 245
Cogen. *See* Total energy system
Coil(s)
    air, 312
    air cooling, 310

ammonia, 68
attemperator, 298
baffled, 68
bare, 67
blower, 68, 235, 302, 308
brine, 298
chilling, 60
defrosting, 177
direct expansion, 300
dry expansion, 43
electrical difference between high/low
    inductance, 156
evaporator, 13
food and temperature of, 230
freezing, 60
frost on, 189
hairpin, 63
headered pipe, 67f
industrial blower, 68
natural convection, 302
operation of, 64
pipe, 60, 61
pressure drop in, 51f
role reversal of, 312
starving, 235
temperatures, 134
temperatures for various room temperatures, 283f
trombone, 63, 66, 69
wall-type blower, 68f
water, 312
Cold-storage warehouses, 305–307
    banks and, 307
    considerations for, 305
    economics of, 306
    leakage load of, 307
    location of, 306
    requirements for, 307
    type/size of, 306
Combustion formulas, 1
Commercial tariffs, 140
Compressor(s), 14, 15
    A-frame, 81
    backpressure and, 239
    belt-driven open-type, 84f
    booster, 266f
    capacity control of, 92–93
    capacity of, actual, 248, 307
    capacity/power for, 169f
    capacity reduction of, 260
    capacity variation of, 252, 253

cascade systems, 268–269, 268f
changing operating speed of, 37, 38
checking operation of, 74
clearance pockets on, 92, 92f
as constant-action pumps, 237
construction details of, 85–93
cooling, 90
cross section of open-type, 75f
discharge temperature, 174
displacement of, actual, 246
displacement per inch stroke per cylinder per 100 rpm
    of, 248
dynamic, 100
efficiency v. horsepower in, 239f
head, 173
heat of, 106
hermetically sealed, 23, 83, 84, 84f
historical development of, 80–85
horizontal double-acting, 80, 81f, 82
horsepower calculation formulas for, 241
horsepower of, 241–242
increasing clearance on, 79f
low-pressure refrigerating, 83f
manifold ammonia, 92f
mechanical efficiency of, 241
modern valve arrangement on, 87f
multicylinder, 92
open-type domestic/commercial, 83, 88, 89
outside mounted, 137
positive-displacement, 100, 115
pressure rise in head of, 271
reciprocating, 23, 93
reduction in capacity of, 271
refrigeration cycles of, 76f
remaining conditions leaving, 241
schematic indicator connected to, 77f
self-regulating characteristics of, 168
shaft seal in, 83, 88–89
shut off controls for, 92
size of, 112, 238–240
sizing, 245–251
speeds, 93
stuffing box on, 81
two-cylinder ammonia, 250f
two-cylinder single-acting, 82f
two-stage, 263, 264f, 265f, 267, 268f
varying temperature of, 170
vertical single-acting, 81–82
volumetric efficiencies of ammonia, 240f
volumetric efficiencies of R-134a/R-22, 240f

Compressor(s), (continued)
    volumetric efficiency of, 78
    V-type, 82, 83f
    VW-type, 82, 83f
    workings of, 74
Compressor(s), centrifugal, 23, 34, 99–101, 99f, 125
    capacity of, 100
    tip speed of, 100
Compressor(s), rotary
    booster, 95f
    features of, 94f
    portable, 95
    rotating-vane, 94
    stationary-vane, 94
    uses of, 95
Compressor(s), screw, 96–99
    advantages of, 96
    ammonia, 97f
    disadvantage of, 96
    economizer port on, 98
    rotors of, 98f
    single-stage, 96
    tapered machined screw-type gears in, 98f
    two-stage, 98
    volumetric efficiency of, 96
Compressor(s), scroll, 101–102
    compression ratio of, 102
    schematic illustration of, 101f
    swing link in, 101
    variable displacement, 102
Concrete, 218
Condenser(s), 14
    actions in, 105
    advantages of cooler liquid in, 106
    air-cooled, 106–108, 107, 108
    air-cooled, remote, 107f
    bent-copper-tubing, 109
    capacities of air-cooled, 249f
    commercial size, 215
    cost of, 225
    cross-connected, 304
    dirty, 184
    double-pipe ammonia, 107
    double-pipe water-cooled, 108–109, 108f
    double-pipe water-cooled, cleanable, 110f
    double-row, double-pass, 106
    double-row, single-pass, 106, 107, 107f
    evaporative, 114, 114f
    heat from, 245f
    heat transfer and, 105
    heat transfer in water-cooled, 223
    as liquid receiver, 110
    mean temperature difference in, 224f
    operation of, 105–106
    performance variable in, 224
    pipe-in-pipe, 108
    ratings/data for, 218f
    remote, 316
    requirements of, 105
    schematic for air cooled, 106f
    shell-and-coil, 110
    shell-and-coil water-cooled, 112f
    shell-and-tube, 109–110, 110f, 295
    shell-and-tube, horizontal, 111f, 225f
    shell-and-tube, vertical, 110, 111f
    single-row, single-pass, 106, 106f
    tube-in-tube, 108
    variables in operation of, 224
    water-cooled, 108
    water distributor, 111f
Conduction, 8, 13, 192
    of common materials, 212f
Congealing tanks, 290–291
    selecting of, 291
Contractors, 157–160
Control(s), 128
    adjustment of, 181
    automatic, 178
    backpressure, 128–130, 131
    changing adjustments on, 129f
    costs, 140
    defrost, 160
    domestic refrigerator, 131f
    electrical, 128, 139
    low-pressure, 128–130
    magnetic, 129
    oil pressure, 132
    pressure, 130f
    schematic of pressure, 129f
    temperature, 128, 130, 130f, 132, 312
    valves, 128
Control cutout point, 133
Controlled pressure receiver, 98
Control panel, 57
Control voltage transformer, 54
Convection, 7, 13, 192
    natural, 67
Cooler-freezer, 221f, 227f
Coolers, commercial walk-in, 198
Cooling towers, 111–114, 137
    deck-type, 113, 113f
    forced-draft, 113, 113f

induced draft, 113, 113f
mechanical-draft, 114
natural-draft, 112
required, 112, 114
spray, 112, 113f
COP. *See* Coefficient of performance
Corliss engines, 161
Counterflow, 109
    lack of, 110
    results of, 109f
Crankhouse heater, 137, 177
Cut-in point, 128
Cut-out point, 128

Dairy farms, 282
Dalton's law of partial pressure, 7f, 7
DDC. *See* Direct digital control
Defrost, 147
    automatic, 148, 149, 151
    coils, 177
    controls, 151
    cycle, 302
    electric heater, 149, 312
    high-speed, 312
    hot-gas, 147–149, 308
    off-cycle, 147
    problem of, 308
    suction pressure control for, 151
    time clock, 151
    water-spray, 149, 150, 151f, 308
    wiring diagram for, 152f
Dehydrators, 122–123
    angle-type, 122f
Demand, 140, 160
    metering, 140
Department of Transportation, U. S., 38
Desiccants, 123
Dew point, 231, 234
Differential, 128
    changing, 129
Direct digital control (DDC), 139, 188, 200
    energy management through, 140
    humidity maintained through, 235
Discharge line, 17, 100f
    indicator, 78f
Doors, 197
    cold-storage, 197
    glass, 197
    insulation of, 197
    losses from, 222

Driers, 122–123
    chemical, 197
    liquid-line, 123
    suction-line, 123
Drift, 112
DuPont, 31
    website, 29, 34, 273
DX valve. *See* Valve(s), thermostatic expansion

Electric float
    liquid solenoid valve and, 55, 56
    TDS and, 56, 57
Electric instruments, 208, 209
Electric tariffs, 140
Electrolux system, 329
    simplified, 330f
Electronic leak detector, 34
Energy, 6, 140
    conversion equivalents, 6
    fuel as economical source of, 327
    management, 140
    off-peak usage of, 140
    peak usage of, 140
Enthalpies, 237
Entropy, 237, 241
    constant lines of, 243
Environmental Protection Agency, U.S., 171
Enzymes, 163, 164
    freezing and, 165
EPR. *See* Evaporator pressure-regulating valve
Ether
    boiling temperature of, 11
    heat transfer and, 11, 12
    pressure-temperature characteristics of, 11f
Eutectic holdover plates, 290–291
    considerations for, 290
    forced-draft, 291
    natural-convection, 291
    storage capacity of, 290
    surface area of, 290
Eutectic point, 285
Eutectic solution, 285
Evaporation, 10–14, 60, 230
    cooling effect of, 29
    heat absorbed through, 230
    latent heat of, 3, 4, 5
Evaporator
    accumulator used in flooded, 64f
    analogy, 41f
    automatic expansion valve and, 47f
    bare-pipe, 217

Evaporator (*continued*)
    blower-type, 68f
    changing temperature of, 253f, 254f
    complete analysis of, 242–243
    controlling liquid level in, 132
    convection, 67–68
    domestic, 62, 62f
    dry, 64–66
    dry expansion, 66, 71
    feeds, 69–70
    finned coil, 61f
    flooded, 43, 62–64
    flooded ammonia, 72
    flooded, full, 48
    flooded halocarbon, 72f
    forced-draft, 67–68, 68
    heat exchanger as, 262
    improved designs of, 66–67
    industrial, 69
    industrial flooded, 62
    large tube, 66f
    liquid circulation in boiler-type, 63f
    nearly flooded, 48
    oil in, 71–73, 143
    pipe coils as, 60
    plate surface, 60
    requirements for, 60–61
    self-regulating characteristics of, 168
    shelf-type, 62f
    sizing, 216–219
    small tube, 65f
    stainless steel, 61f
    starved, 48
    starved dry expansion, 71
    suction temperature and, 50
    temperature, 169f, 174f, 234, 235, 309
    types of, 61–68
Evaporator feed device, 40
Evaporator pressure-regulating valve (EPR), 135, 135f
External equalizer, 51
    thermostatic expansion valve and, 52f

Fahrenheit, 1
Faraday, Michael, 30
Faults, 182f, 183f
Filters, 121
    sand, 315
Fire, 138
    codes, 194
FLA. *See* Full-load amperage
Flake ice machine, 316, 316f
Flare fittings, 118f
Flexible diaphragm, 46
Floodback, 44, 51
    preventing, 132
Flourine, 34
Food
    bacterial spoilage of, 165
    coil temperature and, 230f
    dehydration of, 164, 166, 189
    freezing, 165
    humidity to keep, best, 229, 230f
    loads, 214
    odors in, 190
    "off" flavors in, 190
    plant locker storage periods for, 308f
    problems with storing, 189–190
    ripening process of, 163, 164, 165
    turnover, 307
    unsatisfactory preservation of, 189
    vitamin content of, 164
    volumetric factors of, 214
Freezer burn, 166
Freon, 30
Friction, 277
    increased, 279
Frost, 147, 197
    on coils, 189
    insulating effect of, 147
    line, 174
    preventing, 147, 151
Fuel costs, 161
Full-load amperage (FLA), 157
Fusible plug, 138
Fusion, 4

Gage pressure (psig), 6
    atmospheric pressure and, 6
Gas, 5
    flash, 19, 40, 70, 263, 271
    formulas, 5
    foul, 171
    law, 5
    noncondensable, 32, 124, 171
    passages, 86–88
    phosgene, 34
    superheated discharge, 223
    temperature of discharge, 131
    vapor and, difference between, 4, 5

Halide torch, 34
Halocarbons, 34–36
    properties of, 34

HCFCs. *See* Hydrochlorofluorocarbons
HDA. *See* Horizontal double acting pattern
Headering, 66
Heat
    absorption process generating, 328
    auxiliary, 312
    of compressor, 106
    condenser, 245f
    cycle, 17–18, 18f
    effects of, 1
    electric resistance, 312
    evaporation and, 230
    excessive, 198
    fermentation and, 299
    leakage, 192, 215
    leaks, 211–212
    level of, 2
    quantity of, 2
    removing total, 105
    resistance to flow of, 191
    sensible, 14, 105, 148
    temperature and, difference between, 1
    vital, 164
    waste, 162
Heat exchanger, 72f, 126, 126f
    in absorption systems, 328
    common uses for, 260
    effects of, 260, 261f
    effects of ammonia, 262f
    effects of, different refrigerants and, 263f
    as evaporator, 262
    operating efficiency of, 263
    savings from, 259
    suction vapor from, 126
    thermostatic expansion valve and, 54f
Heat, latent, 3–5, 13, 105
    of ammonia, 32, 33
    of dry ice, 320
    of evaporation, 4, 5, 112
    of fusion, 4
    of ice, 213
    per pound of refrigerant, 26
    in steam drives, 161
    of sublimation, 4
Heat pumps, 55, 162, 311–313
    advantages of, 313
    air-source, 311, 311–312
    in far northern/southern regions of planet, 312
    ground source (geothermal), 311, 312
    water-based ground source (geothermal), 312
Heat, specific, 3

    of common foods, 3f
    of iron, 3
Heat transfer, 7, 13
    circulation and, 61
    compressing vapor, 258
    condensers and, 105
    condition of, 234
    double, 268
    ether and, 11, 12
    in horizontal shell-and-tube condenser, 255f
    radiation, 13
    in water-cooled condenser, 223
HFCs. *See* Hydrofluorocarbons
High level alarm (HLA), 58
High level shutdown (HLSD), 58
High-pressure cut-out, 131
High side, 15, 15f
High side float valve, 44–45, 45f, 69
    capillary tube and, differences between, 45–46
    problems with, 44
    purging, 44
    sensitivity of, 45
    use of, 45
HLA. *See* High level alarm
HLSD. *See* High level shutdown
HMI. *See* Human/machine interface
Horizontal double acting pattern (HDA), 80
Human/machine interface (HMI), 209
Humidity
    absolute, 229
    of air, 231
    cabinet, 189
    control of, 309
    DDC maintaining proper, 235
    drying prevented by, 229
    electronic devices measuring, 236, 236f
    hair measurements of, 236
    measuring, 230–234
    mold increased by, 229
    problem of relative, 229
    relative, 229
    removing, 309
Hunting, 51, 273
Hydrochlorofluorocarbons (HCFCs), 31
    extended schedule to eliminate, 31
Hydrofluorocarbons (HFCs), 30, 31

Ice
    crystal clear, 315
    faster methods for freezing, 316
    freezing tank, 314f

Ice (continued)
    latent heat of, 213
    machines, 316, 317
    making, 313–317
    marble, 315
    storage for holdover capacity, 284f
    storage rooms, 315
    time necessary to freeze, 315
Ice cream packing chart, 321f
Ice, dry, 30, 321–322
    advantages of, 320–321
    box/truck body with bunker for, 322f
    box/truck body with coil and bunker for, 322f
    cost to produce, 319, 320
    disadvantages of, 320–321
    latent heat of, 320
    manufacture of, 318–320
    temperature of, 318, 319, 320
Icemakers, automatic, 313
Ice melting equivalents (IME), 4
Ice plant, refrigeration load for, 315
IME. See Ice melting equivalents
Indicator, 74, 75
    diagram, 75, 77f, 78f, 79
    fundamentals of, 75
Industrial freezer storage plants, 67
Industrial tariffs, 140
Inline pressure regulator (IPR), 135
Inspections, 179
Insulation, 191–195
    board, 193
    board, method of laying up, 193f
    classifying, 193
    concrete, 219
    cost of, 192
    on doors, 197
    foamed-in-place, 194
    foamed plastic, 194, 193f, 194
    glass, 195
    installation difficulty of, 192
    loose fill, 194
    mineral, 195
    nonflammable, 191
    organic, 194–195
    perfect, 192
    of pipes, 198
    plastic, 195
    polystyrene, 195
    polyurethane, 195
    recommended thicknesses of, 223f
    reflective, 195
    requirements, 191
    sawdust, 194
    self-supporting, 191
    synthetic, 194
    of tanks, 198
    thickness, 213
    trends in, present, 197
    weight of, 192
    wet, 196
    wood shaving, 194
Interstate Commerce regulations, 26, 38
IPR. See Inline pressure regulator

Kelvin, 1
Kettering, C. F., 30
K factor, 215, 216
k factor, 206, 211–212
    formula with, 212
    of metal, 213

LCL. See Less-than-container-load
Leaks, 188
    air, 290
    check for, 178
    heat, 211, 215
Less-than-container-load (LCL), 322
Lifting stations, 70
Line sizes, 271
Liquid-feed device, 40
Liquid indicator, 126f
Liquid level control, 58
Liquid receiver, 14
    commercial, 116
    condenser as, 110
    horizontal, 116, 117f
    horizontal ammonia, 117f
    industrial, 116
    orientation of, 116
    vertical, 116, 117f
Liquid solenoid valve
    electric float and, 55, 56
    thermal differential sensor and, 56, 57
    variable-capacitance probe and, 57, 58
Liquid trap in suction line, 148f
Liquid-vapor cycle, 19, 19f
Litmus, 302
LLA. See Low level alarm
LLSD. See Low level shutdown
Load, total, 222
Locked rotor amperage (LRA), 157

Locker plants, 307–308
    diagram of typical, 309f
    food storage periods in, 309f
    size of, 307, 308f
Low level alarm (LLA), 58
Low level shutdown (LLSD), 58
Low-oil cut-out, 131, 132
Low-pressure receiver, 319
Low side, 14, 15f
Low side float valve, 42–44
    advantages of, 42
    for ammonia, 42f
    calibrating, 44
    difficulties with, 43
    disadvantages of, 43
    inside liquid header, 43f
    in separate chamber, 43f
Low-side float valve, 69, 139
Low-water cut-out, 132
LRA. *See* Locked rotor amperage
Lubrication, 89, 90. *See also* Oil
    checking, 179
    force-feed, 90, 131, 143
    problems with, 263
    splash systems for, 143
    watching, 189

Maintenance, 178–179
    contract service, 179
    systematic method for, 185
Meat, aging of, 163
"Mechanical ice box," 30
Mercury column, 207
Metering device, 40
    overfeeding, 184
Methyl chloride, 30, 34, 83
    "methyl drunk" effect of, 30
Microprocessor, 139
Midgely, Thomas, 30
Milk coolers, 296
    flash pasteurizing and, 297
    plate-type, 297f
Moisture, 122
    of oil, 144
    proofing, 196–197
    solubility in air, 230f
Mold, 164
    humidity increasing, 229
    reducing, 165
Mollier diagram. *See* Pressure-enthalpy diagram
Montreal Protocol of 1987, 31, 33

Motor drive, 13
    belted, 159
    capacitor split-phase, 156, 156f
    centrifugal devices in, 157
    compound-wound, 161
    dependability of, 154
    economy of, 154
    fractional-horsepower, 156
    gasoline, 161
    heat, 54, 55
    requirements of, 154
    resistance split-phase, 157
    shunt-wound, 160
    simplicity of, 154
    single-phase induction, 155
    speeds of, standard, 160
    split-phase induction, 156f
    starters, 157
    synchronous, 160
    three-phase, 159, 159f
    torque of, 154
Motor drive, electrical, 154–156
    dependability of, 155
    induction, 155
    single-phase, 155
Motor drive, steam, 161
    latent heat and, 161
    problems with, 161
Multipass brine cooler, 63f
Multiplexing
    commercial, 301–305
    faulty, 189
    rules for, general, 301

National Electric Code, 322
Net refrigeration effect (NRE), 238
    changes in, 258
    formulas for, 239
Noise, knocking, 188
NRE. *See* Net refrigeration effect

ODP. *See* Ozone depletion potential
ODS. *See* Special solder fittings
Oil, 23, 279
    alkabenzene, 142
    cloud point of, 144
    conditions of, 145
    dewaxing, 145
    distribution, 143
    ester-based, 142
    in evaporators, 71–73, 143

Oil (*continued*)
   excess, 267
   flash point of, 144, 145
   foaming, 143–144
   lubricating, 142
   mineral, 142
   moisture of, 145
   PAG, 142
   POE, 142
   pour point of, 144, 145
   pressure control, 132f
   pump, 96
   purifiers, 124
   recommendations for, 145
   refineries, 327
   regenerator, 124f
   return, 271
   seal, 89
   selection of, 144–146
   slugging, 86, 87, 173, 177, 188, 199
   specifications, 146f
   synthetic, 142
   viscosity of, 144, 145f
   wet compression and breakdown of, 258
Oil separator, 71, 123, 143, 267
   ammonia, 124f
   location of, 123
   low-pressure, 123
Oil traps, 123, 124, 143, 267
   location of, 123
"Open on pressure rise" device, 131
Operating problems/solutions, 186f, 187f
Overfeed ratio, 70
Overloaded cabinet, 184
Oxyacetylene torch, 118
Ozone depletion, 31
Ozone depletion potential (ODP), 31

Packaged tube ice maker, 64f
Packing, 88
   metal rod, 89f
PAG. *See* Polyalkaline glycol oil
Pak-Ice machine, 316, 317f
Parallel flow, 109
Phenolphthalein, 286
PH indicators, 286f
Pipes/piping, 116–119
   butt-welded, 118
   friction in, 279
   insulation of, 198
   iron, 118
   rules regarding, 277
   Schedule 85, 118
   seamless wrought iron, 118
   size changes in, 279
   steel, 118
   temperatures of, 200
   valves/fittings and, 279f
   vent, 117f, 118f
   X-heavy, 118
Piston speeds, 93
PLC. *See* Programmable logic controllers
Plug/stoppage, 184
Pneumatic high-side float, 58
   thermal economizer and, 58f
POE. *See* Polyol ester oil
Polyalkaline glycol oil (PAG), 142
Polyol ester oil (POE), 142
Pounds per square inch (psi), 6
Power factor, 160
Pressure, 6, 7
   atmospheric, 6, 10
   carbon dioxide, 318
   condensing, 25, 80f, 106, 206
   constant discharge, 79
   control, 129f, 130f
   crankhouse, 23, 88
   crankhouse, reduction of, 177
   critical, 28
   cycle, 14–16
   evaporating, 26, 206
   excess, 109
   head, 239
   head, sufficient, 106
   importance of, 323
   instantly raised, 74
   intermediate absolute, 264, 265
   low, 14
   measuring, 323
   partial, 7
   rise in compressor hear, 271
   spring, 74
   suction, 14, 23
   in thermostatic expansion valve, 50
   vapor, 74
   variations with altitude, 324f
Pressure, absolute (psia), 6, 323
Pressure drop, 51, 51f, 271
   calculating, 273–277
   effect of, 271, 272
   in lines, 276f

in liquid column, 272f
in liquid line, 271
per foot of lift, 272f
permissible, 272, 272f
variables affecting, 273
Pressure-enthalpy diagram, 243
    ammonia standard ton conditions on, 256f
    compression cycle on, 245f
    compression ratios on, effects of, 260f
    dual-effect compression on, 269f
    R-134a, 244f
    standard ton conditions on, 256f
Pressure gage, 6, 14, 205–208, 206f
    accuracy of, 207
    air pressure and, 324f
    altitude on, effects of, 324f
    compound, 206, 206f
    interior of, 205f
    micron, 206
    protecting, 207
    range of, 206
    recording, 75, 207
    sea-level, 323
    sizes of, 206
    vibrational dampeners for, 207
Pressure-limiting device, 131
Pressure-relief devices, 138
Pressure, saturation, 323
Prest-O-Lite torch, 108
Programmable logic controllers (PLC), 139
psi. *See* Pounds per square inch
psia. *See* Pressure, absolute
psig. *See* Gage pressure
Psychrometric chart
    high temperature, 232f
    low-temperature, 233f
    with problem, 234f
Pump-down cycle, automatic, 134
Purgers, 172f
    automatic, 124, 125, 125f, 171
    centrifugal, 125

Radiation, 7, 13, 192
Range setting, 128
    changing, 129
Rankine, 1
Recirculated system, 70, 71
Recirculator
    package, 70, 71f
    pumps, 70

Refrigerant
    azeotropic, 31
    boiling temperatures of, 25
    carbon dioxide as, 318
    changing, 37–38, 258
    charge, 171
    compression temperature of, 28
    control device, 40
    cost of, 25
    cylinder colors, 188
    detectors, 188
    discharge temperature for various, 176f
    drums, 38, 38f
    evaporator pressure and, 23
    fatal dose of, 26
    feed device, 40
    flammability/explosion hazards of, 26
    flow reversal of, 311
    freezing point of, 25, 26
    graphic displacement of common, 30f
    halocarbon, 60, 71, 72, 118
    handling, 38, 39
    historical development of, 29–32
    insufficient, 185
    latent heat per pound of, 26
    losses, 290
    low-pressure, 66
    mixing, 142
    overcharge in, 185
    physical properties of, 25–28
    properties of common, 28f
    properties of less common, 29f
    removing, 172
    requirements of, 22–25
    saturated, 16
    solubility curve for, 143
    specific gravity of, 27
    stability of, 23
    suitable condensing pressure of, 22, 23
    suitable for cascade system, 267
    superheated, 16
    synthetic, 34
    temperature of, 237
    temperature-pressure relationships for common, 23, 24f
    thermodynamic, 23
    too much, 172
    velocities of, allowable, 272f
    water as, 327
    weight of, 26

Refrigerant 13, 34
    physical properties of, 27f
Refrigerant 35
    critical pressure for, 28
Refrigerant 717. See Ammonia
Refrigerants
    11, 100
    12, 34, 83
    13B1, 34
    21, 35
    22, 35, 83
    30, 34
    40, 35
    113, 35
    114, 35
    123, 35
    401A, 47
    404A, 36
    407C, 36
    500, 35
    502, 36
    503, 36
Refrigeration
    automatic, 177
    commercial, 8
    cost of, 168, 223
    cycles, 20f, 331f
    difficulties of water for, 11
    direct systems of, 282
    domestic, 8
    economy of, 168–172
    erratic, 188
    indirect, 283f
    indirect systems of, 282
    industrial, 8, 168
    losses, 292
    marine, 8
    operational requirements of, 167
    pressure cycle of, 16f
    safety, 172–177
    schematic, 238f
    schematic of, 15f
    self-regulating character of, 167
    service technician, 181
    short-time storage, 168
    small halocarbon systems for, 125
    system faults in, 181
    temperature cycle of, 17f
    ton of, 4, 20
    variation in requirements for, 181–189
    volume of liquid per ton of, 275f

Refrigeration effect. See Net refrigeration effect
Refrigerator, reach in, 213
Relays
    domestic system showing starting, 157f
    potential, 156
    voltage, 156
Remote computer screen. See Direct digital control
Restrictor tube, 45
RLA. See Running load amperage
Running load amperage (RLA), 158
Rupture disk, 138f, 138
R value, 194, 212–214

Safety head, 86
Safety Standard for Refrigeration Systems, 332
Seasonal variation, 235
Settling, 191
Shaft seals, 83, 88–89
    bellows-type, 89, 89f
    diaphragm-type, 89, 89f
    rotary, 89, 90f
Shortcut factors, 210
Short cycling, 44, 120
    preventing, 302
Sight glasses, 125
    liquid, 126f
Silicon thyristor, 159
Sling psychrometer, 230, 231f
Slip, 158
Sodium dichromate, 285
Solder
    fittings, 119f
    hard, 118
    types of, 117
Solenoid valve, 73, 134f
    direct-acting, 133
    in liquid line two-temperature system, 135f
    operated by room thermostat, 134f, 304
    pilot-operated, 133
Special solder fittings (ODS), 117
Standard ton conditions, 21, 20f, 36, 254–256
    on pressure-enthalpy diagram, 256f
    on pressure-enthalpy diagram, ammonia, 256f
Steam jets
    air conditioning and, 103
    as lettuce cooling system, 103
Strainers, 121
Subcooling, 17, 106, 254
    calculations neglecting, 223
    with constant condensing pressure, 257f
    effects of, 256

efficiency systems and, 106
on pressure-enthalpy diagram, 261f
varying, 257f
Sublimation, 4
Sulfur dioxide, 30, 83
Sun effect, 220
Superheating, 126, 254
    added, 175
    calculations neglecting, 223
    suction vapor, 258f, 259f
Surge drum
    arrangement of, 69f
    elimination of, 70

TDS. See Thermal differential sensor
Temperature
    aging, 299
    ambient, 210
    ammonia boiling, 11, 12, 14f, 31
    ammonia discharge, 174
    ammonia evaporating, 300
    ammonia saturation, 14
    breakdown, 23
    brine freezing, 284
    cabinet, raising, 189
    of coil, 136
    compression, 28
    compressor discharge, 174
    condensing, 271
    condensing, changing, 254, 255f
    control, 128, 130, 130f, 313
    correcting design, 220
    cost of holding, 223
    critical, 23, 26, 33
    cycle, 16–17
    design, 210, 211
    discharge, 184
    discharge, change in, 175
    discharge, determining, 257
    discharge gas, 131
    discharge vapor, 257
    drop in outside, 211
    dry-bulb, 112
    dry-bulb, map of outside design, 211f
    of dry ice, 318, 319, 320
    ether boiling, 11
    evaporator, 234, 235, 310
    evaporator, changing, 253f, 254f
    faulty regulation of, 301
    fermenting, 297
    finding discharge, 175

    heat and, difference between, 1
    maintaining, 167
    mean differences in, 227f
    of pipes, 200
    reduction in freezing, 284
    of refrigerant, 237
    refrigerating capacities at various, 173f
    scales, 1, 206
    selecting design, 211
    suction vapor discharge, 175f
    suction vapor, increase in, 55
    in thermostatic expansion valve, 50
    variation in, 301
    variation in power for different, 173f
    various refrigerant discharge, 176f
    water boiling, 2, 10
    wet-bulb, 112
    wet compression reducing discharge, 258
Temperature, suction
    changing, 253, 254
    evaporator conditions and, 49f
Thermal differential sensor (TDS), 56, 57
    electric float and, 56, 57
Thermal economizer, 58
    pneumatic high-side float and, 58f
Thermal electric expansion valve, 54f
Thermal expansion valve. See Valve(s), thermostatic expansion
Thermistor, 208
    reverse-acting, 54, 55
Thermobank system, 149, 150f
Thermocouple, 208
Thermometer
    binoc, 202, 202f
    checking accuracy of, 204
    dry-bulb, 230
    fluid in, 200
    insertion, 203
    limits of error in, 203
    location of, 204
    required points for, 203f
    spirit, 201f
    stem, 200–204
    wet-bulb, 230
Thermometer, dial, 204
    advantages of, 204
    remote, 204
    remote bulb and, 204
Thermometer, mercury
    accuracy of, 200
    disadvantage of, 201

Thermometer, mercury (*continued*)
    improving visibility of, 202
    parallax in reading, 201f
    reflecting-type, 202f
    special designs of, 201
Thermometer, recording, 205, 205f
    memory capability of, 205
Thermometer well, 203
    clip-on, 203
    construction details of welded, 202f
    screw type, 202f
    temporary, 202
Thermostat, 47
    solenoid valve operated by, 134f, 304
Thermostatic bulb, 313
Thread sealer, 118
Total energy system (cogen), 161
Transducer, 208
Transmission, 212, 223
    gain, 213
Tube-Ice machine, 317
Tubing
    ACR, 117
    copper, 116
    flare connections in, 117
    hard-drawn, 117
    size of, 229
    soft-drawn, 116, 117
TX valve. *See* Valve(s), thermostatic expansion

U factor, 212
Usage, 214
    misleading results of, 219
    reduction in, 263

Vacuum, inches of, 6
Valve(s), 128
    ammonia glove, 122f
    arrangement of, 87f
    automatic, 42, 178
    backpressure-regulating, 135
    backseated, 91, 92
    bypass, 91, 92, 93f
    capacity-reduction, 96, 102
    check, 69, 137
    constant-pressure, 135, 136f
    construction of, 91f
    control, 128
    crankhouse pressure-regulating, 136
    crankhouse-responsive, 136
    diaphragm-type, 119

    discharge, 74, 85
    discharge service, 90
    downstream-responsive, 136
    drain, 267
    evaporator pressure-regulating, 136
    evaporator-responsive regulating, 135
    faulty compressor, 184, 188
    feather, 85, 87f
    four-way, 311, 312
    hand, 119–122, 121f
    improperly closed, 290
    insulation, 191
    low-pressure discharge, 86f
    magnetic, 132
    midseated, 91
    needle, 41, 121
    packed low-pressure, 120f
    packless, 119, 121f
    pipes and, 277f
    plug, 41
    poppet, 77, 85, 87f
    positions of, 172
    pressure-reducing, 138
    pressure-relief, 125, 139f
    ring, 85, 87f
    service, 90, 90f, 91
    sliding, 96
    solid-state expansion, 54–55
    suction, 85
    suction, heavy, 77, 79f
    suction pressure, 74, 135
    suction service, 90
    suction service, front-seating, 90, 188
    throttling-type, 135, 284
    water-regulating, 137, 138f
Valve(s), automatic expansion, 46–48, 46f, 132
    action of, 47
    advantages of, 47
    altitude and, 325
    constant temperature characteristics of, 48
    disadvantages of, 47
    evaporator and, 47f
    exceptions to, 133
    multiplexing, 48f
    pressure range of, 48
    relative simplicity of, 48
Valve(s), constant-pressure, 292
    altitude and, 325
Valve(s), expansion, 13, 14, 40, 128
    capacity reduction of, 272
    complete analysis of, 242–243

conditions at, 41
exchanging, 37
function of, 40, 41
hand, 41, 41f, 42, 42f
leaks in, 135
manipulation/adjustment of, 174
open, 181
starved, 184
Valve(s), float, 132
exchanging, 41
Valve(s), thermostatic expansion, 48, 49f, 64, 133
advantages of, 51
capacities of, 52f
commercial applications of, 54
construction variation in, 51
disadvantages of, 51
external equalizer and, 52f
flooded operation of, 53
manual adjustment of, 51
operating, 126
pressures/temperatures in, 50f
Vapor, 5
ammonia, 327
clearance, 74
condition of returning, 174
density of, 237
discharge temperatures of, 175f
dry saturated, 5
gas and, difference between, 5
heat transfer from compressing, 258
pressure, 6, 7
removal of, 74
removal rate of, 237
sources of, 265, 266
suction, 126
superheated, 5, 16, 258f, 259f
superheated suction, 243
temperature of discharge, 257
volume of, 258, 275f
weight discharged, 265
wet, 5
wet suction, 243
Vaporization. *See* Evaporation
Variable-capacitance probe, 57–58, 57f
Variable-frequency drives (VFDs), 92, 160, 161
cost of, 160
Velocity
allowable refrigerant, 272f
calculating, 273–277
flow, 274f

VFDs. *See* Variable-frequency drives
Voltmeters, 209

Wall construction, 214f
Water, 32
boiling temperature of, 2, 10
charged, 296
circulating system, 294f
cooling requirements of, 292
cooling table, 294f
cost of, 103, 112, 170
critical pressure for, 28
drinking, 292, 293f
evaporation of, 231
flow speed of, 293
intercooling with, 267f
latent heat of evaporation of, 112
operating costs of condensing, 170
physical properties of, 26f
pressure-temperature characteristics of, 11f
quantity of, 224
refrigeration difficulties of, 11
seltzer, 296
separator, 125
soda, 296
supply, 111–114
velocity of, 224, 225
Water coolers, 292–295
industrial, 294, 295
instantaneous remote, 293f
outlet of, 293
self-contained, 293f
shell-and-tube, 295
Watt meters, 209
Wax, 142
Weather Bureau, United States, 211, 219
Weld fittings, 119f
Wet compression, 257, 258
oil breakdown and, 258
outline of, 259f
reducing discharge temperature, 258
Wine
aging of, 300
congealing point of, 299
external coolers for, 300
making, 299
Wire drawing, 53

DISCARD

DISCARD

APR 0 5 2010
NOV 1 6 2009
FEB 0 4 2010
FEB 0 6 2010
JUN 1 4 2010
JUN 1 6 2010
JUN 2 1 2010
DEC 0 4 2010
JUL 2 5 2011
AUG 2 2 2011
OCT 1 3 2011
JAN 2 0 2012
SEP 1 0 2012

NOV 0 7 2012
JAN 1 7 2013
MAY 1 0 2013
JAN 2 9 2014
AUG 3 1 2015
MAR 0 7 2016
APR 2 1 2023

TP 492 .L37 2008
Langley, Chris, 1955-
Refrigeration principles,
 practices, and   7/2008